# Mechanisms and Machine Science

Volume 88

This book series establishes a well-defined forum for monographs, edited Books, and proceedings on mechanical engineering with particular emphasis on MMS (Mechanism and Machine Science). The final goal is the publication of research that shows the development of mechanical engineering and particularly MMS in all technical aspects, even in very recent assessments. Published works share an approach by which technical details and formulation are discussed, and discuss modern formalisms with the aim to circulate research and technical achievements for use in professional, research, academic, and teaching activities.

This technical approach is an essential characteristic of the series. By discussing technical details and formulations in terms of modern formalisms, the possibility is created not only to show technical developments but also to explain achievements for technical teaching and research activity today and for the future.

The book series is intended to collect technical views on developments of the broad field of MMS in a unique frame that can be seen in its totality as an Encyclopaedia of MMS but with the additional purpose of archiving and teaching MMS achievements. Therefore, the book series will be of use not only for researchers and teachers in Mechanical Engineering but also for professionals and students for their formation and future work.

The series is promoted under the auspices of International Federation for the Promotion of Mechanism and Machine Science (IFToMM).

Prospective authors and editors can contact Mr. Pierpaolo Riva (publishing editor, Springer) at: pierpaolo.riva@springer.com

Indexed by SCOPUS and Google Scholar.

More information about this series at http://www.springer.com/series/8779

Erwin-Christian Lovasz ·
Inocentiu Maniu · Ioan Doroftei ·
Mircea Ivanescu · Corina-Mihaela Gruescu
Editors

# New Advances in Mechanisms, Mechanical Transmissions and Robotics

## MTM & Robotics 2020

Springer

IFToMM

*Editors*
Erwin-Christian Lovasz
Department of Mechatronics
Polytechnic University of Timişoara
Timişoara, Romania

Inocentiu Maniu
Department of Mechatronics
Polytechnic University of Timişoara
Timişoara, Romania

Ioan Doroftei
Department of Mechanical Engineering,
Mechatronics and Robotics
Gheorghe Asachi Technical
University of Iasi
Iaşi, Romania

Mircea Ivanescu
Mechatronics
University of Craiova
Craiova, Romania

Corina-Mihaela Gruescu
Department of Mechatronics
Polytechnic University of Timişoara
Timişoara, Romania

ISSN 2211-0984 ISSN 2211-0992 (electronic)
Mechanisms and Machine Science
ISBN 978-3-030-60075-4 ISBN 978-3-030-60076-1 (eBook)
https://doi.org/10.1007/978-3-030-60076-1



This Springer imprint is published by the registered company Springer Nature Switzerland AG
The registered company address is: Gewerbestrasse 11, 6330 Cham, Switzerland

# Preface

The present volume displays up-to-date contributions in the field of mechanisms, mechanical transmissions, robotics and mechatronics. The topics developed within this volume are: mechanisms and robotics, mechanical transmissions and machine parts, mechatronics and biomechanics.

The results reported in the papers are going to be discussed at MTM & Robotics 2020 - The Joint International Conference of the XIII International Conference on Mechanisms and Mechanical Transmissions (MTM) and the XXIV International Conference on Robotics (Robotics), in Timisoara, Romania.

The content of the contributions was carefully reviewed and selected for presentation and publication by independent reviewers and members of the International Scientific Committee.

MTM & Robotics 2020 is organized by the Department of Mechatronics, University Politehnica Timisoara, with the support of the Romanian Association for Theory of Machines and Mechanisms (ARoTMM), the Robotics Society of Romania (RSR), IFToMM Technical Committee for Linkages and Mechanical Controls and IFToMM Technical Committee for Robotics and Mechatronics.

The previous editions of the two conferences were held in Romania until 2012, beginning with 1972, every four years—MTM and since 1981, every two years—Robotics. In 2012, the works of the conferences were organized as joint conferences at Pascal Institute, Clermont-Ferrand, France, and then, in 2016, at RWTH Aachen, Germany.

We are grateful to the authors for their contributions and to the reviewers for their recommendations as improvement guidance for the selected papers. We would like to express thanks to the International Federation for the Promotion of Mechanism and Machine Science (IFToMM), ARoTMM and the Robotics Society of Romania (RSR). We also thank the members of the Organizing Committee of

MTM & Robotics 2020 and Springer Publishers for their excellent technical and editorial support.

August 2020

Erwin-Christian Lovasz
Inocentiu Maniu
Ioan Doroftei
Mircea Ivanescu
Corina-Mihaela Gruescu

# Organization

## Conference Chair

Erwin-Christian Lovasz — Politehnica University of Timisoara, Romania

## Conference Co-chairs

Inocentiu Maniu — Politehnica University of Timisoara, Romania
Ioan Doroftei — "Gheorghe Asachi" Technical University of Iasi
Mircea Ivanescu — University of Craiova

## International Scientific Committee

Stelian Alaci, Romania
Cătălin Alexandru, Romania
Mircea-Viorel Bara, Romania
Istvan Biro, Hungary
Karsten Berns, Germany
Stelian Brad, Romania
Cornel Brişan, Romania
Mehmet Ismet Can Dede, Turkey
Giuseppe Carbone, Italy
Maja Čavić, Serbia
Marco Ceccarelli, Italy
Dorian Cojocaru, Romania
Burkhard Corves, Germany
Valer Dolga, Romania
Jean-Christophe Fauroux, France
Dirk Lefeber, Belgium
Carlo Ferraresi, Italy
Alessandro Gasparetto, Italy
Grigore Gogu, France
Alfonso Hernandez, Spain
Tian Huang, China
Mathias Hüsing, Germany
Juan Carlos Jauregui-Correa, Mexico
Andres Kecskemethy, Germany
Gökhan Kiper, Turkey
Chin-Hsing Kuo, Australia
Corneliu Lazăr, Romania
Song Lin, China
Dan Mândru, Romania
Elena Mereuţă, Romania
Calin Micloşina, Romania
Niels Modler, Germany
Andreas Müller, Austria
Mircea Neagoe, Romania
Mircea Niţulescu, Romania
Doru Adrian Pănescu, Romania
Paul Ciprian Patic, Romania
Ivan Petrovič, Croatia
Victor Petuya, Spain
Doina Pîsla, Romania

## Organizing Committee



## With the Support of

International Federation for Promotion of Mechanism and Machine Science—IFToMM
Romanian Association of Mechanism and Machine Science—ARoTMM
Robotics Society of Romania—RSR

# Contents

**Biomechanics**

# Mechanisms and Robotics

# Singularity Loci of a Particular IRI 3-UPU Geometry

Raffaele Di Gregorio(✉)

Department of Engineering, University of Ferrara, 44122 Ferrara, Italy
raffaele.digregorio@unife.it

**Abstract.** Translational parallel manipulators (TPMs) have attracted the interest of the scientific community since the appearance of the DELTA robot. Many TPM architectures have been proposed in the literature, and TPMs of 3-UPU type are among the most studied ones. Nevertheless, only some particular 3-UPU geometries have been studied in depth. The IRI 3-UPUs constitute a new family of recently proposed geometries which could be interesting for industrial applications. Here, the singularity analysis of these novel geometries is addressed with reference to the most promising IRI 3-UPU particular geometry.

**Keywords:** Translational parallel manipulator · Instantaneous DOF · Singularity locus · Constraint singularity

## 1 Introduction

Parallel manipulators (PMs) feature two rigid bodies, one fixed (base) and the other mobile (platform), connected to one another through a number of kinematic chains (limbs). Translational PMs (TPMs) are 3-degrees-of-freedom (DOF) PMs whose platform can perform only spatial translations. The great majority of TPM architectures proposed in the literature [1] exhibit three limbs of the same type with one actuated joint per limb. TPMs of 3-UPU type belong to this class of TPMs: they have three equal limbs of Universal-Prismatic-Universal (UPU) type with the P pair as actuated joint (Fig. 1) [2].

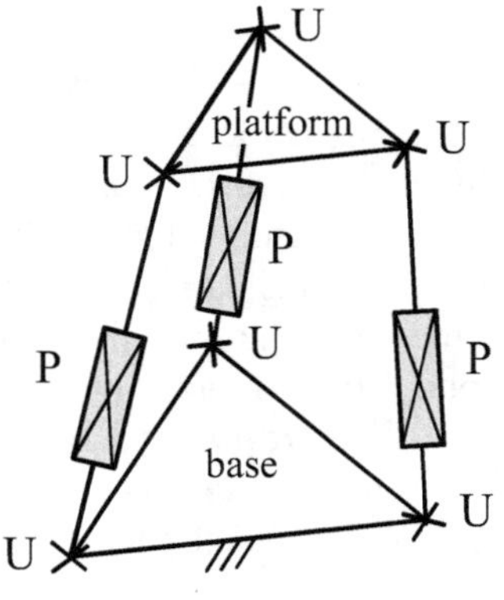


**Fig. 1.** A 3-UPU architecture


E.-C. Lovasz et al. (Eds.): MTM&Robotics 2020, MMS 88, pp. 3–11, 2021.
https://doi.org/10.1007/978-3-030-60076-1_1

The majority of the published papers on translational 3-UPUs [3] deals with the 3-UPU geometries in which the axes of the three revolute (R) pairs adjacent to the base (to the platform) are coplanar, even though the few papers on different geometries (see, for instance, [4–6]) showed that they could be of practical interest.

IRI 3-UPUs [6] are a family of reconfigurable 3-UPUs where the axes of the three R-pairs adjacent to the base (to the platform) share a unique intersection point, but are not coplanar. This feature makes them able to be assembled either as a TPM or as a parallel wrist. In [6], Sarabandi et al. based on the results of the position analysis showed that these architectures might be of interest.

Here, the singularity analysis of the most promising geometry of translational IRI 3-UPUs (Fig. 2) is addressed. In particular, the algebraic equations of the singularity surfaces are deduced and analyzed. The result of this study is that the studied IRI 3-UPU presents wide free-from-singularity regions where the useful workspace could be located.

The paper is organized as follows. Section 2 provides some background concepts and presents the studied geometry. Then, Sect. 3 deduces the singularity loci's equations. Eventually, Sect. 4 discusses the results and draws the conclusions.

## 2 Background and Notations

PMs' singular configurations (singularities) [7–9] are those configurations where the instantaneous input-output relationship fails to state a one-to-one correspondence between instantaneous inputs (i.e., the actuated joints' rates) and outputs (i.e., the platform twist). They are collected into three sets [7]:

- type-I (serial) singularities, which occur when the actuated joints' rates are not determined even though the platform twist is assigned;
- type-II (parallel) singularities, which occur when the platform twist is not determined even though the actuated joints' rates are assigned;
- type-III singularities, which are configurations where both the two previous conditions are satisfied.

From a kinematic point of view, the boundaries of the reachable workspace is the type-I singularity locus; also, at a type-I singularity, the instantaneous-DOFs of the platform are lower than its finite DOFs (i.e., they are stationary configurations [10]). Differently, type-II singularities are usually located inside the reachable workspace and, at a type-II singularity, the instantaneous-DOFs of the platform are greater than its finite DOFs (i.e., they are uncertainty configurations [10]).

From a static point of view, type-I singularities correspond to configurations where the platform can carry large forces (even infinitely large) without using one or more of the actuators' torques. Differently, type-II singularities correspond to configurations where the platform cannot carry forces (even infinitesimal) along one or more directions without requiring that one or more actuators supply infinite generalized torques (i.e., without breaking at least one link). Consequently, type-II singularities must be

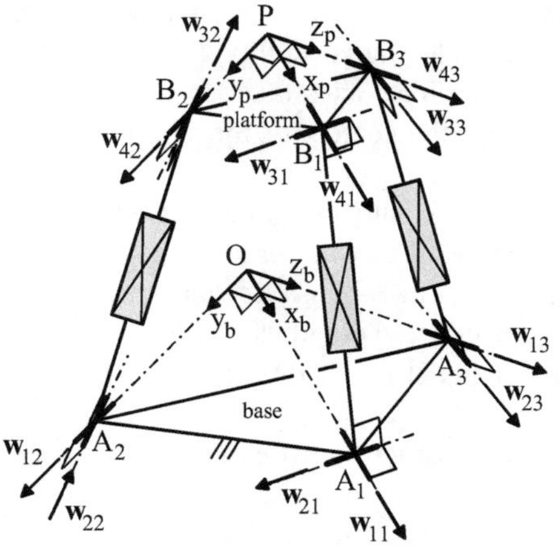


**Fig. 2.** IRI 3-UPU with the R-pair axes that are fixed in the base (platform) mutually orthogonal: notations

identified during design, and avoided during operation by locating the useful workspace in free-from-singularity regions of the operational space. The existence of wide free-from-singularity regions is a necessary condition to state that a new PM architecture could be interesting for actual applications.

In TPMs, when the limb connectivity[1] is greater than 3, the additional instantaneous DOFs gained at a type-II singularity may correspond to instantaneous rotations of the platform (i.e., at that configuration, the platform is not constrained to translate any longer). Such singularities are named "constraint singularities" [11].

Figure 2 shows a particular geometry of IRI 3-UPU assembled to be a TPM. The peculiarity of this architecture is that the axes of the three R-pairs adjacent to the base (to the platform) are mutually orthogonal. Such geometry is the farthest from the two extremes of this family (i.e., the SNU 3-U$\underline{\text{P}}$U [12], which has coplanar axes with a common intersection point and a rotation (constraint) singularity at the home position, and the structural singularity [8] of the translational 3-U$\underline{\text{P}}$U that occurs when the abovementioned R-pair axes are all parallel [13]). Thus, it, intuitively, should provide the wider free-from singularity region around the home position.

Since an UPU limb is a particular RRPRR limb, 3-UPU architectures are special cases of the 3-RRPRR ones. In [14], Di Gregorio and Parenti-Castelli showed that a PM of 3-RRPRR type becomes a TPM, if it is manufactured and assembled, out of constraint singularities, so that, in each RRPRR limb,

(i) the axes of the two intermediate R-pairs are parallel to one another, and
(ii) the axes of the two ending R-pairs are parallel to each other.

With reference to Fig. 2, the following notations are introduced:

[1] According to [10], here, the term "limb connectivity" denotes the DOF number the platform would have if it were connected to the base only through that limb.

- $Ox_by_bz_b$ and $Px_py_pz_p$ are two Cartesian references fixed to the base and the platform, respectively; without losing generality, these two references have been chosen with the homologous coordinate axes that are parallel[2];
- $A_i(B_i)$ for i = 1, 2, 3 are the centers of the U joints adjacent to the base (to the platform);
- in each UPU limb the four R pairs are numbered with an index, j, that increases by moving from the base toward the platform;
- $\mathbf{w}_{ji}$, for j = 1, …, 4, is the j-th R-pair axis' unit vector of the i-th UPU limb, i = 1, 2, 3; the i-th limb is manufactured and assembled so that these unit vectors respect the above-mentioned geometric conditions for the translation [14] as follows: $\mathbf{w}_{1i} = \mathbf{w}_{4i}$, and $\mathbf{w}_{2i} = \mathbf{w}_{3i}$;
- $\mathbf{w}_{11}$, $\mathbf{w}_{12}$, and $\mathbf{w}_{13}$ ($\mathbf{w}_{41}$, $\mathbf{w}_{42}$, and $\mathbf{w}_{43}$) are unit vectors of the coordinate axes $x_b$, $y_b$, and $z_b$ ($x_p$, $y_p$, and $z_p$), respectively, and, at the same time, unit vectors of the three R-pair axes fixed to the base (to the platform);
- $\mathbf{w}_{2i}$ and $\mathbf{w}_{3i}$ are perpendicular to the axis of the i-th limb (i.e., the line through $A_i$ and $B_i$), for i = 1, 2, 3.

Also the following definitions/choices are introduced:

- $d_p = B_1P = B_2P = B_3P$;
- $d_b = A_1O = A_2O = A_3O$;
- $l_i = A_iB_i$, for i = 1, 2, 3;
- $\mathbf{p} = (P - O) = x\mathbf{w}_{11} + y\mathbf{w}_{12} + z\mathbf{w}_{13}$, where $(x, y, z)^T$ collect the coordinates of point P measured in $Ox_by_bz_b$;
- $\mathbf{b}_i = (B_i - O) = \mathbf{p} + d_p\mathbf{w}_{4i} = \mathbf{p} + d_p\mathbf{w}_{1i}$, for i = 1, 2, 3;
- $\mathbf{a}_i = (A_i - O) = d_b\mathbf{w}_{1i}$, for i = 1, 2, 3;
- $\mathbf{g}_i = (\mathbf{b}_i - \mathbf{a}_i)/l_i$ for i = 1, 2, 3;
- $\mathbf{r}_i = \mathbf{w}_{1i} \times \mathbf{w}_{2i}$ for i = 1, 2, 3;
- $\mathbf{h}_i = \mathbf{w}_{3i} \times \mathbf{w}_{4i}$ for i = 1, 2, 3[3];
- $\mathbf{n}_i = [(\mathbf{b}_i - \mathbf{a}_i) \cdot \mathbf{r}_i]\mathbf{h}_i$ for i = 1, 2, 3.

## 3 Singularity Analysis

From an analytic point of view, the instantaneous input-output relationship of a PM [7] is a linear and homogeneous system that relates the platform twist and the actuated-joint rates. In this relationship, the two coefficient matrices (Jacobians) that multiply the platform twist (parallel Jacobian, $\mathbf{J}_p$) and the actuated-joint rates (serial Jacobian, $\mathbf{J}_s$), respectively, depend only on the PM configuration. $\mathbf{J}_p$ is always a 6 × 6 matrix; whereas, $\mathbf{J}_s$ is a 6 × f matrix, where f is the DOF number of the PM. Singularities are PM configurations that make either or both these Jacobians rank-deficient: type-I

[2] The parallelism of the coordinate axes is kept during the motion since the analyzed 3-UPU is translational.

[3] With reference to Fig. 2, note that, if the translation conditions hold, then, $\mathbf{r}_i = -\mathbf{h}_i$.

singularities occur when $\mathbf{J}_s$ is rank-deficient, type-II singularities occur when $\mathbf{J}_p$ is rank-deficient, and type-III singularities occur when $\mathbf{J}_s$ and $\mathbf{J}_p$ are both rank-deficient.

The instantaneous input-output relationship of a translational 3-UPU can be written as follows [3]:

$$\mathbf{J}_p\boldsymbol{\xi} = \mathbf{J}_s\dot{\mathbf{q}} \tag{1}$$

with

$$\boldsymbol{\xi} = \begin{pmatrix} \dot{\mathbf{p}} \\ \boldsymbol{\omega} \end{pmatrix}, \; \dot{\mathbf{q}} = \begin{pmatrix} \dot{l}_1 \\ \dot{l}_2 \\ \dot{l}_3 \end{pmatrix}, \; \mathbf{J}_s = \begin{bmatrix} \mathbf{1}_{3\times3} \\ \mathbf{0}_{3\times3} \end{bmatrix}, \; \mathbf{J}_p = \begin{bmatrix} \mathbf{G}_{3\times3} & \mathbf{K}_{3\times3} \\ \mathbf{0}_{3\times3} & \mathbf{N}_{3\times3} \end{bmatrix} \tag{2}$$

where $\boldsymbol{\omega}$ is the angular velocity of the platform, $\mathbf{1}_{3\times3}$ is the $3 \times 3$ identity matrix, $\mathbf{0}_{3\times3}$ is the $3 \times 3$ null matrix, $\mathbf{G}_{3\times3} = [\mathbf{g}_1, \mathbf{g}_2, \mathbf{g}_3]$, $\mathbf{K}_{3\times3} = [(\mathbf{b}_1 - \mathbf{p}) \times \mathbf{g}_1, (\mathbf{b}_2 - \mathbf{p}) \times \mathbf{g}_2, (\mathbf{b}_3 - \mathbf{p}) \times \mathbf{g}_3]$, and $\mathbf{N}_{3\times3} = [\mathbf{n}_1, \mathbf{n}_2, \mathbf{n}_3]$.

Expression (2) of $\mathbf{J}_p$ reveals that a type-II singularity occurs when

$$\det(\mathbf{J}_p) = \det(\mathbf{G}_{3\times3})\det(\mathbf{N}_{3\times3}) = 0 \tag{3}$$

Equation (3) is the analytic expression of the translational 3-UPUs' singularity locus. Such equation is satisfied if $\det(\mathbf{N}_{3\times3})$ or $\det(\mathbf{G}_{3\times3})$ is equal to zero.

### 3.1 Rotation (Constraint) Singularities

If $\det(\mathbf{N}_{3\times3})$ is not equal to zero, the last three equations of system (1) admit the only solution $\boldsymbol{\omega} = 0$, that is, the platform can only translate. The same three equations reveal that, if $\det(\mathbf{N}_{3\times3}) = 0$, the platform angular velocity, $\boldsymbol{\omega}$, can be different from zero, that is, a rotation (constraint) singularity occurs. According to the adopted notations, the singularity condition $\det(\mathbf{N}_{3\times3}) = 0$ can be written as follows

$$\begin{aligned}\det(\mathbf{N}_{3\times3}) &= \mathbf{n}_1 \cdot (\mathbf{n}_2 \times \mathbf{n}_3) \\ &= [(\mathbf{b}_1 - \mathbf{a}_1)\cdot\mathbf{r}_1][(\mathbf{b}_2 - \mathbf{a}_2)\cdot\mathbf{r}_2][(\mathbf{b}_3 - \mathbf{a}_3)\cdot\mathbf{r}_3]\mathbf{h}_1 \cdot (\mathbf{h}_2 \times \mathbf{h}_3) = 0\end{aligned} \tag{4}$$

The analysis of Eq. (4) brings to the conclusion that a constraint singularity occurs if either of the following conditions is satisfied

a) at least one of the coefficients $[(\mathbf{b}_i - \mathbf{a}_i)\cdot\mathbf{r}_i]$, $i = 1, 2, 3$, is equal to zero;
b) the mixed product $\mathbf{h}_1 \cdot (\mathbf{h}_2 \times \mathbf{h}_3)$ is equal to zero.

Regarding condition (a), it occurs when, in the i-th limb, the limb axis lies on the plane of U-joints' cross links. This condition certainly makes the limb rotation possible around its axis (i.e., it is a limb singularity, see [15]). Nevertheless, the fact that it yields an additional mobility to the platform is questionable since these scalar coefficients are the result of the algebraic manipulations used for eliminating some passive-joint rates from the initial equation and may appear in different positions in the simplified system depending on the adopted elimination procedure. Indeed, these same coefficients, which, in [6], appear in the last three equations (the one related to the rotation singularities), in [15], appear in the equations related to the translation singularities.

Differently, condition (b) occurs when the R-pair axes are all parallel to a unique plane. Such an R-pair arrangement allows platform's instantaneous rotations around any axis parallel to that plane. Therefore, condition (b) clearly identifies constraint singularities. It will be expanded and analyzed in the remaining part of this section.

The expression of $\mathbf{w}_{3i}$ as a function of point P's coordinates (i.e., of the platform position) can be deduced by considering that (see Fig. 2)

$$\mathbf{w}_{3i} = \frac{\mathbf{w}_{4i} \times (\mathbf{b}_i - \mathbf{a}_i)}{|\mathbf{w}_{4i} \times (\mathbf{b}_i - \mathbf{a}_i)|} = \frac{\mathbf{w}_{1i} \times [\mathbf{p} + (d_p - d_b)\mathbf{w}_{1i}]}{|\mathbf{w}_{1i} \times [\mathbf{p} + (d_p - d_b)\mathbf{w}_{1i}]|} = \frac{\mathbf{w}_{1i} \times \mathbf{p}}{|\mathbf{w}_{1i} \times \mathbf{p}|} \tag{5}$$

Formula (5), when introduced into the definition of $\mathbf{h}_i$ yields

$$\mathbf{h}_i = \mathbf{w}_{3i} \times \mathbf{w}_{4i} = \frac{(\mathbf{w}_{1i} \times \mathbf{p}) \times \mathbf{w}_{1i}}{|\mathbf{w}_{1i} \times \mathbf{p}|} = \frac{\mathbf{p} - (\mathbf{w}_{1i} \cdot \mathbf{p})\mathbf{w}_{1i}}{|\mathbf{w}_{1i} \times \mathbf{p}|} \tag{6}$$

which, after the introduction of the explicit expression of $\mathbf{p}$ (i.e., $\mathbf{p} = x\mathbf{w}_{11} + y\mathbf{w}_{12} + z\mathbf{w}_{13}$), gives

$$\mathbf{h}_1 = \frac{y\mathbf{w}_{12} + z\mathbf{w}_{13}}{\sqrt{y^2 + z^2}}, \mathbf{h}_2 = \frac{x\mathbf{w}_{11} + z\mathbf{w}_{13}}{\sqrt{x^2 + z^2}}, \mathbf{h}_3 = \frac{x\mathbf{w}_{11} + y\mathbf{w}_{12}}{\sqrt{x^2 + y^2}} \tag{7}$$

Formulas (7), when introduced into $\mathbf{h}_1 \cdot (\mathbf{h}_2 \times \mathbf{h}_3)$, yield the following equation of the rotation singularity locus

$$\mathbf{h}_1 \cdot (\mathbf{h}_2 \times \mathbf{h}_3) = \frac{2xyz}{\sqrt{(x^2 + z^2)(x^2 + y^2)(y^2 + z^2)}} = 0 \tag{8}$$

The analysis of Eq. (8) reveals that the rotation singularity locus is constituted by the 3 coordinate planes x = 0, y = 0, and z = 0 (see Fig. 3). Also, the analysis of Fig. 2 reveals that

- when point P lies on the $y_b z_b$ coordinate plane (i.e., x = 0), the platform can perform rotations around axes parallel to the $x_b$ axis;
- when point P lies on the $x_b z_b$ coordinate plane (i.e., y = 0), the platform can perform rotations around axes parallel to the $y_b$ axis;
- when point P lies on the $x_b y_b$ coordinate plane (i.e., z = 0), the platform can perform rotations around axes parallel to the $z_b$ axis.

As a consequence, when P coincides with O (i.e., x = y = z = 0) the platform locally acquires 3 rotational DOFs even though the expression at the left-hand side of Eq. (8) becomes indeterminate.

## 3.2 Translation Singularities

Out of constraint singularities (i.e., when $\det(\mathbf{N}_{3\times 3}) \neq 0$) the platform angular velocity, $\boldsymbol{\omega}$, must be equal to zero and the first three equations o system (1) reduce themselves to the following form

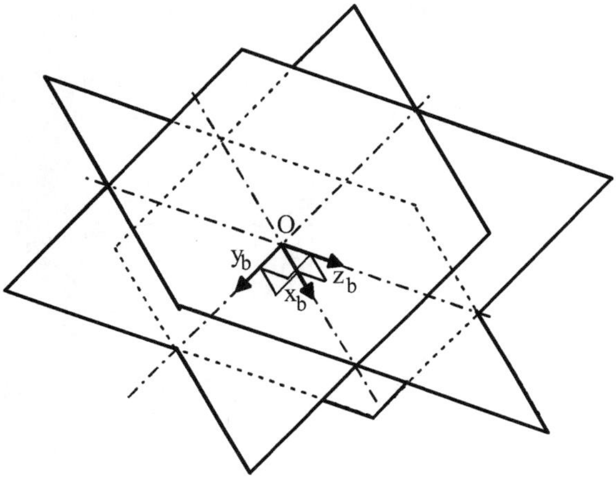


**Fig. 3.** Rotation (constraint) singularity locus

$$\mathbf{G}_{3\times 3}\,\dot{\mathbf{p}} = \dot{\mathbf{q}} \tag{9}$$

System (9) reveals that, when the actuators are locked (i.e., $\dot{\mathbf{q}} = \mathbf{0}$), the platform translation is locked (i.e., $\dot{\mathbf{p}} = \mathbf{0}$), too, if and only if $\det(\mathbf{G}_{3\times 3})$ is not equal to zero; whereas, if $\det(\mathbf{G}_{3\times 3}) = 0$, platform's instantaneous translation (i.e., $\dot{\mathbf{p}} \neq \mathbf{0}$) is possible even though the actuators are locked, that is, a translation singularity occurs. According to the adopted notations, the singularity condition $\det(\mathbf{G}_{3\times 3}) = 0$ can be written as follows

$$\det(\mathbf{G}_{3\times 3}) = \mathbf{g}_1 \cdot (\mathbf{g}_2 \times \mathbf{g}_3) = \frac{1}{l_1 l_2 l_3}(\mathbf{b}_1 - \mathbf{a}_1) \cdot [(\mathbf{b}_2 - \mathbf{a}_2) \times (\mathbf{b}_3 - \mathbf{a}_3)] = 0 \tag{10}$$

The analysis of Eq. (10) brings to the conclusion that a translation singularity occurs if either of the following conditions is satisfied

c) the limb lengths become infinite[4];
d) the mixed product $(\mathbf{b}_1 - \mathbf{a}_1) \cdot [(\mathbf{b}_2 - \mathbf{a}_2) \times (\mathbf{b}_3 - \mathbf{a}_3)]$ is equal to zero.

Regarding condition (c), even though it is physically unreachable, it, anyway, warns that configurations with long limb lengths have a bad kinematic performance in term of positioning precision since they are nearer to a singular configuration [16].

[4] It is worth noting that it is not possible to have only one infinite limb length.

Regarding condition (d), it geometrically occurs when the three limb axes are all parallel to a unique plane. From an analytic point of view, the following relationships hold

$$\mathbf{b}_1 - \mathbf{a}_1 = \left(x + d_p - d_b\right)\mathbf{w}_{11} + y\mathbf{w}_{12} + z\mathbf{w}_{13} \tag{11a}$$

$$\mathbf{b}_2 - \mathbf{a}_2 = x\mathbf{w}_{11} + \left(y + d_p - d_b\right)\mathbf{w}_{12} + z\mathbf{w}_{13} \tag{11b}$$

$$\mathbf{b}_3 - \mathbf{a}_3 = x\mathbf{w}_{11} + y\mathbf{w}_{12} + \left(z + d_p - d_b\right)\mathbf{w}_{13} \tag{11c}$$

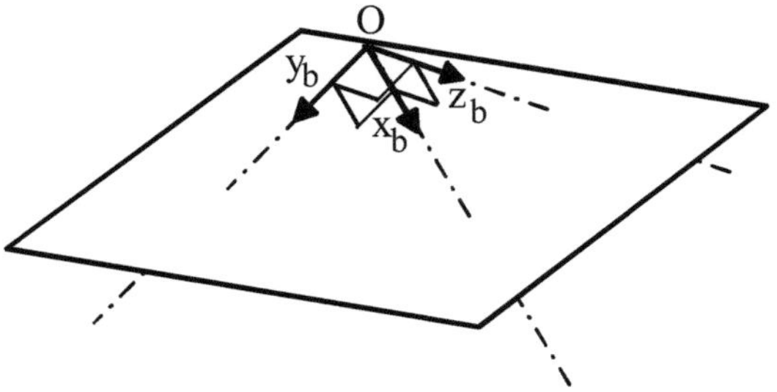


**Fig. 4.** Translation singularity locus

Formulas (11), when introduced into the mixed product $(\mathbf{b}_1 - \mathbf{a}_1) \cdot [(\mathbf{b}_2 - \mathbf{a}_2) \times (\mathbf{b}_3 - \mathbf{a}_3)]$, yield the following equation of the translation singularity locus

$$(\mathbf{b}_1 - \mathbf{a}_1) \cdot [(\mathbf{b}_2 - \mathbf{a}_2) \times (\mathbf{b}_3 - \mathbf{a}_3)] = \left(d_p - d_b\right)^2 \left(x + y + z + d_p - d_b\right) = 0 \tag{12}$$

The analysis of Eq. (12) reveals that, over a structural translation singularity that occurs when $d_p = d_b$, the translation singularity locus is constituted by one plane: the plane $x + y + z + d_p - d_b = 0$ (see Fig. 4).

## 4 Conclusions

The singularity analysis of the most promising geometry of translational IRI 3-UPUs has been addressed. The algebraic equations of the singularity surfaces have been deduced. These equations showed that the constraint-singularity locus consists of three mutually orthogonal planes, and the translation-singularity locus is a fourth plane. Such four planes leave wide free-from-singularity regions where the useful workspace could be located. Therefore, the conclusion is that the analyzed geometry is interesting for practical applications.

Future works will be oriented to find the optimal location of the useful workspace inside the found free-from-singularity regions.

**Acknowledgments.** This work has been developed at the Laboratory of Mechatronics and Virtual Prototyping (LaMaViP) of Ferrara Technopole, supported by FAR2019 UNIFE funds.

## References

1. Kong, X., Gosselin, C.: Type Synthesis of Parallel Mechanisms. Springer, Heidelberg (2007)
2. Tsai, L.W.: Kinematics of a three-DOF platform with three extensible limbs. In: Lenarcic, J., Parenti-Castelli, V. (eds.) Recent Advances in Robot Kinematics, pp. 401–410. Kluwer Academic Publishers, Dordrecht (1996)
3. Di Gregorio, R.: A review of the literature on the lower-mobility parallel manipulators of 3-UPU or 3-URU type. Robotics **9**(1), 5 (2020)
4. Lu, Y., Hu, B.: Analysis of kinematics and solution of active/constrained forces of asymmetric 2UPU+X parallel manipulators. Proc. Inst. Mech. Eng. Part C J. Mech. Eng. Sci. **220**(12), 1819–1830 (2006)
5. Chebbi, A., Parenti-Castelli, V.: The potential of the 3-UPU topology for translational parallel manipulators and a procedure to select the best architecture for a given task. Rom. J. Tech. Sci. Appl. Mech. **58**(1–2), 5–32 (2013)
6. Sarabandi, S., Grosch, P., Porta, J.M., Thomas, F.: A reconfigurable asymmetric 3-UPU parallel robot. In: Proceedings of the 2018 International Conference on Reconfigurable Mechanisms and Robots (ReMAR2018), Deft, Netherlands, pp. 2–9 (2018)
7. Gosselin, C.M., Angeles, J.: Singularity analysis of closed-loop kinematic chains. IEEE Trans. Robot. Autom. **6**(3), 281–290 (1990)
8. Ma, O., Angeles, J.: Architecture singularities of platform manipulators. In: Proceedings of the 1991 IEEE International Conference on Robotics and Automation. Sacramento (CA, USA), pp. 1542–1547 (1991)
9. Zlatanov, D., Fenton, R.G., Benhabib, B.: A unifying framework for classification and interpretation of mechanism singularities. ASME J. Mech. Des. **117**(4), 566–572 (1995)
10. Hunt, K.H.: Kinematic Geometry of Mechanisms. Clarendon Press, Oxford (1990)
11. Zlatanov, D., Bonev, I.A., Gosselin, C.M.: Constraint singularities of parallel mechanisms. In: Proceedings of the IEEE International Conference on Robotics and Automation, Washington, DC, USA, pp. 496–502 (2002)
12. Walter, D.R., Husty, M.L., Pfurner, M.: A complete kinematic analysis of the SNU 3-UPU parallel robot. In: Bates, D.J., Besana, G.M., Di Rocco, S., Wampler, C.W. (eds.) Interactions of Classical and Numerical Algebraic Geometry, Contemporary Mathematics 496, pp. 331–346. AMS, Providence, RI (2009)
13. Zhao, T.S., Li, Y.W., Chen, J., Wang, J.C.: A novel four-DOF parallel manipulator mechanism and its kinematics. In: Proceedings of the 2006 IEEE Conference on Robotics, Automation and Mechatronics (RAM 2006), Bangkok, Thailand, pp. 1–5 (2006)
14. Di Gregorio, R., Parenti-Castelli, V.: A translational 3-DOF parallel manipulator. In: Lenarcic, J., Husty, M.L. (eds.) Advances in Robot Kinematics: Analysis and Control, pp. 49–58. Kluwer, Norwell (1998)
15. Di Gregorio, R., Parenti-Castelli, V.: Mobility analysis of the 3-UPU parallel mechanism assembled for a pure translational motion. ASME J. Mech. Des. **124**(2), 259–264 (2002)
16. Gosselin, C., Angeles, J.: A global performance index for the kinematic optimization of robotic manipulators. ASME J. Mech. Des. **113**(3), 220–226 (1991)

# Optimal Task Placement for Energy Minimization in a Parallel Manipulator

Lorenzo Scalera[1(✉)], Paolo Boscariol[2], Giovanni Carabin[3], Renato Vidoni[3], and Alessandro Gasparetto[1]

[1] University of Udine, Udine, Italy
{lorenzo.scalera,alessandro.gasparetto}@uniud.it
[2] University of Padova, Padua, Italy
paolo.boscariol@unipd.it
[3] Free University of Bozen-Bolzano, Bolzano, Italy
{giovanni.carabin,renato.vidoni}@unibz.it

**Abstract.** This paper presents the optimization of the task placement to achieve energy efficiency in a parallel manipulator. The proposed approach is based on the parametrization of the task whose position and orientation are varied within the workspace of the robot. The energy consumed by the actuators is considered as the cost function for the optimization problem and is computed on the basis of the dynamic and electro-mechanical models of the manipulator. The method is general and can be used as an off-line tool for the optimal placement of different tasks within the workspace of the parallel manipulator.

**Keywords:** Robotics · Dynamics · Parallel robot · Optimization · Energy-efficiency

## 1 Introduction

Parallel manipulators are closed-loop mechanisms whose end-effector is linked to the base by several independent kinematic chains [1]. These manipulators have several advantages with respect to their serial-link counterparts: they can achieve higher velocities and accelerations, better performance in accuracy, rigidity and ability to manipulate large loads, thanks to their lighter arms and greater stiffness. Both rigid-link [1, 2], flexible-link [3] and cable-driven parallel robots [4, 5] have been designed and applied in several different fields. Examples are industrial machines, simulators for flight and entertainment, applications for astronomy, and haptic interfaces.

In the industrial environment, one of the main applications of parallel robots is the high-speed pick and place of small objects for sorting and packaging. These operations are often cyclical and repeated for large volumes of products. Therefore, achieving energy efficiency in such robotic systems is a relevant and actual question in the industrial environment. In this context, several strategies and optimization methods have been investigated [6]. Given a defined parallel manipulator, the energetic performance can be improved for example by properly planning the trajectories that the machine has to follow [7–9], or by exploiting the natural dynamics of the system [10–12].


E.-C. Lovasz et al. (Eds.): MTM&Robotics 2020, MMS 88, pp. 12–22, 2021.
https://doi.org/10.1007/978-3-030-60076-1_2

A different approach relies on the optimization of the task placement with respect to the robot base while satisfying its workspace bounds. Several performance criteria for path location problems have been considered in the literature, such as travel time, kinematic performance, collision and redundancy, as well as dynamic performance. In [13] Nelson and Donath take into account the manipulability measure as the optimization criterion for the optimum location of an assembly task in a manipulator workspace. In [14], Pamanes and Zeghloul consider several kinematic criteria, such as manipulability, condition number, magnitude of force and velocity, for the optimal placement of robotic manipulators. Furthermore, in [15] Feddema describes an algorithm for the optimal robot placement within a work cell to minimize motion time, whereas in [16] Aspragathos and Foussias consider velocity performance of the robot end-effector for the best path location. Kamrani et al. propose in [17] an approach for the optimal robot placement using the response surface method to attain the minimum cycle time. A multi-objective path placement optimization is applied to a parallel manipulator to minimize energy consumption, shaking forces and actuator torques by Ur-Rehman et al. in [18].

In more recent years, the optimal placement problem for a parallel robot has been addressed by Boschetti et al. in [19], where task-dependent and direction-selective performance indexes are presented to account for robot kinematics and task geometrical features. Moreover, Doan and Lin analyze in [20] the optimal robot placement with consideration of redundancy problem for wrist-partitioned robots, whereas Hassan et al. address in [21] the problem of the optimization of multiple base placement for multi-robot collaboration. Other recent examples are given by Valsamos et al. [22], who consider the task placement to minimize the joint velocities, and by Boscariol et al. [23], who present a task-dependent energetic analysis of a 3-DOF robot.

As far as energy reduction in parallel manipulators is considered, only few examples can be found in the present literature. In [2] the dynamic performance analysis of a high-speed parallel manipulator is presented and a dynamic performance index based on the inertia matrix is proposed, whereas in [24] energy consumption maps are computed to provide the more energy-efficient task location for a 4-DOF parallel robot.

In this paper an optimization problem is formulated to define the minimum-energy location of a predefined task in a parallel manipulator. The proposed approach is based on the parametrization of the task whose position and orientation are varied within the workspace of the robot. The energy consumed by the actuators is considered as the cost function for the optimization problem and is computed on the basis of the dynamic and electro-mechanical models of the manipulator.

The paper is organized as follows: Sect. 2 describes the dynamic and the electro-mechanical models of the parallel manipulator. Section 3 presents the optimization problem for the task placement, Sect. 4 the case study, whereas the numerical results are reported and discussed in Sect. 5. Finally, the paper is concluded in Sect. 6.

## 2 Dynamic and Electro-Mechanical Models

In this section the dynamic and electro-mechanical models of the parallel manipulator considered in this work are briefly recalled. The robot, shown in Fig. 1(a), is a four-leg delta-like (4-RUU) robot based on the design reported in [25]. It allows three translations and one rotation of the end-effector around its vertical axis. The end-effector is composed of a mobile platform, which is split into two parts that define a planar four-bar linkage. The rotation of the traveling plate, which allows a motion of the fourth degree of freedom $\vartheta$ within ±45°, is amplified by means of a gearbox up to ±180°. Due to the presence of the gearbox, the mobile platform is not perfectly symmetric with respect to the x and y axis of the robot base reference frame [25].

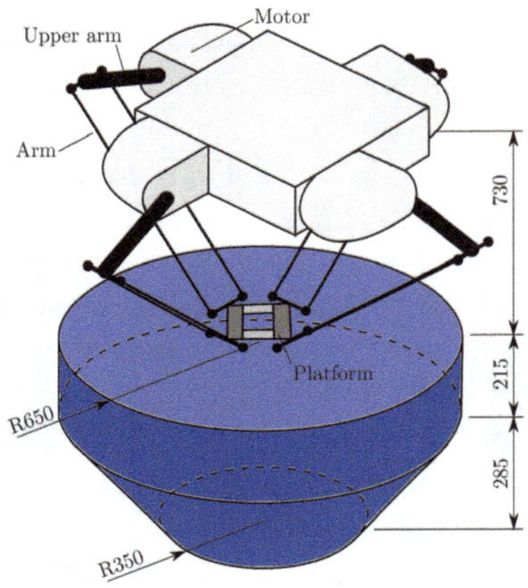


(a)

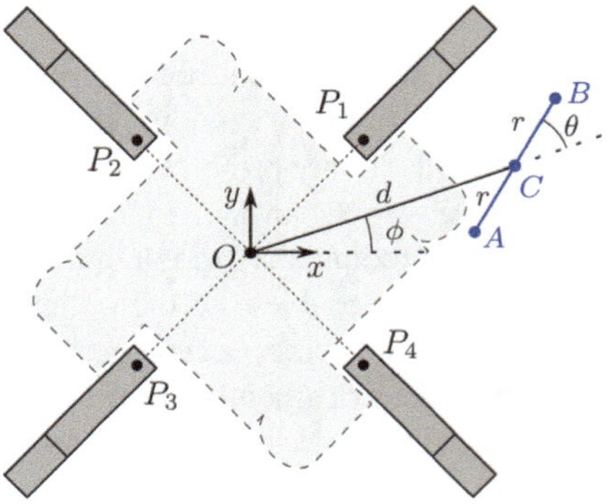


(b)

**Fig. 1.** The 4-DOF parallel manipulator considered in this work with its workspace (dimensions expressed in mm) (a); parametrization of the path, top view (b).

The dynamic equation of motion, which describes the torques at joints $\boldsymbol{\tau}$ as a function of the joint variables $\boldsymbol{q} = [q_1, q_2, q_3, q_4]^T$, are computed using the Lagrangian formalism, leading to the following formulation:

$$\boldsymbol{\tau} = \boldsymbol{M}(\boldsymbol{q})\ddot{\boldsymbol{q}} + \boldsymbol{C}(\boldsymbol{q}, \dot{\boldsymbol{q}}) + \boldsymbol{g}(\boldsymbol{q}) + \boldsymbol{f}_s + \boldsymbol{F}_v\dot{\boldsymbol{q}} \tag{1}$$

where $\boldsymbol{M}(\boldsymbol{q})$ is the mass matrix of the robot, $\boldsymbol{C}(\boldsymbol{q}, \dot{\boldsymbol{q}})$ is the matrix of the centrifugal and Coriolis effects, $\boldsymbol{f}_s = f_s\,\mathrm{sign}(\dot{\boldsymbol{q}})$ accounts for the Coulomb friction force, whereas $\boldsymbol{F}_v = \mathrm{diag}([f_v\ f_v\ f_v\ f_v])$ is the matrix of viscous friction coefficients. The term $\boldsymbol{g}(\boldsymbol{q})$ accounts for the gravitational force. Please refer to [24, 25] for a detailed description of the kinematic and dynamic models of the 4-DOF manipulator.

The electro-mechanical model of the robot is derived assuming that equivalent brushless DC motors are used, as in [23, 24]. The torque provided by the $j$-th motor $\tau_{m,j}(t) = \tau_j / K_r$ is related to the motor armature current $i_j(t)$ by means of the motor torque constant $K_t$ as:

$$\tau_{m,j}(t) = K_t i_j(t) \tag{2}$$

Then, the armature model that describes the voltage drop $v_j(t)$ across the $j$-th motor can be expressed as follows:

$$v_j(t) = R i_j(t) + K_e \dot{q}_{m,j}(t) \tag{3}$$

where $R$ and $K_e$ represent the resistance of motor windings and the back-emf constant, respectively, whereas $\dot{q}_{m,j} = \dot{q}_j K_r$ is the velocity of the $j$-th motor shaft. By indicating with $\eta_d$ the efficiency of the driver, the electro-mechanical model allows to compute the instantaneous electric power $P_{e,j}(t)$ drawn by the $j$-th actuator as the voltage-current product:

$$P_{e,j}(t) = \frac{v_j(t)\, i_j(t)}{\eta_d} \tag{4}$$

Then, the overall energy consumption $E_c$ can be obtained by integrating the electric power of the four actuators of the robot over the time period $T$. Only positive values of the electric power are computed, since energy regeneration in motor drives is not considered. In this case, the current drawn from the actuators to the driver during deceleration is dissipated in a braking resistor. Therefore, the following equation is obtained:

$$E_c = \sum\nolimits_{j=1}^{4} \int_0^T P_{c,j} dt \qquad \text{where} \qquad P_{c,j}(t) = \begin{cases} P_{e,j} & \text{if} \quad P_{e,j} \geq 0 \\ 0 & \text{if} \quad P_{e,j} < 0 \end{cases} \tag{5}$$

## 3 Proposed Method for the Optimal Task Placement

The method proposed in this paper allows to evaluate the optimal task placement within the workspace of the robot to minimize the overall energy consumption $E_c$. Given a general task to be performed in the operative space of the robot, the task can be positioned in the robot workspace in countless ways keeping the relative position of the eventual via-points fixed. Therefore, a parametrization of the task allows to change its position and orientation within the bounds of the workspace. Figure 1(b) shows a graphical overview of the path parametrization. The point $C$ defines the center of the path and its coordinates can be computed in the cylindrical reference frame as $(d, \phi)$: the parameter $d$ is the distance between $C$ and the robot base $O$, whereas $\phi$ represents the orientation of segment $OC$ with respect to the $x$-axis. Then, the angle $\theta$ accounts for the orientation of the path with respect to the segment $d$. With this parametrization, the end-effector pose $P$ in the general points $A$ and $B$ can be defined as follows:

$$P_A = [d\cos(\phi) - r\cos(\theta + \phi), d\sin(\phi) - r\ \sin(\theta + \phi), z_A, \vartheta_A]^T$$
$$P_B = [d\cos(\phi) + r\cos(\theta + \phi), d\sin(\phi) + r\ \sin(\theta + \phi), z_B, \vartheta_B]^T \quad (6)$$

$z_A$ and $z_B$ are the vertical coordinates ($z_A = z_B$) of points $A$ and $B$, and $\vartheta_A = 0, \vartheta_B = \pi/6$ are the orientation of the end-effector at the pick and place points, respectively.

The optimal task placement problem aims at defining the task coordinates $(z, d, \phi, \theta)$ that minimize the overall energy consumption $E_c$ required by the motors of the manipulator to perform the given trajectory. Two different cases are considered:

- Case (1): the optimal task placement is found for discrete values of coordinate $z$, by optimizing $(d, \phi, \theta)$ for each different layer;
- Case (2): the optimal task placement is found by optimizing $(z, d, \phi, \theta)$ over the entire workspace of the parallel manipulator.

The minimum-energy design problem for the more general scenario is formulated as a bounded constrained nonlinear optimization problem that assumes the following form:

$$\begin{aligned} &\min_{x \in \chi}\{f(\boldsymbol{x})\} \\ &f(\boldsymbol{x}) := E_c(\boldsymbol{x}) \\ &\boldsymbol{x} := [z, d, \phi, \theta]^T \\ &\boldsymbol{\chi} := [\boldsymbol{x}^L, \boldsymbol{x}^U]^T \\ &\text{subject to: } \boldsymbol{H}(\boldsymbol{x}) \geq c \end{aligned} \quad (7)$$

$\boldsymbol{H}$ and $c$ define the inequality constraint that accounts for the truncated cone shape of the robot workspace. The lower and upper bounds for the design domain $\boldsymbol{\chi}$ assume the values $\boldsymbol{x}^L := [-1.230,\ 0.000,\ 0.00,\ 0.00]^T$ and $\boldsymbol{x}^U := [-0.730, 0.650, \pi/2, \pi]^T$.

In Case (1) the inequality constraint is replaced by the computation of the upper bound of variable $d$ for each coordinate $z$ prior to starting the optimization solution.

## 4 Case Study

The task taken into account in this work is a common pick-and-place trajectory composed of two vertical and one horizontal spans, as shown in Fig. 2. The motion profile is planned on the basis of the '434' spline algorithm, introduced by Cook and Ho in [26]. This trajectory planning strategy allows to compute a sequence of third and fourth degree polynomial functions through a sequence of via-points by knowing the time distance between each two consecutive via-points. The degrees of the polynomials ensure $C^2$ continuity.

In this work, the sequence of via-points is defined in the operational space in order to keep the length of the path fixed for the purposes of the energetic analysis. As it can be seen in Fig. 2, nine via-points are taken into account: $A$ and $B$ define the pick and place endpoints, $D$, $E$, $F$ and $G$, $H$, $I$ are junction points in the corners of the path, and $M$ is the via-point at the center of the horizontal track. The path is symmetrical with respect to vertical axis that passes through points $M$ and $C$. The height and length of the path respect the 1″-12″-1″ dimensions of the common benchmark task (straight line motion of 25 mm up, 305 mm over, 25 mm down, and back along the same path). A radius of 5 mm blends the path corners. The total execution time is set to 0.400 s.

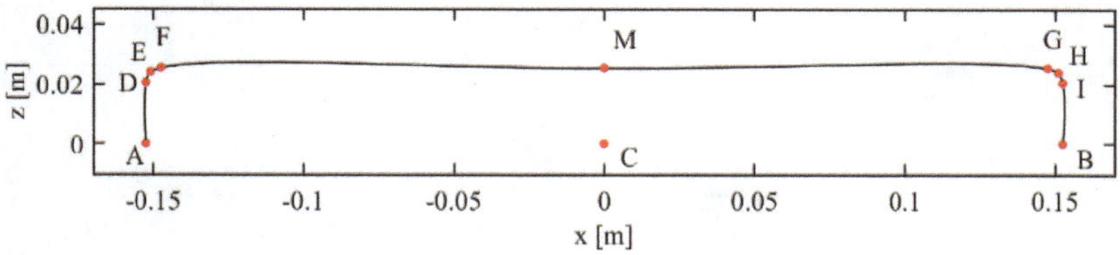


**Fig. 2.** Path of the robot in the operative space (aligned with the $x$-axis).

The dynamic and electro-mechanical parameters of the parallel manipulator implemented in this work are reported in Table 1, and an example trajectory in the operative space is shown in Fig. 3.

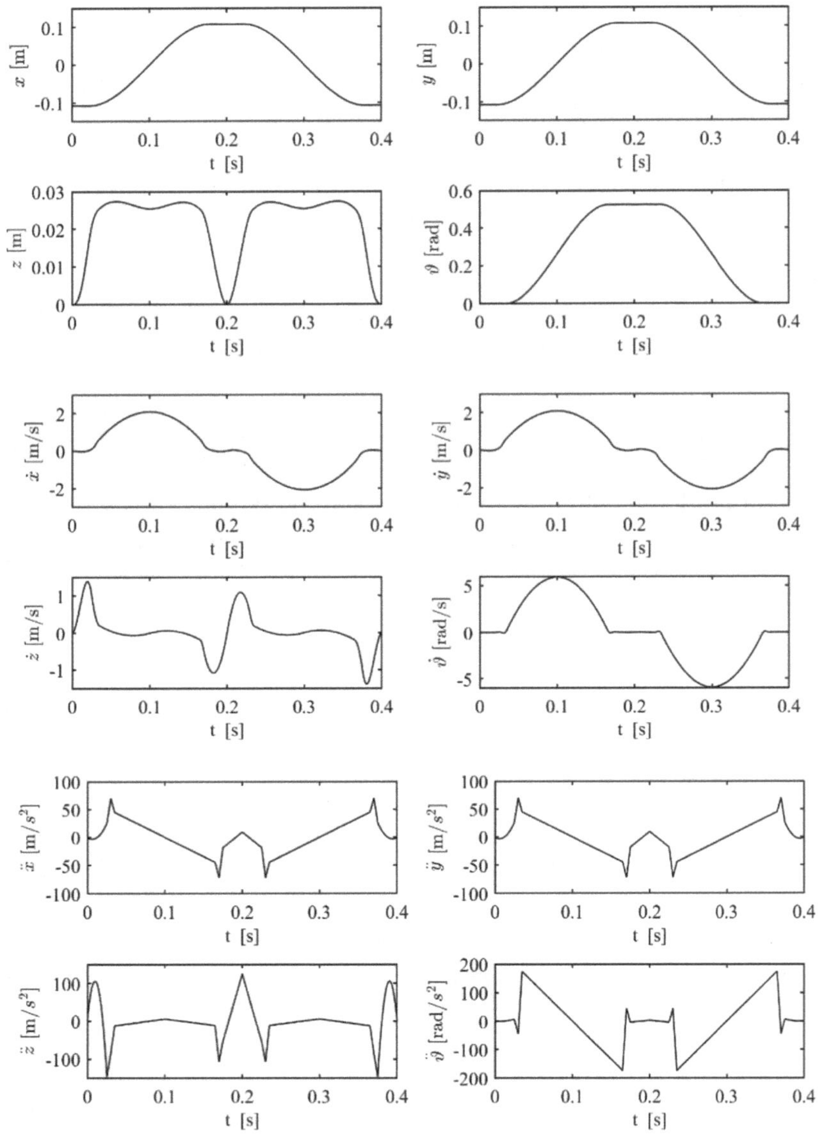


**Fig. 3.** Example trajectory in the operative space ($z = 0.00$, $d = 0.00$, $\phi = 0.00$, $\theta = \pi/4$).

**Table 1.** Dynamic and electro-mechanical parameters of the parallel manipulator.

| Parameter | Symbol | Value |
|---|---|---|
| Static friction coefficient | $f_s$ | 0.109 Nm |
| Viscous friction coefficient | $f_v$ | $1.62 \cdot 10^{-4}$ Nm • s/rad |
| Reduction ratio | $K_r$ | 50 |
| Motor torque constant | $K_t$ | 0.99 Nm/A |
| Motor back-emf constant | $K_e$ | 0.61 V · s/rad |
| Motor winding resistance | $R$ | 2.81 Ω |
| Driver efficiency | $\eta_d$ | 0.90 |

## 5 Results

The optimization problem is solved using the constrained nonlinear multivariable Matlab function *fmincon*, which implements the interior-point method. For the solution of the problem a multi-start optimization is performed, in order to better explore the entire design domain $\chi$. In Case (1) 20 steps for coordinate $z$ are chosen within the vertical bounds of the robot workspace. For each layer, a multi-start optimization is performed starting from 20 points randomly chosen in the design domain, for a total of 400 runs. Several different solutions for each layer are found, often very close to each other due to the numerical stop conditions of the optimization solver in the proximity of a minimum point. For each layer, only the minimum-energy solution within the feasible ones is considered. Figure 4 shows the optimal task placement for each of the 20 layers defined in Case (1). Red dots indicate the starting point $A$ of the path (Fig. 2).

Table 2 reports the optimal solutions for the $z$ coordinates analyzed in Case (1). The overall minimum energy solution in highlighted. From Fig. 4 and from the values reported in Table 2 it can be seen that for values of $z$ greater than $-1.00$ m (upper part of the workspace) the minimum-energy solutions are located near the bound of the workspace for $d \cong 0.50$ m, $\phi \cong 1.52$ rad and $\theta \cong 1.64$ rad.

**Table 2.** Optimal solutions for Case (1). The overall minimum solution is highlighted.

| $z$ [m] | $d$ [m] | $\phi$ [rad] | $\theta$ [rad] | $E_{c,min}$ [J] | $z$ [m] | $d$ [m] | $\phi$ [rad] | $\theta$ [rad] | $E_{c,min}$ [J] |
|---|---|---|---|---|---|---|---|---|---|
| -0.730 | 0.498 | 1.523 | 1.651 | 111.3 | -0.980 | 0.394 | 1.513 | 1.650 | 91.46 |
| -0.755 | 0.498 | 1.521 | 1.646 | 106.4 | -1.005 | 0.254 | 0.004 | 1.564 | 90.74 |
| -0.780 | 0.498 | 1.521 | 1.643 | 102.5 | -1.030 | 0.163 | 0.003 | 1.569 | 89.95 |
| -0.805 | 0.498 | 1.522 | 1.639 | 99.30 | -1.055 | 0.092 | 0.005 | 1.571 | 89.43 |
| -0.830 | 0.498 | 1.525 | 1.637 | 96.78 | -1.080 | 0.077 | 0.008 | 1.573 | 89.36 |
| -0.855 | 0.498 | 1.513 | 1.640 | 94.83 | -1.105 | 0.076 | 0.036 | 1.526 | 89.77 |
| -0.880 | 0.498 | 1.516 | 1.637 | 93.41 | -1.130 | 0.066 | 0.122 | 1.440 | 90.76 |
| -0.905 | 0.498 | 1.522 | 1.634 | 92.47 | -1.155 | 0.064 | 0.103 | 1.433 | 92.46 |
| -0.930 | 0.496 | 1.513 | 1.635 | 92.00 | -1.180 | 0.058 | 0.161 | 1.376 | 95.13 |
| -0.955 | 0.450 | 1.515 | 1.641 | 91.75 | -1.205 | 0.057 | 0.205 | 1.337 | 99.17 |

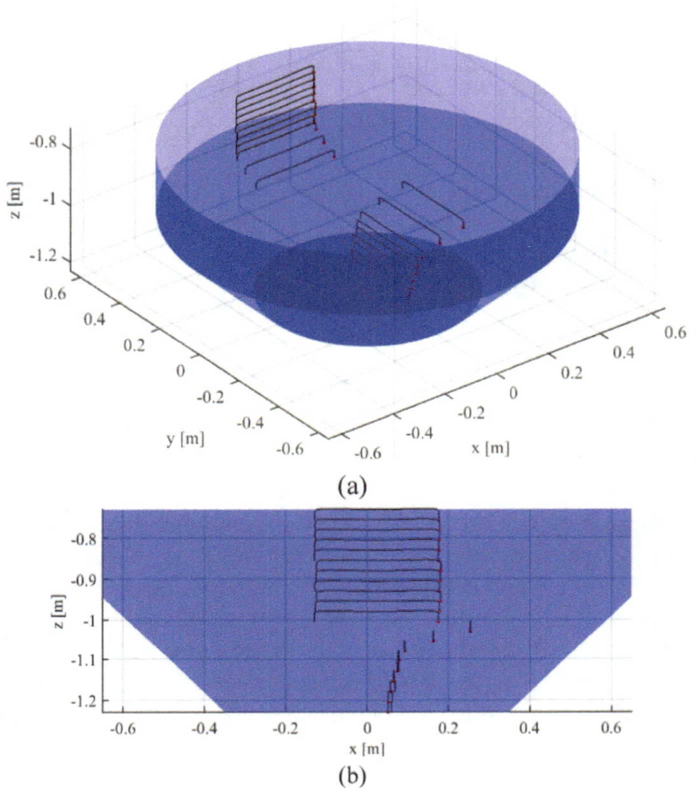


**Fig. 4.** Optimal task placements for Case (1): isometric view (a); front view (b).

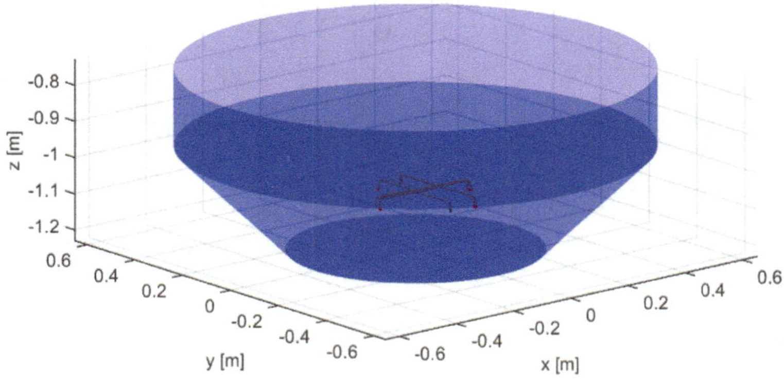


**Fig. 5.** Optimal task placements for Case (2).

As $z$ decreases below $-1.00\,\mathrm{m}$ (lower part of the workspace), the optimal paths move to the center of the working volume (Fig. 4(b)) and are oriented with $\phi$ close to $0.00\,\mathrm{rad}$ and $\theta$ close to $1.50\,\mathrm{rad}$. The overall minimum-energy task is found for

$z = -1.080\,\text{m}$, $d = 0.077\,\text{m}, \phi = 0.008\,\text{rad}$ and $\theta = 1.573\,\text{rad}$, and corresponds to a value of consumed energy equal to $E_{c,min} = 89.36\,\text{J}$. The results obtained for Case (1) could be useful when the relative vertical positioning between robot and working surface is defined a priori.

In Case (2) the optimal placement of the pick-and-place task is found within the entire robot workspace. In this case the multi-start optimization is performed starting from 100 random points within the domain, yielding to 16 different solutions. The optimal task placements for Case (2) are shown in Fig. 5. Within these 16 solutions, the overall minimum-energy one is obtained for $z = -1.072\,\text{m}$, $d = 0.081\,\text{m}, \phi = 0.008\,\text{rad}$ and $\theta = 1.574\,\text{rad}$. This solution corresponds to a value of consumed energy equal to $E_{c,min} = 89.33\,\text{J}$, perfectly in accordance with the results obtained for Case (1). The other 15 solutions found by the optimization solver show energy values that differ by less than 1% from this one.

## 6 Conclusions

This paper presented the optimization of the task placement to achieve energy efficiency in a parallel manipulator. The described approach is based on the parametrization of the task whose position and orientation are varied within the robot workspace. The energy consumed by the actuators is considered as the cost function for the optimization problem and is computed on the basis of the dynamic and electromechanical models of the manipulator. The minimum-energy solution is found either for discrete values of the vertical coordinate, or globally within the entire workspace of the robot. The method is general and can be used as an off-line tool for the optimal placement of different tasks within the workspace of the parallel manipulator. Future developments of this work could include the experimental validation of the system and the development of a toolbox to compute the optimal solution for generic user-defined robot paths.

## References


1. Merlet, J.P.: Parallel Robots, vol. 128. Springer, Cham (2006)
2. Mo, J., Shao, Z.F., Guan, L., Xie, F., Tang, X.: Dynamic performance analysis of the X4 high-speed pick-and-place parallel robot. Robot. Comput. Integr. Manuf. **46**, 48–57 (2017)
3. Vidoni, R., Boscariol, P., Gasparetto, A., Giovagnoni, M.: Kinematic and dynamic analysis of flexible-link parallel robots by means of an ERLS approach. In: ASME 2012 International Design Engineering Technical Conferences and Computers and Information in Engineering Conferences, pp. 1449–1458. ASME Digital Collection (2012)
4. Tang, X.: An overview of the development for cable-driven parallel manipulator. Adv. Mech. Eng. **6**, 823028 (2014)
5. Scalera, L., Gasparetto, A., Zanotto, D.: Design and experimental validation of a 3-DOF underactuated pendulum-like robot. IEEE/ASME Trans. Mechatron. **25**(1), 217–228 (2020)
6. Carabin, G., Wehrle, E., Vidoni, R.: A review on energy-saving optimization methods for robotic and automatic systems. Robotics **6**(4), 39 (2017)

7. Chen, C.T., Liao, T.T.: A hybrid strategy for the time-and energy-efficient trajectory planning of parallel platform manipulators. Robot. Comput. Integr. Manuf. **27**(1), 72–81 (2011)
8. Khoukhi, A., Baron, L., Balazinski, M.: Constrained multi-objective trajectory planning of parallel kinematic machines. Robot. Comput. Integr. Manuf. **25**(4–5), 756–769 (2009)
9. Zhang, X., Ming, Z.: Trajectory planning and optimization for a Par4 parallel robot based on energy consumption. Appl. Sci. **9**(13), 2770 (2019)
10. Scalera, L., Palomba, I., Wehrle, E., Gasparetto, A., Vidoni, R.: Natural motion for energy saving in robotic and mechatronic systems. Appl. Sci. **9**(17), 3516 (2019)
11. Barreto, J.P., Corves, B.: Resonant delta robot for pick-and-place operations. In: IFToMM World Congress on Mechanism and Machine Science, pp. 2309–2318. Springer (2019)
12. Scalera, L., Carabin, G., Vidoni, R., Wongratanaphisan, T.: Energy efficiency in a 4-DOF parallel robot featuring compliant elements. Int. J. Mech. Control **20**(02) (2019)
13. Nelson, B., Donath, M.: Optimizing the location of assembly tasks in a manipulator's workspace. J. Robot. Syst. **7**(6), 791–811 (1990)
14. Pamanes, G.J.A., Zeghloul, S.: Optimal placement of robotic manipulators using multiple kinematic criteria. In: Proceedings of the IEEE International Conference on Robotics and Automation, pp. 933–938 (1991)
15. Feddema, J.T.: Kinematically optimal robot placement for minimum time coordinated motion. In: Proceedings of the IEEE International Conference on Robotics and Automation, vol. 4, pp. 3395–3400 (1996)
16. Aspragathos, N.A., Foussias, S.: Optimal location of a robot path when considering velocity performance. Robotica **20**(2), 139–147 (2002)
17. Kamrani, B., Berbyuk, V., Wäppling, D., Stickelmann, U., Feng, X.: Optimal robot placement using response surface method. Int. J. Adv. Manuf. Technol. **44**(1–2), 201–210 (2009)
18. Ur-Rehman, R., Caro, S., Chablat, D., Wenger, P.: Multi-objective path placement optimization of parallel kinematics machines based on energy consumption, shaking forces and maximum actuator torques: application to the Orthoglide. Mech. Mach. Theory **45**, 1125–1141 (2010)
19. Boschetti, G., Rosa, R., Trevisani, A.: Optimal robot positioning using task-dependent and direction-selective performance indexes: general definitions and application to a parallel robot. Robot. Comput. Integr. Manuf. **29**, 431–443 (2013)
20. Doan, N.C.N., Lin, W.: Optimal robot placement with consideration of redundancy problem for wrist-partitioned 6R articulated robots. Robot. Comput. Integr. Manuf. **48**, 233–242 (2017)
21. Hassan, M., Liu, D., Paul, G.: Collaboration of multiple autonomous industrial robots through optimal base placements. J. Intell. Robot. Syst. **90**(1–2), 113–132 (2018)
22. Valsamos, C., Wolniakowski, A., Miatliuk, K., Moulianitis, V.C.: Optimal placement of a kinematic robotic task for the minimization of required joint velocities. Int. J. Mech. Control **20**, 3–14 (2019)
23. Boscariol, P., Scalera, L., Gasparetto, A.: Task-dependent energetic analysis of a 3 DOF industrial manipulator. In: International Conference on Robotics in Alpe-Adria Danube Region, pp. 162–169. Springer, Cham (2019)
24. Scalera, L., Boscariol, P., Carabin, G., Vidoni, R., Gasparetto, A.: Enhancing energy efficiency of a 4-DOF parallel robot through task-related analysis. Machines **8**(1), 10 (2020)
25. Pierrot, F., Nabat, V., Company, O., Krut, S., Poignet, P.: Optimal design of a 4-DOF parallel manipulator: from academia to industry. IEEE Trans. Robot. **25**, 213–224 (2009)
26. Cook, C.C., Ho, C.Y.: The application of spline functions to trajectory generation for computer-controlled manipulators. In: Computing Techniques for Robots, pp. 101–110. Springer, Boston (1984)

# A Numerical Procedure for Position Analysis of a Robotic Structure. Part I: General Methodology

Stelian Alaci[1(✉)], Ioan Doroftei[2], Florentin Buium[2], Florina-Carmen Ciornei[1], and Ionut-Cristian Romanu[1]

[1] Stefan cel Mare University, Suceava, Romania
stelian.alaci@usm.ro
[2] "Gheorghe Asachi" Technical University of Iasi, Iasi, Romania

**Abstract.** A method for solving position analysis of spatial mechanisms with cylindrical joints was proposed by Hartenberg and Denavit, based on the point coordinates transformation relations when the reference frame is changed. The equivalent scalar equations derived from the matrix closure equation are cumbersome and improbable to be solved analytically. To surpass this inconvenience, Uicker et al. presented a numerical algorithm for solving the matrix equation deduced by Hartenberg and Denavit. This paper proposes a generalization of Uicker's method that can be applied both for robotic systems and for mechanisms with all types of lower pairs. To this end, the spherical and planar pairs from the structure of the robot should be replaced by assemblies of cylindrical pairs. The proposed method assumes that all of the analysed kinematic chain's joints are revolute joints. The initial data to be stipulated are the constructive parameters of the analysed chain together with the homogenous operator which describes the position of the end-link with respect to the ground (the base). Therefore, a set of arbitrary position parameters of the chain are considered and a matrix equation is found; this equation allows for finding the corrections to be applied to the initial set of parameters, in order to obtain a new set of parameters which provide a closer position to the theoretical imposed position. The values of the corrections are found from a linear over-constrained system of 12 equations.

**Keywords:** Spatial kinematics · Robotic chain · Numerical procedure

## 1 Introduction

Due to new technology trends and market developments, the production quantity and variety of industrial robots increases constantly. Robotics is an interdisciplinary field and the progress of every sector leads to a dynamic development of the domain. From digital sensors linked to smart software, human-robot collaboration, or the digital connectivity of robots, all these directions of evolution are reflected in cutting edge medical equipment [1, 2], machining technologies [3], home appliances [4], environmental protection [5, 6], humanitarian demining [7], rehabilitation and health care [8], social robotics [9], co-manipulation and transportation [10, 11], ocean study


E.-C. Lovasz et al. (Eds.): MTM&Robotics 2020, MMS 88, pp. 23–32, 2021.
https://doi.org/10.1007/978-3-030-60076-1_3

machineries [12] or spatial equipment [13]. Most robots are required to move in order to complete a programmed task. The spatial kinematic chains are widely used in robotics and mechanisms applications [14, 15]. A general methodology for the kinematics study of spatial mechanisms with cylindrical joints was offered by Hartenberg and Denavit [16]. The method is based on the remark that the constituent parts of a mechanism create closed chains. A coordinate system is attached to each link of the chain, where one of the links is considered as reference link; on this link, a generic point is taken and the coordinates of this point can be expressed in any of the frames attached to the links. Hartenberg and Denavit proposed that the relation for coordinate transformation should be written as $4 \times 4$ matrices. These matrices describe the relative position of the two frames, by means of the vector which characterises the relative position of the origins, on one side, and of the square type $3 \times 3$ matrix, describing the orientations of the axis of a system versus the other. The elements of the matrix are the projections of the current system versors on the axes of the reference frame, on the other side. For a given mechanism, the base is generally considered reference link. The other links are taken in the succession given by the chain connection, and the transformation relations of a point's coordinates are written from the base until the last link is overlapping the base. The subsequent equation states that the product of all transformation matrices must equal the unit $4 \times 4$ matrix. This matrix equation is equivalent to a system of 16 scalar equations, from which the equations corresponding to the elements of the last row are identically satisfied. From the 12 remaining equations, only 6 are independent, three obtained from the coincidence of displacement vectors, and the latter three resulting from the equality of the rotation matrices.

The problem's complication stems from the selection of the characteristic equations for the rotation matrices identity, and subsequently the solving of the system. The system of equation with six unknowns is difficult to solve because it is a system of trigonometric equations with multiple solutions, corresponding to all the possible positions for assembling the parts, when the positions of the driving links are stipulated. Since the analytical solution of the system is difficult to reach, Uicker et al. [17] proposed a numerical method for finding the position of the mechanism. The method consists in considering all 12 equations of the system and resolving it as an over constrained system. The actual manner assumes accepting a set of parameters characteristic to the position of the mechanism, and then, using the over constrained system, finding the corrections required by the initial position parameters. These values revise the initial set of position parameters and a new calculus of correction starts. The procedure runs until the values of the position parameters become stable. An application of the method is presented in [18, 19] with the mention of rapid convergence as only a few iterations are necessary for stable results.

## 2 Modified Methodology. Problem Statement

Uicker et al.'s proposed method [17], has two main weaknesses: it can only be applied for mechanisms (closed kinematic chains) and for cylindrical pairs. Fischer [20] applies the Hartenberg-Denavit method by expressing the matrices in dual form (as used by

Dimentberg [21] and Yang [22]) and develops the methodology for all types of mechanisms with lower pairs, stipulating the technique of replacing the spherical and planar joints.

This paper applies the Hartenberg-Denavit method for an open chain, characteristic to a robot, Fig. 1. A robot is an open kinematical chain, with controlled motion and with the declared task of placing an object in *a priori* précised position. Both instances are taken into account, when the position of the object is stationary or variable. The problem to be solved is to find the parameters from the driving joints of a given robot having particular constructive characteristics, in otder to obtain the imposed position of the object. The convention used in the Hartenverg-Denavit method is applied for resolving the task: the motions are performed with respect to the axes of the cylindrical joints, denoted by $Oz$, and the geometry of the parts is stipulated by the twisting angle and the length of the common normal of two axes of motion placed on the same link.

## 3 Solving the Problem. Corrections

The robot from Fig. 1 has $n$ links and $n$ cylindrical joints. The object to be placed, fixed to the last $n^{th}$ link has a reference frame denoted "*e*" attached to it (here *e* comes from effector). The position of the object with respect to the base "*1*" is given by the $\boldsymbol{T}_{1e}$ matrix and, according to McCarthy [23], this matrix has the form

$$\boldsymbol{T}_{1e} = \boldsymbol{T}_{12}\boldsymbol{T}_{23}\cdot\ldots\cdot\boldsymbol{T}_{n-1,n}\boldsymbol{T}_{n,e}. \tag{1}$$

Applying the Hartenberg-Denavit convention, a factor form Eq. (1) is expressed as a product of two matrix factors

$$\boldsymbol{T}_{k,k+1} = \boldsymbol{Z}(\theta_k, s_k)\boldsymbol{X}(\alpha_{k,k+1}, a_{k,k+1}), \tag{2}$$

where the matrices

$$\boldsymbol{Z}(\theta, s) = \begin{bmatrix} cos\,\theta & -sin\,\theta & 0 & 0 \\ sin\,\theta & cos\,\theta & 0 & 0 \\ 0 & 0 & 0 & s \\ 0 & 0 & 0 & 1 \end{bmatrix} \tag{3}$$

characterise the rotation of angle $\theta$ and the translation $s$ on $Oz$ direction and

$$\boldsymbol{X}(\alpha, a) = \begin{bmatrix} 1 & 0 & 0 & a \\ 0 & cos\,\alpha & -sin\,\alpha & 0 \\ 0 & sin\,\alpha & cos\,\alpha & 0 \\ 0 & 0 & 0 & 1 \end{bmatrix}, \tag{4}$$

where $\alpha$ describes the twist and $a$ is the length of the common normal between two axes of motion $Oz$ from the same link. The equation of condition, expressed as Eq. (1), can be written under the form

$$\boldsymbol{Z}(\theta_1, s_1)\boldsymbol{X}(\alpha_{12}, a_{12})\ldots..\boldsymbol{Z}(\theta_k, s_k)\boldsymbol{X}(\alpha_{k,k+1}, a_{k,k+1})\ldots..\boldsymbol{Z}(\theta_n, s_n)\boldsymbol{X}(\alpha_{n,e}, a_{ne}) = \boldsymbol{T}_{1e}. \tag{5}$$

Assuming that the same steps are made as in the case of a mechanism (when always $\boldsymbol{T}_{1e} \equiv I_4$), following the method summarized above, the problem is theoretically solved. But the difficulties in analytical calculus require the use of a numerical procedure, similar to the one described by Uicker [17]. Unlike Uicker's methodology, where the revolute joint and the prismatic joint are independently explained, this paper intends a unitary character and considers cylindrical joints for all the pairs of the robot ($\boldsymbol{T}_{1e} \neq I_4$).

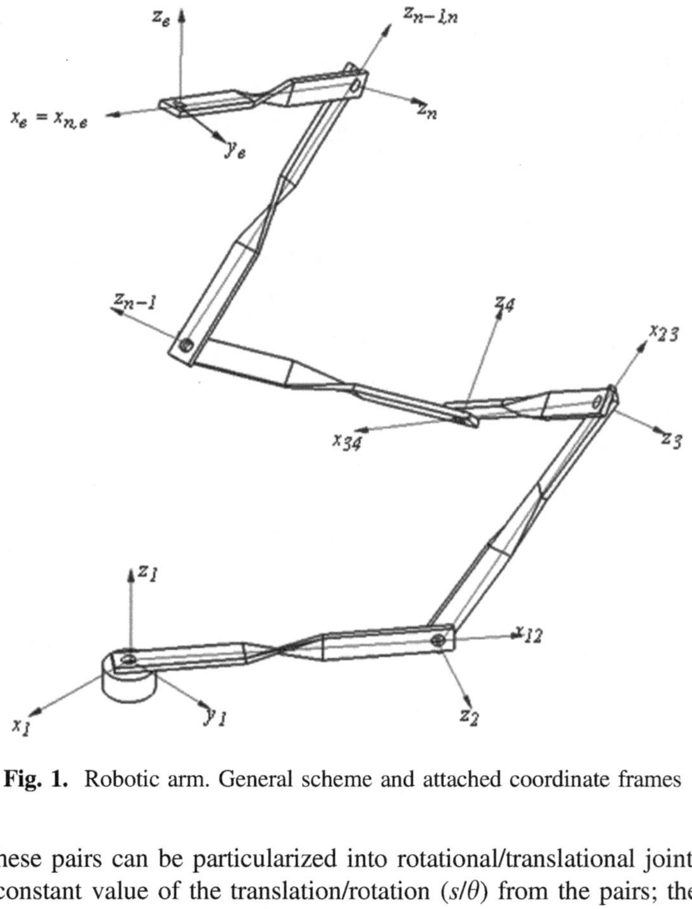


**Fig. 1.** Robotic arm. General scheme and attached coordinate frames

Next, these pairs can be particularized into rotational/translational joints by specifying the constant value of the translation/rotation ($s$/$\theta$) from the pairs; these parameters now describe the geometry of the robot instead of motion. The set of variable parameters of position $\theta_i$ and $s_j$ ($i$ is the index of the revolute joint and $j$ is the index of prismatic joint) verify the Eq. (5). This equation is written as

$$\boldsymbol{Z}(\theta_1, s_1)\boldsymbol{X}(\alpha_{12}, a_{12})\ldots..\boldsymbol{Z}(\theta_k, s_k)\boldsymbol{X}(\alpha_{k,k+1}, a_{k,k+1})\ldots..\boldsymbol{Z}(\theta_n, s_n)\boldsymbol{X}(\alpha_{n,e}, a_{ne}) = \boldsymbol{T}^o_{1e}, \tag{6}$$

where $\boldsymbol{T}^o_{1e}$ is the position matrix imposed to the final link "$e$". For another set of parameters which differ by $d\theta_i$ and $ds_j$ respectively from the first set, the product from the left member of Eq. (6) is a matrix $\boldsymbol{T}^a_{1n}$, different from $\boldsymbol{T}^o_{1e}$,

$$\begin{aligned}&\boldsymbol{Z}(\theta_1 + d\theta_1, s_1 + ds_1)\boldsymbol{X}(\alpha_{12}, a_{12})\ldots..\boldsymbol{Z}(\theta_k + d\theta_k, s_k + ds_k)\boldsymbol{X}(\alpha_{k,k+1}, a_{k,k+1})\ldots..\\&\ldots.\boldsymbol{Z}(\theta_n + d\theta_n, s_n + ds_n)\boldsymbol{X}(\alpha_{n,e}, a_{ne}) = \boldsymbol{T}^n_{1e}\end{aligned}. \tag{7}$$

The $d\theta_i$ and $ds_j$ growths are considered zero in the pairs where the respective motions don't take place. Considering the relations

$$cos\, d\theta \cong 1 \tag{8}$$

$$sin\, d\theta \cong d\theta, \tag{9}$$

which are appropriate for very small $d\theta < 10°$, ($\pi/18 = 0.174$; $sin(10°) = 0.1736$; $cos(10°) = 0.9848$), we write

$$cos(\theta + d\theta) = cos\,\theta\, cos\, d\theta - sin\,\theta\, sin\, d\theta \cong cos\,\theta - sin\,\theta\, d\theta \tag{10}$$

$$sin(\theta + d\theta) = sin\,\theta\, cos\,\theta + cos\,\theta sin\, d\theta = sin\,\theta + d\theta\, cos\,\theta \tag{11}$$

and taking into account the relation

$$Z(\theta, s) = Z(\theta, 0)Z(0, s) \tag{12}$$

then

$$\boldsymbol{Z}(\theta + d\theta, s + ds) = \boldsymbol{Z}(\theta + d\theta, 0)\boldsymbol{Z}(0, s + ds). \tag{13}$$

Or, explicitly

$$\boldsymbol{Z}(\theta + d\theta, 0) = \begin{bmatrix} cos(\theta + d\theta) & -sin(\theta + d\theta) & 0 & 0 \\ sin(\theta + d\theta) & cos(\theta + d\theta) & 0 & 0 \\ 0 & 0 & 1 & 0 \\ 0 & 0 & 0 & 1 \end{bmatrix} = \boldsymbol{Z}(\theta, 0)(\boldsymbol{I}_4 + d\theta\, Q_\theta), \tag{14}$$

where

$$\boldsymbol{Q}_\theta = \begin{bmatrix} 0 & -1 & 0 & 0 \\ 1 & 0 & 0 & 0 \\ 0 & 0 & 0 & 0 \\ 0 & 0 & 0 & 0 \end{bmatrix}. \tag{15}$$

In a similar manner it is shown that

$$\boldsymbol{Z}(0, s+ds) = \begin{bmatrix} 0 & 0 & 0 & 0 \\ 0 & 0 & 0 & 0 \\ 0 & 0 & 0 & s+ds \\ 0 & 0 & 0 & 0 \end{bmatrix} = \boldsymbol{Z}(0, s)(\boldsymbol{I}_4 + ds\boldsymbol{Q}_s), \tag{16}$$

where

$$\boldsymbol{Q}_s = \begin{bmatrix} 0 & 0 & 0 & 0 \\ 0 & 0 & 0 & 0 \\ 0 & 0 & 0 & 1 \\ 0 & 0 & 0 & 0 \end{bmatrix}. \tag{17}$$

Based on Eqs. (14) and (17), it results that

$$\begin{aligned} &\boldsymbol{Z}(\theta + d\theta, s+ds) = \boldsymbol{Z}(\theta + d\theta, 0)\boldsymbol{Z}(0, s+ds) \\ &= \boldsymbol{Z}(\theta, 0)[\boldsymbol{I}_4 + d\theta\boldsymbol{Q}_q)\boldsymbol{Z}(0, s+ds)(\boldsymbol{I}_4 + ds\boldsymbol{Q}_s) \\ &= \boldsymbol{Z}(\theta, 0)\boldsymbol{Z}(0, s)[\boldsymbol{I}_4 + d\theta\,\boldsymbol{Q}_q][\boldsymbol{I}_4 + ds\boldsymbol{Q}_s] \\ &= \boldsymbol{Z}(\theta, 0)\boldsymbol{Z}(0, s)[\boldsymbol{I}_4 + d\theta\,\boldsymbol{Q}_\theta + ds\,\boldsymbol{Q}_s + ds\,d\theta\,\boldsymbol{Q}_q\boldsymbol{Q}_s] \\ &\cong \boldsymbol{Z}(\theta, s)[\boldsymbol{I}_4 + d\theta\,\boldsymbol{Q}_\theta + ds\,\boldsymbol{Q}_s] \end{aligned} \tag{18}$$

In addition, this can be verified by straightforward calculus

$$\boldsymbol{Z}(\theta, s)[\boldsymbol{I}_4 + d\theta\,\mathbf{Q}_\theta + ds\,\boldsymbol{Q}_s] = [\boldsymbol{I}_4 + d\theta\,\boldsymbol{Q}_\theta + ds\,\boldsymbol{Q}_s]\boldsymbol{Z}(\theta, s). \tag{19}$$

The left side of the Eq. (7) is expanded, taking into account the Eq. (19), after the small infinities of degree bigger than one are neglected and becomes

$$
\begin{gathered}
\mathbf{Z}(\theta_1 + d\theta_1, s_1 + ds_1)\boldsymbol{X}(\alpha_{12}, a_{12})\ldots..\boldsymbol{Z}(\theta_k + d\theta_k, s_k + ds_k)\boldsymbol{X}(\alpha_{k,k+1}, a_{k,k+1})\ldots \\
\boldsymbol{Z}(\theta_n + d\theta_n, s_n + ds_n)\boldsymbol{X}(\alpha_{n,e}, a_{ne}) \\
= [\mathbf{I}_4 + d\theta_1\, \boldsymbol{Q}_\theta + ds_1\, \boldsymbol{Q}_s]\mathbf{Z}(\theta_1, s_1)\mathbf{X}(\alpha_{12}, a_{12})\ldots \\
.[\mathbf{I}_4 + d\theta_k\, \boldsymbol{Q}_\theta + ds_k\, \boldsymbol{Q}_s]\boldsymbol{Z}(\theta_k, s_k)\boldsymbol{X}(\alpha_{k,k+1}, a_{k,k+1})\ldots\ldots.. \\
\cdot\ldots..[\mathbf{I}_4 + d\theta_n\, \boldsymbol{Q}_\theta + ds_n\, \boldsymbol{Q}_s]\mathbf{Z}(\theta_n, s_n)\boldsymbol{X}(\alpha_{ne}, a_{ne}) \\
= \boldsymbol{Z}(\theta_1, s_1)\mathbf{X}(\alpha_{12}, a_{12})\ldots..\boldsymbol{X}(\alpha_{k-1,k}, a_{k-1,k})\mathbf{Z}(\theta_k, s_k)\ldots\boldsymbol{Z}(\theta_n, s_n)\boldsymbol{X}(\alpha_{n,e}, a_{n,e}) \\
+\ \boldsymbol{Z}(\theta_1, s_1)(d\theta_1\, \boldsymbol{Q}_\theta + ds_1\, \boldsymbol{Q}_s)\boldsymbol{X}(\alpha_{12}, a_{12})\ldots \\
\mathbf{Z}(\theta_k, s_k)(d\theta_k\, \boldsymbol{Q}_\theta + ds_k\, \boldsymbol{Q}_s)\boldsymbol{X}(\alpha_{k,k+1}, a_{k,k+1})\ldots\ldots. \\
\cdot\boldsymbol{Z}(\theta_n, s_n)(d\theta_n\, \boldsymbol{Q}_\theta + ds_n\, \boldsymbol{Q}_s)\boldsymbol{X}(\alpha_{ne}, a_{ne}) \\
= \boldsymbol{Z}(\theta_1, s_1)\boldsymbol{X}(\alpha_{12}, a_{12})\ldots..\boldsymbol{X}(\alpha_{k-1,k}, a_{k-1,k})\mathbf{Z}(\theta_k, s_k)\ldots\boldsymbol{Z}(\theta_n, s_n)\boldsymbol{X}(\alpha_{n,e}, a_{n,e}) \\
+ \sum_{k=1}^{n} \boldsymbol{W}_k
\end{gathered}
, \tag{20}
$$

where the notation $\boldsymbol{W}_k$ is applied for

$$
\boldsymbol{W}_k = \begin{cases}
\mathbf{Z}(\theta_1, s_1)(d\theta_1 \boldsymbol{Q}_q + ds_1 \boldsymbol{Q}_s)\boldsymbol{X}(\alpha_{1,2}, a_{1,2})\cdot \\
\quad \cdot \prod_{p=k}^{n} \boldsymbol{Z}(\theta_p, s_p)\boldsymbol{X}(\alpha_{p,p+1}, a_{p,p+1}) & k = 1 \\
\left\{\prod_{p=1}^{k-1} \boldsymbol{Z}(\theta_p, s_p)\boldsymbol{X}(\alpha_{p,p+1}, a_{p,p+1})\right\} \cdot \left[d\theta_k \boldsymbol{Q}_q + ds_k \boldsymbol{Q}_s\right] & k > 1, \\
\quad \cdot \left\{\prod_{p=k}^{n} \boldsymbol{Z}(\theta_p, s_p)\boldsymbol{X}(\alpha_{p,p+1}, a_{p,p+1})\right\} & n + 1 \equiv e
\end{cases}
. \tag{21}
$$

With the above considerations, the Eq. (7) is written as

$$
\begin{aligned}
&\boldsymbol{Z}(\theta_1, s_1)\boldsymbol{X}(\alpha_{12}, a_{12})\ldots..\boldsymbol{X}(\alpha_{k-1,k}, a_{k-1,k})\boldsymbol{Z}(\theta_k, s_k)\ldots\boldsymbol{Z}(\theta_n, s_n)\boldsymbol{X}(\alpha_{n,e}, a_{n,e}) \\
&+ \sum_{k=1}^{n} \boldsymbol{W}_k = \boldsymbol{T}^a_{1e}
\end{aligned}
\tag{22}
$$

and by obeying Eq. (6), it results that

$$
\sum_{k=1}^{n} \boldsymbol{W}_k + \ldots\ldots.. = \boldsymbol{T}^a_{1e} - \boldsymbol{T}^o_{1e}. \tag{23}
$$

The matrix products occurring in Eq. (23) can be completed in a straightforward way by rewriting the Eq. (21) under the form

$$\begin{aligned}
\boldsymbol{W}_k &= \left\{\prod_{p=1}^{k-1} \boldsymbol{Z}(\theta_p, s_p)\boldsymbol{X}(\alpha_{p,p+1}, a_{p,p+1})\right\}\left[d\theta_k \boldsymbol{Q}_q + ds_k \boldsymbol{Q}_s\right]\left\{\prod_{p=k}^{n} \boldsymbol{Z}(\theta_p, s_p)\boldsymbol{X}(\alpha_{p,p+1}, a_{p,p+1})\right\} \\
&= \left[\prod_{p=1}^{k-1} \boldsymbol{Z}(\theta_p, s_p)\boldsymbol{X}(\alpha_{p,p+1}, a_{p,p+1})\right]\left[d\theta_k \boldsymbol{Q}_q + ds_k \boldsymbol{Q}_s\right]\left\{\prod_{p=1}^{k-1} \boldsymbol{Z}(\theta_p, s_p)\boldsymbol{X}(\alpha_{p,p+1}, a_{p,p+1})\right\}^{-1} \\
&\quad \left[\prod_{p=1}^{k-1} \boldsymbol{Z}(\theta_p, s_p)\boldsymbol{X}(\alpha_{p,p+1}, a_{p,p+1})\right]\left[\prod_{p=k}^{n} \boldsymbol{Z}(\theta_p, s_p)\boldsymbol{X}(\alpha_{p,p+1}, a_{p,p+1})\right] \\
&= \left[\prod_{p=1}^{k-1} \boldsymbol{Z}(\theta_p, s_p)\boldsymbol{X}(\alpha_{p,p+1}, a_{p,p+1})\right]\left[d\theta_k \boldsymbol{Q}_q + ds_k \boldsymbol{Q}_s\right]\left\{\prod_{p=1}^{k-1} \boldsymbol{Z}(\theta_p, s_p)\boldsymbol{X}(\alpha_{p,p+1}, a_{p,p+1})\right\}^{-1} \cdot \boldsymbol{T}^o_{1e}
\end{aligned} \tag{24}$$

Next, taking into account the following transformation

$$\boldsymbol{T}^o_{1,k} = \prod_{p=1}^{k-1} \boldsymbol{Z}(\theta_p, s_p)\boldsymbol{X}(\alpha_{p,p+1}, a_{p,p+1}), \tag{25}$$

which represents the matrix of position of frame $k$ with respect to the ground.

Therefore, Eq. (23) can be written under the form

$$\sum_{k=1}^{n} \boldsymbol{T}^o_{1,k}\left[d\theta_k \boldsymbol{Q}_q + ds_k \boldsymbol{Q}_s\right]\boldsymbol{T}^{o-1}_{1,k}\boldsymbol{T}^o_{1,e} = \boldsymbol{T}^a_{1,e} - \boldsymbol{T}^o_{1,e}. \tag{26}$$

The Eq. (26) is multiplied by $\boldsymbol{T}^{o-1}_{1e}$ at the right, resulting

$$\sum_{k=1}^{n} \boldsymbol{T}^o_{1,k}\left[d\theta_k \boldsymbol{Q}_q + ds_k \boldsymbol{Q}_s\right]\boldsymbol{T}^{o-1}_{1,k} = \boldsymbol{T}^a_{1,e}\boldsymbol{T}^{o-1}_{1e} - \boldsymbol{I}_4. \tag{27}$$

The Eq. (27) allows for finding the corrections $d\theta_i$ and $ds_j$ to be applied to the vector of the initial displacements; a better approximation of the solution of Eq. (7) results

$$\theta_i = \theta_{0i} + d\theta_i,\ s_j = s_{0j} + ds_j. \tag{28}$$

As stated above, the Eq. (27) is equality between two $4 \times 4$ matrices. From the resulting 16 scalar equations, 4 equations are identically satisfied (the elements from the last row). Uicker [17] considers the remaining 12 equations as a constrained system with $i + j \leq 12$ which requires a numerical procedure to be solved.

## 4 Conclusions

The paper presents a generalized methodology for the kinematics analysis of spatial mechanisms. The principle of the method was offered by Hartenberg and Denavit who employed the homogenous operators ($4 \times 4$ matrices) to describe the coordinate transformations of a point, passing successively through all the frames attached to the links of a mechanism until reaching the frame from the initial link. The result is a matrix product that must be equal to the unity matrix $\boldsymbol{I}_4$. Equating the corresponding components of the rotation matrices and the components of sliding vectors, a system of 12 scalar equations is obtained; the analytical solving of this system is less probable.

Uicker et al. give a numerical method for solving the equation described by Hartenberg-Denavit, for the case of a mechanism with cylindrical pairs exclusively and the cases of rotation pairs and translation pairs are considered distinctly. The present paper approaches an expanded form of Uicker's method suitable for spatial kinematic chains (closed or open) with any type of lower pairs. With this aim, the spherical and cylindrical pairs are replaced by combinations of cylindrical, revolute and prismatic pairs.

The proposed method considers that all the pairs are cylindrical and may be particularized into revolute/prismatic pairs by blocking the sliding/rotation in a cylindrical pair. As input data, the values of the constructive parameters (that characterize the relative position of the axes of the pairs belonging to the same link) must be stated, as well as the operator describing the position of the last link referring to the ground.

Firstly, an initial set of parameters is accepted for the displacements from the cylindrical pairs. Then, the corrections to be applied to these initial values are found, in order to attain a new set of values, providing the robot with a position closer to the one theoretically imposed. Consequently, running a relatively small number of iterations, the set of desired parameters (equal to the theoretical ones) is obtained.

Evidently, for the general case, the homogenous operator characterizing the position of the last link can be a function of time, and the final values obtained by applying the method should be time functions, representing the relative motions from the driving pairs.

## References


1. Varrasi, S., Lucas, A., Soranzo, A., McNamara, J., Di Nuovo, A.: IBM cloud services enhance automatic cognitive assessment via human-robot interaction. new trends in medical and service robotics. In: Carbone, G., Ceccarelli, M., Pisla, D. (eds.) Advances in Theory and Practice, pp. 169–176 (2019)
2. Pîslă, D., Gherman, B., Vaida, C., Plitea, N.: Kinematic modeling of a 5 DOF parallel hybrid robot designed for laparoscopic surgery. Robotica **30**(07), 1095–1107 (2012)
3. Hoffman, P.J., et al.: Precision Machining Technology, 1st edn., p. 628. Cengage Learning (2011)
4. Prassler, E., Kosuge, K.: Domestic robotics. In: Siciliano, B., Khatib, O. (eds.) Springer Handbook of Robotics, pp. 1253–1281. Springer Heidelberg (2008)

5. Atanasoae, P., Pentiuc, R.: Indices for the power quality monitoring in the romanian power transmission system. In: ICHQP, Book Series: International Conference on Harmonics and Quality of Power, pp. 68–71 (2014)
6. Atanasoae, P., Pentiuc, R.: Considerations on the green certificate support system for electricity production from renewable energy sources. Procedia Eng. **181**, 796–803 (2017)
7. Doroftei, I., Baudoin, Y.: A concept of walking robot for humanitarian demining. Ind. Robot. Int. J. **39**(5), 441–449 (2012)
8. Racu (Cazacu), C.M., Doroftei, I.: Ankle rehabilitation device with two degrees of freedom and compliant joint. In: IOP Conference Series: Materials Science and Engineering, vol. 95, p. 012054 (2015)
9. Doroftei, I., Adascalitei, F., Lefeber, D., Vanderborght, B., Doroftei, I.A.: Facial expressions recognition with an emotion expressive robotic head. In: IOP Conference Series: Materials Science and Engineering, vol. 147, p. 012086 (2016)
10. Hichri, B., Fauroux, J.-C., Adouane, L., Doroftei, I., Mezouar, Y.: Design of cooperative mobile robots for co-manipulation and transportation tasks. Robot. Comput. Integr. Manuf. **57**, 412–421 (2019)
11. Hichri, B., Fauroux, J.-C., Adouane, L., Mezouar, Y., Doroftei, I.: Flexible co-manipulation and transportation with mobile multi-robot system. Assem. Autom. **39**(3), 422–431 (2019)
12. Fossum, T.O., et al.: Toward adaptive robotic sampling of phytoplankton in the coastal ocean. Sci. Robot. **4**(27), eaav304 (2019). https://doi.org/10.1126/scirobotics.aav3041
13. Launius, R.G.: Robots in Space: Technology, Evolution, and Interplanetary Travel (New Series in NASA History), p. 334. The Johns Hopkins University Press, Baltimore (2012)
14. Gogu, G.: Structural Synthesis of Parallel Robots, Part 1: Methodology. Series: Solid Mechanics and Its Applications, vol. 149, p. 706 (2008)
15. Lenarcic, J., Khatib, O. (eds.): Advances in Robot Kinematics, p. 561. Springer, Cham (2014)
16. Hartenberg, R.S., Denavit, J.: Kinematic Synthesis of Linkages. McGraw-Hill Series in Mechanical Engineering, p. 435. McGraw-Hill, New York (1965)
17. Uicker, J.J., Denavit, J.J., Hartenberg, R.S.: An iterative method for the displacement analysis of spatial mecahnisms. J. Appl. Mech. **31**, ASME Trans. Ser. E **86**, 309–314 (1964)
18. Ciornei, F.C., Alaci, S., Pentiuc, R.D., Doroftei, I.: Use of dual numbers in kinematical analysis of spatial mechanisms. Part I: principle of the method. In: IMT Oradea, IOP-MSE, vol. 568, p. 012033 (2019). https://doi.org/10.1088/1757-899X/568/1/012033
19. Alaci, S., Pentiuc, R.D., Doroftei, I., Ciornei, F.C.: Use of dual numbers in kinematical analysis of spatial mechanisms. Part II: applying the method for the generalized Cardan mechanism. In: IMT Oradea IOP-MSE, vol. 568, p. 012032 (2019). https://doi.org/10.1088/1757-899X/568/1/012032
20. Fisher, I.S.: Dual Number Methods in Kinematics, Statics and Dynamics, p. 242. CRC Press, Boca Raton (1998)
21. Dimentberg, F.M.: The Screw Calculus and Its Applications in Mechanics. PN ASIN: B00ANZ592Q, 1 January 1968
22. Yang, A.T., Freudenstein, F.: Application of dual-number quaternion algebra to the analysis of spatial mechanisms. Trans. ASME J. Appl. Mech. E **86,** 300–308 (1964)
23. McCarthy, J.M.: Dual orthogonal matrices in manipulator kinematics. Int. J. Robot. Res. **5** (2), 45–51 (1986)

# A Numerical Procedure for Position Analysis of a Robotic Structure. Part II: 3C Robotic Arm Illustration

Stelian Alaci[1(✉)], Ioan Doroftei[2], Florentin Buium[2], Florina-Carmen Ciornei[1], and Ionut-Cristian Romanu[1]

[1] Stefan cel Mare University, Suceava, Romania
stelian.alaci@usm.ro
[2] "Gheorghe Asachi" Technical University of Iasi, Iasi, Romania

**Abstract.** The paper considers a robotic arm with stipulated geometry and imposed position. In the first part of the work, we deduced the matrix equation that allows for establishing the required corrections for an initial set of displacement values from the pairs of the robot. With the amended set of parameters, a new set of parameters is obtained, providing the robot a position closer to the desired one. This process can be regarded as a sequence from an iterative process that aims a final position of the robotic arm with predetermined accuracy. The second part of the paper applies the relations deduced in the first part, for a certain case, namely a robotic arm that has three cylindrical pairs in the structure. The straightforward application of the equations obtained in the first part develops an over-constrained system of linear equations that necessitates the use of generalized inverse matrix Moon Penrose. This requires, in turn, the application of several matrix operations upon matrices with 12 rows. Next, the equation from the first part is written for the robotic arm with three cylindrical pairs, resulting in the occurrence of several special matrices. The properties of the special matrices are considered and a new simpler method is proposed for finding the corrections necessary to amend the initial parameters. This method reduces the calculus to a system of six equations.

**Keywords:** Matrix method · Iterative procedure · 3C robotic arm · Positional analysis

## 1 General Remarks

A robotic arm is considered with imposed final position. The displacements from the pairs must be found, and in the first part of the paper, the matrix equation allowing for finding the correction to be applied to an initial set of positional parameters was developed. It aimed to attain a new set of positional parameters which better satisfy the conditional equation, expressed in a general form:

$$\boldsymbol{Z}(\theta_1^o, s_1^o)\boldsymbol{X}(\alpha_{12}, a_{12})\ldots\ldots\boldsymbol{Z}(\theta_k^o, s_k^o)\boldsymbol{X}(\alpha_{k,k+1}, a_{k,k+1})\ldots\ldots\boldsymbol{Z}(\theta_k^o, s_k^o)\boldsymbol{X}(\alpha_{n,e}, a_{ne}) = \boldsymbol{T}_{1e}^o \quad (1)$$


E.-C. Lovasz et al. (Eds.): MTM&Robotics 2020, MMS 88, pp. 33–42, 2021.
https://doi.org/10.1007/978-3-030-60076-1_4

For a set of initial parameters $\theta_i$ and $s_j$, the Eq. (1) is written:

$$\boldsymbol{Z}(\theta_1, s_1)\boldsymbol{X}(\alpha_{12}, a_{12})\ldots\ldots\boldsymbol{Z}(\theta_k, s_k)\boldsymbol{X}(\alpha_{k,k+1}, a_{k,k+1})\ldots\ldots\boldsymbol{Z}(\theta_n, s_n)\boldsymbol{X}(\alpha_{n,e}, a_{ne}) = \boldsymbol{T}_{1e} \tag{2}$$

and the corrections $d\theta_i$ and $ds_j$ to be applied to the initial set $\theta_i$ and $s_j$ are given by solving the equation

$$\sum_{k=1}^{n} \boldsymbol{T}_{1,k}\left[d\theta_k \boldsymbol{Q}_q + ds_k \boldsymbol{Q}_s\right] \boldsymbol{T}_{1,k}^{-1} \boldsymbol{T}_{1,e} = \boldsymbol{T}_{1,e}^{o} - \boldsymbol{T}_{1,e}, \tag{3}$$

where $\boldsymbol{T}_{1,k}$ is the matrix of position of the $k$ link with respect to the fixed reference frame 1, having the expression:

$$\boldsymbol{T}_{1,k} = \prod_{p=1}^{k-1} \boldsymbol{Z}(\theta_p, s_p)\boldsymbol{X}(\alpha_{p,p+1}, a_{p,p+1}) \tag{4}$$

The form of the matrices $\boldsymbol{Z}(\theta, s)$, $\boldsymbol{X}(\alpha, a)$, $\boldsymbol{Q}_\theta$ and $\boldsymbol{Q}_s$ involved in Eqs. (3–4), are given in Appendix. The $\boldsymbol{T}_{1,k}$ matrices have the general form

$$\boldsymbol{T}_{1,k} = \begin{bmatrix} & \boldsymbol{R}_{1,k} & & \boldsymbol{d}_{1,k} \\ 0 & 0 & 0 & 1 \end{bmatrix}, \tag{5}$$

where the $\boldsymbol{R}_{1,k}$ matrix is the rotation matrix of the frame $k$ with respect to the frame 1 and $\boldsymbol{d}_{1,k}$ is the translational vector of the origin of the frame $k$ with respect to the origin of the system 1. The matrix $\boldsymbol{T}_{1,k}^{-1}$ is obtained using the relation

$$\boldsymbol{T}_{1,k}^{-1} = \begin{bmatrix} & \boldsymbol{R}_{1,k}^{T} & & -\boldsymbol{d}_{1,k} \\ 0 & 0 & 0 & 1 \end{bmatrix} \tag{6}$$

Another necessary matrix from Eq. (3) is $d\theta_k \boldsymbol{Q}_q + ds_k \boldsymbol{Q}_s$ and it can be written under the form

$$d\theta_k \boldsymbol{Q}_\theta + ds_k \boldsymbol{Q}_s = \begin{bmatrix} & \boldsymbol{Q} & & \boldsymbol{q} \\ 0 & 0 & 0 & 1 \end{bmatrix}, \boldsymbol{Q} = \begin{bmatrix} 0 & -1 & 0 \\ 1 & 0 & 0 \\ 0 & 0 & 0 \end{bmatrix}, \boldsymbol{q} = \begin{bmatrix} 0 \\ 0 \\ 1 \end{bmatrix}, \tag{7}$$

where $\boldsymbol{Q}$ is a skew-symmetric $3 \times 3$ matrix and the $\boldsymbol{q}$ matrix is a column vector of $3 \times 1$ type. Performing the product $\boldsymbol{T}_{1,k}\left[d\theta_k \boldsymbol{Q}_q + ds_k \boldsymbol{Q}_s\right] \boldsymbol{T}_{1,k}^{-1}$, it results that

$$\boldsymbol{T}_{1,k}[d\theta_k \boldsymbol{Q}_\theta + ds_k \boldsymbol{Q}_s]\, \boldsymbol{T}_{1,k}^{-1} = \begin{bmatrix} \boldsymbol{R}_{1,k}\boldsymbol{Q}\boldsymbol{R}_{1,k}^{T} d\theta_k & -\boldsymbol{R}_{1,k}\boldsymbol{Q}\boldsymbol{d}_{1,k} d\theta_k + \boldsymbol{R}_{1,k}\boldsymbol{q}\, ds_k \\ 0 \;\; 0 \;\; 0 & 0 \end{bmatrix} \tag{8}$$

Since the $\boldsymbol{Q}$ matrix is skew-symmetric, the $\boldsymbol{R}_{1,k}\boldsymbol{Q}\boldsymbol{R}_{1,k}^T$ matrix will also be $3 \times 3$ skew-symmetric (as shown in Appendix). Denoting by

$$\boldsymbol{H}^k = \boldsymbol{R}_{1,k}\boldsymbol{Q}\boldsymbol{R}_{1,k}^T d\theta_k \tag{9}$$

and

$$\boldsymbol{h}^k = -\boldsymbol{R}_{1,k}\boldsymbol{Q}\boldsymbol{d}_{1,k}d\theta_k + \boldsymbol{R}_{1,k}\boldsymbol{q}d_{s_k}, \tag{10}$$

the Eq. (3) can now be expressed under the form

$$\begin{bmatrix} \sum_{k=1}^{n} \boldsymbol{H}^k & \sum_{k=1}^{n} \boldsymbol{h}^k \\ 0 \;\; 0 \;\; 0 & 1 \end{bmatrix} \boldsymbol{T}_{1e} = \boldsymbol{T}_{1e}^o - \boldsymbol{T}_{1e}. \tag{11}$$

The above matrix equation is equivalent to an over-constrained system of 12 linear equations (with a number of $i+j$ unknowns) namely

$$\boldsymbol{A}\left[d\theta_1, \ldots., d\theta_i, ds_1 \ldots ..ds_j\right]^T = \boldsymbol{T}, \tag{12}$$

where $\boldsymbol{A}$ is a $12 \times (i+j)$ type matrix, and $\boldsymbol{T}$ is a $(12 \times 1)$ type vector. The solution of the system Eq. (12) is obtained by means of generalized Moon Penrose inverse $\boldsymbol{A}^I$ [1] given by

$$\boldsymbol{A}^I = (\boldsymbol{A}^T\boldsymbol{A})^{-1}\boldsymbol{A}^T. \tag{13}$$

In conclusion, the solution of the system is written as

$$\left[d\theta_1, \ldots., d\theta_i, ds_1 \ldots ..ds_j\right]^T = (\boldsymbol{A}^T\boldsymbol{A})^{-1}\boldsymbol{A}^T\boldsymbol{T}. \tag{14}$$

## 2 Applying the Methodology for a Robotic Arm with Three Cylindrical Pairs

A robotic arm is presented in Fig. 1, where the ground is denoted 1, the mobile parts are 2, 3 and the final part (gripper attached to the part 3) is $e = 4$. The cylindrical pairs $C_{12}$, $C_{23}$ and $C_{34}$ occur between these links.

A set of constructive parameters is accepted (angles in *rad* and translations in *mm*):

$$\alpha_{12} = \pi/6, \alpha_{23} = \pi/3, \alpha_{3e} = \pi/4, a_{12} = 25, a_{23} = 15, \alpha_{3e} = 35. \tag{15}$$

The position of the final link is characterised by the motion: rotations $\theta_1, \theta_2$ and $\theta_3$ and translations $s_1$, $s_2$ and $s_3$ from pairs, and by the geometrical parameters of the parts

$\alpha_{12}$, $a_{12}$, $\alpha_{23}$, $a_{23}$, $\alpha_{3e}$, $a_{3e}$. The theoretical displacements are considered with the values (angles in *rad* and translations in *mm*):

$$\theta_1^o = 0.3, \theta_2^o = 0.5, \theta_3^o = 1.1, s_1^o = 20, s_2^o = 30, s_3^o = 40. \tag{16}$$

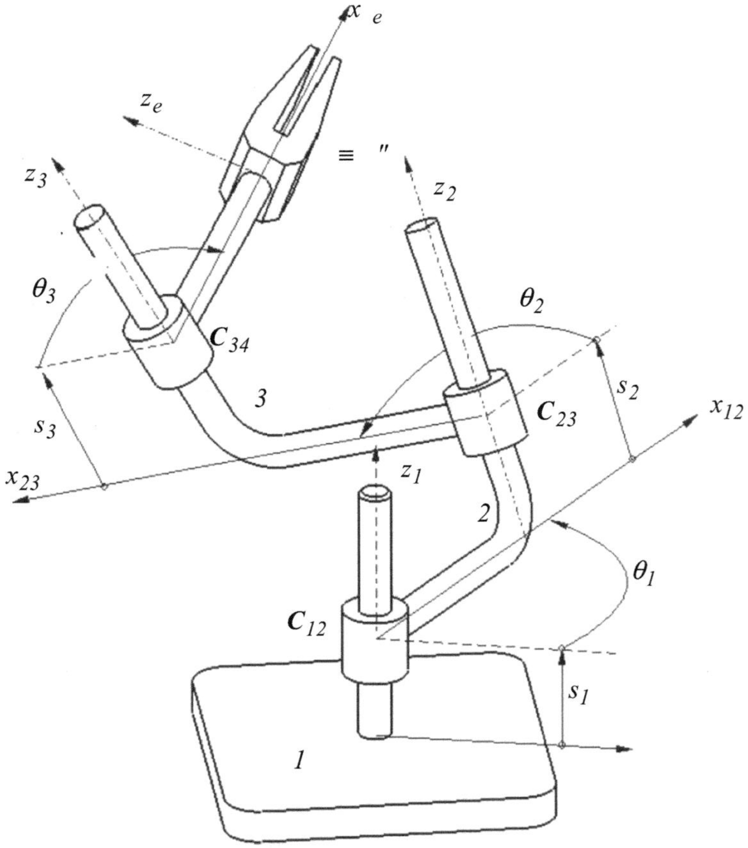


**Fig. 1.** Robotic arm with three cylindrical pairs

The matrix $\boldsymbol{T}_{1e}^o$ is found using the Eq. (1) and has the form

$$\boldsymbol{T}_{1e}^o = \begin{bmatrix} 0.135 & -0.049 & 0.99 & 70.361 \\ 0.189 & -0.979 & -0.074 & -20.274 \\ 0.973 & 0.197 & -0.122 & 85.74 \\ 0 & 0 & 0 & 1 \end{bmatrix}. \tag{17}$$

The initial set of values for the robotic arm pairs' displacements was chosen (angles in *rad* and translations in *mm*):

$$\theta_1 = 0.5, \theta_2 = 0.7, \theta_3 = 1.5, s_1 = 40, s_2 = 70, s_3 = 60. \tag{18}$$

The set of corrections required to be applied to the initial parameters (18) was found using the Eq. (12). The amended parameters are in fact a new set of parameters to be corrected in the next step. Table 1 presents the results obtained for 5 successive iterations applied for refining the initial parameters.

**Table 1.** Positional parameters of the robotic arm in five steps

| | 0 | 1 | 2 | 3 | 4 | 5 |
|---|---|---|---|---|---|---|
| $\theta_1$ | 0.500 | 0.549 | 0.190 | 0.290 | 0.300 | 0.300 |
| $\theta_2$ | 0.700 | 0.244 | 0.624 | 0.511 | 0.500 | 0.500 |
| $\theta_3$ | 1.500 | 1.246 | 1.035 | 1.094 | 1.100 | 1.100 |
| $s_1$ | 40.000 | 13.339 | 48.326 | 25.342 | 20.112 | 20.112 |
| $s_2$ | 70.000 | 41.489 | -3.960 | 23.862 | 29.874 | 29.874 |
| $s_3$ | 60.000 | 37.558 | 53.922 | 43.162 | 40.067 | 40.067 |

Analysing the data from Table 1, we note that for a small number of iterations (five) the values of the corrected parameters practically coincide to the exact values Eq. (16), proving the strong convergence of the method.

## 3 Improved Methodology for the Correction of Initial Parameters

The methodology practiced in the previous paragraph generally follows the iterative calculation method proposed by Uicker et al. [2], based on the homogenous operators methodology, due to Hartenberg and Denavit [3]. As noticed from Eq. (14), a series of matrix operations (product, transpose, inverse calculus) is required for acquiring the corrections of the initial parameters' values. The computing resources will evidently increase with the complexity of the analysed kinematic chain or mechanism. Next, an improved solution for obtaining the correction vector is proposed with the purpose of reducing computing time. To this end the Eq. (3) is written in expanded form

$$\begin{bmatrix} \boldsymbol{Z}(\theta_1, s_1)(d\theta_1 \boldsymbol{Q}_\theta + ds_1 \boldsymbol{Q}_s)\boldsymbol{Z}(-\theta_1, -s_1) + .. \\ \boldsymbol{Z}(\theta_1, s_1)\boldsymbol{X}(\alpha_{1,2}, a_{1,2})\boldsymbol{Z}(\theta_2, s_2)(d\theta_1 \boldsymbol{Q}_\theta + ds_1 \boldsymbol{Q}_s)\cdot \\ \boldsymbol{Z}(-\theta_2, -s_2)\boldsymbol{X}(-\alpha_{12}, -a_{12})\boldsymbol{Z}(-\theta_1, -s_1) + .. \\ \boldsymbol{Z}(\theta_1, s_1)\boldsymbol{X}(\alpha_{1,2}, a_{1,2})\boldsymbol{Z}(\theta_2, s_2)\boldsymbol{X}(\alpha_{23}, a_{23})\boldsymbol{Z}(\theta_3, s_3)(d\theta_1 \boldsymbol{Q}_\theta + ds_1 \boldsymbol{Q}_s) \\ \cdot \boldsymbol{Z}(-\theta_3, -s_3)\boldsymbol{X}(-\alpha_{23}, -a_{23})\boldsymbol{Z}(-\theta_2, -s_2)\boldsymbol{X}(-\alpha_{21}, -a_{21})\boldsymbol{Z}(-\theta_1, -s_1) \end{bmatrix} \boldsymbol{T}_{1e} = \tag{19}$$

$$= \boldsymbol{T}^o_{1e} - \boldsymbol{T}_{1e}$$

Applying the identity

$$\boldsymbol{Z}(\theta, s)(d\theta\,\boldsymbol{Q}_\theta + ds\boldsymbol{Q}_s)\boldsymbol{Z}(-\theta, -s) = d\theta\,\boldsymbol{Q}_\theta + ds\boldsymbol{Q}_s, \tag{20}$$

which is proved by direct calculus, the Eq. (20) takes a simpler form

$$\begin{aligned}
&d\theta_1\boldsymbol{Q}_\theta + ds_1\boldsymbol{Q}_s + ..\\
&\boldsymbol{Z}(\theta_1, s_1)\boldsymbol{X}(\alpha_{1,2}, a_{1,2})(d\theta_2\boldsymbol{Q}_\theta + ds_2\boldsymbol{Q}_s)\boldsymbol{X}(-\alpha_{12}, -a_{12})\boldsymbol{Z}(-\theta_1, -s_1) + ..\\
&\boldsymbol{Z}(\theta_1, s_1)\boldsymbol{X}(\alpha_{1,2}, a_{1,2})\boldsymbol{X}(\alpha_{23}, a_{23})(d\theta_3\boldsymbol{Q}_\theta + ds_3\boldsymbol{Q}_s)\\
&\cdot \boldsymbol{X}(-\alpha_{23}, -a_{23})\boldsymbol{Z}(-\theta_2, -s_2)\boldsymbol{X}(-\alpha_{21}, -a_{21})\boldsymbol{Z}(-\theta_1, -s_1) = \boldsymbol{T}^o_{1e}\boldsymbol{T}^{-1}_{1e} - \boldsymbol{I}_4
\end{aligned} \tag{21}$$

and is equivalent to Eq. (11) expressed as

$$\begin{bmatrix} \sum\limits_{k=1}^{n} \boldsymbol{H}^k & \sum\limits_{k=1}^{n} \boldsymbol{h}^k \\ 0 \quad 0 \quad 0 & 0 \end{bmatrix} = \boldsymbol{T}^o_{1e}\boldsymbol{T}^{-1}_{1e} - \boldsymbol{I}_4. \tag{22}$$

The Eq. (22) in expanded form is

$$\begin{bmatrix} 0 & \sum\limits_{k=1}^{n} H^k_{12} & \sum\limits_{k=1}^{n} H^k_{13} & \sum\limits_{k=1}^{n} h^k_1 \\ -\sum\limits_{k=1}^{n} H^k_{12} & 0 & \sum\limits_{k=1}^{n} H^k_{23} & \sum\limits_{k=1}^{n} h^k_2 \\ -\sum\limits_{k=1}^{n} H^k_{13} & -\sum\limits_{k=1}^{n} H^k_{23} & 0 & \sum\limits_{k=1}^{n} h^k_3 \\ 0 & 0 & 0 & 0 \end{bmatrix} = \boldsymbol{T}^o_{1e}\boldsymbol{T}^{-1}_{1e} - \boldsymbol{I}_4. \tag{23}$$

The matrices $\boldsymbol{H}^k$ from the above relation are skew symmetric matrices and therefore, any linear combination $\sum\limits_{k=1}^{n} \boldsymbol{H}^k d\theta_k$ will be on its turn a skew symmetric matrix. The right member of Eq. (23) is denoted by

$$\boldsymbol{B} = \boldsymbol{T}^o_{1e}\boldsymbol{T}^{-1}_{1e} - \boldsymbol{I}_4. \tag{24}$$

The first three elements from the principal diagonal of the two members are equated and the three equations obtained can be satisfied only for the ideal case when

$$\boldsymbol{T}_{1e} = \boldsymbol{T}^o_{1e}. \tag{25}$$

But in such a case, a procedure for correcting the initial values is not necessary. This suggests that, from the 6 equations resulting from equating the non-diagonal elements (1, 2), (1, 3), (2, 3), (2, 1), (3, 1), (3, 2) from the two sides, only the upper triangular matrix elements should be considered. In conclusion, from the matrix Eq. (23), only the next equations must be maintained

$$\begin{array}{l}\sum_{k=1}^{n} \boldsymbol{H}_{12}^{k} = \boldsymbol{B}_{12}, \sum_{k=1}^{n} \boldsymbol{H}_{13}^{k} = \boldsymbol{B}_{13}, \sum_{k=1}^{n} \boldsymbol{H}_{23}^{k} = \boldsymbol{B}_{23} \\ \sum_{k=1}^{n} \boldsymbol{h}_{1}^{k} = \boldsymbol{B}_{14}, \sum_{k=1}^{n} \boldsymbol{h}_{2}^{k} = \boldsymbol{B}_{24}, \sum_{k=1}^{n} \boldsymbol{h}_{3}^{k} = \boldsymbol{B}_{34}\end{array}. \tag{26}$$

The first three equations from the system of Eqs. (26) are, according to Eq. (11), linear combinations only of the angular corrections ($d\theta_1$, $d\theta_2$ and $d\theta_3$) while the last three equations, are linear combinations of all the corrections ($d\theta_1$, $d\theta_2$, $d\theta_3$, $ds_1$, $ds_2$, $ds_3$ and $ds_3$).

The system of Eq. (26) is written under the form:

$$\begin{bmatrix} A_{11} & A_{12} & A_{13} & 0 & 0 & 0 \\ A_{21} & A_{22} & A_{23} & 0 & 0 & 0 \\ A_{31} & A_{32} & A_{33} & 0 & 0 & 0 \\ A_{41} & A_{42} & A_{43} & A_{44} & A_{45} & A_{46} \\ A_{51} & A_{52} & A_{53} & A_{54} & A_{55} & A_{56} \\ A_{61} & A_{62} & A_{63} & A_{64} & A_{65} & A_{66} \end{bmatrix} \begin{bmatrix} d\theta_1 \\ d\theta_2 \\ d\theta_3 \\ ds_1 \\ ds_2 \\ ds_3 \end{bmatrix} = \begin{bmatrix} B_{12} \\ B_{13} \\ B_{23} \\ B_{14} \\ B_{24} \\ B_{34} \end{bmatrix}. \tag{27}$$

Equation (27) illustrates that the system is a linear algebraic one whose unknowns are the corrections $d\theta_1$, $d\theta_2$, $d\theta_3$, $ds_1$, $ds_2$ and $ds_3$. The system of six equations with six unknowns is compatible. It can also be noticed that the first three equations are independent from the last three and so can be solved separately.

As shown in [3–5], for this system to be compatible, it is required that the structure of the robotic arm contains at least three rotational pairs; the equivalent consequence of this necessity is that if in the structure of a robot with six DOF there are more than three prismatic pairs, the final link cannot reach every imposed position.

With the Eq. (27), the correction vector is obtained rapidly using the relation

$$\begin{bmatrix} d\theta_1 \\ d\theta_2 \\ d\theta_3 \\ ds_1 \\ ds_2 \\ ds_3 \end{bmatrix} = \begin{bmatrix} A_{11} & A_{12} & A_{13} & 0 & 0 & 0 \\ A_{21} & A_{22} & A_{23} & 0 & 0 & 0 \\ A_{31} & A_{32} & A_{33} & 0 & 0 & 0 \\ A_{41} & A_{42} & A_{43} & A_{44} & A_{45} & A_{46} \\ A_{51} & A_{52} & A_{53} & A_{54} & A_{55} & A_{56} \\ A_{61} & A_{62} & A_{63} & A_{64} & A_{65} & A_{66} \end{bmatrix}^{-1} \begin{bmatrix} B_{12} \\ B_{13} \\ B_{23} \\ B_{14} \\ B_{24} \\ B_{34} \end{bmatrix}. \tag{28}$$

In this case, the corrections are found via reduced number of matrix operations. Additionally, the matrices involved in these calculus are of smaller dimensions compared to the ones from Eq. (14).

To illustrate the applicability of the method, for the robotic arm with the same constructive data from Eq. (14) and having the same imposed position, the problem starts with a set of parameters distant from the solution of Eq. (16) compared to the set from Eq. (18).

From Table 2 we note that the new proposed method converges rapidly, in the same manner as the original method, with the advantage of much reduced volume of calculus, since the overconstrained system was replaced by a linear system.

**Table 2.** Positional parameters of the robotic arm in five steps for the simplified method

| | 0 | 1 | 2 | 3 | 4 | 5 |
|---|---|---|---|---|---|---|
| $\theta_1$ | 0.700 | −0.522 | −0.261 | 0.201 | 0.294 | 0.300 |
| $\theta_2$ | 0.1 | 1.799 | 1.018 | 0.607 | 0.506 | 0.500 |
| $\theta_3$ | 0.2 | 0.239 | 0.872 | 1.009 | 1.096 | 1.100 |
| $s_1$ | 70 | 1662 | 1087 | 231.8 | 56.66 | 20.35 |
| $s_2$ | 80 | −1795 | −708.8 | −253.9 | −8.294 | 29.60 |
| $s_3$ | 100 | 927.1 | 118.6 | 256.9 | 62.61 | 40.22 |

The data from Table 2 are plotted in Fig. 2 to highlight the rapid convergence of the method.

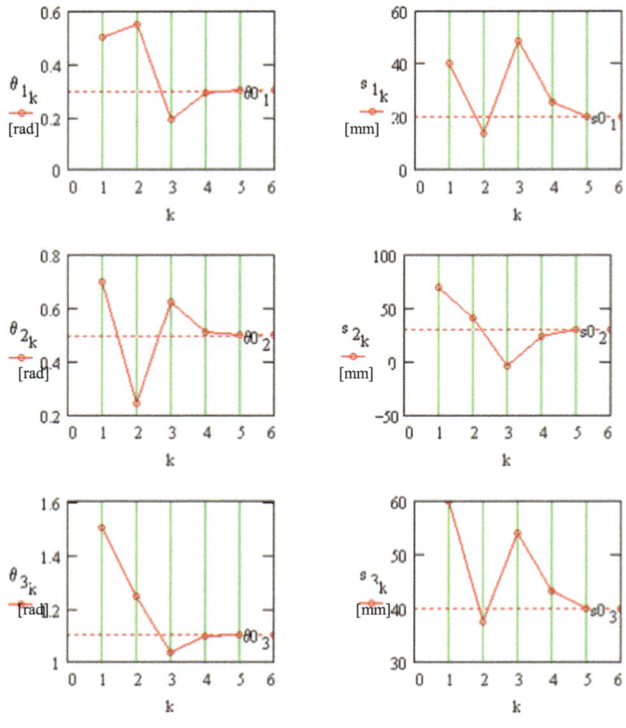

**Fig. 2.** The displacements from the cylindrical pairs of the robot versus iteration number

## 4 Conclusions

The general methodology for obtaining the corrections to be applied to an initial set of positional parameters of a robotic structure in order to determine a new set of parameters, corresponding to a nearer position referred to the desired one, was deduced in the first part of the work. The relations obtained in the first part are applied in the second part for a 6 DOF robotic structure with three cylindrical pairs, in order to find the corrections of the initial positional parameters. In an initial stage, the general equations from the first part are applied straightforwardly, and an overconstrained system of 12 linear equations must be solved. The procedure is an iterative one and presents a very good convergence. For the proposed example, after only six iterations, the values imposed for the parameters were practically attained. Another section of the paper presents a more convenient manner of writing the matrix equation deduced in the first part, and by taking into account a series of properties of special matrices occurring in this equation, a new methodology is proposed for finding the corrections for the initial parameters. These corrections are here the unknowns of a compatible determined linear system of six equations. The convergence of the new presented method is comparable to the one of the initial method, but the calculus is substantially reduced. The two methods are illustrated by presenting the values of positional parameters of the robot corresponding to five iterations for each method; the theoretical prescribed position is reached in both alternatives.

## Appendix

Proposition: if $\theta$ is a skew symmetric matrix ($\theta^T = -\theta$), then $H = M\theta M^T$ is a skew symmetric matrix, too.
Demonstration:

$$H^T = (M\theta M^T)^T = (M^T)^T\theta^T M^T = M(-\theta)M^T = -M\theta M^T = -H, \text{ q.e.d.} \tag{A1}$$

$$\mathbf{Z}(\theta, s) = \begin{bmatrix} \cos\theta & -\sin\theta & 0 & 0 \\ \sin\theta & \cos\theta & 0 & 0 \\ 0 & 0 & 0 & s \\ 0 & 0 & 0 & 1 \end{bmatrix}; \boldsymbol{X}(\alpha, a) = \begin{bmatrix} 1 & 0 & 0 & a \\ 0 & \cos\alpha & -\sin\alpha & 0 \\ 0 & \sin\alpha & \cos\alpha & 0 \\ 0 & 0 & 0 & 1 \end{bmatrix}. \tag{A2}$$

$$\boldsymbol{Q}_\theta = \begin{bmatrix} 0 & -1 & 0 & 0 \\ 1 & 0 & 0 & 0 \\ 0 & 0 & 0 & 0 \\ 0 & 0 & 0 & 0 \end{bmatrix}; \boldsymbol{Q}_s = \begin{bmatrix} 0 & 0 & 0 & 0 \\ 0 & 0 & 0 & 0 \\ 0 & 0 & 0 & 1 \\ 0 & 0 & 0 & 0 \end{bmatrix}. \tag{A3}$$

## References

1. Angeles, J.: Rational Kinematics, Springer Tracts in Natural Philosophy (34), p. 121. Springer, New York (1988)
2. Uicker, J.J., Denavit, J.J., Hartenberg, R.S.: An iterative method for the displacement analysis of spatial mecahnisms. J. Appl. Mech. **31**, ASME Trans. Ser. E **86**, 309–314 (1964)
3. Hartenberg, R.S., Denavit, J.: Kinematic Synthesis of Linkages. McGraw-Hill Series in Mechanical Engineering, p. 435. McGraw-Hill, New York (1965)
4. Ciornei, F.C., Alaci, S., Pentiuc, R.D., Doroftei, I.: Use of dual numbers in kinematical analysis of spatial mechanisms. Part I: principle of the method. In: IMT Oradea, IOP-MSE, vol. 568, p. 012033 (2019). https://doi.org/10.1088/1757-899X/568/1/012033
5. Alaci, S., Pentiuc, R.D., Doroftei, I., Ciornei, F.C.: Use of dual numbers in kinematical analysis of spatial mechanisms. Part II: applying the method for the generalized Cardan mechanism. In: IMT Oradea IOP-MSE, vol. 568, p. 012032 (2019). https://doi.org/10.1088/1757-899X/568/1/012032

# The Effect of Position and Orientation Characteristic on the Forward Position Solution of Parallel Mechanisms

Huiping Shen[1(✉)], Qing Xu[1], Boxiong Zeng[1], Guanglei Wu[2], and Tingli Yang[1]

[1] Research Center for Advanced Mechanism Theory, Changzhou University, Changzhou 213016, China
shp65@126.com

[2] School of Mechanical Engineering, Dalian University of Technology, Dalian 116024, China
gwu@dlut.edu.cn

**Abstract.** The author's extensive experience on forward position solution (FPS) of numerous spatial parallel mechanisms (PMs) leads to an important observation that the type and number of position and orientation characteristics (POC) of the moving platform significantly affect the forward position analysis of PMs. Some specific type of POCs not only makes the forward kinematic modelling of the PM simple but also leads to symbolic FPS. The reason is that these specific POCs introduce some auxiliary constrained position equations. Therefore, simple mathematical method can be used to find the symbolic solutions. A three-translation (3T) PM is taken as an example to derive its symbolic forward position solutions for illustration.

**Keywords:** Position and orientation characteristic · Forward position solution · Coupling degree · Parallel mechanism

## 1 Introduction

The forward position solution (FPS) is one of the important and complicated problems in the field of parallel mechanisms (PMs). It is the foundation for the analysis of velocity/acceleration, workspace, singularity, error analysis, dimension synthesis, and dynamic analysis. Generally, analytical solutions of forward position mean symbolic solutions or closed-form solutions. So-called symbolic FPS are the formula solutions, which are more accurate than closed-form solutions or numerical solutions. In particular, the symbolic FPS are very beneficial to the dimension synthesis, error compensation, motion control, and dynamic analysis of a PM. However for complex planar mechanisms and general space PMs with multiple loops, to obtain symbolic solutions is still an open issue in the mechanism community [1–4].

The type and number of position and orientation characteristic (POC) of a PM refer to the type and number of output translation and/or rotation motion of the moving platform [5, 6]. Let the output motion of the moving platform be $a$T$b$R (where T and R


E.-C. Lovasz et al. (Eds.): MTM&Robotics 2020, MMS 88, pp. 43–52, 2021.
https://doi.org/10.1007/978-3-030-60076-1_5

stand for translation and rotation motion, respectively, and a $\leq$ 3, b $\leq$ 3). According to the dual nature of motion and constraint, it means that the topological structure of the PM constrains both (3-*a*) translation components and (3-*b*) rotation components. Therefore, only *a* translation components and *b* rotation components need to be solved for forward position analysis.

From this point of view, it is much easier to solve the forward position of planar mechanisms than that of spatial PMs since the output motions of the planar mechanisms include only two translations and one rotation (2T1R). Therefore, only two translations (2T) and one rotation (1R) that is perpendicular to the motion plane are needed to be solved. That is, only three variables (x, y, $\gamma$) need to be solved, which is relatively simple.

While for the output motions of the spatial PMs with degree-of-freedom (DOF) of from 3 to 6 are *a*T*b*R($0 \leq a, b \leq 3$), as an example, for the 3T3R PM, six variables ($x, y, z, \alpha, \beta, \gamma$) need to be solved. Hence, it is more complicated.

Thus, for the spatial PMs with 3 to 6 DOFs, POC analysis is very important since it directly affects how many components among six components ($x, y, z, \alpha, \beta, \gamma$) to be obtained, which affect the efficiency of FPS.

In the past year, the authors calculated the FPS of a large number of spatial PMs [7] and found that: 1) the more complete the POC types (translation, rotation), the more variables required to be solved, and the more difficult it is to be solved. On the other hand, the fewer types of POC the PM has, the fewer motion variables to be solved it has, which simplify the kinematic modelling. 2) the types and number of POCs play an important role in the modelling and solving of forward positions. In other words, the types and number of POCs will form topological constraints, while the topological constraints can form more auxiliary position constraint equations that are conducive to modelling and solving forward positions. 3) In addition, when the moving platform has six special POC characteristics such as 1T, 2T, 3T 1T1R, 2T1R and 3T1R, only elementary mathematics are adopted to obtain the symbolic FPS directly.

Based on the previous observations, this paper takes the symbolic FPS of a three-translation (3T) PM as an example to illustrate the fact that how the POC affects the modelling and solving forward positions. This work provides a new perspective for the efficient calculation of FPS for complex PMs.

The organization of the remaining paper is organized as follow. In the Sect. 2, topology analysis of the 3T PM proposed is introduced. Position analysis including FPS and inverse solutions the of the PM is performed in the Sect. 3. The conclusions are drawn in the last section.

## 2 Topology Analysis of PM

### 2.1 Topological Design

The 3T PM proposed in this paper is illustrated in Fig. 1. The base platform 0 is connected to the moving platform 1 by two hybrid chains (HC) that contain $loop_{(s)}$ [6], their topological constraints are depicted as follows.

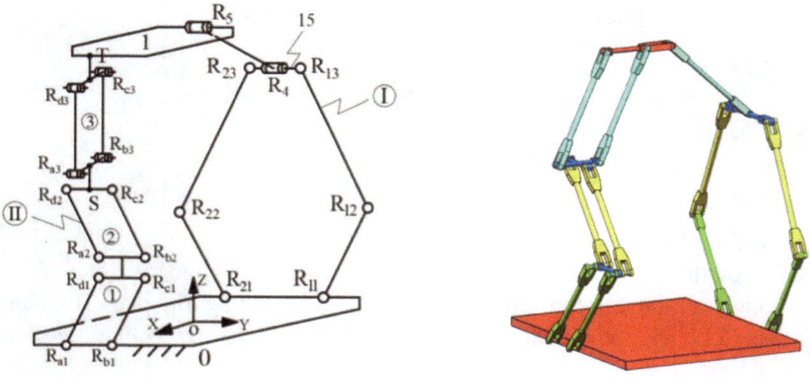


(a) Schematic diagram of mechanism (b) 3D CAD design

**Fig. 1.** Kinematic structure of the 3T PM

(1) On the intermediate link 15 of the right 6R planar mechanism, two parallel axes revolute joints $R_4$ and $R_5$ are connected in series, where $R_4$ is connected to link 15 and $R_5$ is connected to the moving platform 1, to obtain the first hybrid chain I, i.e., $HC_{I}$.

(2) The left side branch is made up of three 4R parallelogram mechanisms connected in series, and the three parallelograms connected from the base platform 0 to the moving platform 1 are recorded as ①, ② and ③ respectively, after the parallelograms ① and ②are rigidly connected in the same plane, they are connected to the parallelogram ③ in their orthogonal plane to obtain the second hybrid chainII, i.e., $HC_{II}$.

(3) The revolute joints $R_{11}$, $R_{21}$, $R_{a1}$ and $R_{b1}$ are connected to the base platform 0, and $R_{11}$ is parallel to $R_{a1}$. When the PM is moving, the 6R planar mechanism is always parallel to the plane of the parallelograms ① and ②.

## 2.2 Analysis of Topology Characteristics

**Analysis of the POC Set.** The POC set equations for serial and parallel mechanisms are expressed, respectively, as follows [5–7]:

$$M_{bi} = \bigcup_{i=1}^{m} M_{Ji} \tag{1}$$

$$M_{Pa} = \bigcap_{i=1}^{n} M_{bi} \tag{2}$$

Where,

$M_{Ji}$ - POC set generated by the $i$-$^{\text{th}}$ joint.
$M_{bi}$ - POC set generated by the end link of $i$-$^{\text{th}}$ branched chain.
$M_{Pa}$ - POC set generated by the moving platform of PM.
$\cup$ - union operation.
$\cap$ - intersection operation.

Obviously, the output motions of the intermediate link 15 of the 6R planar mechanism on the hybrid chain I involve two translations and one rotation (2T1R). The output motions of the link S of the parallelogram ② on the hybrid chain II include two translations (2T), and the output motions of the link T of the parallelogram ③ contain three translations (3T). Therefore, the topological architecture of the hybrid chain I and II of the PM can be equivalently depicted as, respectively:

$$HC_1\{-\Diamond(R_{11}||R_{12}||R_{13}||R_{23}||R_{22}||R_{21})\perp R_4||R_5-\}$$

$$HC_2\left\{-P^{(4R)}-P^{(4R)}-P^{(4R)}-\right\}$$

The POC sets of the end link of the two $HC_S$ are determined according to Eqs. (1) and (2) as follows:

$$M_{HC_1}=\begin{bmatrix} t^2(\perp R_{13}) \\ r^1(||R_{13}) \end{bmatrix}\cup\begin{bmatrix} t^2(\perp R_4) \\ r^1(||R_4) \end{bmatrix}=\begin{bmatrix} t^3 \\ r^2(||\Diamond(R_{13},R_4)) \end{bmatrix}$$

$$M_{HC_2}=\bigcup_{i=1}^{3}\begin{bmatrix} t^1(||\Diamond(R_{ai}R_{bi}R_{ci}R_{di})) \\ r^0 \end{bmatrix}=\begin{bmatrix} t^3 \\ r^0 \end{bmatrix}$$

The POC set of the moving platform of this PM is determined from Eq. (2) by

$$M_{Pa}=M_{HC_1}\cap M_{HC_2}=\begin{bmatrix} t^3 \\ r^0 \end{bmatrix}$$

This formula indicates that the moving platform 1 of the PM produces only three translational motion. It is further known that the hybrid chain II in the PM itself can realize the design requirement of three translations, which simultaneously constrains the two rotational outputs of the hybrid chain I.

It is calculated that the DOF of the PM is 3. Therefore when the revolute joints $R_{11}$, $R_{21}$ and $R_{a1}$ on the base platform 0 are selected as three actuated joints, the moving platform 1 can realize three translational motion outputs.

**Determining the Coupling Degree.** According to the composition principle of PM based on single-opened-chains (SOC) units [6, 7], any PM can be decomposed into groups of Assur kinematics chains (AKC), and an AKC with $v$ independent loops can be decomposed into $v$ SOC. The constraint degree of the $j^{th}$ SOC is defined by

$$\Delta_j = \sum_{i=1}^{m_j} f_i - I_j - \xi_{L_j} = \begin{cases} \Delta_j^- = -5, -4, -2, -1 \\ \Delta_j^0 = 0 \\ \Delta_j^+ = +1, +2, +3, \cdots \end{cases} \quad (3)$$

Where,

$m_j$ - the number of joints contained in the $j^{th}$ $SOC_j$. $f_i$ - DOF of the $i^{th}$ joints.
$I_j$ - the number of actuated joints in the $j^{th}$ $SOC_j$. $\xi_{L_j}$ - the number of independent equations of the $j^{th}$ loop.

Then, the coupling degree of AKC [6, 7] is expressed as

$$\kappa = \frac{1}{2}\min\left\{\sum_{j=1}^{v} |\Delta_j|\right\} \quad (4)$$

The coupling degree $\kappa$ reflects the correlation and dependence between kinematic variables of each independent loop of the PM. It has been proved that the higher $\kappa$ it is equal to, the higher the complexity of the kinematic and dynamic solutions of the PM will be.

The independent displacement equations of $loop_1(R_{11} - R_{12} - R_{13} - R_{23} - R_{22} - R_{21})$ and $loop_2(P^{(4R)} - P^{(4R)} - P^{(4R)})$ have been calculated as $\xi_{L_1} = 3$, $\xi_{L_2} = 5$, thus, the constraint degree of the two independent loops are calculated by Eq. (3), respectively, as follows.

$$\Delta_1 = \sum_{i=1}^{m_1} f_i - I_1 - \xi_{L_1} = 6 - 2 - 3 = 1$$

$$\Delta_2 = \sum_{i=1}^{m_2} f_i - I_2 - \xi_{L_2} = 5 - 1 - 5 = -1$$

The coupling degrees of the AKC is calculated by Eq. (4) as

$$k = \frac{1}{2}\sum_{j=1}^{v} |\Delta_j| = \frac{1}{2}(|+1| + |-1|) = 1$$

Thus, the PM contains only one AKC, and its coupling degrees equals to 1. Therefore, when solving the FPS of the PM, it is necessary to set only one virtual variable in the $loop_1$ whose constraint degree is positive ($\Delta_1 = 1$). Then, a position constraint equation with this virtual variable is established in the $loop_2$ whose constraint degree is negative ($\Delta_2 = -1$). The real value of the virtual variable can be obtained.

Due to the special POC, i.e., 3-translation constraint, of the PM, the $loop_2$ with the negative constraint degree directly applied to the topological constraint of the $loop_1$ with the positive constraint degree ($\Delta_1 = 1$), i.e., the motion of the link 15 is always parallel to the base platform 0, from which the virtual variable is easily obtained. Therefore, the symbolic FPS of the PM can be directly obtained in the following section, which significantly simplifies the FPS. This method for FPS based on special topological constraints is of general [8].

## 3 Position Analysis

### 3.1 The Coordinate System and Parameterization

The kinematic modelling of the PM is shown in Fig. 2. The base platform 0 and the moving platform 1 are both in the shape of rectangle. The frame coordinate system O-XYZ is established with the origin located at the geometric center of the base platform 0, the X- and Y-axis being perpendicular and parallel to the line $A_1A_2$, and the Z-axis is determined by the right hand Cartesian coordinate rule. The moving coordinate system O′-X′Y′Z′ is attached to the center of the moving platform 1, of which the X′ and Y′ axes are parallel and perpendicular to the line $D_2G_3$, and the Z′ axis is determined the right hand Cartesian coordinate rule.

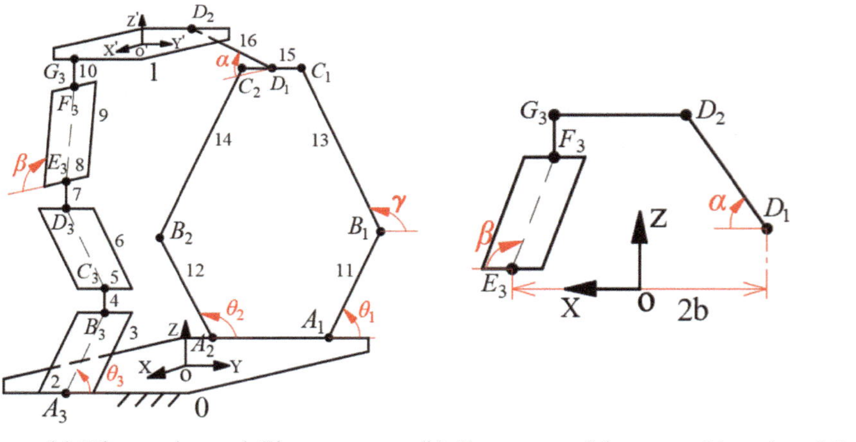


**Fig. 2.** Kinematic modelling of the 3T PM

The length of the line $A_1A_2$ on the hybrid chain I is equal to $2a$, and the lengths of the driving links 11 and 12 are both equal to $l_1$. Moreover, the lengths of the connecting links 13 and 14 are both equal to $l_2$, and the lengths of the intermediate links 15 and 16 are equal to $l_3$ and $l_4$ respectively.

The length of the parallelogram short links 2, 5, 8 on the hybrid chain II is equal to $l_5$. The point $A_3$, $B_3$, $C_3$, $D_3$, $E_3$ and $F_3$ are the midpoint of the short edge, and the length of the long links 3, 6, 9 is equal to $l_6$. The length of the connecting links 4, 7 between the parallelograms is equal to $l_7$. The length of the connecting link 10 is $l_8$, the distance from point $A_3$ to line $A_1A_2$ on the base platform 0 is equal to $2b$, and the length of the line $D_2G_3$ on the moving platform 1 is equal to $2d$.

The three input angles between the vectors $A_1B_1$, $A_2B_2$, $A_3B_3$ and the Y-axis are set to $\theta_1$, $\theta_2$ and $\theta_3$, respectively. The angle between the vectors $B_1C_1$ and the Y-axis is denoted as $\gamma$, and the angle $\gamma$ is supposed to a virtual variable. The angles between the vectors $D_1D_2$, $E_3F_3$ and the X-axis are $\alpha$ and $\beta$, respectively.

## 3.2 Forward Kinematics

To obtain the FPS, i.e., it is to compute the position $O'(x, y, z)$ of the moving platform with the known actuated joints $\theta_1$, $\theta_2$ and $\theta_3$.

1) **Solving the first loop with positive constraint degree**

$$\text{Loop}_1 : A_1 - B_1 - C_1 - C_2 - B_2 - A_2$$

The points $A_1$, $A_2$ and $A_3$ on the base platform 0 are known as below, respectively $A_1 = (-b, a, 0)^T, A_2 = (-b, -a, 0)^T, A_3 = (b, 0, 0)^T$.
The coordinates of points $B_1$, $B_2$, $B_3$, and $C_3$ are readily calculated as

$$B_1 = (-b, a + l_1 \cos\theta_1, l_1 \sin\theta_1)^T,\ B_2 = (-b, -a + l_1 \cos\theta_2, l_1 \sin\theta_2)^T,$$

$$B_3 = (b, l_6 \cos\theta_3, l_6 \sin\theta_3)^T,\ C_3 = (b, l_6 \cos\theta_3, l_6 \sin\theta_3 + l_7)^T$$

Due to the special constraint of the three-translation of the moving platform 1, during the motion of the PM, the intermediate link 15 of the 6R planar mechanism is always parallel to the base platform 0, namely, $C_1C_2||A_1A_2$, then we have

$$z_{C_1} = z_{C_2} \tag{5}$$

Therefore, the coordinates of points $C_1$ and $C_2$ are calculated as

$$C_1 = (-b, a + l_1 \cos\theta_1 + l_2 \cos\gamma, l_1 \sin\theta_1 + l_2 \sin\gamma)^T$$

$$C_2 = (-b, a + l_1 \cos\theta_1 + l_2 \cos\gamma - l_3, l_1 \sin\theta_1 + l_2 \sin\gamma)^T$$

With the link length constraints defined by $B_2C_2 = l_2$, the constraint equation can be deduced as

$$A \sin\gamma + B \cos\gamma + C = 0 \tag{6}$$

$$\gamma = 2 \arctan \frac{A \pm \sqrt{A^2 + B^2 - C^2}}{B - C} \tag{7}$$

Where, $A = 2Dl_2, B = 2Fl_2, C = D^2 + F^2$,

$$D = l_1 \sin\theta_1 - l_1 \sin\theta_2, F = 2a + l_1 \cos\theta_1 - l_1 \cos\theta_2 - l_3.$$

Here, the second loop with negative constraint degree ($\Delta_2 = -1$) applies to the special topological constraint, i.e., Eq. (5), on the first loop with positive constraint degree ($\Delta_2 = 1$), which is the key issue to directly find the symbolic solutions of the virtual variable $\gamma$. This is why the symbolic FPS based on the topological constraint analysis of the PM can be easily obtained.

2) **Solving the second loop with negative constraint degree**

$$Loop_2 : D_1 - D_2 - G_3 - F_3 - E_3 - D_3 - C_3 - B_3 - A_3$$

The coordinates of points $D_1$ and $D_2$ obtained from points $C_1$ and $C_2$ are derived as

$$D_1 = (-b, a + l_1 \cos\theta_1 + l_2 \cos\gamma - l_3/2, l_1 \sin\theta_1 + l_2 \sin\gamma)^T$$

$$D_2 = [-b + l_4 \cos\alpha, a + l_1 \cos\theta_1 + l_2 \cos\gamma - l_3/2, l_1 \sin\theta_1 + l_2 \sin\gamma + l_4 \sin\alpha]$$

Additionally, the coordinates of point $O'$ can be calculated as:

$$O' = \begin{bmatrix} x \\ y \\ z \end{bmatrix} = \begin{bmatrix} -b + l_4 \cos\alpha + d \\ a + l_1 \cos\theta_1 + l_2 \cos\gamma - l_3/2 \\ l_1 \sin\theta_1 + l_2 \sin\gamma + l_4 \sin\alpha \end{bmatrix} \quad (8)$$

Further, the coordinates of points $G_3$, $F_3$, $E_3$ and $D_3$ are represented by the coordinates of point $O'$ defined as:

$$G_3 = (x + d, y, z)^T, F_3 = (x + d, y, z - l_8)^T$$

$$E_3 = (b, y, z - l_8 - l_6 \sin\beta)^T, D_3 = (b, y, z - l_8 - l_6 \sin\beta - l_7)^T \quad (9)$$

Due to the link length constraints defined by $B_3C_3 = l_6$, the constraint equation can be deduced as below.

$$(x_{C_3} - x_{B_3})^2 + (y_{C_3} - y_{B_3})^2 + (z_{C_3} - z_{B_3})^2 = l_6^2 \quad (10)$$

Let

$$l_4 \sin\alpha - l_6 \sin\beta = t \quad (11)$$

there exists

$$t = -H_1 \pm \sqrt{H_2} \tag{12}$$

Where,
$H_1 = l_1 \sin\theta_1 + l_2 \sin\gamma - l_6 \sin\theta_3 - l_8 - 2l_7$, $H_2 = l_6^2 - (y_{D_3} - y_{C_3})^2$.

During motion, the 6R planar mechanism is always parallel with the plane of the parallelograms ① and ②, therefore, there are always two topological relations as

$$y_{D_1} = y_{D_3} \tag{13}$$

$$l_4 \cos\alpha + 2d - l_6 \cos\beta = 2b \tag{14}$$

Eliminating $\beta$ from Eqs. (11) and (14) yields

$$\alpha = 2\arctan\frac{J_1 \pm \sqrt{J_1^2 + J_2^2 - J_3^2}}{J_2 - J_3} \tag{15}$$

Where, $J_1 = 2l_4t$, $J_2 = 4l_4(b - d)$, $J_3 = l_6^2 - l_4^2 - t^2 - 4(b - d)^2$.

Finally, substituting the values of $\gamma$ obtained by Eq. (7) and $\alpha$ by Eq. (15) into Eq. (8), the coordinates of point $O$' of the moving platform 1 can be easily obtained.

Further, from Eq. (7), $\gamma = f_1(\theta_1, \theta_2)$, it is known the PM has partial input-output motion decoupling.

It can be seen that the topological constraint Eqs. (5), (13) and (14) are key steps to derive the symbolic FPS of the PM.

## 4 Conclusions

The type and number of POC of the moving platform have a significant influence to the FPS of PMs. Some specific POCs not only make the FPS analysis of the PM simple and easy, but also make the PMs have symbolic FPS.

The symbolic FPS is illustrated with a 3T PM to verify this observation. The method needs fewer mathematical operations and knowledge, but results in clear physical and geometric meaning.

**Acknowledgement.** The support by National Natural Science Foundation of China (No.51975062, 51375062) is greatly appreciated.

## References

1. Nanua, P., Waldron, K.J.: Direct kinematic solution of a Stewart platform. In: Proceedings of the IEEE International Conference on Robotics and Automation, 14–19 May 1989, pp. 431–437 (1989)

2. Husain, M., Waldron, K.J.: Direct position kinematics of the 3-1-1-1 Stewart platforms. J. Mech. Des. **116**(4), 1102–1107 (1994)
3. Wampler, C.W.: Forward displacement analysis of general six-in-parallel SPS (Stewart) platform manipulator using soma coordinates. Mech. Mach. Theory **31**(3), 311–337 (1996)
4. Husty, M.L.: Algorithm for solving the direct kinematic of Stewart-Gough-type platforms. Mech. Mach. Theory **31**(4), 365–380 (1996)
5. Yang, T., Liu, A., Luo, Y., et al.: Theory and Application of Robot Mechanism Topology. Science Press (2012)
6. Yang, T., Liu, A., Shen, H., et al.: Topology Design of Robot Mechanism. Springer, Berlin (2018)
7. Shen, H., Yang, T., Li, J., Zhang, D., Deng, J., Liu, A.: Evaluation of topological properties of parallel manipulators based on the topological characteristic indexes. Robotica **11** (2019)
8. Shen, H., Chablat, D., Zen, B., et al.: A translational three-degrees-of-freedom parallel mechanism with partial motion decoupling and analytic direct kinematics. ASME J. Mech. Robot. **12**(2), 021112 (2020)

# The Effect of the Type of Sub-kinematic Chains on the Forward Position Solutions of Parallel Mechanisms

Huiping Shen(✉), Qing Xu, Ju Li, Tao Li, and Ting-li Yang

Research Center for Advanced Mechanism Theory, Changzhou University, Changzhou 213164, China
shp65@126.com

**Abstract.** Sub-kinematic chain (SKC) refers to the smallest analytical unit that contains only one Assur Kinematic Chain with zero degree-of-freedom (DOF), and any parallel mechanism (PM) can be decomposed into several SKCs. This paper proposes an SKC-based kinematic modelling method for forward position solution (FPS), that is, the position analysis of a complex PMs can be decomposed into the modelling of several SKCs. Meanwhile, the modelling method for position analysis of a SKC based on single-open-chain units are also given. Therefore, the difficulty and complexity of solving the position problem of a PM depend on the type and number of SKCs contained in the PM itself. A 5-DOF two-translation and three-rotation PM is used as an example to illustrate this method. This kinematic modelling method has a clear road-map that greatly reduces the difficulty of the FPS. This method is general and also easily conducive to automatic computer generation.

**Keywords:** Sub-kinematic chain · Forward position solutions · Parallel mechanism · Coupling degree

## 1 Introduction

Forward position solutions (FPS) of a PM is one of the most important and basic problems in PMs community.

At present, most scholars use the vector loop method [1, 2] to establish the input-output position equations of a PM, and then use numerical or algebraic methods to find out its solutions. The numerical methods [3, 4] can be used directly to solve the position equations. Its advantage is that the real number solutions can be obtained, whereas its disadvantage are that the iteration is easy to diverge and the calculations are large. By eliminating the unknowns in the position equations, the advanced algebraic methods [5–10] gives a one-variable higher-order equation, which can be adept to search all possible solutions.

The establishment of the above-mentioned position equations based on the vector loop method does not consider the effect of the topological characteristics on the kinematics of PM. Therefore, the position equation has a large number of variables.


E.-C. Lovasz et al. (Eds.): MTM&Robotics 2020, MMS 88, pp. 53–65, 2021.
https://doi.org/10.1007/978-3-030-60076-1_6

Furthermore, the mathematical derivation and elimination processes are complicated when the algebraic methods are used.

This paper proposes a kinematic modelling method for the FPS of PM based on the SKC units. The paper is organized as follows. In Sect. 2, a method for FPS based on SKC is introduced. The idea is to decompose a complex PM into $SKC_{(s)}$, and then to solve the kinematics of each SKC. Thus the complexity of PM kinematics depends on the type and number of the $SKC_{(s)}$ itself. Especially when the coupling degree of a single SKC is lower or zero, its symbolic FPS can be easily obtained. Take the FPS of a 5-DOF two-translation and three-rotation PM with zero coupling degree as an example to illustrate this method in Sect. 3. Conclusions are given in Sect. 4.

## 2 Method for FPS Based on SKC

### 2.1 Basic Principle for FPS Based on SKC

The SKC-based method for FPS proposed in the paper includes the following three steps:

1) Topological analysis [13, 14]. When a PM is decomposed into a series of single-open-chain (SOC) with three constraint degrees of positive, zero and negative, denoted as SOC ($\Delta_j^+$), SOC ($\Delta_j^0$), SOC ($\Delta_j^-$), respectively, the constraint degree ($\Delta$) of each SOC is obtained, based on which these SOCs can be further grouped into one or several $SKC_{(s)}$, and then the coupling degree $\kappa_i\left(\kappa_i = \Delta_j^+ = |\Delta_j^-|\right)$ of each $SKC_i$ is calculated. At this time, it can be determined whether the PM contains only one SKC or several SKCs.
2) Establishing the position equations. For each SKC with different coupling degree, both the geometric constraint characteristic and/or topological constraint characteristic are used to generate its position equations with minimal (just $\kappa_i$) virtual variables.
3) Mathematical solving. The position of each SKC can be easily solved by using simple mathematical methods, which finally leads to the FPS of the PM. For most PMs, the symbolic solutions or closed-form solutions can be obtained.

It can be known from the foregoing statement that the position analysis of SKCs unit is the basis for FPS of PMs.

### 2.2 Position Solution of SKC Unit

According to [11], the most common SKC units for PMs are: $SKC^0(0)$, $SKC^1(1, -1)$, $SKC^1(1, 0, -1)$, $SKC^1(1, 0, 0, -1)$, $SKC^1(1, 0, 0, 0, -1)$, $SKC^2(2, -2)$, $SKC^2(2, -1, -1)$. Different combinations of these different SKCs units can form a number of PMs with different topology. Now take $SKC^0(0)$ and $SKC^1(1, -1)$ as examples to illustrate the kinematic modelling and solution of the SKC unit.

**Modelling and Solution of $SKC^0(0)$**

$SKC^0(0)$ means that the SKC contains only one loop and its constraint degree is zero, which indicates that its motion is self-determined. For example, for all planar mechanisms with Assur group II, there is no need to assign virtual variables, the vector method can be used to directly solve the position equation.

When the number of constraint position equations of a loop is greater than or equal to the number of independent displacement equations (NIDE) [13, 14] of the loop, the symbolic solutions or closed-form solutions can be obtained.

**Modelling and Solution of $SKC^k$ (1, −1)**

① $SKC^1(1, -1)$ means that the SKC contains two loops, and the constraint degree of the first and second loop are 1 and −1, respectively. For a $SKC^1(1, -1)$, the motion of the first loop is uncertain. To determine its motion, assign one ($\Delta_1 = 1$) virtual variable $\alpha_1^*$ for the first loop. The form of the virtual variable can be an angle or a distance. In this way, the positions of all the joints in the first loop can be determined and should be the functions of the virtual variable $\alpha_1^*$;

② Next step is to analyze the position and orientation characteristics (POC) of the sub-moving platform and/or moving platform of the PM, and judge whether this one virtual variables $\alpha_1^*$ can be obtained directly from the geometric constraint equations in the first loop. If not, the topological constraint equations introduced by the topological characteristic of the second loop should be searched and founded. This topological constraint characteristic condition generally comes from the second loop adjacent to the first loop. Because the first and second loops are coupled, the second loop with a negative constraint degree ($\Delta_2 = -1$) may apply a special POC topological constraint to the first loop with a positive constraint degree ($\Delta_1 = 1$). This topological constraint can form additional auxiliary constraint equations that is applied in the first loop. In this way, when the total number of constraint position equations in the first loop is greater than or equal to the NIDE of the loop, i.e., $\xi_L$. While $\xi_L$ is obtained during the topology analysis stage. Thus, the symbolic solution of the virtual variable $\alpha_1^*$ in the first loop can be obtained by using the elementary mathematics of one or two linear or quadratic equations.

③ If the auxiliary topological constrain equation can not be found in the first loop, then a suitable geometric constraint condition, for example, the link length condition must be found in the second loop with a negative constraint degree ($\Delta_2 = -1$). From this, a position equation containing the virtual variables $\alpha_1^*$ is established below.

$$f(\alpha*) = f(\theta_i, l_j, \alpha*) = 0 \tag{1}$$

It is easy to know that when $\Delta_j^+ = \Delta_j^- = 1(\kappa = 1)$, Eq. (1) is a position equation with one virtual variable $\alpha_1^*$, where $\theta_i$ is the known value of the actuated joints and $l_j$ are link parameters.

④ It is preferred to use mathematical software tools and/or algebraic transformation to perform symbol processing on Eq. (1), and derive an one-variable higher-order equation, and find the closed-form solutions of Eq. (1), or directly use the conventional numerical method to find the numerical solutions of Eq. (1). Furthermore, by substituting the real value into the position calculation process in each of the aforementioned loops, the FPS of the PM can be finished.

In the same way, the modelling and solving of $SKC^k(1, 0, -1)$ unit can be handled.

# 3 Case Analysis

## 3.1 Topological Analysis of 5-DOF 2T3R PM

The 5-DOF 2T3R PM proposed in this paper is shown in Fig. 1. It consists of a base platform 0, a moving platform 1 and three branches. Among them, the first branch I is a hybrid chain (HC) that includes a 2-DOF planar mechanism that can be equivalently denoted as $P_1$-$P^{(4R)}$-$R_1||R_2\perp P_2$, which further can be expressed as $HC_I$. It is easy to know that the output of the 2-DOF planar mechanism is two-translation motion. The 2-DOF planar mechanism is further connected to the moving platform 1 by the spherical joint $S_1$ and to the base platform 0 by the two prismatic joints $P_1$ and $P_2$.

The second branch II is also a hybrid chain, denoted as $HC_{II}$, which is connected to the moving platform 1 through the spherical joint $S_5$, and two spherical joints $S_2$ and $S_3$ are connected to the base platform 0. The branch III is a simple chain $S_6$-$P_5$-$S_7$, denoted as $C_{III}$, which is connected to the moving platform 1 by the spherical joint $S_7$, and to the base platform 0 by the spherical joint $S_6$. Among them, $P_1$, $P_2$, $P_3$, $P_4$, and $P_5$ will be used as actuated joints.

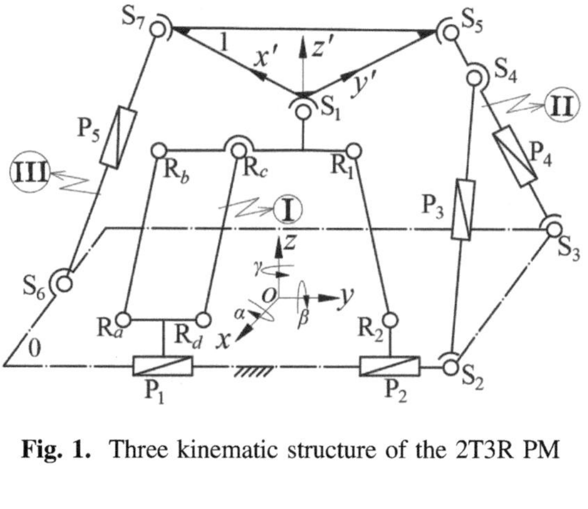


**Fig. 1.** Three kinematic structure of the 2T3R PM

**Analysis of the POC Set.** The POC set equations for serial and parallel mechanisms are expressed respectively as follows [12]:

$$M_{b_i} = \bigcup_{i=1}^{m} M_{J_i} \tag{2}$$

$$M_{\mathrm{P_a}} = \bigcap_{i=1}^{n} M_{\mathrm{b}_i} \tag{3}$$

where $M_{Ji}$ is POC set generated by the $i$-$^{\text{th}}$ joint. $M_{bi}$ is POC set generated by the end link of $i^{\text{th}}$ branched chain. $M_{Pa}$ is POC set generated by the moving platform of PM.

Obviously, the topological architecture of the chain I, II and III of the PM can be denoted as, respectively:

$HC_{\mathrm{I}}\{\text{-}\diamond(\mathrm{P_1\text{-}P^{(4R)}\text{-}R_1||R_2\perp P_2)\text{-}S_1\text{-}}\}$, $HC_{\mathrm{II}}\{\text{-}(\mathrm{S_2\text{-}P_3\text{-}S_4\text{-}P_4\text{-}S_3)\text{-}S_5\text{-}}\}$,
$C_{\mathrm{III}}\{\text{-}\mathrm{S_6\text{-}P_5\text{-}S_7}\text{-}\}$

By using Eq. (2), the POC sets of the three chains are determined as follows, respectively.

$$M_{HC_{\mathrm{I}}} = \begin{bmatrix} t^2(\perp R_1) \\ r^0(||R_1) \end{bmatrix} \cup \begin{bmatrix} t^0 \\ r^3 \end{bmatrix} = \begin{bmatrix} t^2(\perp R_1) \\ r^3 \end{bmatrix},\ M_{HC_{\mathrm{II}}} = \begin{bmatrix} t^3 \\ r^3 \end{bmatrix},\ M_{C_{\mathrm{III}}} = \begin{bmatrix} t^3 \\ r^3 \end{bmatrix}.$$

By using Eq. (3), the POC set of the moving platform of this PM is determined as

$$M_{pa} = M_{HC_{\mathrm{I}}} \cap M_{HC_{\mathrm{II}}} \cap M_{C_{\mathrm{III}}} = \begin{bmatrix} t^2(\perp R_1) \\ r^3 \end{bmatrix} \cap \begin{bmatrix} t^3 \\ r^3 \end{bmatrix} \cap \begin{bmatrix} t^3 \\ r^3 \end{bmatrix} = \begin{bmatrix} t^2(\perp R_1) \\ r^3 \end{bmatrix}$$

This equation indicates that the moving platform 1 of the PM can produce two-translation and three-rotation motion outputs.

**Determining the DOF.** The general full-cycle DOF formula for PMs proposed in authors' work [12] is given below:

$$F = \sum_{i=1}^{m} f_i - \sum_{i=1}^{v} \xi_{\mathrm{L}j} \tag{4}$$

$$\xi_{\mathrm{L}_j} = \dim.\left\{ \left(\bigcap_{i=1}^{j} M_{\mathrm{b}_i}\right) \bigcup M_{\mathrm{b}_{(j+1)}} \right\} \tag{5}$$

where
$F$ is DOF of PM. $f_i$ is DOF of the $i^{\text{th}}$ joint. $m$ - number of all joints of the PM.
$v$ is number of independent loops, and $v = m - n + 1$.
$\xi_{L_j}$ is the number of independent displacement equations (NIDE) of the $j^{\text{th}}$ loop.
$\bigcap_{i=1}^{j} M_{b_i}$ is POC set generated by the sub-PM formed by the former $j$ branches.
$M_{b(j+1)}$ is POC set generated by the end link of $j+1$ sub-chains.

The PM can be decomposed into four independent loops, and their NIDE are calculated as follows:

① The first independent loop is composed of 2-dof planar mechanism in the hybrid chain I, i.e., loop $P_1$-$P^{(4R)}$-$R_1||R_2\perp P_2$, which can be denoted as:

$$Loop_1\{\text{-}P_1\text{-}P^{(4R)}\text{-}R_1||R_2\perp P_2\text{-}\}.$$

Obviously, the NIDE of the first loop is $\xi_{L_1} = 3$.

② The second independent loop is composed of $S_2$-$P_3$-$S_4$-$P_4$-$S_3$ in the hybrid chain II, which can be denoted as:

$$Loop_2\{\text{-}S_2\text{-}P_3\text{-}S_4\text{-}P_4\text{-}S_3\text{-}\}.$$

Obviously, the NIDE of the second loop is $\xi_{L_2} = 6$.

③ The third loop is composed of $R^{(S_2-S_3)}$-$S_5$-$S_1$, which can be denoted as:

$$Loop_3\{\text{-}R^{(S_2-S_3)}\text{-}S_1\text{-}S_5\text{-}\}.$$

Here, $R^{(S_2-S_3)}$ is a local passive rotational DOF around the line $S_2$-$S_3$. Obviously, the NIDE of the third loop is $\xi_{L_3} = 6$.

④ The fourth loop is consisted of another local passive rotational DOF, i.e., $R^{(S_1-S_5)}$, and branch $S_6$-$P_5$-$S_7$, which can be denoted as:

$$Loop_4\{\text{-}R^{(S_1-S_5)}\text{-}S_6\text{-}P_5\text{-}S_7\text{-}\}.$$

Obviously, the NIDE of the fourth Loop is $\xi_{L_4} = 6$.

Thus, the DOF of the PM is calculated from Eq. (4) as:

$$F = \sum_{i=1}^{m} f_i - \sum_{i=1}^{v} \xi_{L_j} = (5+8+6+7) - (3+6+6+6) = 5$$

Therefore, when the prismatic joints $P_1$, $P_2$, $P_3$, $P_4$ and $P_5$ are selected as the actuated joints, the moving platform 1 can realize 2-translation and 3-rotation motion outputs.

**Determining the Coupling Degree.** According to the composition principle of mechanism based on SOC units [12, 13], any PM can be decomposed into a series of SKC that contain only one Assur kinematics chains (AKC) with zero DOF. Further, each SKC with $v$ independent loops can also be further decomposed into $v$ $SOC_s$. The constraint degree of the $j^{th}$ SOC is defined by [12].

$$\Delta_j = \sum_{i=1}^{m_j} f_i - I_j - \xi_{L_j} = \begin{cases} \Delta_j^- = -5, -4, -2, -1 \\ \Delta_j^0 = 0 \\ \Delta_j^+ = +1, +2, +3, \cdots \end{cases} \tag{6}$$

where

$m_j$ is the number of joints contained in the $j^{th}$ $SOC_j$. $f_i$ is DOF of the $i^{th}$ joints. $I_j$ is the number of actuated joints in the $j^{th}$ $SOC_j$. $\xi_{L_j}$ is the NIDE of the $j^{th}$ loop. For any SKC, the following equation must be satisfied

$$\sum_{j=1}^{v} \Delta_j = 0 \tag{7}$$

Then, the coupling degree of a SKC [12] is defined by

$$\kappa = \frac{1}{2}\min\left\{\sum_{j=1}^{v} |\Delta_j|\right\} \tag{8}$$

The coupling degree $\kappa$ reflects the correlation and dependence between kinematic variables of each independent loop of the mechanism. It has been proved that the kinematics and dynamics of the PMs with higher $\kappa$ could be more complex.

Since the NIDEs of $Loop_1$, $Loop_2$, $Loop_3$ and $Loop_4$ have already been calculated in the previous section, i.e., $\xi_{L_1} = 3$, $\xi_{L_2} = 6$, $\xi_{L_3} = 6$, $\xi_{L_4} = 6$, thus, the constraint degree of the four loops are calculated by Eq. (6), respectively, and can be obtained as follows,

$$\Delta_1 = \sum_{i=1}^{m_1} f_i - I_1 - \xi_{L_1} = 5 - 2 - 3 = 0, \Delta_2 = \sum_{i=1}^{m_2} f_i - I_2 - \xi_{L_2} = 8 - 2 - 6 = 0,$$

$$\Delta_3 = \sum_{i=1}^{m_3} f_i - I_3 - \xi_{L_3} = 6 - 0 - 6 = 0, \Delta_4 = \sum_{i=1}^{m_4} f_i - I_4 - \xi_{L_4} = 7 - 1 - 6 = 0.$$

Thus, from Eq. (7), it can be known that this PM contains four SKCs. That is, $Loop_1$, $Loop_2$, $Loop_3$, and $Loop_4$ constitutes $SKC_1$, $SKC_2$, $SKC_3$, and $SKC_4$, respectively. There are actuated joints $P_1$ and $P_2$ in $SKC_1$, actuated joints $P_3$ and $P_4$ in $SKC_2$, and actuated joint $P_5$ in $SKC_4$, respectively. Therefore, at this moment, it can be judged that the PM has input-output motion decoupling [12–14], which will also be proved in the following Sect. 3.2.

The coupling degree of each SKC is calculated by Eq. (8) as,

$$\kappa_1 = \kappa_2 = \kappa_3 = \kappa_4 = \frac{1}{2}\sum_{j=1}^{v} |\Delta_j| = \frac{1}{2}(0) = 0$$

Zero coupling degree means that symbolic FPS of the PM can be directly derived from each SKC.

### 3.2 Position Analysis

**Direct Kinematics.** The kinematic modelling of the PM is shown in Fig. 2. Let the base platform 0 be a square with a side length $2a$. The point O on the base platform is selected as the origin of the frame coordinate system, with the $x$-axis parallel to $A_4A_3$ and the $y$-axis parallel to $A_1A_2$. Set moving platform 1 as an isosceles right-angled triangle with a right-angled side length $b$. The O′ point on the moving platform is selected as the origin of the moving coordinate system, the $x'$ axis is along the $O'E_2$ direction, and the $y'$ axis is along the $O'E_3$ direction. The axes $z$ and $z'$ are determined by the right-hand system. Let $A_iB_i = l, B_iC_i = l_b(i = 1 \sim 2)$, $C_1D = l_1, DC_2 = l_2$, $A_3B_3 = l_3, A_4B_3 = l_4$, $A_5E_3 = l_5$, $DO' = l_6$, $B_3E_2 = l_7$. The five input parameters are $y_{A_1}, y_{A_2}$, $l_3$, $l_4$, $l_5$, respectively. The attitude angle α, β and $\gamma$ are defined as the rotation angles of the moving platform about the $x$, $y$, and $z$ axes, respectively.

Solving the direct kinematics is to compute the position $O'(x, y, z)$ and attitude angle $(\alpha, \beta, \gamma)$ of the moving platform when setting the five input parameters of the five actuated joints $P_1$, $P_2$, $P_3$, $P_4$ and $P_5$ (with two coordinates $y_{A_1}, y_{A_2}$ and three length $l_1$, $l_2$, $l_3$).

It is easy to know that the coordinates of points $A_1$, $A_2$, $A_3$, $A_4$, $A_5$, $B_1$ and $B_2$ are respectively given as

$A_1 = (a, y_{A_1}, 0)^T$, $A_2 = (a, y_{A_2}, 0)^T$, $A_3 = (a, a, 0)^T$, $A_4 = (-a, a, 0)^T$, $A_5 = (0, -a, 0)^T$, and $B_1 = (a, y_{A_1}, l)^T$, $B_2 = (a, y_{A_2}, l)^T$.

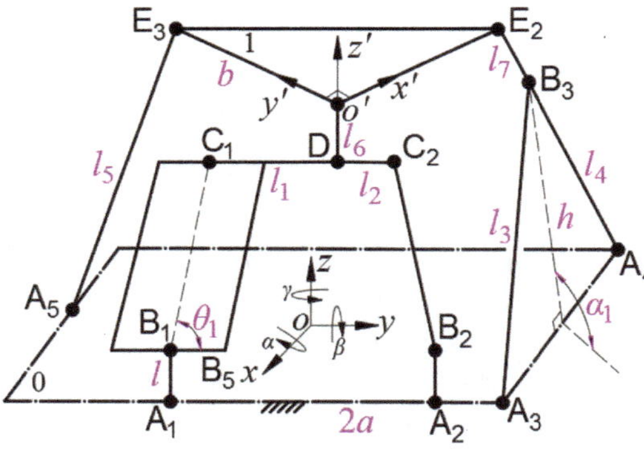


**Fig. 2.** Kinematic modelling of the 2T3R PM

*Find the Coordinates of Point O′ from SKC1.* In the first loop $A_1$-$B_1$-$C_1$-$D$-$C_2$-$B_2$-$A_2$ (i.e.SKC$_1$), the angle $\theta_1$ between $B_1C_1$ and $B_1B_5$ is set as the intermediate variable. The coordinates of the points $C_1$, $C_2$ and O′ are given respectively below. $C_1 = (a, y_{A_1} + l_b\cos\theta_1, l + l_b\sin\theta_1)^T$, $C_2 = (a, y_{A_1} + l_b\cos\theta_1 + l_1 + l_2, l + l_b\sin\theta_1)^T$

$$O' = \begin{bmatrix} x \\ y \\ z \end{bmatrix} = \begin{bmatrix} a \\ y_{A_1} + l_b\cos\theta_1 + l_1 \\ l + l_b\sin\theta_1 \end{bmatrix} \tag{9}$$

According to the length constraint condition defined by $C_2B_2 = l_b$, $\theta_1$ can derived as,

$$\theta_1 = \arccos(\frac{y_{A_1} + l_1 + l_2 - y_{A_2}}{-2l_b}) \tag{10}$$

Substituting Eq. (10) into Eq. (9) leads the coordinates of the point O′.

*Find the Coordinates of Point E2 from SKC2 and SKC3.* In the second loop $A_3$-$B_3$-$A_4$ (i.e., $SKC_2$), let $h$ be the vertical distance from $B_3$ to $A_3A_4$ and let $\alpha_1$ be the angle between the plane $A_3B_3A_4$ and the base plane. Then, the coordinates of the point $E_2$ are calculated by

$$E_2 = \begin{bmatrix} -a + \frac{l_4^2 + 4a^2 - l_3^2}{4a}\frac{l_4 + l_7}{l_4} \\ a + \frac{l_4 + l_7}{l_4} h\cos\alpha_1 \\ \frac{l_4 + l_7}{l_4} h\sin\alpha_1 \end{bmatrix} \tag{11}$$

In the third loop $B_3$-$E_2$-$O'$ (i.e., $SKC_3$), according to the length constraint condition defined by $E_2O' = b$, we can get,

$$\alpha_1 = 2\arctan(\frac{-I_2 \pm \sqrt{I_1^2 + I_2^2 - I_3^2}}{I_3 - I_1}) \tag{12}$$

where $I_1 = 2h(a - y)\frac{l_4 + l_7}{l_4}, I_2 = -2hz\frac{l_4 + l_7}{l_4}$, $I_3 = (x_{E_2} - x)^2 + (a - y)^2 + z^2 - b^2 + (h\frac{l_4 + l_7}{l_4})^2$.

Substituting Eq. (12) into Eq. (11), the coordinates of point $E_2$ can be obtained.

*Find the Coordinates of Point $D_3$ from $SKC_4$.* In the fourth loop $A_5$-$E_3$ (i.e., $SKC_4$), let the coordinates of point $E_3$ be $E_3 = (x_{E_3}, y_{E_3}, z_{E_3})$. According to the two length constraint conditions defined by $A_5E_3 = l_5$ and $E_3O' = E_3E_2 = b$, we can get,

$$\begin{cases} x_{E_3}^2 + (y_{E_3} + a)^2 + z_{E_3}^2 = l_5^2 \\ (x_{E_3} - x)^2 + (y_{E_3} - y)^2 + (z_{E_3} - z)^2 = b^2 \\ (x_{E_3} - x_{E_2})^2 + (y_{E_3} - y_{E_2})^2 + (z_{E_3} - z_{E_2})^2 = b^2 \end{cases} \tag{13}$$

From Eq. (13), we can get

$$\begin{cases} z_{E_3} = \frac{-S_2 \pm \sqrt{S_2^2 - 4S_1S_3}}{2S_1} \\ y_{E_3} = T_1 z_{E_3} + T_2 \\ x_{E_3} = W_1 z_{E_3} + W_2 \end{cases} \tag{14}$$

where $M = -\frac{a+y}{x}, N = -\frac{z}{x}, L = -\frac{a^2 - l_5^2 - x^2 - y^2 - z^2 + b^2}{2x}, P = 2(x_{E_2} - x)M + 2(y_{E_2} - y)$,

$$U = 2(x_{E_2} - x)N + 2(z_{E_2} - z),\ R = 2(x_{E_2} - x)L + x^2 + y^2 + z^2 - x_{E2}^2 - y_{E2}^2 - z_{E2}^2 + b^2,$$
$$T_1 = -\frac{U}{P}, T_2 = -\frac{R}{P}, W_1 = N - \frac{MU}{P}, W_2 = L - \frac{MR}{P},\ S_1 = W_1^2 + T_1^2 + 1,$$
$$S_2 = 2W_1W_2 + 2T_1T_2 + 2aT_1, S_3 = W_2^2 + T_2^2 + 2aT_2 + a^2 - l_5^2.$$

*Solving the Attitude Angle of the Moving Platform from the Transformation Matrix.* The coordinates in the frame coordinate system for any point $S$ on the moving platform 1 can be expressed as:

$$S = QS' + K \tag{15}$$

where $\boldsymbol{Q}$ *is* the transformation matrix between the moving and frame coordinate systems, i.e., $\mathbf{Q} = \begin{bmatrix} \cos\beta\cos\gamma & -\cos\beta\sin\gamma & \sin\beta \\ \sin\alpha\sin\beta\cos\gamma + \cos\alpha\sin\gamma & \cos\alpha\cos\gamma - \sin\alpha\sin\beta\sin\gamma & -\sin\alpha\cos\beta \\ \sin\alpha\sin\gamma - \cos\alpha\sin\beta\cos\gamma & \cos\alpha\sin\beta\sin\gamma + \sin\alpha\cos\gamma & \cos\alpha\cos\beta \end{bmatrix}$.

$S'$ is the coordinate of point $S$ in the moving coordinate system.

$K$ is the coordinate of the origin of the moving coordinate system in the frame coordinate system.

According to the Eq. (15), the coordinates of the points $E_2$ and $E_3$ in the frame coordinate system, with the relative coordinates $E_2$ = ($b$, 0, 0) and $E_3$ = (0, $b$, 0) in the moving coordinate system, are

$$\begin{aligned} E_2 = {} & (b\cos\beta\cos\gamma + a, b(\cos\alpha\sin\gamma + \sin\alpha\sin\beta\cos\gamma) + y, \\ & b(\sin\alpha\sin\gamma - \cos\alpha\sin\beta\cos\gamma) + z)^{\mathrm{T}} \end{aligned} \tag{16}$$

$$\begin{aligned} E_3 = {} & (-b\cos\beta\sin\gamma + a, b(\cos\alpha\cos\gamma - \sin\alpha\sin\beta\sin\gamma) + y, \\ & b(\sin\alpha\cos\gamma + \cos\alpha\sin\beta\sin\gamma) + z)^{\mathrm{T}} \end{aligned} \tag{17}$$

To solve Eqs. (11), (14), (16) and (17) simultaneously, the attitude angle of the moving platform can be obtained as,

$$\begin{cases} \alpha = 2\arctan\left(\dfrac{-N_2 \pm \sqrt{N_2^2 + N_1^2 - N_3^2}}{N_3 - N_1}\right) \\ \beta = \arccos\left(\dfrac{x_{E_2} - a}{b\cos\gamma}\right) \\ \gamma = \arctan\left(\dfrac{a - x_{E_3}}{x_{E_2} - a}\right) \end{cases} \tag{18}$$

where $N_1 = -b\sin\beta\sin\gamma, N_2 = b\cos\gamma,\ N_3 = z - z_{E_3}$.

From Eqs. (10) and (18), we can know,

$$\begin{cases} y = f_2(y_{A_1}, y_{A_2}) \\ z = f_3(y_{A_1}, y_{A_2}) \end{cases}, \begin{cases} \alpha = f_{11}(y_{A_1}, y_{A_2}, l_3, l_4, l_5) \\ \beta = f_{12}(y_{A_1}, y_{A_2}, l_3, l_4, l_5) \\ \gamma = f_{13}(y_{A_1}, y_{A_2}, l_3, l_4, l_5) \end{cases}$$

That is, the PM has partial input-output motion decoupling, which is beneficial to trajectory planning and motion control of the PM.

**Inverse Kinematics.** Solving the inverse kinematics is to compute the position values of the prismatic joints $P_1$, $P_2$, $P_3$, $P_4$ and $P_5$ (with the coordinates $y_{A_1}$, $y_{A_2}$ and the length of $l_1$, $l_2$, $l_3$), when setting the position O′($x$, $y$, $z$) and attitude angle ($\alpha$, $\beta$, $\gamma$) of the moving platform.

From O′($x$, $y$, $z$), the coordinates of points $C_1$ and $C_3$ can be obtained as $C_1 = (x, y - l_1, z - l_6)^T$, $C_2 = (x, y + l_2, z - l_6)^T$. According to the two length constraint conditions defined by $C_1B_1 = C_2B_2 = l_b$, we can get,

$$y_{A_1} = y - l_1 \pm \sqrt{l_b^2 - (l_6 + l - z)^2} \tag{19}$$

$$y_{A_2} = y + l_2 \pm \sqrt{l_b^2 - (l_6 + l - z)^2} \tag{20}$$

From Eqs. (14) and (16), the coordinates of points $E_2$ and $E_3$ can be obtained. Thus, according to the length constraint condition defined by $E_2A_4 = l_4 + l_7$, we can get,

$$l_4 = -l_7 + \sqrt{(x_{E_2} + a)^2 + (y_{E_2} - a)^2 + z_{E_2}^2} \tag{21}$$

Since the points $E_2$, $B_3$, and $A_4$ are on a straight line, we know that the coordinates of point $B_3$ are,$B_3 = (\frac{l_4}{l_4 + l_7}(x_{E_2} + a) - a, \frac{l_4}{l_4 + l_7}(y_{E_2} - a) + a, \frac{l_4}{l_4 + l_7} z_{E_2})^T$.

According to the length constraint condition $A_3B_3 = l_3$, we can get,

$$l_3 = \sqrt{(x_{B_3} - a)^2 + (y_{B_3} - a)^2 + z_{B_3}^2} \tag{22}$$

According to the length constraint condition $A_5E_3 = l_5$, we can get,

$$l_5 = \sqrt{x_{E_3}^2 + (y_{E_3} + a)^2 + z_{E_3}^2} \tag{23}$$

So far,the input values of the five prismatic joints $P_1$, $P_2$, $P_3$, $P_4$ and $P_5$ have been solved.

**Numerical Valuation for Forward and Inverse Solutions.** Set the parameters of the PM as: $a = 80, b = 90, l = 10, l_1 = 40, l_2 = 20,\ l_6 = 10, l_7 = 10, l_b = 100,\ y_{A_1} = -45, y_{A_2} = 20,\ l_3 = 133.3, l_4 = 144.3, l_5 = 122.5$ (unit: mm).

From Eqs. (9) to (18), four sets of real solutions are obtained, as shown in Table 1.

**Table 1.** The values of direct solutions

| No. | y/mm | z/mm | α/(°) | β/(°) | γ/(°) |
|---|---|---|---|---|---|
| 1* | −2.5 | 119.9687 | 179.4162 | 179.9713 | −44.4458 |
| 2 | −2.5 | 119.9687 | 0.5274 | 179.9713 | −44.4558 |
| 3 | −2.5 | 119.9687 | −83.4946 | 135.7687 | 4.8979 |
| 4 | −2.5 | 119.9687 | −89.6637 | 135.7687 | 4.8979 |

Substituting the forward solution data No. 1* in Table 1 into Eqs. (19)–(23), and one set of solutions is gotten as $y_{A_1} = -45, y_{A_2} = 20$, $l_3 = 133.2515$, $l_4 = 144.2521$, $l_5 = 122.5622$, which is the same as the five position values of the prismatic joints $P_1$, $P_2$, $P_3$, $P_4$ and $P_5$ when solving the forward solution, and the maximum relative error is 0.050776%. Therefore, all equations of the forward solution and the inverse solution are derived correctly.

## 4 Conclusions

The FPS of the foregoing 5-DOF two-translation and three-rotation PM and other many complex PMs the authors have performed show that the SKC-based modelling and solution method is general and has the following advantages.

1) The difficulty and complexity of FPS of PMs depend on the type and number of SKC itself. In particular, when the coupling degree of a single SKC is lower or zero, its position equation is very easy, and the symbolic solutions can be obtained [15].
2) The method presented in this paper has a clear analysis road-map, which can greatly reduce the difficulty of FPS of complex PMs. In addition, the SKC-based modelling method is easy to be programmed into automatic computer generation [16].

## References


1. Pennock, G.R., Hasan, A.: A polynomial equation for a coupler curve of the double butterfly linkage. J. Mech. Des. **124**(1), 39–46 (2002)
2. Kong, X., Gosselin, C.M.: Generation and forward displacement analysis of RPR-PR-RPR analytic planar parallel manipulators. J. Mech. Des. **8**(18), 2195–2304 (2001)
3. Liu, A., Yang, T.: Kinematics Design of Mechanical System. China Petrochemical Press, Beijing (1999)
4. Liu, A., Yang, T.: Finding all positive solutions of a general 6-SPS parallel robot mechanism. Mech. Sci. Technol. **15**(4), 543–546 (1996)
5. Liao, Q., Seneviratne, L.D., Earles, S.W.E.: Forward positional analysis for the general 4-6 In parallel platform. Proc. Inst. Mech. Eng. Part C J. Mech. Eng. Sci., **209**, 55–67 (1995)
6. Nanua, P., Waldron, K.J.: Direct kinematic solution of a Stewart platform. In: Proceedings of the IEEE International Conference on Robotics and Automation 1989, 14–19 May 1989 pp. 431–437 (1989)
7. Husain, M., Waldron, K.J.: Direct position kinematics of the 3-1-1-1 Stewart platforms. J. Mech. Des. **116**(4), 1102–1107 (1994)

8. Wampler, C.W.: Forward displacement analysis of general six-in-parallel Sps (Stewart) platform manipulator using soma coordinates. Mech. Mach. Theory **31**(3), 311–337 (1996)
9. Husty, M.L.: Algorithm for solving the direct kinematic of Stewart-Gough-type platforms. Mech. Mach. Theory **31**(4), 365–380 (1996)
10. Dasgupta, B., Mruthyunjaya, T.S.: A constructive predictor-corrector algorithm for the direct position kinematics problem for a general 6-6 Stewart platform. Mech. Mach. Theory **31**(6), 799–811 (1996)
11. Shen, H.: Topological Characteristics-Based Kinematics. Changzhou University, January 2020
12. Yang, T., Liu, A., Luo, Y., et al.: Topological Design of Robot Mechanism. Science Press, Beijing (2012)
13. Yang, T., Liu, A., Shen, H., et al.: Topology Design of Robot Mechanism. Springer, Singapore (2018)
14. Shen, H., Yang, T., Li, J., Zhang, D., et al.: Evaluation of topological properties of parallel manipulators based on the topological characteristic indexes. Robotica **11** (2019)
15. Shen, H., Chablat, D., Zen, B., et al.: A translational three-degrees-of-freedom parallel mechanism with partial motion decoupling and analytic direct kinematics. J. Mech. Robot. **12**, 021112-1–021112-7 (2020)
16. Zuo, S., Shen, H.: Position Analysis and Computer Automatic Generation of Parallel Mechanism. Changzhou University, June 2016

# Development of a Total Inspection System for Peaches

Kazuyoshi Ishida(✉), Koji Makino, Hiromi Watanabe, Yutaka Suzuki, Tsutomu Tanzawa, Shinji Kotani, and Hidetsugu Terada

University of Yamanashi, Kofu, Yamanashi 400-8511, Japan
{isawa, kohjim, hwatanabe, yutakas, ttanzawa, kotani, terada}@yamanashi.ac.jp

**Abstract.** To inspect the peach moth larvae for peach fruit, the X-ray inspection system has been developed. However, it is insufficient to test using that system, so that, to detect the generated hole by on the surface of peach, the surface inspection system is developed. Furthermore, to remove trichome and moth eggs from the surface, an air blower system with rotation mechanism is developed. As a result, 0.2 mm diameter drilled holes on the peach surface can be detected by the inspection camera device at a rate of 94%. We have confirmed that this peach surface inspection system can be executed stably. In addition, it was found that peaches were not damaged by the improved end effector.

**Keywords:** Automatic inspection · Peach transfer · Flexible handling · Peach fruit moth · Quarantine control

## 1 Introduction

The export of agricultural products is important for many countries [1]. However, most products must be often checked in quarantine. This study focuses on peaches exported from Japan to Taiwan because peaches are very popular in Taiwan [2] and the amount of exported peaches is third grade in Japan [3]. Peaches should not contain any larvae of the peach moth [4]. If any larvae are detected in Taiwan, then exports from a prefecture or the whole of Japan are halted. This stoppage is a very heavy penalty for the exporting country and a serious problem for many farmers. Therefore, to avoid exporting any peaches that contain larvae, the farmers visually inspect all peaches by searching for holes that are less than 0.2 mm in diameter bored by the larvae enter the fruit. Such inspection is very difficult and time-consuming. Therefore, it is desirable to develop an automatic inspection system, but it is difficult to develop the system for agricultural products because such products are available only for a few months at a time.

Many types of inspection machinery have been developed and studied [5, 6]. We have also developed a peach inspection system that uses X-rays to detect the larvae [7, 8]. It is difficult to detect all the larvae because there is a tradeoff between the accuracy and rate of detection. Therefore, it is necessary to employ various sensors or systems. There are many


E.-C. Lovasz et al. (Eds.): MTM&Robotics 2020, MMS 88, pp. 66–79, 2021.
https://doi.org/10.1007/978-3-030-60076-1_7

studies on the integration of inspection systems. Air is blown over each peach to remove any moth eggs and loose hairs from the skin. After that, visual inspection looks for any larva holes. There are two pre-processes used for checking. Therefore, we believe that it is effective to integrate various sensors and systems to check fruit and other farm products.

The peculiarity of export grade peaches for Taiwan that are a unique Japanese variety makes peach inspection system development more difficult. The surface of the peach is soft and easily damaged, therefore it is necessary to evaluate the handling using real peaches. As the dummy peach cannot reproduce the surface condition of the real peach, the correct evaluation of handling must use the real peach. In addition, there are no peaches peculiar to Japan used for handling evaluation in other countries. Furthermore, export grade peaches are available only in Japan at a limited time and place.

In this paper, a total inspection system for peach moth larvae in peaches is described. Our system is illustrated in Fig. 1. This system consists of devices for scanning the interior of the peaches, detecting any larva holes on the skins, and removing any moth eggs clinging to the skins. The air blowing and hole detection are performed automatically and sequentially. Section 2 of this paper introduces the actual processes of searching for holes and larvae, as well as removing eggs. Section 3 describes the devices, which are a camera device, an air blower, and a handling unit, that compose the peach surface inspection system. Section 4 discusses the experiments conducted on the peach surface inspection system. Section 5 presents the conclusions of this study.

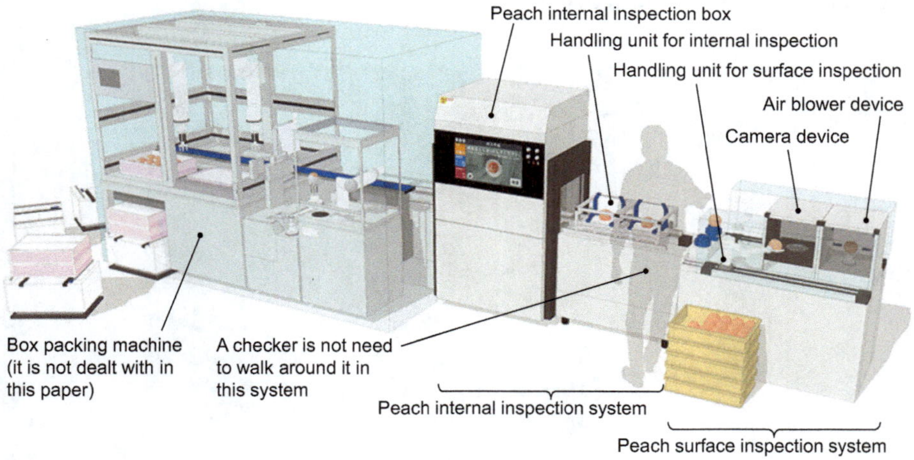


**Fig. 1.** Image illustration of total inspection system.

## 2 Peach Inspection

### 2.1 Current Checking Procedure

A checker that means a staff who checks peaches performs two procedures: blowing air and searching for holes. First, the checker blows air to remove the trichome (fine hair covering the skin of a peach). Blowing air accomplishes two goals at once: it is easier to find holes if the trichome is removed and any moth eggs laid on the skin are removed with the trichome. An egg that remains would hatch into a larva that bores a hole into the interior of the peach.

Second, the checker uses a large magnifying glass to search for any holes. The holes are about 0.2 mm in diameter. Even with a magnifying glass, it is usually very difficult to find such a small hole. This difficulty motivated us to develop our system to search the interior of a peach to detect any larva.

### 2.2 Peach Internal Inspection System

We developed our inspection system to use X-rays and search the interior of a peach, as shown in Fig. 2.

The checker places two peaches onto the handling unit and pushes the start button on the display. The handling unit enters the inspection box automatically. Images are obtained from the X-ray machine and analyzed by our detection algorithm [9]. If a larva or void in the peach is not detected, then the handling unit exits at the opposite side. If a larva is detected, the handling unit returns the peach to the initial position.

The peach internal inspection system cannot inspect the peach surface. Therefore, we designed and produced the peach surface inspection system consisting of various devices shown in the next section, and the details of the system are described in this paper. In particular, this end effector that is built in the handling unit is an important mechanical element in peach handling.

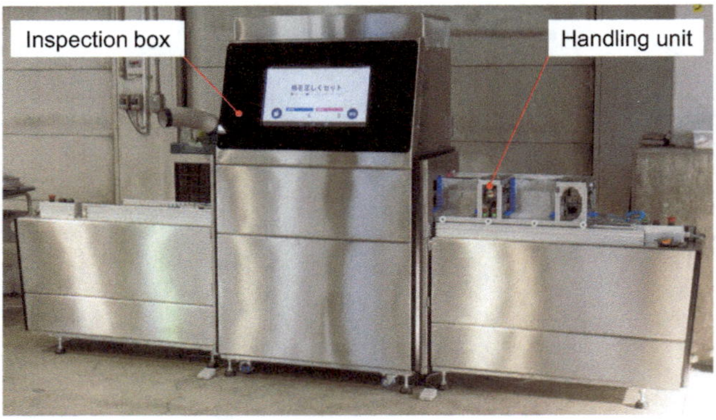


**Fig. 2.** Peach internal inspection system.

## 3 New Devices for Peach Surface Inspection System

### 3.1 Arrangement of Devices

We considered arrangements of the devices and the handling units [10]. The schematic diagram of the round-type arrangement of the devices is shown in Fig. 3, which is a schematic diagram of the system shown in Fig. 1. Two handling units are employed. The checker places a peach on the access table and presses the start button to transfer the peach to the air blower device. After blowing, the handling unit A grasps the peach again to transfer it from the air blower device to the camera device, and returns to its initial position. At the same time, the handling unit B transfers the peach that is placed at the camera device to the access table. The checker can place the next peach on the access table.

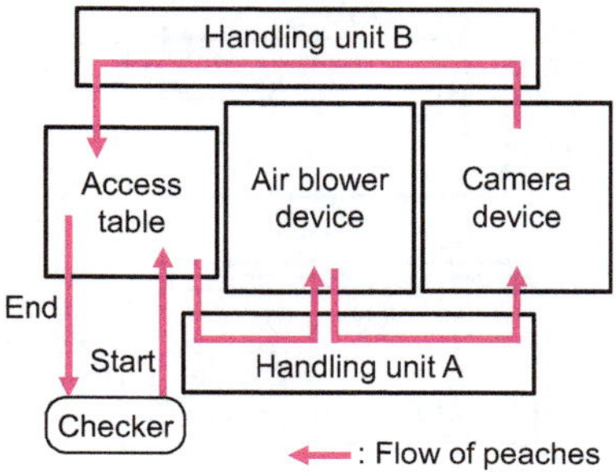


**Fig. 3.** Arrangement of devices.

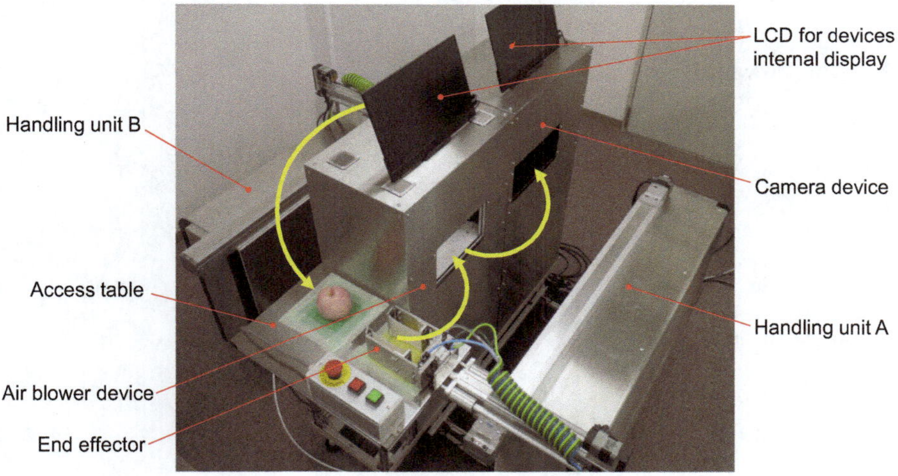


**Fig. 4.** Peach surface inspection system (W:1320 × H:1365 × D:1645 [mm]).

We developed the peach surface inspection system shown in Fig. 4. Each device was sealed and had an LCD monitor for the observation of its inner workings. Start, stop, and emergency stop buttons were employed. The checker stood at the end effector, placed a peach on the access table, and pushed the start button. Handling unit A grasped the peach from the access table. After all processes were completed, handling unit B grasped and transferred the peach back onto the access table.

The configuration diagram for the peach surface inspection system is shown in Fig. 5. The timings of the devices were controlled by a PC (EPSON, MR8100).

The camera and air blower were controlled by a microprocessor (Mbed: NXP Semiconductors, LPC1768) and the handling unit was controlled by a programmable logic controller (PLC: Mitsubishi Electric, FX3G-60MT/DS). These devices communicated with the PC via wired serial communication.

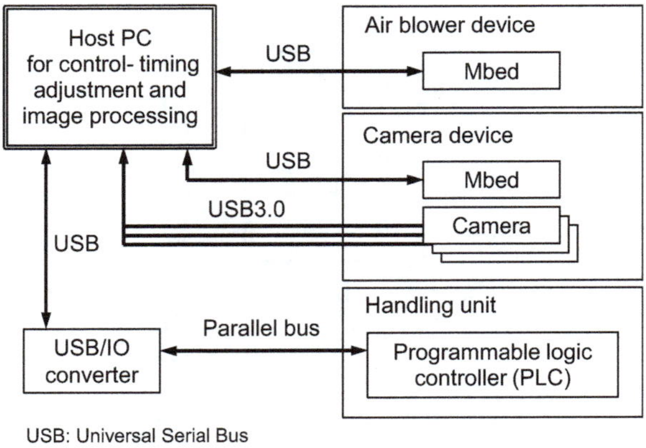


**Fig. 5.** Configuration diagram for the peach surface inspection system.

## 3.2 Air Blower Device

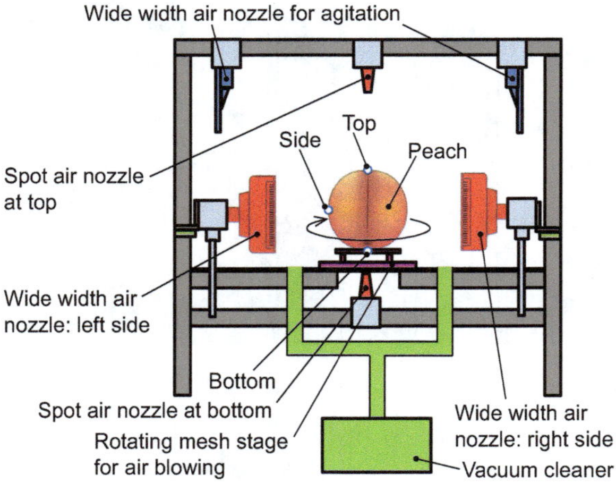


**Fig. 6.** Arrangement of air and vacuum nozzles.

The skin of a peach is covered by trichome, to which the moth eggs are attached. The trichome of all peaches exported to Taiwan from Japan is removed by the air blower.

We experimented to check if the amount, velocity, and direction of air blown were sufficient to remove all the trichome on a peach. The air nozzles (depicted in red) and the peach were arranged as shown in Fig. 6. One spot air nozzle was set at the top and another at the bottom of the box while a wide air nozzle (depicted in blue) were set to each side of the top nozzle and another to each side of the peach to agitate the surrounding air in order to ensure that the removed trichome did not fall back onto the peach, which was placed onto the mesh stage and subjected to air blown from the bottom spot nozzle. The peach was rotated at 104°/s. In the pre-experiments, we confirmed that any eggs attached to the skin were removed if the air velocity was over 10 m/s. Therefore, we set the air velocity of the top nozzle to 26 m/s and those of the others to approximately 18 m/s. The removed trichome was sucked away by a vacuum nozzle. Then, the peach was placed in a sealed box. When the peach was placed in or taken out, a shutter attached to the device was opened.

The air blower device is shown in Fig. 7. The outer plate is removed, a peach is placed on the rotating stage, the door is closed, and the air blower starts. The blowing time, including the shutter action, takes 7 s. The evaluation whether the trichomes are removed or not is performed in visual inspection of human who has received a lecture for the peach inspection. There is not the quantitative basis for the evaluation. The checker judged that the removal result by the air blow device is enough.

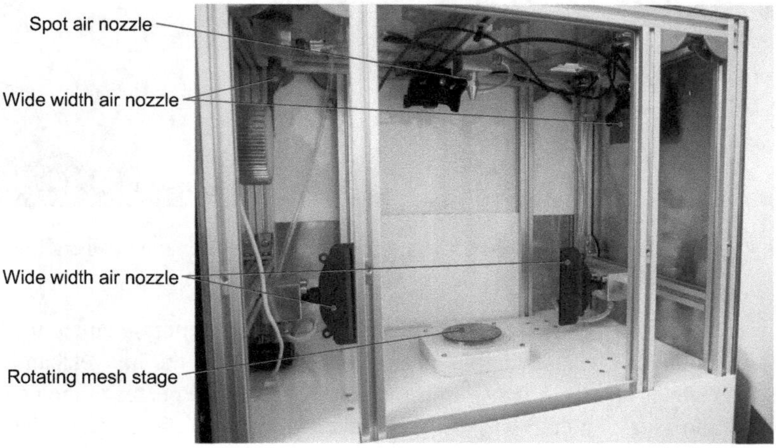


**Fig. 7.** Air blower devices (outer plate and shutter not shown).

### 3.3 Camera Device

Three cameras are used to detect any holes on the skins of the peaches. The larvae normally bore holes that are less than 0.2 mm in diameter, so it is difficult to find them under normal conditions. Therefore, a special lighting condition is used to detect the edges of any hole and an analytical method is used to convert the camera image.

The arrangement of the cameras is shown in Fig. 8. Three cameras are placed at the top, side, and bottom. The top and bottom cameras are set at ±60° from the side camera. Each camera has autofocus with 24-degree horizontal and 18-degree vertical viewing angles.

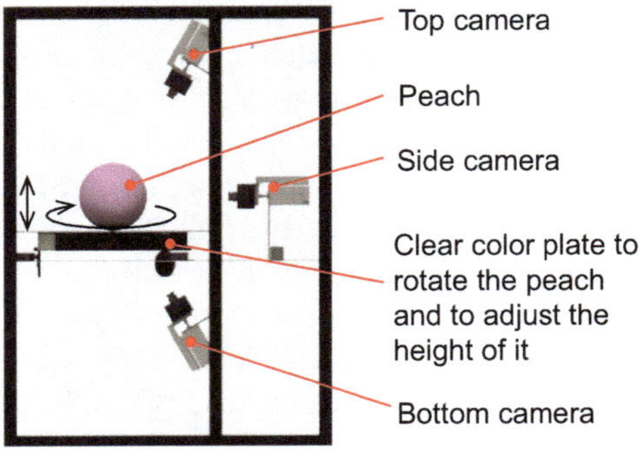


**Fig. 8.** Arrangement of three cameras.

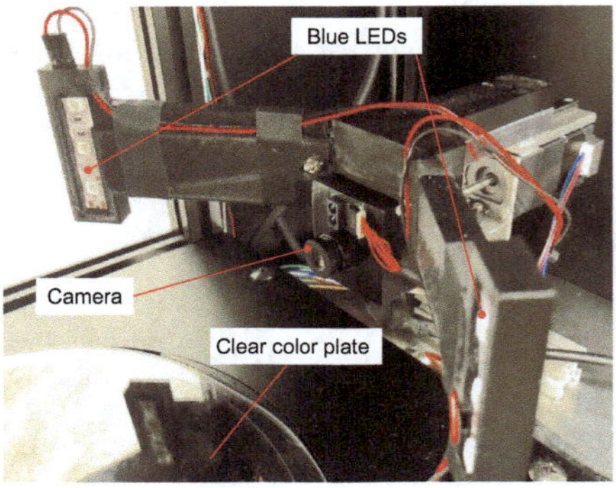


**Fig. 9.** Arrangement of light against camera.

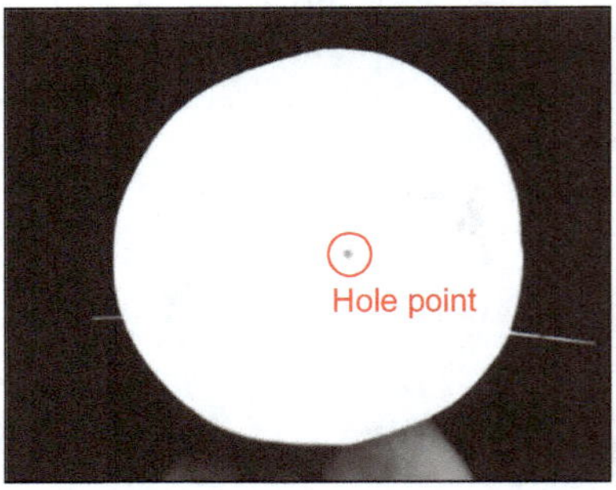


**Fig. 10.** Converted camera image.

A peach is placed on a clear color plate. The bottom camera enables the bottom image to be obtained because the plate can be rotated so that the images can be taken from various angles and the height can be adjusted to align the center of the peach with the axis of the side camera. An image is taken at every 45°. In total, 24 (3 × 8 directions) images are obtained. The arrangement of the cameras and light is shown in Fig. 9.

The inside of the device is a dark box. When the peach is placed within or taken out, a shutter attached to the device is opened similarly to the air blower. The images are converted by algorithms. A converted camera image is shown in Fig. 10. The hole can be expressed as a black point.

### 3.4 Handling Unit

Consisting of an end effector and two slide rails made by the linear actuator, as shown in Fig. 11, the handling unit transfers the peach from the initial position to the air blower, then to the camera, and back to the initial position.

The handling unit consists of an end effector and two slide rale difficult to use a normal robotic hand. The end effector grasps the peach by using a special concavo-convex sheet [11] as shown in Fig. 12. The end effector can grasp a peach of dimensions 94 (width) × 80 (height) × 89 (depth) mm. The effector is controlled by two air cylinders whose strokes in the side and upward directions are 40 and 10 mm, respectively. The sheet is pulled by a hinge with a spring.

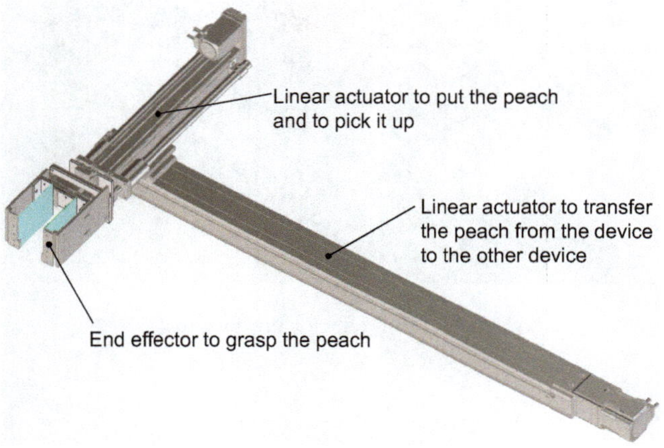


**Fig. 11.** Structure of the handling unit.

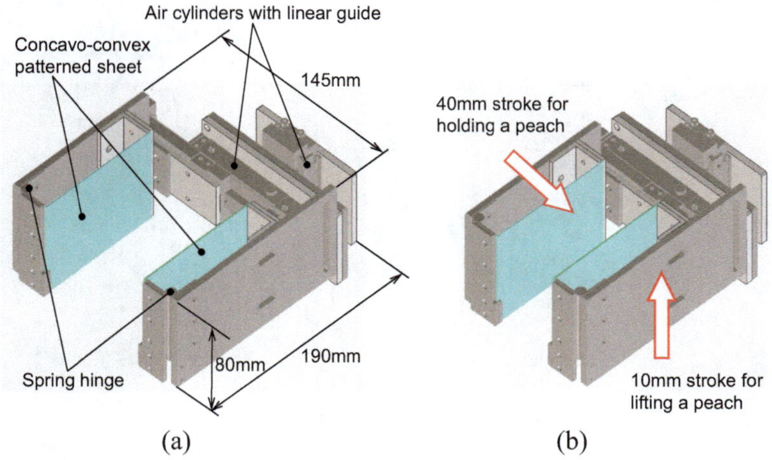


**Fig. 12.** End effector using special sheet to grasp peach: (a) Release state, (b) Holding state.

The end effector is shown in Fig. 13. The accuracy and the velocity of the linear slider were 0.2 mm and 100 mm/s, respectively. We investigated the performance of the end effector and determined that it was able to place a peach within 10 mm in error diameter.

After that, a performance test was conducted using the real peaches.

**Fig. 13.** End effector grasps peach.

# 4 Experiments

## 4.1 Detection Performance of an Inspection Camera Device

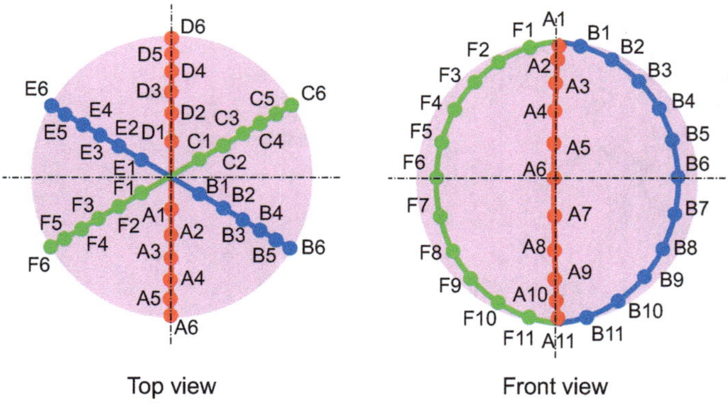


**Fig. 14.** Position of holes on peach.

Using a 0.2-mm drill, we bored 66 holes into a peach to investigate the performance of the camera device. The positions of the hole are shown in Fig. 14. The results are shown in Table 1. Of the peaches, 94% were detected. The time to check each peach was 24 s.

**Table 1.** Results of detection by camera device.

| Latitude position | Longitudinal positions | | | | | |
|---|---|---|---|---|---|---|
| | A | B | C | D | E | F |
| 1 | √ | √ | √ | √ | NG | NG |
| 2 | √ | √ | √ | √ | √ | √ |
| 3 | √ | √ | √ | √ | √ | √ |
| 4 | √ | √ | √ | √ | √ | √ |
| 5 | √ | √ | √ | √ | √ | √ |
| 6 | √ | √ | √ | √ | √ | √ |
| 7 | √ | √ | √ | √ | √ | √ |
| 8 | NG | √ | √ | √ | √ | √ |
| 9 | √ | √ | √ | √ | √ | √ |
| 10 | √ | √ | √ | √ | √ | √ |
| 11 | NG | √ | √ | √ | √ | √ |
| | | | | √: Detectable<br>NG: Undetectable | | |

## 4.2 Evaluation of a Handling Motion

The handling motion test was performed for the peach surface inspection system. First, the series motion was confirmed. Figure 15 illustrates each process. Figure 15(a) and (c) show the transfer by handling unit A. Figure 15(b) shows the wait during air blowing. Figure 15(d) shows the handling unit A's return to the initial position during camera detection. Figure 15(e) and (f) illustrate the peach's being removed from the camera and returned to the initial position. We confirmed that all processes were connected and there were no motion errors. The time of the series motion was 127 s. The details of each process are shown in Table 2. The velocity of the handling unit was kept slow to maintain safety throughout the experiments. If the handling unit is covered by a safety fence, then the velocity can be doubled. Since two peaches are transferred simultaneously, the time per peach is 40 s.

Second, the series motions were performed 30 times continuously. No errors occurred. We confirmed that this system could be executed stably.

**Table 2.** Time of each process.

| Process name | Process time [s] |
|---|---|
| Transfer motion to air blower | 24 |
| Air blower | 7 |
| Transfer motion to camera | 34 |
| Hole detection | 24 |
| Remove peach from camera and return to initial position | 38 |
| **Total** | 127 |

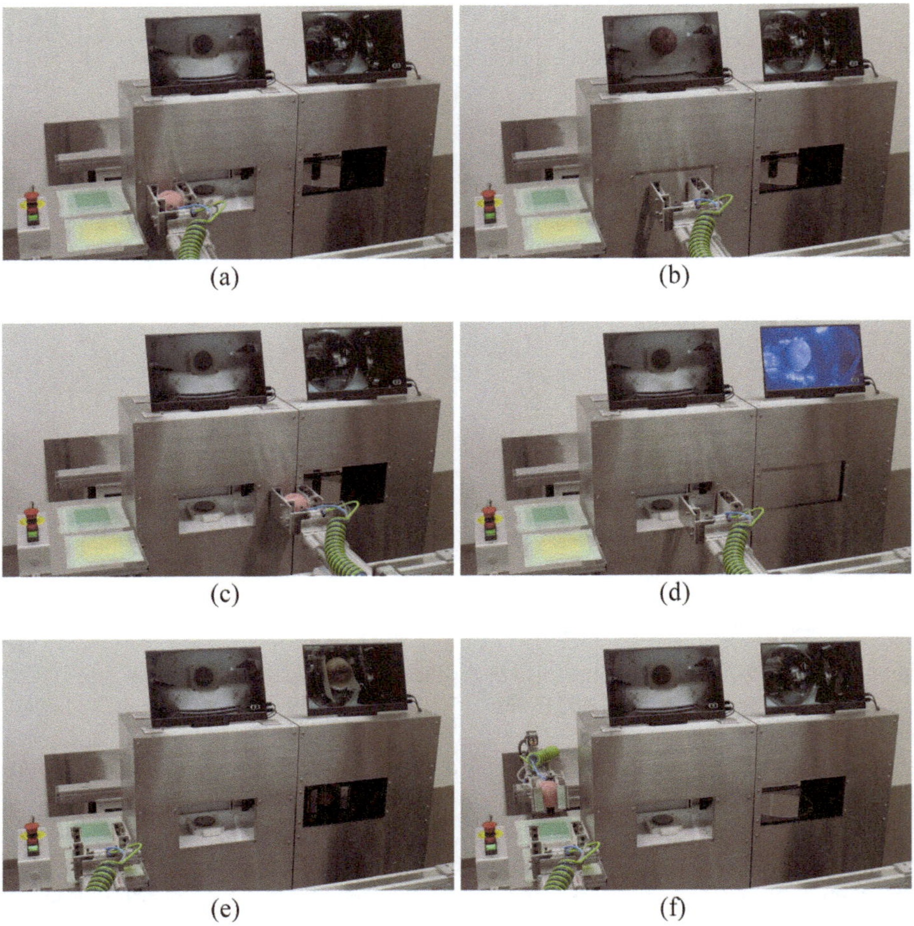

**Fig. 15.** Procedure for checking: (a) Transfer motion to air blower device, (b) During air blower, (c) Transfer motion to camera device, (d) During hole detection, (e) Remove peach from camera device, (f) Transfer motion to initial position.

## 4.3 Improvement of End Effector and Its Performance Test

The end effector used in the handling unit may damage the peach. To solve this problem, the contact area between the end effector and the peach was improved. Figure 16(a) shows the improved end effector. The contact area with the peach is composed of two 10 mm thick cushioning blocks covered with the concavo-convex sheet (Fig. 16(b)). The cushioning block on the peach contact side has a hole (0 to 60 mm in diameter, flat for 0 mm) to reduce the contact pressure with peach.

Next, to confirm the peach lifting operation, the holding force was measured using the capacitive force sensor (WACOH-TECH, WEF-6A200-4-RC5). The shape of the force sensor is a sphere, and the diameter can be selected to 75, 80, 85, 90 and 95 mm. The size of the sensor is similar to the size of the peach. Figure 17 shows the relationship between the holding force and the hole diameter. When the sphere diameter of the force sensor is in the range of 75 to 90 mm, the holding force tends to decrease as the hole diameter increases. In addition, when the sphere diameter of the force sensor was 75 mm, the force sensor could not be lifted with the hole diameters of 40 mm and 60 mm. Since the diameter range of the target peach was 87 to 94 mm, it was found that the peach could be grasped with a cushioning block hole diameter of 60 mm or less. From Fig. 17, we confirmed that the holding force of the hole diameter of 0 to 40 mm was 11.8 to 21.0 N under the condition of the force sensor diameter of 85 to 95 mm.

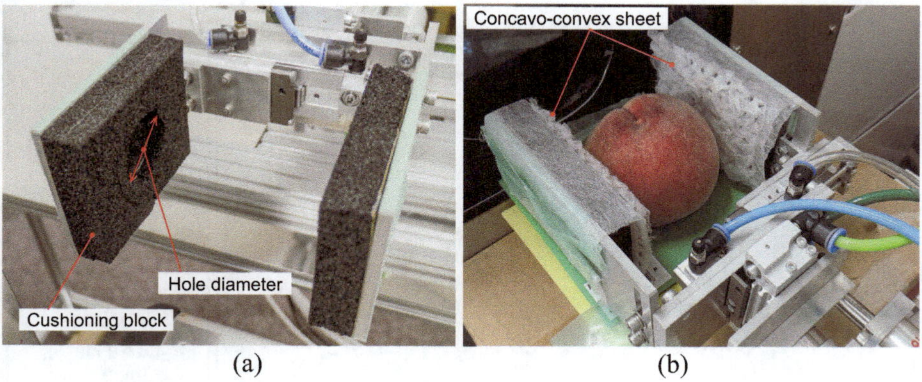


**Fig. 16.** End effector using the cushioning block with concavo-convex sheet: (a) Appearance without concavo-convex sheet, (b) Peach grasp test.

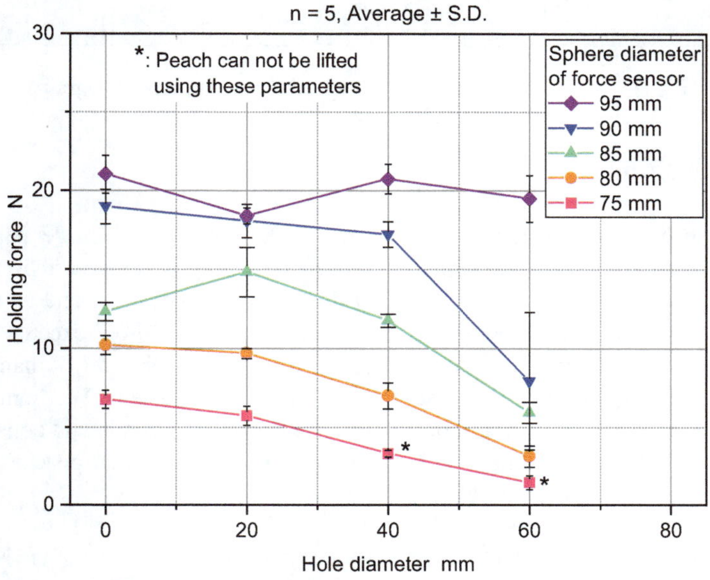


**Fig. 17.** Relationship between holding force and hole diameter on the cushioning block covered with concavo-convex sheet.

Finally, to confirm the damage to the peach, the performance of the improved end effector was evaluated by the grasp test using the export grade peaches. This size of the export grade peaches is from 87 to 94 mm. The cushioning block hole diameters were three types (0, 20 and 40 mm), and the number of grasp at each diameter was 2, 20 and 100 times. In addition, the grasping time per one trial is 5 s. The peaches after the grasp test were stored in a room out of direct sunlight (room temperature 25 ± 1 °C, humidity 63 ± 9%). Figure 18 shows the surface condition of the peaches after 8 days in the grasp test. It became clear that the improved end effector can be grasped without damaging the export grade peaches.

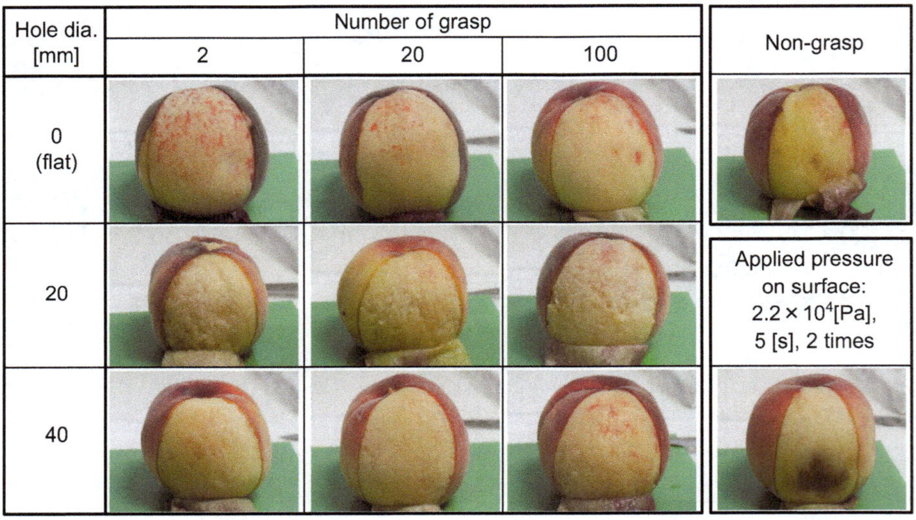


**Fig. 18.** Surface condition of each peach (grasping time: 5 s/grasp).

# 5 Conclusions

We developed a total system for detecting peach moth larvae inside peaches and removing moth eggs from the skins of the peaches. This paper describes the development of two devices: a camera to inspect the skins for larva holes and an air blower to remove the eggs from the skins. The camera and air blower execute the same functions performed during a visual inspection by a human checker. A handling unit was developed to automate the peach surface inspection system. We confirmed the performance of each device and series motion. This system is stable and consumes less time than would a human checker. Furthermore, it was found that peaches were not damaged by the improved end effector.

**Acknowledgment.** This research was supported by grants from the Project of the NARO Bio-oriented Technology Research Advancement Institution (the special scheme project on regional developing strategy).

## References


1. Summary of the Basic Plan for Food, Agriculture and Rural Areas, WEB site on the Ministry of Agriculture, Forestry and Fisheries. http://www.maff.go.jp/e/policies/law_plan/attach/pdf/index-2.pdf. Accessed 13 Mar 2020
2. Fruits and Vegetables Grown in Japan Are Safe and of the Finest Quality, WEB site on the Ministry of Agriculture, Forestry and Fisheries. http://www.maff.go.jp/j/export/e_info/vege_fruit/pdf/fruit_e_all.pdf. Accessed 13 Mar 2020
3. Japan Fruit Association: Introduction to Fruit Export and Trade, Agriculture Study Group Report (2017)
4. Development of AFFriinovation for Agriculture, Forestry, and Fisheries, WEB site on the Ministry of Agriculture, Forestry and Fisheries. http://www.maff.go.jp/e/policies/food_ind/attach/pdf/index-4.pdf. Accessed 13 Mar 2020
5. Ihara, F., et al.: Non-destructive observation of peach fruit moth, Carposina sasakii Matsumura (Lepidoptera: Carposinidae), in young apple fruits by MRI. Jpn. J. Appl. Entomol. Zool. **52**(3), 123–128 (2008)
6. Toyoshima, S., et al.: The ability of a double-sensored NIR device to detect apples ('Fuji' Cultivar) injured by the peach fruit moth, Carposina sasakii Matsumura (Lepidoptera: Carposinidae). Bull. Natl. Inst. Fruit Tree Sci. **7**, 13–20 (2008)
7. Kotani, S., Suzuki, Y.: Development of an X-ray penetration imaging system for detecting peaches injured by the peach fruit moth. J. Jpn. Soc. Precis. Eng. **79**(11), 995–998 (2013)
8. Ishida, K., et al.: Development of a peach fruit moth inspection system. In: SICE, pp. 1397–1402 (2018)
9. Watanabe, H., et al.: Research and verification tests of detecting system using X-ray for peaches injured by peach fruit moth. IEEJ Trans. Electron. Inf. Syst. **139**(9), 993–1000 (2019)
10. Makino, K., et al.: Motion planning of a rotation type peach fruit moth inspection system. In: Proceedings of the 7th European Conference on Mechanism Science, pp. 173–180 (2018)
11. Makino, K., et al.: Development of a concavo-convex non-woven cloth to reduce shock to fruit. In: 26th IEEE International Symposium on Robot and Human Interactive Communication (RO-MAN), pp. 1210–1215 (2017)

# Position Analysis of a Novel Translational 3-URU with Actuators on the Base

Raffaele Di Gregorio(✉)

Department of Engineering, University of Ferrara, 44122 Ferrara, Italy
raffaele.digregorio@unife.it

**Abstract.** A recent patent application of the author has presented a translational parallel manipulator (TPM) with three equal limbs of Universal-Revolute-Universal (URU) type, with only one actuated joint per limb, which has overall size and characteristics similar to the DELTA robot. Translational 3-URU architectures are not novel in the literature, but the presented one has the actuators on the frame (base) even though the actuated joints are not on the base, and a particular geometry, which makes it novel. Putting the actuators on the base allows a substantial reduction of the mobile masses, thus promising good dynamic performances, and makes the remaining part of the limb a simple chain constituted by only passive R-pairs. Here, the position analysis of this novel 3-URU is studied.

**Keywords:** Lower-mobility manipulator · Translational parallel manipulator · Kinematic analysis · Position analysis

## 1 Introduction

Parallel manipulators (PMs) feature two rigid bodies, one fixed (base) and the other mobile (platform), connected to one another through a number of kinematic chains (limbs). Translational PMs (TPMs) are 3-degrees-of-freedom (DOF) PMs whose platform can perform only spatial translations. This author, in a recent patent application [1], has presented a TPM with three equal limbs of Universal-Revolute-Universal (URU) type, with only one actuated joint per limb.

3-URU architectures are not a novelty in the literature [2–6]. Nevertheless, the ones proposed in the literature [2–4] have the R-pairs, adjacent to the base, as actuated joints or, when presented as a spatial mechanism without actuated joints [5], have the axes of the three R-pairs adjacent to the base (to the platform) that are coplanar and with a common intersection. Changing the actuated joints and/or modifying the base (the platform) geometry affect the behavior of the machine in a substantial manner as regard both to the load redistribution among the links and to the functional aspects (e.g., useful workspace sizes and location).


E.-C. Lovasz et al. (Eds.): MTM&Robotics 2020, MMS 88, pp. 80–90, 2021.
https://doi.org/10.1007/978-3-030-60076-1_8

The novelty of the translational 3-URU proposed in [1], hereafter named LaMaViP 3-URU, stands in the fact that:

(i) the actuators are on the base even though the actuated joints are not on the base,
(ii) in each URU limb, the actuated R-pair is the one not adjacent to the base in the U-joint adjacent to the base, and
(iii) it has a particular base (platform) geometry (see Fig. 1) where the axes of the three R-pairs adjacent to the base (to the platform) share a common intersection point, but are not coplanar.

Putting the actuators on the base allows a significant reduction of the mobile masses, thus promising good dynamic performances, and makes the remaining part of the limb a simple chain constituted by only passive R-pairs.

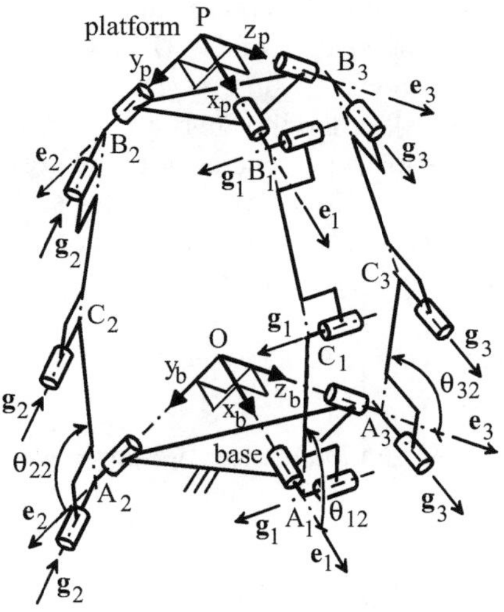


**Fig. 1.** LaMaViP 3-URU with the R-pair axes that are fixed in the base (platform) mutually orthogonal: scheme and notations

This paper deals with the position analysis of the LaMaViP 3-URU. The organization of the paper is as follows. Section 2 presents the LaMaViP 3-URU together with some background concepts and the adopted notations. Section 3 analyzes the kinematics of one type of mechanical transmission suitable for connecting the actuator on the base to the actuated joint in each limb. Then, Sect. 4 addresses the position analysis of the novel TPM and Sect. 5 draws the conclusions.

## 2 Description of the LaMaViP 3-URU

Out of constraint singularities [5], a 3-URU architecture is a TPM if it is manufactured and assembled so that in each URU limb the axes of the two ending R-pairs are parallel to one another and the axes of the three intermediate R-pairs are all parallel [6].

Figure 1 shows the reference geometry for a LaMaViP 3-URU[1]. For the geometry of Fig. 1, in [1], this author proved the existence of wide free-from-singularity regions of the platform operational space, where the useful workspace can be located. The geometry of Fig. 1 has the axes of the three R-pairs adjacent to the base (to the platform) that are mutually orthogonal and share a common intersection point.

With reference to Fig. 1,

- $Ox_by_bz_b$ and $Px_py_pz_p$ are two Cartesian references fixed to the base and to the platform, respectively; without losing generality, these two references have been chosen with the homologous coordinate axes that are parallel to one another[2];
- $A_i(B_i)$ for i = 1, 2, 3 are the centers of the U joints adjacent to the base (to the platform);
- without losing generality [8], in the i-th limb, i = 1, 2, 3, the points $A_i$ and $B_i$ are assumed to lie on the same plane perpendicular to the axes of the three intermediate R-pairs; such plane intersects at $C_i$ the axis of the R-pair between the two U-joints;
- $\mathbf{e}_1$, $\mathbf{e}_2$, and $\mathbf{e}_3$ are unit vectors of the coordinate axes $x_b$, $y_b$, and $z_b$($x_p$, $y_p$, and $z_p$), respectively, and, at the same time, unit vectors of the three R-pair axes fixed to the base (to the platform);
- $\mathbf{g}_i$, i = 1, 2, 3, is the unit vector parallel to the axes of the three intermediate R-pairs of the i-th limb.

Moreover, the following definition/choices are introduced:

- $d_p = B_1P = B_2P = B_3P$;
- $d_b = A_1O = A_2O = A_3O$;
- in each URU limb, the five R-pairs are numbered with an index, j, that increases by moving from the base toward the platform; the actuated joint is the second R-pair;
- $\theta_{ij}$, for i = 1, 2, 3, and j = 1,..., 5, is the joint variable of the j-th R-pair of the i-th limb; the actuated-joint variables are the angles $\theta_{i2}$, i = 1, 2, 3 (see Fig. 1);

[1] The patent application [1] covers a wider family of 3-URU geometries that are obtained by slightly modifying the geometry of Fig. 1. Since the singularity surfaces [6, 7] are geometric representations of algebraic equations whose coefficients continuously vary with the variations of the geometric constants, the results reported in [1] for the reference geometry of Fig. 1 does not significantly change for the other geometries of the family.

[2] The parallelism of the coordinate axes is kept during the motion since the analyzed 3-URU is translational.

- $\theta_{iM}$, for i = 1, 2, 3, is the rotation angle of the motor shaft (see Fig. 2) of the actuator of the i-th limb;
- $f_i = A_iC_i$, for i = 1, 2, 3; $r_i = B_iC_i$, for i = 1, 2, 3;
- $\mathbf{p} = (P - O) = x\mathbf{e}_1 + y\mathbf{e}_2 + z\mathbf{e}_3$, where $(x, y, z)^T$ collects the coordinates of point P in $Ox_by_bz_b$; such coordinates also identify the platform pose during motion since the studied 3-URU is translational;
- $\mathbf{a}_i = (A_i - O) = d_b\mathbf{e}_i$, for i = 1, 2, 3;
- $\mathbf{b}_i = (B_i - O) = \mathbf{p} + d_p\mathbf{e}_i$, for i = 1, 2, 3;
- $\mathbf{c}_i = (C_i - O) = \mathbf{a}_i + f_i\mathbf{u}_i$, for i = 1, 2, 3;
- $\mathbf{h}_i = \mathbf{g}_i \times \mathbf{e}_i$, for i = 1,2,3;
- $\mathbf{u}_i = (C_i - A_i)/f_i = \cos\theta_{i2}\,\mathbf{e}_i + \sin\theta_{i2}\,\mathbf{h}_i$, for i = 1, 2, 3;
- $\mathbf{v}_i = (B_i - C_i)/r_i = \cos\theta_{i3}\,\mathbf{u}_i + \sin\theta_{i3}\,(\cos\theta_{i2}\,\mathbf{h}_i - \sin\theta_{i2}\,\mathbf{e}_i)$ for i = 1, 2, 3;

Figure 2 shows a possible mechanical transmission, based on a bevel gearbox that actuates the second R-pair of the i-th limb by keeping the actuator on the base. Figure 3 shows a constructive scheme of the i-th URU limb. Figures 2 and 3 highlight that the actual construction of the proposed type of URU limb is quite simple.

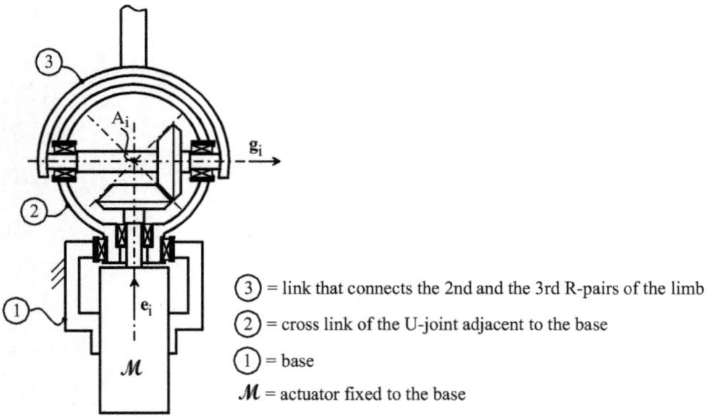


**Fig. 2.** A possible mechanical transmission, based on a bevel gearbox, for actuating the 2nd R-pair of the i-th limb by keeping the actuator fixed to the base.

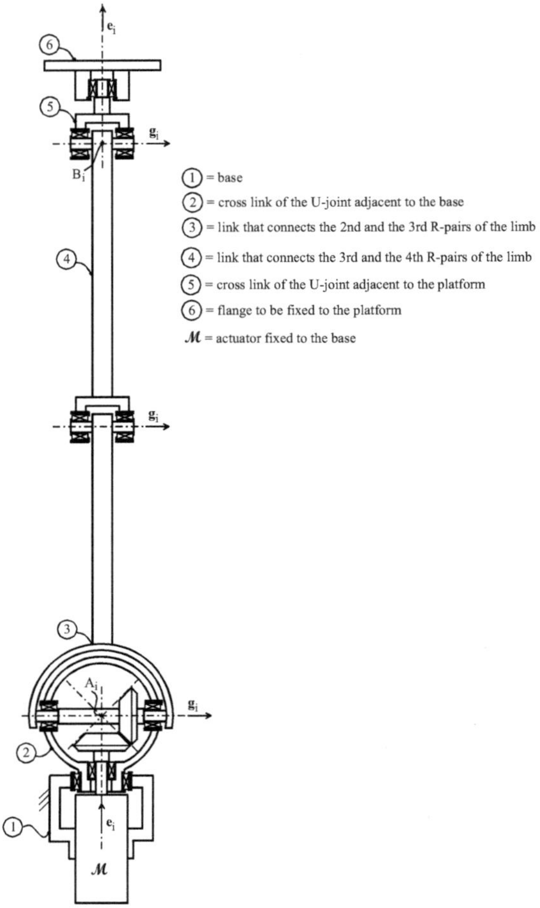


**Fig. 3.** Constructive scheme of the i-th URU limb

## 3 Kinematic Analysis of the Actuation Device

The analysis of Figs. 2 and 3 reveals that the following relationships hold:

$$^{i}\boldsymbol{\omega}_{21} = \dot{\theta}_{i1}\mathbf{e}_{i}, {}^{i}\boldsymbol{\omega}_{32} = \dot{\theta}_{i2}\mathbf{g}_{i}, {}^{i}\boldsymbol{\omega}_{M1} = \dot{\theta}_{iM}\mathbf{e}_{i} \qquad i = 1, 2, 3 \tag{1}$$

where $^{i}\boldsymbol{\omega}_{pq}$ denotes the angular velocity of link p with respect to link q in the i-th limb, and the index M denotes the motor shaft.

Moreover, the relative motion theorems [9] states that

$$^{i}\boldsymbol{\omega}_{M2} = {}^{i}\boldsymbol{\omega}_{M1} - {}^{i}\boldsymbol{\omega}_{21} = (\dot{\theta}_{iM} - \dot{\theta}_{i1})\mathbf{e}_{i} \qquad i = 1, 2, 3 \tag{2}$$

Let $k_i$ be the speed ratio of the bevel gearbox of the i-th limb, the following relationship must hold:

$$k_i = \frac{{}^{i}\boldsymbol{\omega}_{32} \cdot \mathbf{g}_i}{{}^{i}\boldsymbol{\omega}_{M2} \cdot \mathbf{e}_i} = \frac{\dot{\theta}_{i2}}{\dot{\theta}_{iM} - \dot{\theta}_{i1}} \qquad i = 1, 2, 3 \tag{3}$$

Equation (3) yields the following expressions of the actuated-joint rates as a function of the angular velocities of the motor shafts

$$\dot{\theta}_{i2} = k_i(\dot{\theta}_{iM} - \dot{\theta}_{i1}) \qquad i = 1, 2, 3 \tag{4}$$

Equation (4) can be integrated to give the relationship

$$\theta_{i2} = k_i[(\theta_{iM} - \theta_{i1}) - (\theta_{iM|0} - \theta_{i1|0})] \qquad i = 1, 2, 3 \tag{5}$$

where $\theta_{iM|0}$ and $\theta_{i1|0}$ are the values of $\theta_{iM}$ and $\theta_{i1}$, respectively, when $\theta_{i2}$ is equal to zero.

## 4 Position Analysis

The notations introduced in Sect. 2 yield (see Fig. 1)

$$\mathbf{g}_1 = \cos\theta_{11}\mathbf{e}_2 + \sin\theta_{11}\mathbf{e}_3,\ \mathbf{g}_2 = -\cos\theta_{21}\mathbf{e}_1 + \sin\theta_{21}\mathbf{e}_3,\ \mathbf{g}_3 = \cos\theta_{31}\mathbf{e}_1 + \sin\theta_{31}\mathbf{e}_2 \tag{6}$$

whose introduction into the definition of $\mathbf{h}_i$ gives

$$\mathbf{h}_1 = -\cos\theta_{11}\mathbf{e}_3 + \sin\theta_{11}\mathbf{e}_2,\ \mathbf{h}_2 = -\cos\theta_{21}\mathbf{e}_3 - \sin\theta_{21}\mathbf{e}_1,\ \mathbf{h}_3 = -\cos\theta_{31}\mathbf{e}_2 + \sin_{31}\mathbf{e}_1 \tag{7}$$

Moreover, the analysis of the i-th limb reveals that the following relationships hold

$$\mathbf{b}_i = \mathbf{p} + d_p\mathbf{e}_i = d_b\mathbf{e}_i + f_i\mathbf{u}_i + r_i\mathbf{v}_i \qquad i = 1, 2, 3 \tag{8}$$

which, by replacing the above-reported explicit expressions of $\mathbf{p}$ and of the unit vectors $\mathbf{u}_i$ and $\mathbf{v}_i$, become

$$\begin{aligned} &\{x + d_p - d_b - [f_1\cos\theta_{12} + r_1\cos(\theta_{12} + \theta_{13})]\}\mathbf{e}_1 \\ &+ \{y - \sin\theta_{11}[f_1\sin\theta_{12} + r_1\sin(\theta_{12} + \theta_{13})]\}\mathbf{e}_2 \\ &+ \{z + \cos\theta_{11}[f_1\sin\theta_{12} + r_1\sin(\theta_{12} + \theta_{13})]\}\mathbf{e}_3 = 0 \end{aligned} \tag{9a}$$

$$\begin{aligned} &\{x + \sin\theta_{21}[f_2\sin\theta_{22} + r_2\sin(\theta_{22} + \theta_{23})]\}\mathbf{e}_1 \\ &+ \{y + d_p - d_b - [f_2\cos\theta_{22} + r_2\cos(\theta_{22} + \theta_{23})]\}\mathbf{e}_2 \\ &+ \{z + \cos\theta_{21}[f_2\sin\theta_{22} + r_2\sin(\theta_{22} + \theta_{23})]\}\mathbf{e}_3 = 0 \end{aligned} \tag{9b}$$

$$\{x - \sin\theta_{31}[f_3\sin\theta_{32} + r_3\sin(\theta_{32}+\theta_{33})]\}\mathbf{e}_1 + \{y + \cos\theta_{31}[f_3\sin\theta_{32} + r_3\sin(\theta_{32}+\theta_{33})]\}\mathbf{e}_2 + \{z + d_p - d_b - [f_3\cos\theta_{32} + r_3\cos(\theta_{32}+\theta_{33})]\}\mathbf{e}_3 = 0 \quad (9c)$$

The dot products of Eqs. 9 respectively by $\mathbf{e}_1$, $\mathbf{e}_2$, and $\mathbf{e}_3$ yield nine equations in twelve unknowns: the platform pose parameters, $(x, y, z)^T$, and the joint variables $(\theta_{i1},\theta_{i2}, \theta_{i3})^T$ for i = 1, 2, 3. This system is the reference closure-equation system for the solution of the position analysis problems.

### 4.1 Inverse Position Analysis

The inverse position analysis (IPA) consists in the determination of the actuated-joint variables for assigned platform-pose parameters. In the LaMaViP 3-URU, the particular actuation system (Fig. 2) relates (see Eq. (5)) the actuated-joint variables, $\theta_{i2}$ for i = 1, 2, 3, to the rotation angles, $\theta_{iM}$ for i = 1, 2, 3, of the motor shafts and the joint variables $\theta_{i1}$ for i = 1, 2, 3. Consequently, the IPA has to be the computation of ($\theta_{i1}$, $\theta_{i2}$), i = 1, 2, 3, as functions of $(x, y, z)^T$.

The joint variables $\theta_{i1}$ for i = 1, 2, 3 can be expressed as functions of $(x, y, z)^T$ by comparing Eqs. (6) and the following alternative expression (see Fig. 1)

$$\mathbf{g}_i = \frac{\mathbf{e}_i \times (\mathbf{b}_i - \mathbf{a}_i)}{|\mathbf{e}_i \times (\mathbf{b}_i - \mathbf{a}_i)|} = \frac{\mathbf{e}_i \times [\mathbf{p} + (d_p - d_b)\mathbf{e}_i]}{|\mathbf{e}_i \times [\mathbf{p} + (d_p - d_b)\mathbf{e}_i]|} = \frac{\mathbf{e}_i \times \mathbf{p}}{|\mathbf{e}_i \times \mathbf{p}|} \qquad i = 1,2,3, \quad (10)$$

which yield

$$\mathbf{g}_1 = \frac{y\mathbf{e}_3 - z\mathbf{e}_2}{\sqrt{y^2+z^2}},\ \mathbf{g}_2 = \frac{-x\mathbf{e}_3 + z\mathbf{e}_1}{\sqrt{x^2+z^2}},\ \mathbf{g}_3 = \frac{x\mathbf{e}_2 - y\mathbf{e}_1}{\sqrt{x^2+y^2}} \quad (11)$$

Such a comparison provides the following explicit formulas

$$\sin\theta_{11} = \frac{y}{\sqrt{y^2+z^2}},\ \cos\theta_{11} = \frac{-z}{\sqrt{y^2+z^2}},\ \sin\theta_{21} = \frac{-x}{\sqrt{x^2+z^2}},\ \cos\theta_{21} = \frac{-z}{\sqrt{x^2+z^2}} \quad (12a)$$

$$\sin\theta_{31} = \frac{x}{\sqrt{x^2+y^2}},\ \cos\theta_{31} = \frac{-y}{\sqrt{x^2+y^2}} \quad (12b)$$

which yield

$$\theta_{11} = \mathrm{ATAN2}(y, -z),\ \theta_{21} = \mathrm{ATAN2}(-x, -z),\ \theta_{31} = \mathrm{ATAN2}(x, -y) \quad (13)$$

Formulas (12) make it possible to simplify the closure-equation system coming from Eqs. (9) as follows

$$f_1\cos\theta_{12} + r_1\cos(\theta_{12}+\theta_{13}) = x + d_p - d_b \tag{14a}$$

$$f_1\sin\theta_{12} + r_1\sin(\theta_{12}+\theta_{13}) = \sqrt{y^2+z^2} \tag{14b}$$

$$f_2\cos\theta_{22} + r_2\cos(\theta_{22}+\theta_{23}) = y + d_p - d_b \tag{14c}$$

$$f_2\sin\theta_{22} + r_2\sin(\theta_{22}+\theta_{23}) = \sqrt{x^2+z^2} \tag{14d}$$

$$f_3\cos\theta_{32} + r_3\cos(\theta_{32}+\theta_{33}) = z + d_p - d_b \tag{14e}$$

$$f_3\sin\theta_{32} + r_3\sin(\theta_{32}+\theta_{33}) = \sqrt{x^2+y^2} \tag{14f}$$

By isolating $\cos(\theta_{i2} + \theta_{i3})$ and $\sin(\theta_{i2} + \theta_{i3})$ for i = 1, 2, 3 in Eqs. (14) and, then, imposing that the trigonometric identities $\cos^2(\theta_{i2} + \theta_{i3}) + \sin^2(\theta_{i2} + \theta_{i3}) = 1$, i = 1, 2, 3, must hold, the following simplified closure-equation system comes out

$$\alpha_i^2 + \beta_i^2 + f_i^2 - r_i^2 - 2\,f_i(\alpha_i\cos\theta_{i2} + \beta_i\sin\theta_{i2}) = 0 \qquad i = 1,2,3 \tag{15}$$

with

$$\alpha_1 = x + d_p - d_b,\ \alpha_2 = y + d_p - d_b,\ \alpha_3 = z + d_p - d_b, \tag{16a}$$

$$\beta_1 = \sqrt{y^2+z^2},\ \beta_2 = \sqrt{x^2+z^2},\ \beta_3 = \sqrt{x^2+y^2} \tag{16b}$$

The introduction of the trigonometric identities $\cos\theta_{i2} = (1-t_i^2)/(1+t_i^2)$ and $\sin\theta_{i2} = 2t_i/(1+t_i^2)$, where $t_i = \tan(\theta_{i2}/2)$, into Eqs. (15) transforms them into quadratic equations whose solutions are

$$t_i = \frac{2\,f_i\beta_i \mp \sqrt{4\,f_i^2(\alpha_i^2 + \beta_i^2) - (\alpha_i^2 + \beta_i^2 + f_i^2 - r_i^2)^2}}{(\alpha_i + f_i)^2 + \beta_i^2 - r_i^2} \qquad i = 1,2,3 \tag{17}$$

Formulas (17) provide up to two values of $\theta_{i2}$. From a geometric point of view, these two solutions per limb correspond to the up to two intersections of two circumferences that lie on the plane perpendicular to the unit vector $\mathbf{g}_i$ and passing trough $A_i$ and $B_i$, one with center at $A_i$ and radius $f_i$ and the other with center at $B_i$ and radius $r_i$. These intersections are the possible positions of point $C_i$ (see Fig. 1).

Therefore, the conclusion is that there are up to eight solutions for an assigned platform pose (i.e., for one assigned value of $(x, y, z)^T$). They are obtained by combining the up to two limb configurations compatible with that pose. Such limb configurations are identified by the unique value of $\theta_{i1}$, computed with formula (13), when combined with the two values, say $\theta_{i2,I}$ and $\theta_{i2,II}$, of $\theta_{i2}$ computed with formula (17) to give the two analytic solutions, $(\theta_{i1}, \theta_{i2,I})$ and $(\theta_{i1}, \theta_{i2,II})$, of the i-th limb IPA.

### 4.2 Direct Position Analysis

The direct position analysis (DPA) consists in the determination of the platform-pose parameters compatible with one assigned value of the actuated-joint variables. In the LaMaViP 3-URU, the complexity of the computation of the platform poses compatible with assigned values of $\theta_{iM}$ for i = 1, 2, 3, of the motor shafts depends on the values of the speed ratios, $k_i$, of the bevel gearboxes that appear in Eqs. (5). Anyway, if the actuated-joint variables, $\theta_{i2}$ for i = 1, 2, 3, are assigned and the $\theta_{iM}$ for i = 1, 2, 3, are computed after the computation of the platform poses compatible with the assigned values of $\theta_{i2}$ for i = 1, 2, 3, the computation is possible without assuming any value of speed ratio. This computation is illustrated below.

If the values of $\theta_{i2}$ for i = 1, 2, 3 are assigned, Eqs. (15) will constitute a system of three equations in the three platform-pose parameters $(x, y, z)^T$. Unfortunately, these three equations are irrational since they contain the $\beta_i$ terms, i = 1, 2, 3, (see, formulas (16b)). Nevertheless, by replacing $\alpha_i$ and $\beta_i$ with their explicit expressions (16), such equations, after suitable algebraic manipulations, can be transformed into the following form

$$\left(x^2+y^2+z^2\right)^2+2\left(x^2+y^2+z^2\right)(\gamma_1 x+\delta_1)+(\gamma_1 x+\delta_1)^2-\varepsilon_1\left(y^2+z^2\right)=0 \tag{18a}$$

$$\left(x^2+y^2+z^2\right)^2+2\left(x^2+y^2+z^2\right)(\gamma_2 y+\delta_2)+(\gamma_2 y+\delta_2)^2-\varepsilon_2\left(x^2+z^2\right)=0 \tag{18b}$$

$$\left(x^2+y^2+z^2\right)^2+2\left(x^2+y^2+z^2\right)(\gamma_3 z+\delta_3)+(\gamma_3 z+\delta_3)^2-\varepsilon_3\left(x^2+y^2\right)=0 \tag{18c}$$

where, $\gamma_i = 2\left(d_p - d_b - f_i\right)$, $\delta_i = -2f_i\left(d_p - d_b\right)\cos\theta_{i2} + \left(d_p - d_b\right)^2 + f_i^2 - r_i^2$, and $\varepsilon_i = 4f_i^2\sin^2\theta_{i2}$, for i = 1, 2, 3.

System (18) is constituted by three quartic equations in the three platform-pose parameters, x, y, and z. Since the quartic terms are the same in all the equations, they can be eliminated from two of them by replacing, for instance, Eqs. (18b) and (18c) with the equations obtained by subtracting Eq. (18a) from them. Such algebraic manipulations yield the following system

$$\left(x^2+y^2+z^2\right)^2+2\left(x^2+y^2+z^2\right)(\gamma_1 x+\delta_1)+(\gamma_1 x+\delta_1)^2-\varepsilon_1\left(y^2+z^2\right)=0 \tag{19a}$$

$$\begin{aligned}&2\left(x^2+y^2+z^2\right)\left[(\gamma_2 y+\delta_2)-(\gamma_1 x+\delta_1)\right]+(\gamma_2 y+\delta_2)^2-(\gamma_1 x+\delta_1)^2\\&-\varepsilon_2\left(x^2+z^2\right)+\varepsilon_1\left(y^2+z^2\right)=0\end{aligned} \tag{19b}$$

$$2\left(x^2+y^2+z^2\right)\left[(\gamma_3 z+\delta_3)-(\gamma_1 x+\delta_1)\right]+(\gamma_3 z+\delta_3)^2-(\gamma_1 x+\delta_1)^2 \\ -\varepsilon_3\left(x^2+y^2\right)+\varepsilon_1\left(y^2+z^2\right)=0 \tag{19c}$$

System (19) is constituted by two cubic and one quartic equations, which yield a Bézout number equal to 36. Such a number just gives a limit to the actual maximum number of solutions. The analytic solution of system (19) and the determination of the actual maximum number of DPA solutions will be investigated in future works. Anyway, system (19) is always numerically solvable. Once system (19) has been solved, for each solution $(x, y, z)^T$, the corresponding values of $\theta_{i1}$, i = 1, 2, 3, can be computed through formulas (13). Then, the assigned values of $\theta_{i2}$, i = 1, 2, 3, together with the computed values of $\theta_{i1}$, i = 1, 2, 3, can be used to compute the motor shaft angles, $\theta_{iM}$ for i = 1, 2, 3, thus stating a correspondence between assigned values of $\theta_{i2}$, i = 1, 2, 3, and motor shaft angles.

## 5 Conclusions

The position analysis of a novel translational 3-URU, named LaMaViP 3-URU, has been addressed. The peculiarities of the studied LaMaViP 3-URU are that (i) the actuators are on the base even though the actuated joints are not on the base, (ii) in each URU limb, the actuated R-pair is the one not adjacent to the base in the U-joint adjacent to the base, and (iii) it has a particular base (platform) geometry where the axes of the three R-pairs adjacent to the base (to the platform) share a common intersection point, and are mutually orthogonal.

The particular choice of actuated joints makes the position analysis of the LaMaViP 3-URU different from those of other translational 3-URUs presented in the literature.

The results of this study are as follows. The inverse position analysis of LaMaViP 3-URU has eight solutions that can be determined in explicit form. Differently, its direct position analysis requires the solution of an algebraic non-linear system constituted by two cubic equations plus one quartic equation in the three platform-pose parameters, which can be numerically solved.

Future works on this subject will try to get an analytic solution of the direct position analysis and to determine the maximum number of real solutions of this analysis.

**Acknowledgments.** This work has been developed at the Laboratory of Mechatronics and Virtual Prototyping (LaMaViP) of Ferrara Technopole, supported by FAR2019 UNIFE funds.

## References

1. Di Gregorio, R.: Meccanismo Parallelo Traslazionale, Italy, 23 March 2020. Patent Application No. 102020000006100
2. Huda, S., Takeda, Y.: Kinematic analysis and synthesis of a 3-URU pure rotational parallel mechanism with respect to singularity and workspace. J. Adv. Mech. Des. Syst. Manuf. **1**(1), 81–92 (2007)

3. Huda, S., Takeda, Y.: Kinematic design of 3-URU pure rotational parallel mechanism with consideration of uncompensatable error. J. Adv. Mech. Des. Syst. Manuf. **2**(5), 874–886 (2008)
4. Carbonari, L., Corinaldi, D., Palpacelli, M., Palmieri, G., Callegari, M.: A novel reconfigurable 3-URU parallel platform. In: Ferraresi, C., Quaglia, G. (eds.) Advances in Service and Industrial Robotics, pp. 63–73. Springer, Dordrecht (2018)
5. Zlatanov, D., Bonev, I.A., Gosselin, C.M.: Constraint singularities as C-space singularities. In: Lenarčič, J., Thomas, F. (eds.) Advances in Robot Kinematics: Theory and Applications, pp. 183–192. Springer, Dordrecht (2002)
6. Di Gregorio, R.: A review of the literature on the lower-mobility parallel manipulators of 3-UPU or 3-URU type. Robotics **9**(1), 5 (2020)
7. Di Gregorio, R., Parenti-Castelli, V.: Influence of the geometric parameters of the 3–UPU parallel mechanism on the singularity loci. In: Proceedings of the International Workshop on Parallel Kinematic Machines – PKM 1999, Milan, Italy, 30 November, pp. 79–86 (1999)
8. Di Gregorio, R., Parenti-Castelli, V.: A translational 3-DOF parallel manipulator. In: Lenarcic, J., Husty, M.L. (eds.) Advances in Robot Kinematics: Analysis and Control. Kluwer, Norwell, MA, USA, pp. 49–58 (1998)
9. Ardema, M.D.: Newton-Euler Dynamics. Springer, New York (2005)

# Design and Development of a Robotic Hexapod Platform for Educational Purposes

Mihai Steopan(✉), Bogdan Mocan, Popister Florin, Marcela Bucur, and Virgil Ispas

Technical University of Cluj-Napoca, Cluj-Napoca, Romania
mihai.steopan@muri.utcluj.ro

**Abstract.** The main objective of the work presented in this paper was to design and develop a hexapod robotic platform for educational purposes using performance planning methods. In this respect Design for Six Sigma approach was used for concept development, 3D CAD software and finite element analysis for the modeling and structural optimization while 3D printing technology was used for the manufacturing of the structural components.

The developed robotic hexapod platform consists of two platforms connected with six articulated legs. The upper core platform houses the Arduino control board together with 3 ultrasonic based distance sensors and a fast vision sensor. The lower core platform houses 12 servomotors, a driver board and the power supply. An Arduino based configuration of controller-sensor-drivers-motors was used for the control of the robotic platform.

To exemplify the locomotion modes and functionality of the robotic platform, we highlighted the results of a series of experiments performed on the manufactured prototype.

**Keywords:** Robotic hexapod platform · 3D printing parts · Arduino based configuration · Environment friendly material

## 1 Introduction

Since ancient times man has tried to build mechanisms to mimic as closely as possible the behavior of living things in order to facilitate his work in areas having irregular patterns or materials, or specifically designed for humans.

Mobile robotics is the area of robotics which deals with the design of hardware and software in such a way that the resulting robot can work in a dynamic environment relaying on inconsistent sensor information [1]. Mobile robots may be fully autonomous or remotely controlled. Mobile robotics is usually considered to be a subfield of robotics and information engineering [2]. As stated in reference [3] the mobile robots mainly include wheeled robots, tracked robots, legged robots, snake robots, and spherical robots. Thus, a mobile robot is a robot that is capable of locomotion. Mobile robot needs locomotion mechanisms that enable it to move unbounded throughout its environment. There are a large variety of possible ways to move, and so the selection of the robot locomotion system is a challenge for the engineering team which design the


E.-C. Lovasz et al. (Eds.): MTM&Robotics 2020, MMS 88, pp. 91–103, 2021.
https://doi.org/10.1007/978-3-030-60076-1_9

mobile robot [4]. For the purpose of this study we chose a hexapod design for the robotic platform.

Many mobile robots are already used today in different applications, from simple operations such as vacuuming in private homes, to very complex and dangerous operations, such as Unmanned Air Vehicles (UAV) used by U.S. army [5]. Experts estimate that the field of mobile robotics is one that will experience major changes in the next few years [6]. In order to reach the desired results, scientists will also need to achieve better results in areas which are strongly interconnected with mobile robotics, like new ways of locomotion, algorithms of Artificial Intelligence (AI), new ways of electricity storing devices, data processing algorithms, etc. [7, 8].

The performance of a mobile robot is not only the result of programming skills, but also the ability to integrate and adapt to these three basic components [9]: 1) The hardware components of the robot: body, motors, sensors, battery, etc.; 2) The environment conditions; 3) The developed software for controlling the robot. Any modification of one or more of the three basic components will influence the performance of the robot. It is obvious that the design of locomotion mechanism greatly affects the controllability and performance of the robot [10, 11]. Mechanical and electrical design process is a creative activity which involves assessment and logical decisions using principles of energy, material and mechanics [12, 13].

In this paper we presented the design and development of a 6-Legged robotic platform (robotic hexapod) for testing and simulation of different movement patterns in the laboratory. The robotic hexapod platform was designed using various performance planning methods and for the control of the robot movements an Arduino based hardware was used. The robotic hexapod platform was equipped with multiple sensors for obstacle detection and distance measurement.

The structure of this paper is as follows: Sect. 1 makes the introduction and presents different performance characteristics of the mobile robots, then Sect. 2 presents the preliminaries, and the design mechanical and electrical concept of the proposed robotic platform. Section 3 describes the experimental results: the developed prototype, functional simulation, test results and limitations. Section 4 highlights the conclusions and discussions.

## 2 Design and Development of the Mechanical and Electrical Concept

There are many different approaches to develop products in a such competitive global market. Typically, we use engineering to stage competitive development of complex products such as mechatronic equipment, industrial machineries or different domestic goods. In the case of products with reduced complexity it is a challenge which of the tools/methods for planning the product performance bring value in design process.

One of the conventional approaches [9] is that by which the first step in developing a competitive product is to visually develop the product. This *first phase* (Ideation) is strictly a creative one; very few new ideas will become final products. A simple technique commonly used to formulate ideas for a new concept is brainstorming [13]. The second step is to achieve a detailed segmentation of the market potential (Fig. 1 – the Define phase). This is a significant step because it provides a better product customization, usually leads to better results on the market.

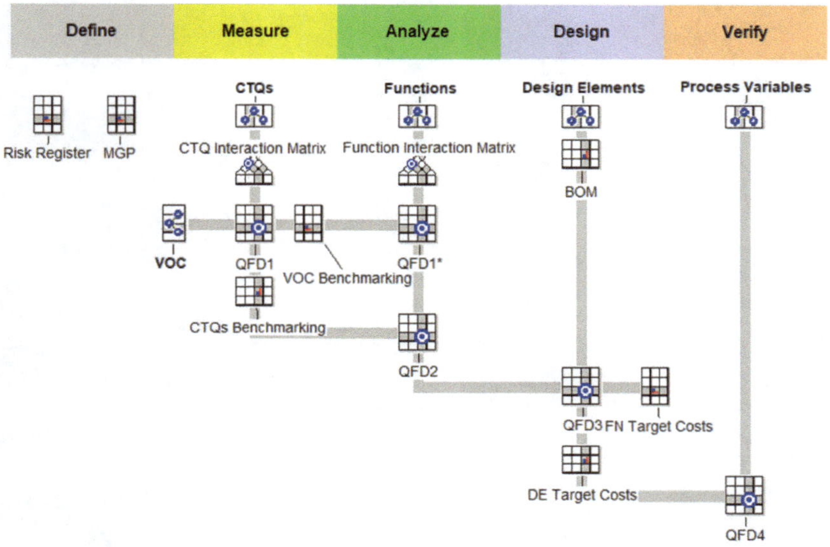


**Fig. 1.** The methodology overview for designing the robotic platform.

The *third phase* (Fig. 1 – the Measure phase) focuses on identifying customer needs and expectations in accordance with our product. One of the best techniques used to identify needs and expectations is the method called VOCT (voice of customer) [13].

The methods used to support this process must be very precise, because, usually, customers have difficulty expressing what they really want. The next step is to sort by reformulating and systemizing needs. Methods such as Mind-Map or morphological diagrams are often used to achieve the desired results.

The next step includes an analysis of potential conflicts of needs. If a potential conflict is identified, we should eliminate that need and use innovative techniques to resolve the conflict. In order to do this, we used TRIZ method [14]. In the same phase we will establish a ranking of needs.

Classification can be achieved by applying different techniques. It can be done after a market research for a specific industry segment or application on a target group, and using, for example, the AHP method. The next step is to define and analyze Key Performance Characteristics associated with that product.

Next step, as presented in Fig. 1 – the Analyze phase, is to define the key features of the product and their interconnections. This involves planning functions, making the product architecture, planning modules and interfaces, making product design, design a prototype (Fig. 1 – the Design phase) to allow further testing and evaluation and optimization of the results. Validation of the manufacturing process is done in the last phase.

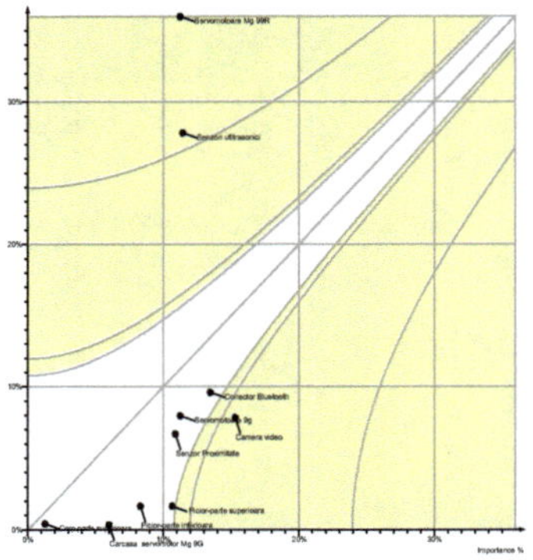


**Fig. 2.** Parts importance vs. Parts cost weight.

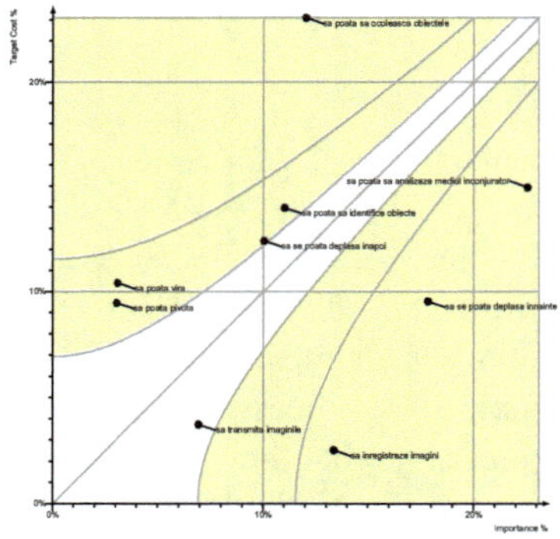


**Fig. 3.** Functions importance vs. Functions cost weight.

To plan the performance of the product, in our case a robotic hexapod platform, we only need to complete certain steps, within a specific methodology, in order to achieve the desired result. In line with this, to plan the performance of the robotic platform we used several methods and tools, like [13]: Voice of Customer Table (VOCT I and II); Analytic

Hierarchy Process (AHP); Quality Function Deployment (QFD I, II, III); Theory of Inventive Problem Solving (TRIZ); Pugh Concept Selection method (PUGH).

AHP is a structured technique that deals with complex decisions, rather than prescribing a correct decision. The four most important stakeholders' needs, ranked using AHP approach, were: 1) To be able to send and receive data - 14%; 2) To be able to generate different movements patterns - 14%; 3) To be able to detect obstacles in the vicinity – 10%; 4) User Safety – 8%.

The method used next was QFD [13]. By successively applying the QFD method in conjunction with other planning tools, the concept of the product is developed form requirements into technical characteristics, product functions, product subsystems and finally individual parts. The Figs. 2 and 3 illustrate the sensitivity cost diagram based on manufacturing cost of the components used for the mobile robot. The robotic hexapod platform must be cost-effective. So that, through this sensitivity cost analyses we evaluated each component part of the robotic hexapod by the point of view of the value added that it brings to the robot's in contrast with how much does it cost (Fig. 2). In each robotic platform component, there must be a balance between value added and its cost. From this diagram the designer knows what the maximum cost for a component part is – that's why for some components from the robotic platform we chose a low-cost part. For a perfect balance between the value added and the cost of a part, it should be placed on the main diagonal of the graph. Thus, the component parts that are below the oblique their cost is greater than the added value brought to the robotic platform, and those that are above the oblique incorporate more added value than their cost. In order to have an idea of how much value added the modules of the robotic platform, by their functions, have, a sensitive diagram (Fig. 3) - Functions importance vs. Functions cost – was done.

It can be seen from the Fig. 4 that the MG 996R servomotor and the Sharp ultrasonic sensors do not meet the desired expectations, thus they will be replaced with similar elements (i.e. MG 995 for the final concept), in the final concept. Thus, Fig. 3 presents the wiring diagram and the components used in the final concept of the robotic hexapod platform. A pixy Cam is mounted on a servomotor and connected to the ICSP port of the Arduino board. It is used for motion detection and object tracking – allowing for multiple scenarios to be programmed into the robotic platform. The PING proximity sensors, connected to ports 2–7, can also be used to identify "obstacles". For the remote command and control a HC-06 Bluetooth module was used. The 12 servomotors used to move the 6 legs are connected to an Adafruit 16-Channel 12-bit PWM/Servo Driver - I2C interface - PCA9685, allowing for simultaneous control of all the servomotors.

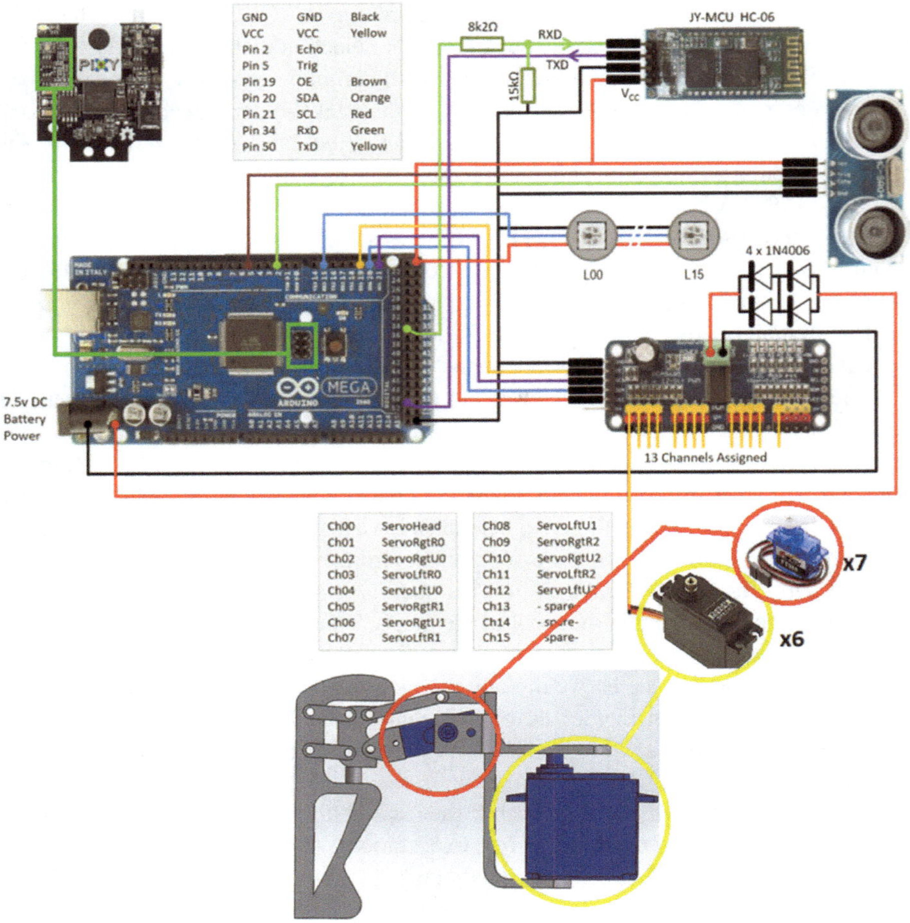


**Fig. 4.** Robotic platform hardware (electrical and mechanical) connections diagram.

## 3 Structural Optimization of the Robotic Hexapod Platform

For the dimensioning of the main mechanical components FEA (finite element analysis) was used to optimize the mass with respect to a given deflection value (using SolidWorks software package provided by Vegra Info S.R.L.)

The final concept for the hexapod platform can be seen in the Fig. 5. It is made from a core consisting of two platforms, and six articulated legs. The upper core platform houses the Arduino command and control board together with 3 ultrasonic based distance sensors and a Pixycam. The lower core platform, Fig. 5, houses 6 MG995 servomotors (the grey boxes), 6 MG90 servomotors (the blue boxes), a driver board (16-CH Servo Motor Controller/ PWM Servo Driver Board for Arduino) and the power supply. One of the main components of the robot are the legs. These are 6 in number to ensure both its stability and strengthen.

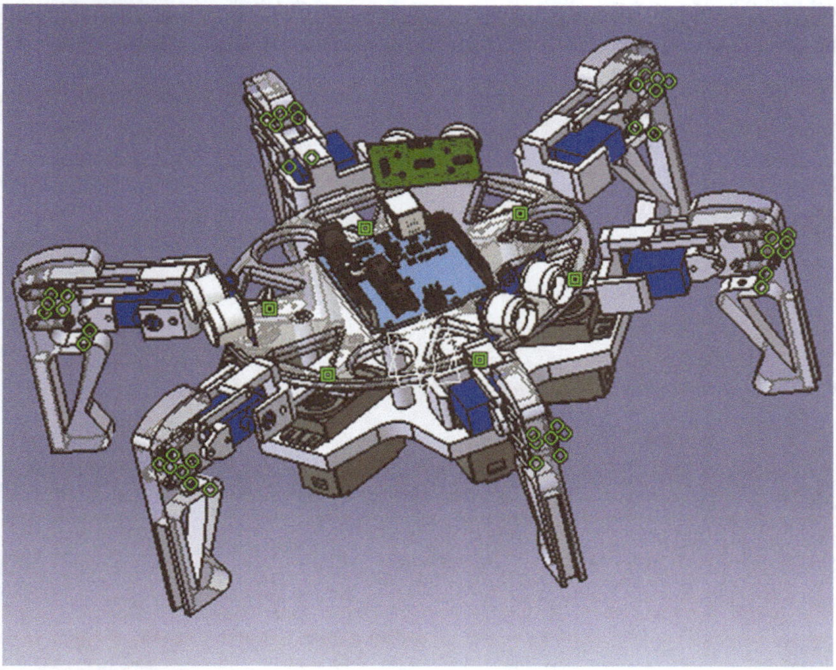

**Fig. 5.** Designed hexapod platform mechanical structure.

### 3.1 Experimental Results: The Developed Prototype, Functional Simulation, Test Results and Limitations

In order to determine the most adequate infill parameters FEA (finite element analysis) was employed [15]. Three cases were taken under consideration (Fig. 6). In the first case a 45° angled straight rib was used; the second case consisted in a X crossed rib pattern; the third case used a double arched structure.

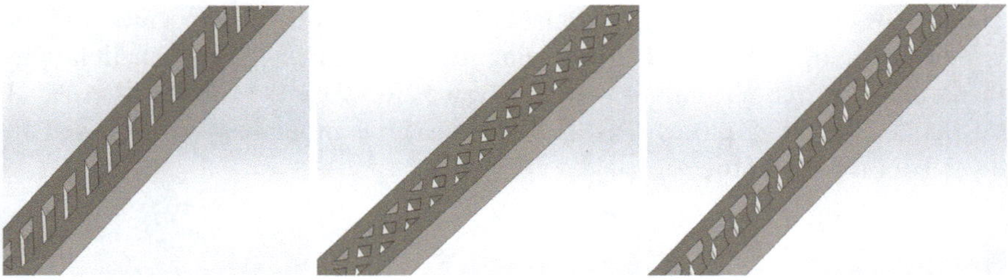

a. b. c.

**Fig. 6.** Internal test bar structure: a. straight; b. crossed; c. arched.

For each of the selected structures 3 sets of parameters were considered for analysis: the external wall thickness (1.5 ÷ 5 mm), the rib thickness (0.5 ÷ 3 mm) and the rib density (0.5 ÷ 5 mm) with a step of 0.5 mm. Considering this, FEA analysis was used in conjunction with design study tool from SolidWorks. The results can be graphically seen in the Fig. 7.

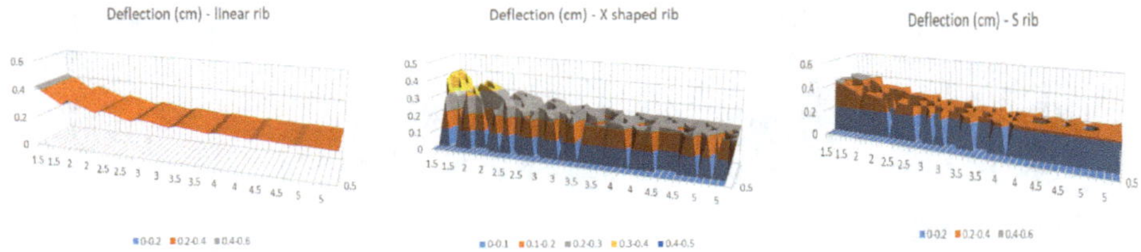


**Fig. 7.** Element deflection based on rib shape

As can be seen from the Fig. 7, considering the parameters mentioned above, the best deflection is attained using a double arched structure with a wall thin of 3 mm, a rib of 0.5 mm and a density of 3 mm. The final structure of the foot inside is specific to 3D Prusa printer models (Fig. 8 – Internal leg infill).

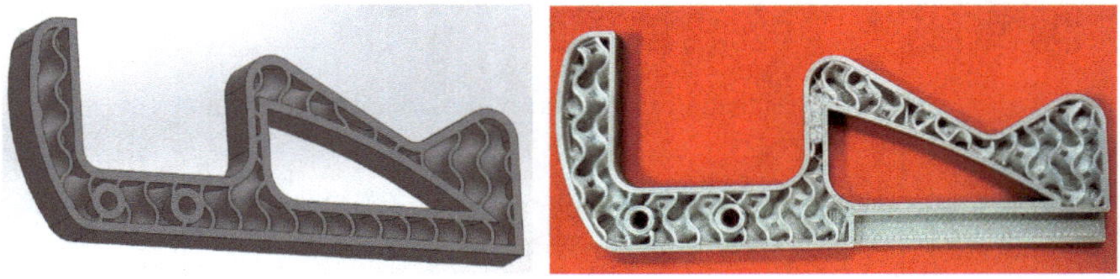

**Fig. 8.** Internal leg infill left – 3D model, right – 3d printed part.

As can be seen from the Fig. 8, the infill of the leg has an atypical structure – most 3D printed parts having a triangular or honeycomb structure. Even though the inner structure is at only 15% infill, due to the geometry of the infill a good enough resilience is attained. As can be seen the deformation of the foot with the specific infill has a maximum value of 1,552e + 06 $N/m^2$, being much lower than the deformation of a leg with normal honeycomb infill, having the value of 5,216e + 06 $N/m^2$ (Fig. 9). The force applied to the 2 pieces was 10N, the material properties were chosen from matweb.com for PLA [16].

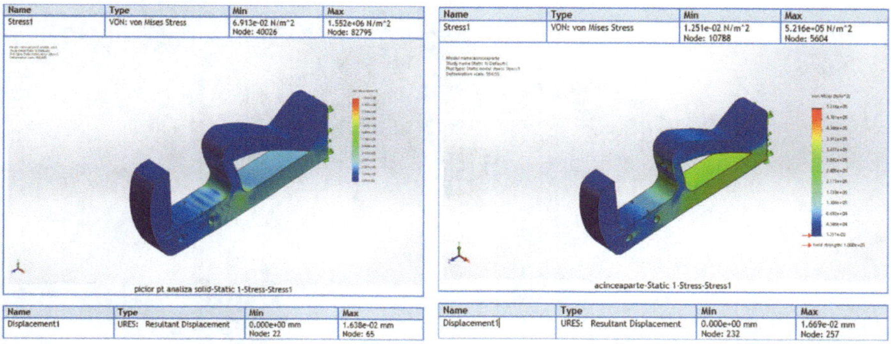


**Fig. 9.** FEA for the robot leg final design.

## 3.2 Final Design

The mechanical structure of the robotic hexapod platform (Fig. 10) was designed using SolidWorks software, and printed on a Prusa Mk3i printer. Several dimensional adjustments were necessary due to the printed parts dimensional differences from the virtual 3D model. It mostly consisted in printing several test probes to determine the variation in joint holes regarding the 3D printed parts.

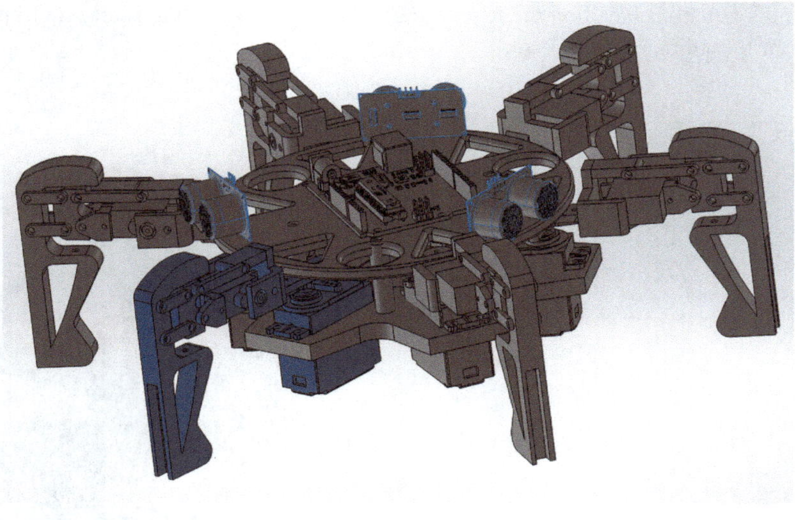

**Fig. 10.** Final 3D model prepared for 3D printing (dark blue – the leg structure).

As mentioned above, some re-dimensioning was required for the joints to fit together because of the 3D printing variations. The final leg structure can be seen in the Fig. 11. The design consists in a base servomotor (1) that swings in a side movement the element 2. Attached to the second element there is a servomotor (3) that swings in an up-down motion the element (4) thus transmitting the movement to element (5). Using (6) the movement is transmitted to the final leg element (7).

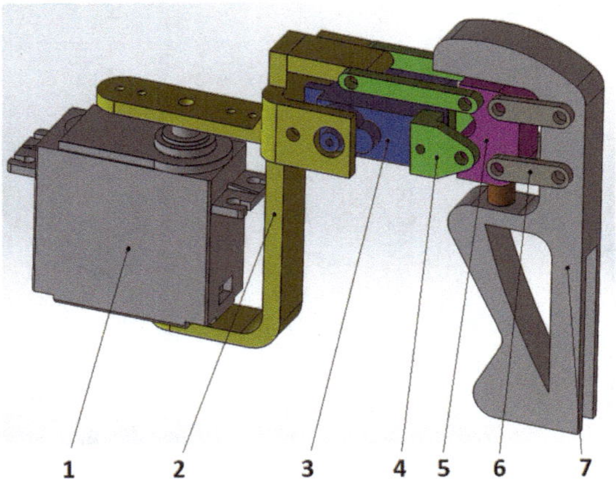


**Fig. 11.** Virtual prototype of the leg structure.

Using the motion study command from the SolidWorks software package, the leg structure can be simulated and analyzed. Each joint can be given individual control commands in a large variety of ways. One such option (Segments) is presented in the figure below. The user can control a virtual actuator through the use of a time table and a given relative displacement at any given time as can be seen in Fig. 12, where (1) is the virtual actuator placed on the leg structure and (2) is the virtual rotary motion drive placed in the animation bar.

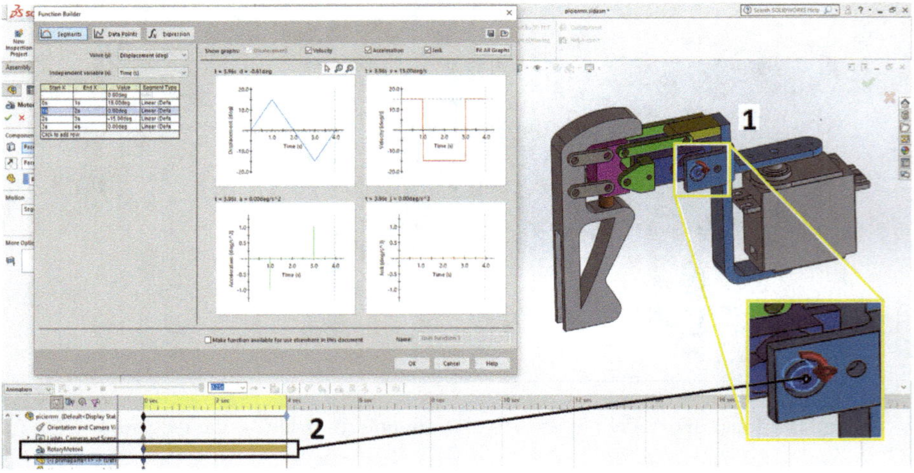


**Fig. 12.** The sequence option definition for the second (dark blue) actuator.

In the same manner a second motor can be attached to the first joint (Fig. 13). A different timetable can be used in order to correlate the movements between the joints.

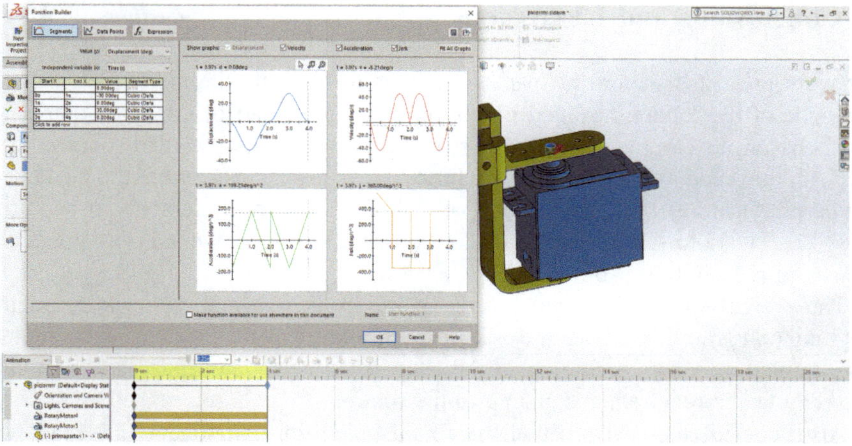

**Fig. 13.** The sequence option definition for the first actuator.

The motion profile functions can then be exported to be used in the programming of the individual motors of the mobile platform. Thus, besides having a simulated model, one can also observe the movement and interaction of various joints.

For example, the sequence for the second actuator is defined by the following functions:

[1] xbar $= (x - x1)/(x2 - x1)$
[2] ybar $= 3 * xbar^2 - 2 * xbar^3$
[3] $y = y1 + (y2 - y1) * ybar$

where: "$x_i$" represents the time needed to get to "$y_i$" position.

These functions can be programmed in the functioning algorithm of the physical platform as motion subroutines for each individual motor (Fig. 14).

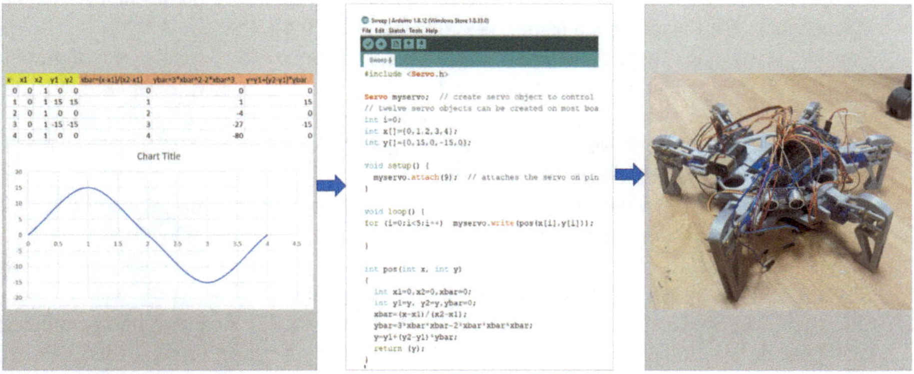


**Fig. 14.** Transferring individual functions into the robot controller.

## 4 Conclusions and Discussions

The work presented within this paper, that highlight mainly the mechanical and electrical side of the robotic hexapod platform, represents a part of a much larger project that focus on development of the smart mobile robots. The obtained robot is a prototype of hexapod robotic platform whose structure was 3D printed. The hexapod robotic platform has been realized by assembling 3D printed elements, 6 legs, robot body, 12 servomotors, Arduino Mega 2560 board, a shield engine, a Pixy camera and 3 ultrasound proximity sensors PING.

The weight of the hexapod robotic platform is reduced, the whole assembly weighing 689 grams. The parts are assembled in a way that the robot can be completely disassembled very fast according to the established needs and thus can be modified according to the requirements arising in the future.

Also, considering that most of the robotic platform components are made from Polylactic acid (PLA), this being a biodegradable plastic obtained from maize, the final robot is environmentally friendly. In addition to this, it was tried to avoid wasting material, choosing alternative solutions even if the assembly time was higher.

Using FEA analysis the mechanical structure was optimized regarding the deflection of the parts in respect to weight considering several types of ribs for the infill of the parts. Part wall thickness, rib thickness and rib density were put through several variations, resulting in a lighter but sufficiently strong structure.

## References

1. Moldovan, C.E., Craciun, A., Dolga, V., Lovasz, E.C., Maniu, I., Sticlaru, C.: On the development of a voice and gesture based HMI for the control of a mobile robot. Appl. Mech. Mater. **762**, 201–204 (2015)
2. Mocan, B., Fulea, M., Brad, E., Brad, S.: State-of-the-art and proposals on reducing energy consumption in the case of industrial robotic systems. In: Proceedings of the 2014 International Conference on Production Research – Regional Conference Africa, Europe and the Middle East; 3rd International Conference on Quality and Innovation in Engineering and Management, Cluj-Napoca, Romania, 1–5 July, pp. 328–334 (2014)
3. Yim, M., Roufas, K., Duff, D., Zhang, Y., Eldersha, C., Homans, S.: Modular reconfigurable robots in space applications. Auton. Robot. **14**, 225–237 (2003)
4. Wu, J.X., Yao, Y.A., Ruan, Q., Liu, X.P.: Design and optimization of a dual quadruped vehicle based on whole close-chain mechanism. ARCHIVE Proc. Inst. Mech. Eng. Part C J. Mech. Eng. Sci. **203–210**, 1989–1996 (2016)
5. Bartha, L., Fulea, M., Mocan, B.: Express-car-hire - competitive solution for electric vehicle hiring within airport terminals. Environ. Eng. Manag. J. **17**(2), 273–283 (2018)
6. Wang, Y.J., Wu, C.L., Yu, L.Q., Mei, Y.Y.: Dynamics of a rolling robot of closed five-arc-shaped-bar linkage. Mech. Mach. Theory **121**, 75–91 (2018)
7. Shirzadeh, M., Shojaeefard, M.H., Amirkhani, A., Behroozi, H.: Adaptive fuzzy nonlinear sliding-mode controller for a car-like robot. In: 5th Conference on Knowledge-Based Engineering and Innovation, pp. 686–691 (2019)
8. Tamas, L., Murar, M.: Smart CPS: vertical integration overview and user story with a cobot. Int. J. Comput. Integr. Manuf. **32**(4–5), 504–521 (2019)

9. Siciliano, B., Khatib, O.: Handbook of Robotics. Springer, Heidelberg (2008). ISBM: 978-3-540- 23957-4
10. Siegwart, R., Nourbakhsh, I.R.: Autonomous Mobile Robots, A Bradford Book. The MIT Press, Cambridge (2004)
11. Moldovan, C.E, Lovasz, E.C., Perju, D., Modler, K.H., Maniu, I.: On the kinematic analysis of the fourth-class, mechanisms. In: Proceedings of the 14th IFToMM World Congress, pp. 482–487 (2015)
12. Yu, L.Q., Mei, Y.Y., Li, L., Wang, Y.J., Wu, C.L.: Capability analysis and motion planning for obstacle negotiation of a closed five-bow-shaped-bar linkage. J. Mech. Eng. **53**, 69–75 (2017)
13. Popescu, S., et al.: Managementul calitãii – Bazele managementului calitătii, vol. I, ed. Casa cărtii de stiintă (1999). ISBN 973-9404-61-8
14. TRIZ40 Homepage. http://www.triz40.com/TRIZ_GB.php. Accessed 17 Mar 2020
15. Păcurar, A., Păcurar, R., Beata, E., Popişter, F., Oţel, C.: Decreasing of the manufacturing time for a thermoforming mold by applying the DFM principles. MATEC Web Conf. **137**, 01008 (2017)
16. MatWeb: Your Source for Materials Information. http://matweb.com/. Accessed 14 Mar 2020

# Dual Least Squares and the Characteristic Length: Applications to Kinematic Synthesis

Bruno Belzile([✉]) and Jorge Angeles

McGill University, 17 rue Sherbrooke Ouest, room 270, Montreal, QC H3A 0C3, Canada
{bruno,angeles}@cim.mcgill.ca

**Abstract.** Least-square problems arise in cases involving arrays with entries bearing different units, rather common in the analysis and synthesis of mechanical systems. Indeed, this is the case with rigid bodies whereby both rotation and translation occur. Dual numbers have been extensively used to take into account this feature. However, the primal part of the system of equations is still usually solved first, followed by the dual part. In this paper, we propose the use of the concept of *characteristic length* to be able to obtain one single dimensionless system of equations in problems of kinematic synthesis where the least-square solution to a system with both rotation and translation is to be found.

**Keywords:** Dual numbers · Kinematics · Synthesis

## 1 Introduction

The formulation of mathematical models of multibody systems is known to be eased when dual numbers are introduced to represent the scalar, vector and matrix quantities that arise in the presence of both rotation and translation. For instance, on the one hand, problems in dynamics usually involve as many equations of motion, derived from the Newton-Euler equations, as independent displacement variables, both equal in number to the degree of freedom of the given system. On the other hand, design problems sometimes lead to overdetermined systems of equations, linear or nonlinear, with a number of scalar equations greater than that of the scalar unknowns.

The natural way of approaching these problems is by minimization of the least-square error involved in the approximation of the overdetermined system of equations thus arising. In the case of purely translational or purely rotational systems, the least-square problem thus resulting is classical. Not so in the case in which both translation and rotation are present, whereby variables representing translations and their counterparts representing rotations bear disparate units. The use of *dual quantities* eases the formulation, for this allows the analyst to represent the motion relations with the same format, as if only rotations were involved. Clifford wrote one of the earliest papers on dual numbers [6] that can be found in the literature, expanding the concept of quaternions to both rotations and translations, which he dubbed *bi-quaternions*. Afterwards,

Supported by NSERC (Canada's Natural Sciences and Engineering Research Council) through grant no. 4532-2010 and the Postdoctoral Fellowship Program.


E.-C. Lovasz et al. (Eds.): MTM&Robotics 2020, MMS 88, pp. 104–113, 2021.
https://doi.org/10.1007/978-3-030-60076-1_10

several researchers used them for many applications, not only in kinematics [5, 8, 18, 19] but also in dynamics [1, 10, 25]. Some have also worked on the synthesis of linkages, including the early work of Levitskii [20], Yang and Freudenstein [26] and Sarkisyan *et al.* [23]. More recently, the authors of this paper reported an algorithm based on a dual version of *Householder reflections* for the approximate synthesis of spatial four-bar linkages [4]. Other researchers proposed a dual version of the Moore-Penrose generalized inverse for kinematic synthesis [9, 21]. The concept of *hyper-dual numbers* was also introduced [13]. The formulation of mechanical problems with dual numbers is closely linked to the *Principle of Transference*, due to Kotel'nikov and Study [14]: *All identities of ordinary trigonometry hold true for dual angles*. More recent work on this principle, including a proof and its extension to hyper-dual numbers, can be found in the literature [7, 22].

Two theoretical issues arise here: the first is the very definition of the minimum error, as all error variables are dual numbers, that can be considered as two-dimensional arrays, just as complex numbers. Now, arrays are not *ordered sets*, in that it is not possible to order a set of dual numbers—just as in the case of complex numbers—from smallest to largest. Here we provide a framework, based on the *context* in which the error arises, to define the minimum error. In our previous paper [4], the primal and dual errors were considered sequentially, the latter after the former.

The second theoretical issue is the definition of the dual error, which should be a dual number itself. In the case of least squares, the problem is eased by realizing that the error is defined as the Euclidean norm of the array of errors. Here we define the dual error as the Euclidean norm of the array of scalar dual errors. Such Euclidean norm is known to be a dual number. The theoretical issue here is that, while the primal part of the norm at stake is positive-definite, and can play the role of an error to minimize, the dual part is sign-indefinite.

We provide one alternative approach to the solution to the least-square problem at hand, with the introduction of the characteristic length [2], which is problem-dependent, casting the overdetermined system in a dimensionally homogeneous form, i.e., all entries bearing either the same physical units, or no units at all. The characteristic length can, for example, be used to obtain a homogeneous Jacobian matrix for a system undergoing both translations and rotations. Without this normalization, the condition number of the matrix coefficients involved is meaningless and an optimization would be impossible [11].

## 2 Background on Dual Numbers

Dual numbers are similar to complex numbers, in that they are represented as an "addition" of two parts, the primal part and the dual part, the latter pre-multiplied by the dual unit $\varepsilon$. This unit mimicks the role of the imaginary unit in the field of complex numbers. A dual number $\hat{x}$ is defined as

$$\hat{x} = x + \varepsilon x_o, \quad x,\, x_o \in \mathbf{R}, \quad \varepsilon^2 = 0 \tag{1}$$

where $x$ and $x_o$ are the real *primal* and *dual* parts of $\hat{x}$, respectively, $\varepsilon$ being the *dual unit*. The zero dual number is naturally defined as the one with *both* its primal and its dual parts equal to zero.

Here, each part has different units, those of the dual part equal those of the primal part times units of length. It should be noted that dual numbers do not constitute a *mathematical field*, but rather a *ring*, because an inverse under multiplication cannot be defined in the case of a *pure* dual number[1]. Rules regarding the sum, the difference, the product and the quotient of dual numbers can be found in books on the subject [14].

To be able to compute any other dual expression, we can use the definition given by Kotel'nikov [19] for a dual *analytical* function and its derivative with respect to its dual argument:

$$\hat{f}(\hat{x}) = f(x+\varepsilon x_o) = f(x) + \varepsilon x_o \frac{\mathrm{d}f(x)}{\mathrm{dx}}, \quad \frac{\mathrm{d}\hat{f}}{\mathrm{d}\hat{x}} = \frac{\mathrm{d}f}{\mathrm{d}x} + \varepsilon \frac{\mathrm{d}f_o}{\mathrm{d}x} \tag{2}$$

The above equation is obtained with the series expansion of $f(x+\varepsilon x_o)$ about $x$, knowing that any power of $\varepsilon$ higher than the first one vanishes.

## 3 Application of the Characteristic Length

Let us first recall the *overdetermined linear system of equations* over the real field:

$$\mathbf{Ax} = \mathbf{b}, \quad \mathbf{A} \in \mathbf{R}^{m\times n}, \, m > n \tag{3}$$

The matrix $\mathbf{A}$ is assumed to be of *full rank*, i.e., $\mathrm{rank}(\mathbf{A}) = n$. The above system does not admit a solution, but, rather, an optimum *approximation* that minimizes the Euclidean norm of the error, with the error $\mathbf{e}$ defined as

$$\mathbf{e} \equiv \mathbf{b} - \mathbf{Ax} \tag{4}$$

To have a norm, $\mathbf{e}$ is assumed to be dimensionally homogeneous—i.e., all entries of $\mathbf{e}$ bear the same physical units. Thus, the *optimum solution* sought minimizes *a norm* of $\mathbf{e}$. Of the various norms available, the *Euclidean norm* is the most favored.

The dual version of the real overdetermined system of linear equations is obtained by simple dualization of the foregoing equation, thereby obtaining

$$\hat{\mathbf{A}}\hat{\mathbf{x}} = \hat{\mathbf{b}}, \quad \hat{\mathbf{A}} \in \mathbb{D}^{m\times n}, \, m > n, \, \hat{\mathbf{b}} \in \mathbb{D}^{m}, \, \hat{\mathbf{x}} \in \mathbb{D}^{n} \tag{5}$$

where $\mathbb{D}$ denotes the *dual ring*. As in the real case, we assume, for brevity, that $\hat{\mathbf{A}}$ has full rank, i.e., that its primal part does; its dual part needn't be so.

The unboundedness of the dual part of the vector norm notwithstanding, we can always equate to zero the derivative of the Euclidean norm-squared of the dual error $\hat{\mathbf{e}} \equiv \hat{\mathbf{b}} - \hat{\mathbf{A}}\hat{\mathbf{x}}$ w.r.t. $\hat{\mathbf{x}}$, which readily leads to

$$\frac{\mathrm{d}\|\hat{\mathbf{e}}\|^2}{\mathrm{d}\hat{\mathbf{x}}} = -2\hat{\mathbf{A}}^T\hat{\mathbf{e}} = \mathbf{0}, \quad \hat{\mathbf{e}} \equiv \hat{\mathbf{b}} - \hat{\mathbf{A}}\hat{\mathbf{x}} \tag{6}$$

---

[1] This is a number whose primal part vanishes.

Upon expansion into its primal and dual parts, Eq. (5) reveals two *decoupled* overdetermined systems of real linear equations, one for the primal, one for the dual part of the unknown $\hat{\mathbf{x}}$, namely,

$$\mathbf{A}\mathbf{x} = \mathbf{b} \tag{7a}$$

$$\mathbf{A}\mathbf{x}_o = \mathbf{b}_o - \mathbf{A}_o\mathbf{x} \tag{7b}$$

In the case of an overdetermined system, the above set of equations can be solved for $\hat{\mathbf{x}}$ in two different ways. First, by sequentially solving first Eq. (7a) for $\mathbf{x}$, then using the least-square value of $\mathbf{x}$ thus resulting to solve Eq. (7b) for $\mathbf{x}_o$. The primal matrix $\mathbf{A}$ only needs to be factored once, with the QR decomposition [24], for instance. The latter is the factoring of matrix $\mathbf{A}$ into the product of the orthogonal matrix $\mathbf{Q}$ and the upper triangular matrix $\mathbf{R}$, which can be done efficiently and robustly with Householder reflections [17].

Another way to accomplish this task is to factor both the primal and the dual parts of $\hat{\mathbf{A}}$ with the dual version of Householder reflections (DHHR), as described in an earlier paper [4]. The numerical solution to an overdetermined system of linear equations can also be computed by means of the singular value decomposition (SVD) [15]. Pennestrì and Stefanelli discussed the SVD of tall rectangular dual matrices [12]. As stated elsewhere [4], we focus on HHR, because "(i) HHR obviate the SVD and (ii) the SVD is iterative by necessity and, hence, unsuitable for online computations."

However, these methods still consider this as a decoupled problem, since the primal and dual parts have different units. To circumvent this issue, the concept of *characteristic length* (CL) can be invoked [2]. The CL is a *normalizing* length that minimizes the condition number of the overall $2m \times n$ matrix of the system (7a & b). First, let us rewrite Eq. (7b) as

$$\mathbf{A}\frac{1}{L}\mathbf{x}_o + \frac{1}{L}\mathbf{A}_o\mathbf{x} = \frac{1}{L}\mathbf{b}_o, \quad L \in \mathbf{R}^+ \tag{8}$$

Then, Eqs. (7a & 8) can be cast in the form

$$\mathbf{N}\mathbf{w} = \mathbf{s} \tag{9}$$

where

$$\mathbf{N} \equiv \begin{bmatrix} \mathbf{A} & \mathbf{O} \\ (1/L)\mathbf{A}_o & \mathbf{A} \end{bmatrix}, \quad \mathbf{w} \equiv \begin{bmatrix} \mathbf{x} \\ (1/L)\mathbf{x}_o \end{bmatrix}, \quad \mathbf{s} \equiv \begin{bmatrix} \mathbf{b} \\ (1/L)\mathbf{b}_o \end{bmatrix} \tag{10}$$

with $\mathbf{O}$ defined as the $m \times n$ zero matrix. We now have a system of $2m$ dimensionally-homogeneous linear equations (dimensionless). The characteristic length $L$ is to be determined so as to minimize a condition number of $\mathbf{N}$, $\kappa(\mathbf{N})$. Because of its geometric significance, which lends itself to a geometric interpretation, we choose the condition number based on the matrix Frobenius norm. Let

$$\mathbf{P}(L) \equiv \mathbf{N}^T\mathbf{N} = \begin{bmatrix} \mathbf{A}^T\mathbf{A} + (1/L)\mathbf{A}_o^T\mathbf{A}_o & (1/L)\mathbf{A}_o^T\mathbf{A} \\ (1/L)\mathbf{A}^T\mathbf{A}_o & \mathbf{A}^T\mathbf{A} \end{bmatrix} \tag{11}$$

Then

$$\kappa^2(\mathbf{N}) = \|\mathbf{N}\|_F^2 \, \|\mathbf{N}^{-1}\|_F^2 = \mathrm{tr}(\mathbf{P})\mathrm{tr}(\mathbf{P}^{-1}) \tag{12}$$

with $||\mathbf{N}||_F$ representing the Frobenius norm of the matrix $\mathbf{N}$. The optimum value of $L$ can be obtained upon zeroing the partial derivative of $\kappa^2(\mathbf{N})$ with respect to $L$, i.e.

$$f(L) \equiv \frac{\partial \kappa^2}{\partial L} = 0 \tag{13}$$

The above expression can be expanded into

$$f(L) = \mathrm{tr}\left(\frac{\partial \mathbf{P}}{\partial L}\right)\mathrm{tr}(\mathbf{P}^{-1}) + \mathrm{tr}(\mathbf{P})\mathrm{tr}\left(\frac{\partial \mathbf{P}^{-1}}{\partial L}\right) \tag{14a}$$

with

$$\mathrm{tr}(\mathbf{P}) = 2\mathrm{tr}(\mathbf{A}^T\mathbf{A}) + (1/L^2)\mathrm{tr}(\mathbf{A}_o^T\mathbf{A}_o) \tag{14b}$$

and

$$\mathrm{tr}\left(\frac{\partial \mathbf{P}}{\partial L}\right) = -(2/L^3)\mathrm{tr}(\mathbf{A}_o^T\mathbf{A}_o) \tag{14c}$$

By partitioning $\mathbf{P}$, its inverse is computed[2] to obtain the remaining two terms of Eq. (14a):

$$\mathbf{P}^{-1} \equiv \begin{bmatrix} \mathbf{W} & \mathbf{Y} \\ \mathbf{X} & \mathbf{Z} \end{bmatrix} \tag{15a}$$

where

$$\mathbf{W} \equiv (\mathbf{A}^T\mathbf{A} + (1/L^2)\mathbf{A}_o^T\mathbf{A}_o - (1/L^2)\mathbf{A}^T\mathbf{A}_o(\mathbf{A}^T\mathbf{A})^{-1}\mathbf{A}_o^T\mathbf{A})^{-1} \tag{15b}$$

and

$$\mathbf{Z} \equiv (\mathbf{A}^T\mathbf{A} - (1/L^2)\mathbf{A}_o^T\mathbf{A}(\mathbf{A}^T\mathbf{A} + (1/L^2)\mathbf{A}_o^T\mathbf{A}_o)^{-1}\mathbf{A}^T\mathbf{A}_o)^{-1} \tag{15c}$$

The matrices $\mathbf{X}$ and $\mathbf{Y}$ needed not to be defined, as they are not used in the sequel, since only the trace of $\mathbf{P}^{-1}$ needs to be computed. Furthermore, both $\mathbf{A}^T\mathbf{A}$ and $\mathbf{A}^T\mathbf{A} + (1/L^2)\mathbf{A}_o^T\mathbf{A}_o$ must be non singular. The last two terms are

$$\mathrm{tr}(\mathbf{P}^{-1}) = \mathrm{tr}(\mathbf{W}) + \mathrm{tr}(\mathbf{Z}) \tag{16a}$$

and

$$\mathrm{tr}\left(\frac{\partial \mathbf{P}^{-1}}{\partial L}\right) = \frac{\partial \mathrm{tr}(\mathbf{W})}{\partial L} + \frac{\partial \mathrm{tr}(\mathbf{Z})}{\partial L} \tag{16b}$$

## 4 Application to Kinematic Synthesis

The procedure detailed above is now applied to the same example of the synthesis of a spatial four-bar linkage reported in an earlier paper [4].

The RCCC mechanism[3] under synthesis, shown in Fig. 1, is to operate as closely as possible to *homokinetic*[4] in the pair $(\psi, \phi)$, which are the input and output angles, respectively.

[2] See 2018. CRC Standard Mathematical Tables and Formulas, 33rd Edition. Chapman and Hall/CRC, p. 93.

[3] R and C stand for revolute and cylindrical joints, respectively.

[4] The *italicized* term means a one-to-one velocity transmission ratio.

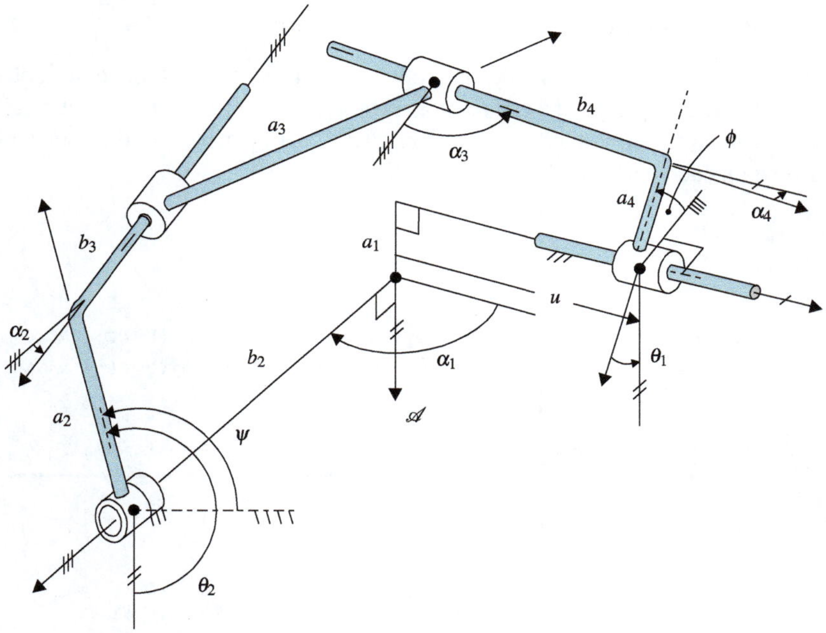


**Fig. 1.** Generic RCCC spatial four-bar linkage.

For the sake of brevity, the input-output (I/O) equation is not derived here, as it is available elsewhere [3]. The input is angle $\hat{\psi}$, a "dual angle" with zero dual part; the output being $\hat{\phi} \equiv \phi + \varepsilon u$. Moreover, the I/O equation is

$$\hat{F}(\hat{\psi}, \hat{\phi}) \equiv \hat{k}_1 + \hat{k}_2 \cos\hat{\psi} + \hat{k}_3 \cos\hat{\psi} \cos\hat{\phi} - \hat{k}_4 \cos\hat{\phi} + \sin\hat{\psi} \sin\hat{\phi} = 0 \tag{17}$$

The primal and dual parts of the *dual Freudenstein parameters* (DFP) $\{\hat{k}_i\}_1^4$ are defined, respectively, as

$$k_1 \equiv \frac{\lambda_1\lambda_2\lambda_4 - \lambda_3}{\mu_2\mu_4}, \quad k_2 \equiv \frac{\lambda_4\mu_1}{\mu_4}, \quad k_3 \equiv \lambda_1, \quad k_4 \equiv \frac{\lambda_2\mu_1}{\mu_2} \tag{18}$$

and

$$k_{o1} \equiv -\frac{a_1\lambda_2\lambda_4\mu_1\mu_2\mu_4 + a_2(\lambda_1\lambda_4 - \lambda_2\lambda_3)\mu_4}{\mu_2^2\mu_4^2} - \frac{-a_3\mu_2\mu_3\mu_4 + a_4(\lambda_1\lambda_2 - \lambda_3\lambda_4)\mu_2}{\mu_2^2\mu_4^2} \tag{19a}$$

$$k_{o2} \equiv \frac{a_1\lambda_1\lambda_4\mu_4 - a_4\mu_1}{\mu_4^2}, \quad k_{o3} \equiv -a_1\mu_1, \quad k_{o4} \equiv \frac{a_1\lambda_1\lambda_2\mu_2 - a_2\mu_1}{\mu_2^2} \tag{19b}$$

with

$$\lambda_i \equiv \cos\alpha_i, \quad \mu_i \equiv \sin\alpha_i \neq 0 \tag{20}$$

## 4.1 Solutions and Comparison

To formulate the approximate-synthesis problem, two sets of input-output variables are given, $\{\psi_i\}_1^m$ and $\{\phi_i, u_i\}_1^m$, where $m > n$, $n$ being the number of unknown linkage parameters. An overdetermined system of $m$ dual linear equations in the four DFPs is obtained when each pair of I/O values is applied to Eq. (17)[5]:

$$\hat{\mathbf{S}}\hat{\mathbf{k}} = \hat{\mathbf{b}} \tag{21a}$$

with

$$\hat{\mathbf{S}} = \underbrace{\begin{bmatrix} 1 & c\psi_1 & c\psi_1 c\phi_1 & -c\phi_1 \\ 1 & c\psi_2 & c\psi_2 c\phi_2 & -c\phi_2 \\ \vdots & \vdots & \vdots & \vdots \\ 1 & c\psi_m & c\psi_m c\phi_m & -c\phi_m \end{bmatrix}}_{\mathbf{S}} + \varepsilon \underbrace{\begin{bmatrix} 0 & -b_2 s\psi_1 & -u_1 c\psi_1 s\phi_1 - b_2 s\psi_1 c\phi_1 & u_1 s\phi_1 \\ 0 & -b_2 s\psi_2 & -u_2 c\psi_2 s\phi_2 - b_2 s\psi_2 c\phi_2 & u_2 s\phi_2 \\ \vdots & \vdots & \vdots & \vdots \\ 0 & -b_2 s\psi_m & -u_m c\psi_m s\phi_m - b_2 s\psi_m c\phi_m & u_m s\phi_m \end{bmatrix}}_{\mathbf{S}_o} \tag{21b}$$

$$\hat{\mathbf{k}} = \underbrace{\begin{bmatrix} k_1 \\ k_2 \\ \vdots \\ k_m \end{bmatrix}}_{\mathbf{k}} + \varepsilon \underbrace{\begin{bmatrix} k_{o1} \\ k_{o2} \\ \vdots \\ k_{om} \end{bmatrix}}_{\mathbf{k}_o}, \quad \hat{\mathbf{b}} = \underbrace{\begin{bmatrix} -s\psi_1 s\phi_1 \\ -s\psi_2 s\phi_2 \\ \vdots \\ -s\psi_m s\phi_m \end{bmatrix}}_{\mathbf{b}} - \varepsilon \underbrace{\begin{bmatrix} -u_1 s\psi_1 c\phi_1 - b_2 c\psi_1 s\phi_1 \\ -u_2 s\psi_2 c\phi_2 - b_2 c\psi_2 s\phi_2 \\ \vdots \\ -u_m s\psi_m c\phi_m - b_2 c\psi_m s\phi_m \end{bmatrix}}_{\mathbf{b}_o} \tag{21c}$$

For this example, a large set of prescribed data triads is used ($m = 501$). This is done to minimize the structural error, since the design error is known to converge toward the structural error when the prescribed data are increased [16]. Moreover, $\psi_i$ and $\phi_i$ are, respectively, uniformly distributed in the intervals $86^\circ \leq \psi \leq 206^\circ$ and $-26^\circ \leq \psi \leq 94^\circ$. The values $u_i$ are symmetrically distributed around $u = 0$, with $u_1 = -a_1/10$ and $u_m = a_1/10$. All lengths are normalized with respect to $a_1$ to be dimensionless. In light of the homokinetic condition imposed on the mechanism, a *cycloidal motion program* for the translation of the output joint is chosen. This motion is defined by a signed distance $u_i$, calculated from axis $\mathscr{A}$ indicated in Fig. 1. The set $\{u_i\}_1^m$ is computed with the distribution given below:

$$u_i = -\frac{a_1}{10} + U\left(\frac{i-1}{m-1} - \frac{1}{2\pi}\sin\frac{2\pi(i-1)}{m-1}\right), \quad i = 1, \ldots, m \tag{22}$$

where $U$ is the amplitude of the motion, set as $a_1/5$. The number of unknowns is reduced to only two in the primal part, to obtain the desired homokinetic behavior and a symmetric mechanism ($k_3 = 0$ and $k_4 = k_2$). The parameter $b_2$ is chosen to be equal to $a_1$. The shafts to be coupled being normal to each other, $\alpha_1$ is equal to $90^\circ$. The number of unknowns in the dual part of the system of equations also reduces to two, given that $k_{o3} = -1$ as per Eq. (19) and $k_{o4} = k_{o2}$, again, by symmetry.

[5] $c\psi_i$, $c\phi_i$, $s\psi_i$ and $s\phi_i$ denote, respectively, $\cos\psi_i$, $\cos\phi_i$, $\sin\psi_i$ and $\sin\phi_i$.

The condition numbers of $\mathbf{S}$ and $\mathbf{S}_o$ are 11.6079 and unbounded, respectively. An unbounded condition number is associated with a rank-deficient rectangular matrix—singular, in the case of square matrices. In our case, the rank-deficiency of $\mathbf{S}_o$ is immaterial, as only $\mathbf{S}$ is to be inverted in the Moore-Penrose generalized sense [4].

As a reminder, using the dual Householder reflections [4], the least-square solution[6,7] is obtained as

$$\begin{aligned} k_1 &= 1.275, \quad k_2 = k_4 = 0.9439 \\ k_{o1} &= 1.3275, \quad k_{o2} = k_{o4} = 0.6008 \end{aligned} \tag{23}$$

The corresponding linkage parameters, computed with Eqs. (18) and (19), are:

$$\begin{aligned} \alpha_2 &= \alpha_4 = 46.65^\circ, \quad \alpha_3 = 132.4^\circ \\ a_2 &= a_4 = -0.3178, \quad a_3 = 1.0408 \end{aligned} \tag{24}$$

The RMS primal and dual errors are, respectively, 0.0194 rad and 0.1234.

Now, by casting the system of equations in the form of a homogeneous system as described in Sect. 3, then solving Eq. (13) for $L$, one can obtain the characteristic length, i.e. $L = 0.7970$ with $\kappa(\mathbf{N}) = 18.9876$. The corresponding least-square solution can now be obtained:

$$\begin{aligned} k_1 &= 1.0190, \quad k_2 = k_4 = 0.7573 \\ k_{o1} &= 1.3554, \quad k_{o2} = k_{o4} = 0.6796 \end{aligned} \tag{25}$$

The corresponding linkage parameters, computed with Eqs. (18) and (19), are:

$$\begin{aligned} \alpha_2 &= \alpha_4 = 52.84^\circ, \quad \alpha_3 = 130.4^\circ \\ a_2 &= a_4 = -0.4321, \quad a_3 = 1.0529 \end{aligned} \tag{26}$$

The design error thus resulting, normalized with the characteristic length, is 0.0689. The latter is dimensionless, because, as previously mentioned, a system of dimensionless homogeneous linear equations is used for the synthesis. For comparison, with the parameters obtained with the DHHR, displayed in Eq. (23), the design error, normalized with the same characteristic length, is 0.1103, thereby showing a major advantage of the novel method.

## 5 Conclusions

Least-square problems are common in the field of mechanisms. Situations involving both rigid-body rotations and translations make dual numbers, and dual least squares, an idea tool. In this paper, a novel approach on problems involving dual least squares was

[6] In the approximate-synthesis problem, the design error $\hat{\mathbf{e}}$ is minimized with respect to its Euclidean norm, not the structural error $\hat{\mathbf{s}} \equiv (\varphi - \phi) + \varepsilon(\mathbf{v} - \mathbf{u})$, with $\varphi \equiv [\varphi_1 \quad \dots \quad \varphi_m]^T$, $\phi \equiv [\phi_1 \quad \dots \quad \phi_m]^T$, $\mathbf{v} \equiv [v_1 \quad \dots \quad v_m]^T$ and $\mathbf{u} \equiv [u_1 \quad \dots \quad u_m]^T$.

[7] The terms $(\varphi_i, v_i)$ are, respectively, the generated values of the outputs $\phi$ and $u$ attained at $\psi = \psi_i$ and $(\phi_i, u_i)$ are the prescribed values of the output angle and position at $\psi = \psi_i$.

proposed. The concept of the characteristic length was used to obtain a homogeneous system of equations. The former is computed to minimize the condition number of the matrix defining the system of equations, which is based on the matrix Frobenius norm. Naturally, another criterion would result in different results. The procedure was applied to overdetermined systems. A numerical example related to kinematics was given and discussed to illustrate the methodology proposed.

## References


1. Agrawal, S.K.: Multibody dynamics: a formulation using Kane's method and dual vectors. J. Mech. Des. Trans. ASME **115**(4), 833–838 (1993). https://doi.org/10.1115/1.2919276
2. Angeles, J.: Fundamentals of Robotic Mechanical Systems, Mechanical Engineering Series, vol. 124. Springer, Cham (2014). https://doi.org/10.1007/978-3-319-01851-5
3. Bai, S.P., Angeles, J.: A unified input-output analysis of four-bar linkages. Mech. Mach. Theory **43**, 240–251 (2008)
4. Belzile, B., Angeles, J.: Reflections over the dual ring—applications to kinematic synthesis. J. Mech. Des. **141**(7), 072302 (2019). https://doi.org/10.1115/1.4043204
5. Bottema, O., Roth, B.: Theoretical Kinematics. North-Holland Publishers Co., Amsterdam (1978)
6. Clifford, W.K.: Preliminary sketch of bi-quaternions. Proc. London Math. Soc. **4**, 381–395 (1873)
7. Cohen, A., Shoham, M.: Principle of transference - an extension to hyper-dual numbers. Mech. Mach. Theory **125**, 101–110 (2018). https://doi.org/10.1016/j.mechmachtheory.2017.12.007
8. Condurache, D.: Higher-order relative kinematics of rigid body motions: a dual lie algebra approach. In: Lenarcic, J., Parenti-Castelli, V. (eds.) Advances in Robot Kinematics 2018 (ARK 2018), pp. 83–91. Springer, Cham (2019). https://doi.org/10.1007/978-3-319-93188-3_10
9. De Falco, D., Pennestrì, E., Udwadia, F.E.: On generalized inverses of dual matrices. Mech. Mach. Theory **123**, 89–106 (2018). https://doi.org/10.1016/j.mechmachtheory.2017.11.020
10. Dimentberg, F.M.: The Screw Calculus and Its Applications in Mechanics. Izdat. Nauka. Moscow. English Translation: AD680993, Clearinghouse for Federal and Scientific Technical Information (1965)
11. Duffy, J.: The fallacy of modern hybrid control theory that is based on "orthogonal complements" of twist and wrench spaces. J. Robot. Syst. **7**(2), 139–144 (1990). https://doi.org/10.1002/rob.4620070202
12. Pennestrì, E., Stefanelli, R.: Linear algebra and numerical algorithms using dual numbers. Multibody Syst. Dyn. **18**, 323–344 (2007). https://doi.org/10.1007/s11044-007-9088-9
13. Fike, J.A., Jongsma, S., Alonso, J.J., van der Weide, E.: Optimization with gradient and Hessian information calculated using hyper-dual numbers. In: 29th AIAA Applied Aerodynamics Conference 2011. American Institute of Aeronautics and Astronautics Inc. (2011). https://doi.org/10.2514/6.2011-3807
14. Fischer, I.S.: Dual-Number Methods in Kinematics, Statics and Dynamics. CRC Press LLC, Boca Raton (1999)
15. Golub, G.H., Loan, C.F.V.: Matrix Computations. The Johns Hopkins University Press, Baltimore (1983)
16. Hayes, M.J.D., Parsa, K., Angeles, J.: The effect of data-set cardinality on the design and structural errors of four-bar function-generators. In: Proceedings of the 10th World Congress on the Theory of Machines and Mechanisms, pp. 437–442 (1999)

17. Householder, A.S.: Unitary Triangularization of a nonsymmetric matrix. J. ACM **5**(4), 339–342 (1958). https://doi.org/10.1145/320941.320947
18. Keler, M.L.: Kinematics and statics including friction in single-loop mechanisms by screw calculus and dual vectors. J. Manuf. Sci. Eng. Trans. ASME **95**(2), 471–480 (1973). https://doi.org/10.1115/1.3438179
19. Kotel'nikov, A.P.: Screw calculus and some of its applications to geometry and mechanics. Annals of The Imperial University of Kazan (1895)
20. Levitskii, N.I.: Design of Pane Mechanisms with Lower Pairs. Izdatelstvo Akademii Nauk SSSR, Moscow-Leningrad (1950). (in Russian)
21. Pennestrì, E., Valentini, P.P., de Falco, D.: The Moore-Penrose dual generalized inverse matrix with application to kinematic synthesis of spatial linkages. J. Mech. Des. **140**(10), 102303 (2018). https://doi.org/10.1115/1.4040882
22. Rico Martínez, J.M., Duffy, J.: The principle of transference: history, statement and proof. Mech. Mach. Theory **28**(1), 165–177 (1993). https://doi.org/10.1016/0094-114X(93)90055-Z
23. Sarkisyan, Y.L., Gupta, K.C., Roth, B.: Kinematic geometry associated with the least-square approximation of a given motion. ASME J. Eng. Ind. **95**(2), 503–510 (1973). https://doi.org/10.1115/1.3438183
24. Stoer, J., Bulirsch, R.: Introduction to Numerical Analysis. Springer, New York (2002)
25. Study, E.: Geometrie der Dynamen. Druck und Verlag von B.G., Teubner, Leipzig (1903)
26. Yang, A.T., Freudenstein, F.: Application of dual-number quaternion algebra to the analysis of spatial mechanisms. J. Appl. Mech. **31**(2), 300 (1964). https://doi.org/10.1115/1.3629601

# Origami-Inspired Design of a Deployable Wheel

John Berre(✉), François Geiskopf, Lennart Rubbert, and Pierre Renaud

ICube, INSA Strasbourg, University of Strasbourg, CNRS, 67100 Strasbourg, France
john.berre@insa-strasbourg.fr

**Abstract.** In this paper, we introduce the design of a variable-diameter wheel for mobile robots. The goal is to improve maneuverability and crossing capacity of such robots by actively controlling the diameter of wheels. Origami patterns have been recently considered in robotics for the design of compliant and deployable structures. In our context, compliance can limit the load capacity of mobile robots. We then propose an origami-inspired design to benefit at the same time from the high deployability of origami patterns and the stiffness of articulated mechanisms. The proposed deployable wheel is based on a flasher structure. Kinematic architecture is determined to allow the transition from a flat configuration to a folded configuration for maximum deployability. The original arrangement of kinematic joints to get the flasher motions with a rigid-body mechanism is presented. In addition, an actuation mechanism is designed to control the wheel radius using a single actuator. Finally, design of peripheral shape is discussed to maintain proper rolling capacity. The implementation of the prototype is then presented with validation of kinematics properties.

**Keywords:** Origami · Origami-inspired mechanism · Variable-diameter wheel · Mobile robot · Deployable mechanism

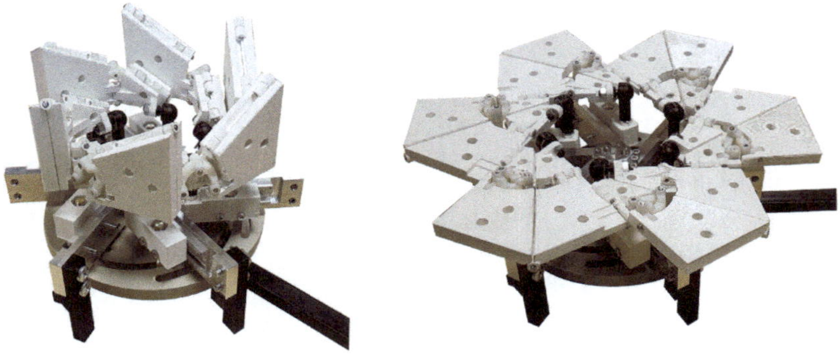

**Fig. 1.** Folded (left) and deployed (right) configurations of the origami-inspired mechanism for deployable wheel


E.-C. Lovasz et al. (Eds.): MTM&Robotics 2020, MMS 88, pp. 114–126, 2021.
https://doi.org/10.1007/978-3-030-60076-1_11

# 1 Introduction

Design of mobile robots for exploration tasks is challenging because of the unstructured environment they can encounter. Bypassing or crossing various obstacles such as holes, stairways or walls requires a high maneuverability as well as crossing abilities. Large diameter wheels help to improve the capability to cross holes or obstacles, and also to enhance the traction performance [1]. Going through narrow passages or tunnels however constrain the maximum allowable size of wheels as observed in [2, 3]. One way to satisfy all requirements is to transform wheels into active components with controllable diameters, as introduced in [2] where the robot is composed of a body and two variable diameter wheels. The overall decrease of diameter then helps to go through narrow spaces, and increase of diameters improves obstacle crossing capability. In addition, steering of the robot can be achieved by variation of the wheel diameters. The interest of such an approach is linked to the achievable range of diameter for the wheel. In this paper, we are therefore interested in the design of wheels of variable diameter, trying to maximize the range of variation. In addition, the wheels have to support the vehicle weight. This constitutes a second design objective. As a result, we are interested in the design of wheels with variable diameter that offer, at the same time high load capacity and large range of diameter.

Several designs have been recently introduced for variable diameter wheels. They can be distinguished by the adopted design principle. Origami patterns have been considered to create wheels using folding of flexible materials. The ratio between maximum and minimum wheel diameter can then reach 2.2 in [2], around 3.1 in [4] and 9.2 in [5]. Other designs have been introduced using rigid-body articulated mechanisms. The ratios are, in these cases, equal to 2 in [6], 1.5 in [7] and 1.3 in [8]. The well-known deployability of origami patterns is clearly an advantage to create wheels with large diameter variations. However, kinematics of origamis is generally only possible thanks to the compliance of the material used for the pattern. Origami-based solutions are also restricted by their load capacity. For example, in [2] the wheel deformation is around 5% under a 1 N radial force. This constitutes a strong limitation for the design of deployable wheels as origami-based wheels will remain compliant, making it difficult to ensure high load capacity. With current solutions, a contradiction exists between using compliant design to remain foldable and using articulated mechanisms to bear high load capacity. The approach we propose is then to consider the interesting deployability of origami patterns, but designing our deployable wheel as an origami-inspired solution that is a rigid-body mechanism. This constitutes our main contribution, illustrated in Fig. 1, with the introduction of a wheel o f variable diameter that is designed following this approach to offer at the same time deployability and load capacity.

The requirements for rolling and crossing capacities may depend in exploration robotics on the ground properties. Different outer profiles of wheels have then been considered [7, 9]. In [6] the proposed shape of the deployable wheel varies with the diameter. One configuration is dedicated to stair crossing, another one to rolling. In [1, 7, 8] or [9] the shape of wheels is chosen to favor obstacle crossing. It is then not possible to improve rolling efficiency without modifying the deployable mechanism. Our proposition is to ensure wheel deployability independently from the wheel outer shape. With the presented design, different types of wheel contact can be obtained for the same deployability. This constitutes a second contribution of the presented work.

Origami-inspired design is an active field, with several recent approaches reviewed in [14]. Building an origami-inspired mechanism requires to select and adapt one of these approaches in order to determine the proposed kinematic architecture. This will be presented as the first part of the work. The interest of the system is also linked to our objective to actuate it with a single actuator. The design of an actuation mechanism using a single actuator is then a second part. The elaboration of the variable-diameter wheel architecture is introduced in Sect. 2, with the description of the selected origami structure, the design of the associated rigid-body mechanism and the actuation mechanism. Section 3 is focused on the analysis of the wheel properties in terms of contact management and deployability through the implementation of a proof of concept for experimental assessment. Finally, conclusions and perspectives are submitted in Sect. 4.

## 2 Design

### 2.1 Introduction of the Selected Origami Pattern

The flasher origami considered in [2] and depicted in Fig. 2 is of interest as it can be switched from a flat unfolded configuration to a fully folded configuration with a large radial variation, as represented in Fig. 3. In [5], a first implementation of this origami using rigid elements was investigated for a deployable solar panel, which confirmed the interest of such origami pattern. The proposed kinematics is however complex to implement with several actuators needed to control the deployment. In the following, we thus consider the flasher origami and derive an associated origami-inspired rigid-body mechanism that can be deployed using a single actuator.

Origami patterns are best described by the planar arrangement of folds, represented graphically by the location and type of folds within the pattern (Fig. 4). The colors are assigned to the fold direction. A red (respectively blue) fold designates a so-called mountain (valley) fold, where the direction leads the fold to come up (down). The facets delimited by the folds are the panels of the origami pattern. Their upper surface is represented in white, their lower surface in grey.

The pattern of the flasher origami is represented in Fig. 2. It is the repetition of an elementary pattern depicted in Fig. 5, composed of 4 panels connected with three mountain folds and one valley fold. Two additional valley folds are used to connect the elementary pattern with the neighbouring panels belonging to other elementary patterns. The folded configuration of the flasher is obtained when the facets B and B′ are superimposed, as well as facets A and A′. We therefore need to reproduce this motion with the rigid and thick panels that compose a rigid-body mechanism.

**Fig. 2.** The pattern of the flasher origami. In red, the elementary pattern and in orange the associated neighboring panels. Blue background is added to outline the panel boundaries.

**Fig. 3.** Sequence showing origami flasher deployment, from flat configuration to folded configuration.

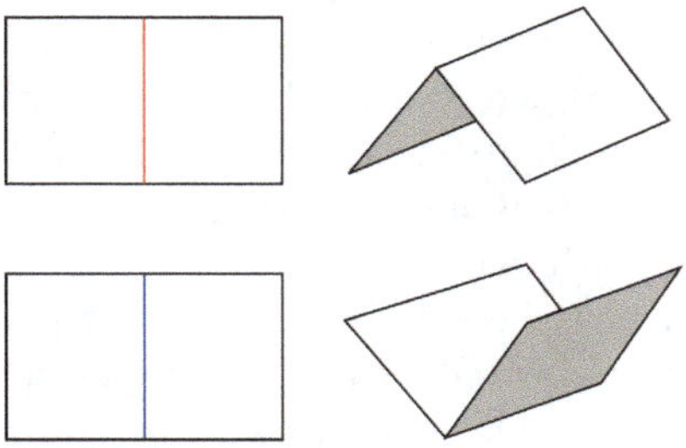

**Fig. 4.** Illustration of mountain and valley folds for a simple origami pattern, before (left) and after (right) folding.

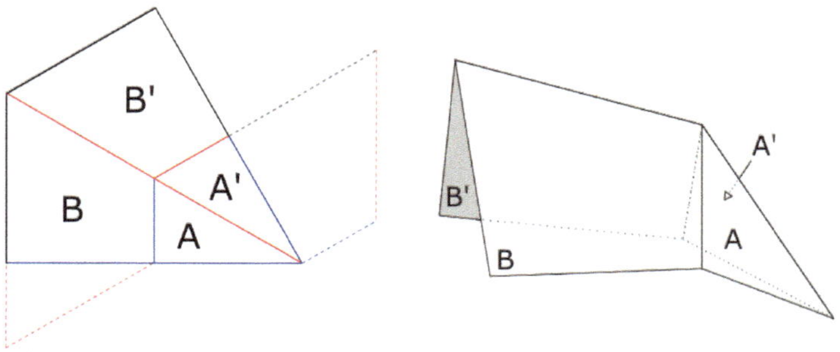


**Fig. 5.** Elementary pattern of the flasher origami in unfolded configuration (left) and during folding (right).

## 2.2 Deployable Mechanism Design

Building an origami-inspired mechanism requires to deal with the thickness of solids, negligible when paper is considered as in origami folding. Several design methods dealing with thick origami have been introduced such as shifting of folding axis [10, 11], membrane folding [5], strained joint area fold [12], the double hinge method [13], tapered panels method [13] or shift panels method [16]. They have been reviewed and discussed in [14] and [15].

All methods have restricted fields of applications defined by conditions on the origami patterns. The flat foldability condition defined in [11], i.e. the existence of flat configurations in folded and unfolded states, is necessary for the double hinge and shifting folding axis methods. Most methods require to have an origami with admissible kinematics, which means the rigid foldability condition [11] has to be fulfilled: it must be possible to fold rigid panels without thickness according to the origami pattern. This is not the case for the flasher origami. As outlined in [14], interference between the panels is difficult to avoid when following [16]. This method is therefore not valid in our case. The methods of [5] and [12] are based on the introduction of additional degrees of freedom to have admissible kinematics. Internal mobilities are therefore generally introduced [14]. This is not acceptable in our context as the mechanism deployment has to be controlled with a single actuator.

No existing method directly fits our situation. The design of the deployable rigid-body mechanism is therefore achieved with a specific approach composed of two steps. In a first step, the impact of panel thickness is investigated experimentally. The objective is to determine the condition that allows the flasher to be kinematically admissible with thick and rigid panels. The flasher kinematics is strongly linked to the behavior of the elementary pattern. So, in this step, only this pattern is considered. A prototype of a flasher elementary pattern is manufactured with rigid thick panels manufactured with 3-mm PMMA polymer sheets. Membranes composed of standard tape are added to implement the creases between the panels. The tape is located on the top layer for the valley folds and bottom layer for the mountain folds. No backlash has been left between panels to ensure that the link replacing the crease can be considered as a revolute joint. Using a trial-and-error approach, it is then identified that the suppression of the crease between panels A′ and B′ (Fig. 4) is the only possible modification to get foldability of the elementary pattern.

In a second step, the suppressed fold is replaced by a kinematic chain that allows us to get admissible kinematics without any internal mobility. The determination process is conducted using a rigid-body simulation software. Kinematic joints are added to constrain the displacement of the panel B′ when the elementary pattern is folded. The adequate kinematic chain is composed of five revolute joints (Fig. 6) connecting panels A′ and B′. The revolute joints 5, 6, 7 allow the planar displacement between A′ and B′ panels which is needed for thickness accommodation. Then the revolute joints 4 and 8 are connected with the panels so they can move like in the origami. Joints 1, 2, 3 and 9 are designed to have mountain folds replaced by revolute joints on bottom edges and valley folds by joints on top edges. The mechanism is obtained after duplicating the elementary pattern to follow the initial flasher arrangement (Fig. 7) and tested using the same software. Kinematics remains admissible, with possibility to fold the mechanism from the flat configuration represented in Fig. 7 to the folded configuration of Fig. 9 passing through partially folded configuration in Fig. 8. This process validates the kinematic scheme of the proposed origami-inspired mechanism.

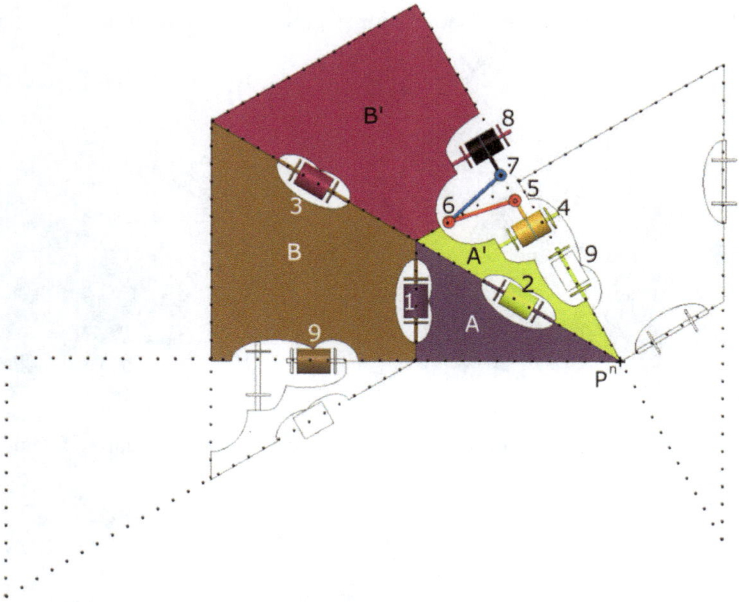


**Fig. 6.** Kinematic scheme of the elementary pattern composing the proposed origami-inspired mechanism.

**Fig. 7.** 3D kinematic illustration of the obtained rigid-body mechanism in flat configuration.

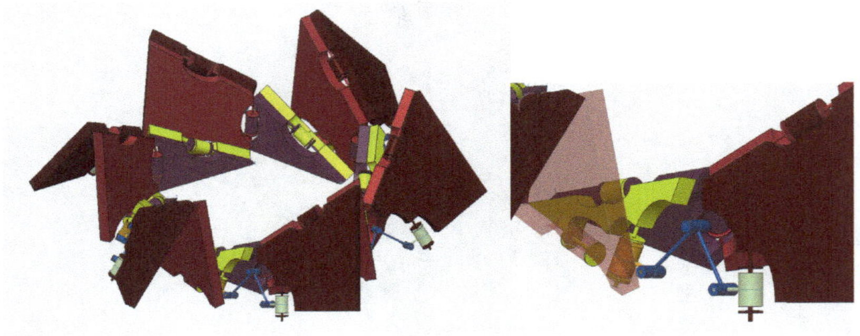

**Fig. 8.** 3D kinematic illustration of the obtained rigid-body mechanism during folding (left) and a zoom on small elementary pattern (right).

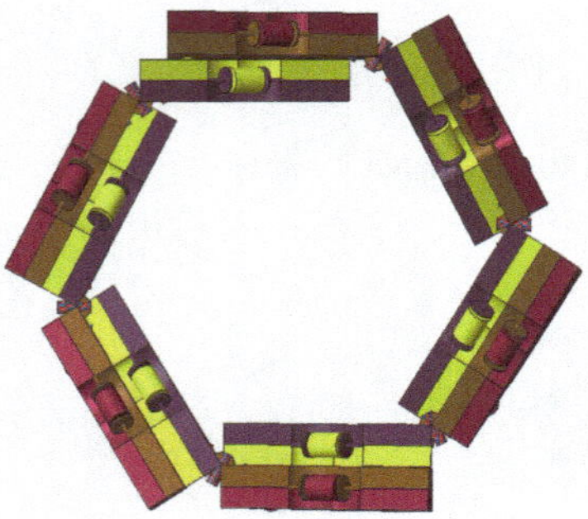

**Fig. 9.** 3D kinematic illustration of the obtained rigid-body mechanism folded configuration.

## 2.3 Actuation Mechanism Design

In order to determine how to actuate the mechanism shown in Fig. 7, the numerical simulation is used to analyze first the relative motions of the components. The folding sequence can then be simulated by placing virtual actuators on the creases. It is determined that it is possible to constrain the motion of points $P_1$ to $P_6$ along axis $x_1$ to $x_6$ (Fig. 10) to get the mechanism deployment. This constraint can be achieved by placing a spherical joint connected to the base by mean of a prismatic joint.

Prismatic joints are connected to the mechanism frame numbered 0 on Fig. 10. For a pattern i, i ∈ [1, 6], the body 2.i is connected with this ground through a prismatic joint of axis $x_i$ and with the flasher-like mechanism through a spherical joint of center $P_i$. The translation of the body 2.i induces the motion of the mechanism connected to the spherical joint. The body 1 is connected with the frame through a revolute joint. Its main functional surfaces are spiral-shaped grooves. Each groove is defined to ensure a radial displacement of the body 2.i constrained by the prismatic joint. The linear motions of points $P_i$, need to be synchronized during deployment to ensure the simultaneous deployment of all elementary patterns. This the aim of the axial repetition of the spiral shaped grooves. Finally, the rotation of the body 1 induces the motion of points $P_i$, and then the mechanism deployment.

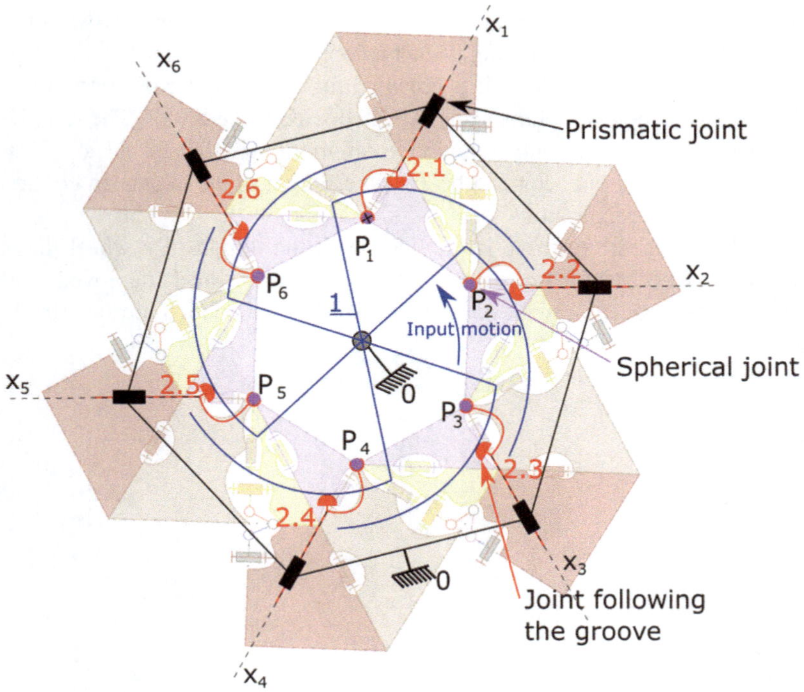


**Fig. 10.** Kinematic scheme of the actuation mechanism.

# 3 Implementation

The analysis and assessment of the wheel properties in terms of contact management and deployability requires the implementation of a proof of concept. In the following, we consider the design of a variable-diameter wheel in the context of a research project dedicated to mobile robots for exploration[1]. The robot design has an impact on the desired wheel size and inner diameter in folded configuration. Because of project specification, the minimal diameter of the wheel has to be higher than 150 mm. According to this value, the dimensions of the mechanism have been adapted. The design is achieved considering manufacturing with 3D printing techniques applied to the panels and machining processes to all other components.

## 3.1 Wheel Profile Management

Exploration robots evolve in various environments. A wheel that can be designed with various profiles has a better versatility and therefore a larger number of use cases.

In its flat configuration, the outer shape of the flasher origami can be cropped. Thus, for our deployable wheel, it is possible to adapt the wheel shape in its deployed configuration. In the folded configuration, as paper has no thickness, the shape of a flasher origami cannot however be modified. On the contrary, the proposed mechanism can be modified thanks to its thickness. The specific shape to adapt the mechanism is integrated to the thickness without consequence on the deployment. In Fig. 11 and 12, a first example is provided, removing the actuation mechanism for sake of clarity. Here, a circular wheel profile is desired. It is obtained for the two extreme configurations. In Fig. 10, the added specific volumes are indicated with dashed lines. The ratio between the diameters of the folded and unfolded configuration is then equal to 170%. This means that the external diameter can be 1.7 times larger in the expanded configuration than in the folded configuration.

In Fig. 13, a second example is provided with the design of a wheel that can be used as a claw wheel (folded configuration) or a circular wheel (deployed configuration). Such a component can be used to maximize the robot motricity in different terrains.

[1] http://www.origabot.cnrs.fr.

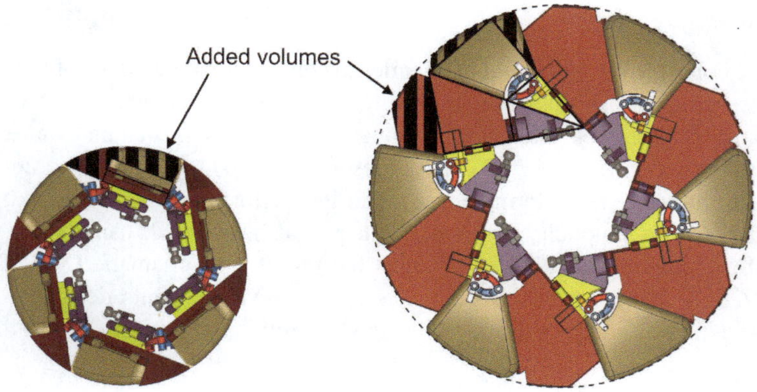


**Fig. 11.** Implementation of a quasi-circular wheel. Side view in folded (left) and expanded (right) configurations at the same scale.

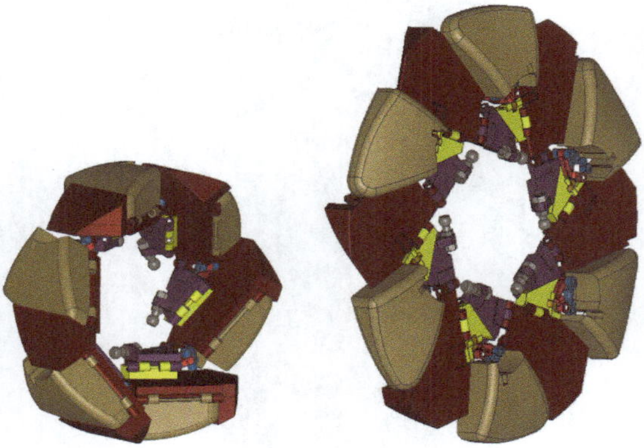

**Fig. 12.** Implementation of a quasi-circular wheel. Three-quarter view in folded (left) and expanded (right) configurations at the same scale.

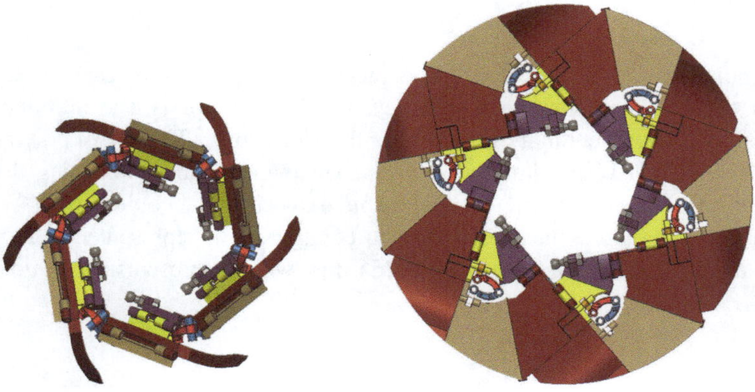

**Fig. 13.** Side view of adaptable wheel with a claw wheel in folded configuration (left) and a circular profile wheel in expanded configuration (right) at the same scale.

### 3.2 Prototype

A proof of concept was built (Fig. 1) in order to confirm the kinematics of the proposed mechanism, and to assess the reachable range of wheel diameters. 3D printing is favoured as it simplifies the manufacturing of the kinematic chain connecting the panels determined in Sect. 2.2. The CAD design (Fig. 14) was performed considering the project requirements in terms of size, and the performance of FDM printing with PLA material. The simplicity of joint design was also favoured, using standard components for the spherical (WGRM-05, Igus Inc.) joints and prismatic (NW-02-17 and NS-01-17, Igus Inc.) and 3-mm steel axes for the revolute joints designed as barrel hinges. The guiding slots are produced using CNC milling to limit friction in the joints.

**Fig. 14.** CAD view of the prototype with the actuation mechanism. Two elementary patterns have been removed for better visibility.

The number of joints and parts makes the assembly process delicate. With this first proof of concept, the geometrical accuracy of 3D printed parts also introduces some defects in joint alignment that overconstrain the mechanism. The use of polymer limits the stiffness of panels and links used in the kinematics chain connecting them. The deployability of the proposed origami-inspired mechanism could still be observed experimentally, as represented with the two configurations represented in Fig. 1. The external diameter could vary from 160 to 265 mm, which represents a variation ratio of 1.65.

## 4 Conclusion

In this paper, we have developed an origami-inspired mechanism to be used for variable-diameter wheels. Such wheels can be of interest in the design of mobile robots as they can contribute to steering or crossing capacity. The proposed system is built as a rigid-body mechanism to try to increase the load capacity of the wheel and therefore of the robot. An approach combining experimental evaluation and simulation was adopted to determine the kinematic architecture, starting from the flasher origami pattern. A kinematic scheme is proposed to actuate the deployable mechanism thanks to a single actuator. The solution is of particular interest for mobile robots as the mechanism for diameter variation can be designed independently from the surfaces needed for a proper contact with the terrain. Two examples are provided with a quasi-circular wheel or a wheel with claws. A proof of concept was implemented. The deployment sequence is confirmed, and the achievable variation of diameter is about 100 mm which corresponds to a ratio of 1.65.

The achievable deployment ratio is dependent on the implementation conditions, such as the minimum component dimension and their manufacturing process. Further work is required to assess the maximum range that can be reached with the proposed kinematic architecture as well as load capacity. Origami-inspired mechanisms are composed of a large number of components and kinematic joints. The implementation of revolute joints constrains the minimal size of the panels. The development of solutions for simplifying the whole manufacturing and assembly process, and in the end improving the size and mechanical efficiency of the component, is another perspective of this work.

**Acknowledgement.** This work was supported by the French National Agency for Research (ORIGABOT ANR-18-CE33-0008), and the Investissements d'Avenir program (Robotex ANR-10EQPX-44, Labex CAMI ANR-11-LABX-0004). We also acknowledge C. Brice for his work on the prototype and B. Wach for his technical support on 3D printing parts.

## References

1. Zeng, W., Gao, F., Jiang, H., Huang, C., Liu, J., Li, H.: Design and analysis of a compliant variable-diameter. Mech. Mach. Theory **125**, 240–258 (2018)
2. Lee, D.-Y., Jung, G.-P., Sin, M.-K., Ahn, S.-H., Cho, K.-J.: Deformable wheel robot based on origami structure. In: 2013 IEEE International Conference on Robotics and Automation (ICRA), pp. 5612–5617 (2013)
3. Lee, D.-Y., Kim, S.-R., Kim, J.-S., Park, J.-J.: Origami wheel transformer: a variable-diameter wheel drive robot using an origami structure. Soft Robot. **4**(2), 163–180 (2017)
4. Felton, S.M., Lee, D.-Y., Cho, K.-J., Wood, R.: A passive, origami-inspired, continuously variable transmission. In: IEEE International Conference on Robotics and Automation (ICRA), pp. 2913–2918 (2014)
5. Zirbel, S.A., Magleby, S., Lang, R., Thomson, M., Sigel, D., Walkemeyer, P., Trease, B., Howell, L.L.: Accommodating thickness in origami-based deployable arrays. ASME J. Mech. Des. **135**(11), 111005-1–111005-11 (2013)

6. Nagatani, K., Kuze, M., Yoshida, K.: Development of transformable mobile robot with mechanism of variable wheel diameter. J. Robot. Mechatron. **19**(3), 252–257 (2007)
7. Ming, H., Bao-Long, S., Wen-Hua, C., Dan-Min, H., De-Bei, L., Ya-Di, H.: Structure synthesis & configuration transformation of variable topology repeated foldable wheel. In: IEEE International Conference on Robotics and Biomimetics (ROBIO), pp. 85–90 (2013)
8. Xiaojiang, Y., Guoyang, X., Gaofeng, Ying, C.: Variable-diameter mobile platform chassis characteristics analysis. In: International Conference on New Technology of Agricultural Engineering (ICAE), pp. 166–169 (2011)
9. Sell, R., Aryassov, G., Petritshenko, A., Kaeeli: Kinematics and dynamics of configurable wheel-leg. In: 8th International DAAAM Baltic Conference (2012)
10. Tachi, T.: Rigid-foldable thick origami. In: Origami 5: Fifth International Meeting of Origami Science, Mathematics, and Education, pp. 253–264 (2011)
11. Chen, Y, Peng, R., You, Z.: Origami of thick panels. Science **349**, 396–400 (2015)
12. Pehrson, N.A., Lang, R.J., Magleby, S.P., Howell, L.L.: Introduction of monolithic origami with thick-sheet materials. In: International Association for Shell and Spatial Structures Annual Symposium (2016)
13. Abel, Z., Cantarella, J., Demaine, E.D., Eppstein, D., Hull, T.C., Ku, J.S., Lang, R.J., Tachi, T.: Rigid origami vertices: conditions and forcing sets. J. Comput. Geom., **7**, 171–184 (2015)
14. Tolman, K.A.: Developing hybrid thickness-accommodation techniques for new origami-inspired engineered systems. Thesis submitted to the faculty of Brigham Young University (2017)
15. Morgan, M.R., Lang, R.J., Magleby, S.P., Howell, L.L.: Towards developing product applications of thick origami using the offset panel technique. Mech. Sci. **7**(1), 69–77 (2016)
16. Edmondson, B.J., Lang, R.J., Magleby, S.P., Howell, L.L.: An offset panel technique for thick rigidly foldable origami. In: ASME 2014 International Design Engineering Technical Conferences and Computers and Information in Engineering Conference (2014)

# An Experimental Survey on the Odometric Error of Mecanum Wheeled Mobile Robots

Amir Shahidi(✉), Mathias Hüsing, and Burkhard Corves

Institute of Mechanism Theory, Machine Dynamics and Robotics, RWTH Aachen University, Aachen, Germany
shahidi@igmr.rwth-aachen.de

**Abstract.** To compensate the odometric errors of mobile robots in motion control tasks in known environments, the measured errors in pre-defined experiments can be used as feed forward of the control loop. In an experimental study on a mecanum-wheeled mobile robot, the odometric errors have been measured and are presented in this paper. The experiments are designed to represent the joint effects of systematic (robot specific) as well as non-systematic (environment specific) errors in the odometry of the robot. Different surfaces are chosen to perform the experiments and the robot is driven with a range of velocities of typical indoor mobile robotic application. The odometric errors of the mobile robot are measured via iGPS system with an accuracy of 200 μm. The results of the experiments are statistically analyzed and the mean values of the translational and rotational errors are reported within a confidence interval of 90%. The presented results may provide the readers with an overview of the proportion of the odometric errors in the translational workspace of holonomic mobile robots. A feed forward control scheme is suggested to use the gathered data to compensate the odometry errors.

**Keywords:** Mobile robotic · Odometry error · Odeomtry calibration

## 1 Introduction

An important part of mobile robot navigation in known environments is localization and motion control. A common way of localization is Dead-reckoning, that is the measurements of the odometry of the robot from an origin in the environment. Different sensoric systems were used in different researches to study localization tasks, among which range sensors and RGB-D camera based localization can be named [4,6]. Nevertheless such localization methods can be prone to accumulative errors. The emergent errors in mobile robotics can be categorized as systematic and non-systematic errors [10]. A model estimating both the systematic and non-systematic translational and rotational error of a synchronous drive mobile robot odometry is introduced in [11].


E.-C. Lovasz et al. (Eds.): MTM&Robotics 2020, MMS 88, pp. 127–134, 2021.
https://doi.org/10.1007/978-3-030-60076-1_12

A possible approach to eliminate the failure and estimate the correct position of the mobile robot is using auxiliary sensors which detect predefined landmarks in the environment and estimate the position of the robot with respect to them [5]. Among these researches the utilization of cameras for detecting the odometric error and Kalman filtering for pose estimation can be named [7]. Particle filtering methods were applied on information from sonar and laser range sensors to estimate the pose of a mobile robot in [2]. In [14] a real time position estimation via deep learning algorithms applied to the gathered data from point clouds generated from cameras is presented. The estimation methods, *e.g.* extended/unscented Kalman filter or particle filter, are based on probabilistic techniques, mostly with background in Bays method [20]. A comparison between these methods can be found in [21]. Almost every known algorithm in this field is based on common feature of probabilistic nature [17]. There has been recent researches for improving accuracy of the localization tasks by modeling the uncertainty of the odometric parameters. For instance, in [1] the efficient incorporation of RGB-D point features in graph-based localization has been investigated and the spatial uncertainty model of the feature is modeled to achieve improvements in the estimation of the trajectories.

Common assumption in these models involve the rigidity of wheels and absence of slippage [9,12]. The effects of slippage have been studied in [13] by comparison of the measurements of the driven and non-driven wheels. A similar study is carried out in [16] using a gyroscope to detect slippage. The effects of loading the mobile robot in position error and slippage is studied in [18]. These effects gain importance, when the robot wheels are equipped with free rolling rollers to facilitate the kinematics of holonomic movement.

In this paper the odometric error of a holonomic mobile robot determined from 450 indoor experiments is presented. To heighten the effects of non-systematic error different environmental condition (*i.e.* different surfaces) have been used for the experiments, namely carpet (high friction), vinyl (low friction) and plank (middle friction). Moreover the robot is driven in the velocity range of 0.25 to 0.75 $\frac{m}{s}$ as normally are used in common indoor applications. The odometry errors are measured through an iGPS system, a laser-based indoor system with optical detectors and transmitters [3,15]. The accuracy of the measurements in the conducted experiments is 200 $\mu$m.

# 2 Mathematical Formulation and Experiment Design

## 2.1 Statistical Analysis

In this paper the pose vector of the robot is defined as

$$^{G}\zeta = [x,\, y,\, \theta]^T \tag{1}$$

in the global coordinate system $G$ and the direction of the movement is defined via variable $\varphi$. The kinematic of holonomic robots enables them to freely move in different translational directions in the planar space. Considering the local coordinate system of the robot defined as $x_R$-axis pointing towards front of the robot (*cf.* Fig. 1) and $z_R$-axis pointing outward from the robot's surface, the movement in $x_R$-direction corresponds to $\varphi = 0$ and the movement in $y_R$-direction corresponds to $\varphi = \frac{\pi}{2}$. The aim of the experiments designed in this study is to measure the mean $\mu$ amount of odometric errors of the robot, in its translational workspace, *i.e.* $x_{err}$ and $y_{err}$ as well as rotation $\theta_{err}$. The translational kinematic space of the holonomic mobile robot is linearly spaced with incremental value of $\frac{\pi}{8}$. Due to the symmetry of the movement, just the first quarter of the space $(0 < \varphi < \frac{\pi}{2})$ is considered, *i.e.* $\varphi \in \{0, \frac{\pi}{8}, \frac{2\pi}{8}, \frac{3\pi}{8}, \frac{4\pi}{8}\}$.

The parameters $x_{err}$, $y_{err}$ and $\theta_{err}$ are assumed to be independent and identically distributed (i.i.d.) continuous random variables and have some $\mathcal{N}(\mu, \sigma^2)$. It is aimed to measure the mean values $\mu$ of their distribution with a certain confidence from $n$ conducted experiments. To achieve this goal, Student's t-distribution is utilized [8]. The variable

$$T_\chi = \sqrt{n}\frac{\bar{\chi} - \mu_\chi}{S}, \tag{2}$$

is defined having t-distribution with degree of freedom $n - 1$, where

$$\begin{cases} \bar{\chi} = \frac{1}{n}\sum_{i=1}^{n} \chi_i, \\ \\ S^2 = \frac{1}{n-1}\sum_{i=1}^{n}(\chi_i - \bar{\chi})^2. \end{cases} \qquad \chi \in \{x_{err}, y_{err}, \theta_{err}\} \tag{3}$$

The interval

$$\left[\bar{\chi} - t_{1-\frac{\alpha}{2};n-1}\frac{S}{\sqrt{n}}, \bar{\chi} + t_{1-\frac{\alpha}{2};n-1}\frac{S}{\sqrt{n}}\right] \tag{4}$$

can be defined as $100\,(1 - \alpha)\%$ confidence interval for $\mu_\chi$. For the sake of the experiments $n = 10$ (the number of experiments for each set up), *i.e.* the degree of freedom for t-distribution is equal to 9. Besides, $\alpha = 0.1$ is chosen to determine 90% confidence intervals for the $\mu_\chi$. Thereof, the value of the variable $t$ can be read from t-distribution tables as

$$t_{1-\frac{\alpha}{2};n-1} = t_{0.95;9} = 1.833. \tag{5}$$

### 2.2 Experimental Setup

The holonomic mobile robot Summit XL Steel of the company Robotnik is used in the experiments in this study. The robot weighs 90 kg and its wheel radius is 125 mm. The experiments are conducted on three different types of surfaces (carpet, vinyl and plank) with three different maximum velocities $(v_{max} \in \{0.25, 0.50, 0.75\}\frac{m}{s})$ in 5 different directions $(\varphi \in \{0, \frac{\pi}{8}, \frac{2\pi}{8}, \frac{3\pi}{8}, \frac{4\pi}{8}\})$. Figure 1 schematically shows the set of experiment in the direction $\varphi = \frac{\pi}{8}$ and $x_{err}$, $y_{err}$ and $\theta_{err}$ in the global coordinate system. In this figure, the robot is supposed to travel from its start position $([x_{R,S}, y_{R,S}]^T)$ to a goal position $([x_{R,G}, y_{R,G}]^T)$ in the direction of $\varphi = \frac{\pi}{8}$. According to the errors ($x_{err}$, $y_{err}$

and $\theta_{err}$), the robot finishes the trajectory in ($[x_{R,A}, y_{R,A}]^T$). The distance that the robot is paving in each experiment is held to be constant as $4\,m$. The measured odometric errors $x_{err}$, $y_{err}$ and $\theta_{err}$ via the iGPS system is transferred to the robot's coordinate system

$$^{R}\boldsymbol{\zeta} = {^{RG}\boldsymbol{T}}\,{^{G}\boldsymbol{\zeta}}, \tag{6}$$

with $^{RG}\boldsymbol{T}$ declaring the transformation from global to robot's coordinate system. The gathered data are post processed using the statistical analysis program R [19].

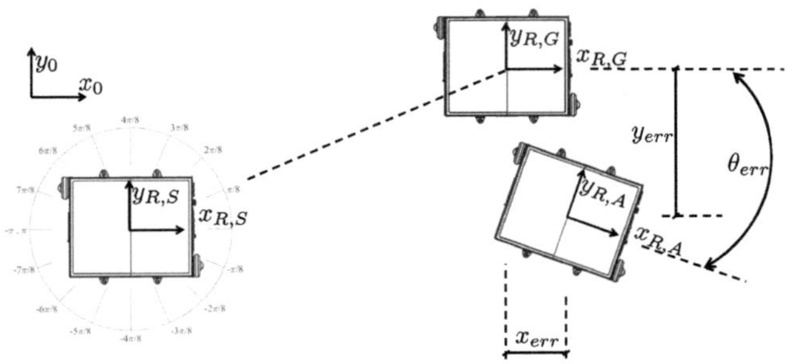


**Fig. 1.** Definition of Robots coordinate system and odometric errors

# 3 Results and Discussion

The complete results of the conducted experiments are presented numerically in Table 1. In the presented table, the following abbreviations are made for the sake of simplicity:

- $\mu_{\chi,0.9}$: The mean value of the 90% confidence interval of the variable $\chi \in \{x_{err}, y_{err}, \theta_{err}\}$
- $\mu_{\chi,LCI}$: Lower confidence interval of the mean value of $\chi \in \{x_{err}, y_{err}, \theta_{err}\}$
- $\mu_{\chi,UCI}$: Upper confidence interval of the mean value of $\chi \in \{x_{err}, y_{err}, \theta_{err}\}$

The translational error of the odometry is shown to be much less when the robot is moving along its $x$-axis than when the robot is moving transversely. The confidence interval has its widest range, when the robot is heading in $\varphi = \frac{\pi}{4}$ direction.

A contour plot of the results in the translational workspace of the robot is presented in Fig. 2. The effect of different maximum velocities of the robot is represented in different radius of a polar plot, *i.e.* in different directions of movement the radial values alongside the chord of the polar contour plots represent the amount of the error in that direction when the robot moves with the corresponding maximum velocity proportional to the length of the chord (max $0.75\,\frac{m}{s}$).

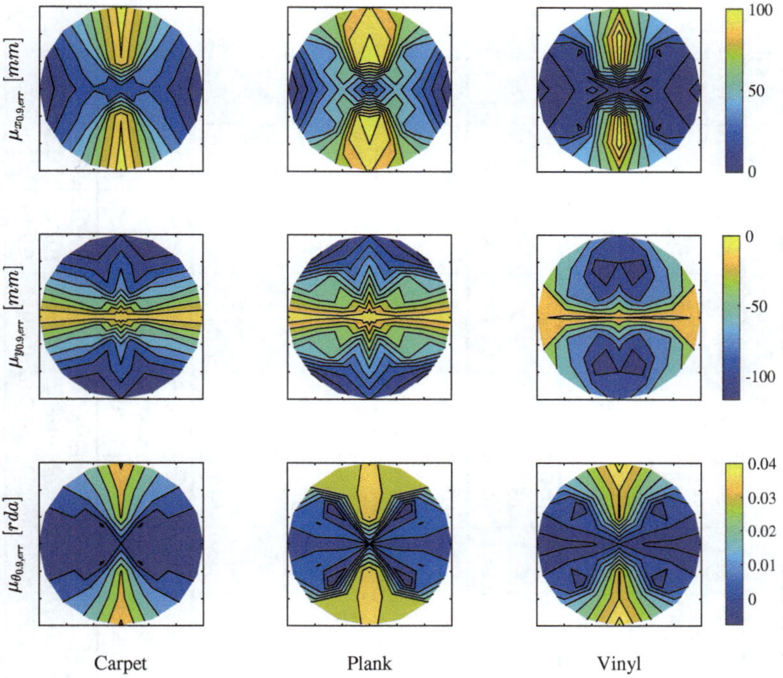


**Fig. 2.** $\mu_{\chi,0.9}$: The mean value of the 90% confidence interval of the variable $\chi \in \{x_{err},\, y_{err},\, \theta_{err}\}$

The modeled data can be used in the feed forward pose control of the holonomic mobile robot. Consider the vector $\boldsymbol{\Omega}$ representing the vector of control variables of a 4-wheel-drive robot in joint-space:

$$\boldsymbol{\Omega} = [\omega_1, \omega_2, \omega_3, \omega_4]^T . \tag{7}$$

The jacobian matrix $\boldsymbol{J}$ (varies depending on the kinematic properties of the robot) projects the joint-space control variables $\boldsymbol{\Omega}$ to the work-space control variables $\boldsymbol{v}$. The functions $\mathcal{F}$ and $\mathcal{K}$ can be introduced such that:

$$\boldsymbol{\zeta} = \mathcal{F}(\boldsymbol{v}), \tag{8a}$$

$$\boldsymbol{v} = \mathcal{K}(\boldsymbol{\zeta}). \tag{8b}$$

A table lookup block composed of the presented results from Table 1 would be introduced with inputs $\mathcal{S}$ (introducing the level of friction or surface type: $\mathcal{S} \in \{\text{carpet}, \text{vinyl}, \text{plank}\}$), $\varphi$ and the velocity. Feeding the read $\boldsymbol{\zeta}_{err}$ into the function $\mathcal{K}$ results in a compensating velocity vector $\boldsymbol{v}_{comp}$ which shall contribute in reduction of the expected amount of the odometric error.

**Table 1.** Statistical conclusion of odometric error of mecanum wheeled mobile robot

| | | | $0.25\ \frac{m}{s}$ | | | | | $0.5\ \frac{m}{s}$ | | | | | $0.75\ \frac{m}{s}$ | | | | |
|---|---|---|---|---|---|---|---|---|---|---|---|---|---|---|---|---|---|
| | | | 0 | $\frac{\pi}{8}$ | $\frac{2\pi}{8}$ | $\frac{3\pi}{8}$ | $\frac{4\pi}{8}$ | 0 | $\frac{\pi}{8}$ | $\frac{2\pi}{8}$ | $\frac{3\pi}{8}$ | $\frac{4\pi}{8}$ | 0 | $\frac{\pi}{8}$ | $\frac{2\pi}{8}$ | $\frac{3\pi}{8}$ | $\frac{4\pi}{8}$ |
| Odometric error $x_{err}$ $[mm]$ | Carpet | $\mu_{x_{err,LCI}}$ | 16.0 | 31.7 | 12.8 | 103.0 | 136.0 | -9.09 | 7.49 | -4.02 | 83.0 | 139.0 | -34.7 | -21.9 | 0.303 | 75.3 | 147.0 |
| | | $\mu_{x_{err,0.9}}$ | 16.9 | 33.4 | 30.8 | 105.0 | 139.0 | -8.3 | 9.87 | 29.0 | 90.0 | 144.0 | -33.8 | -20.9 | 24.4 | 77.0 | 153.0 |
| | | $\mu_{x_{err,UCI}}$ | 17.8 | 35.1 | 48.8 | 108.0 | 143.0 | -7.52 | 12.2 | 62.1 | 97.0 | 149.0 | -32.9 | -20.0 | 48.5 | 78.7 | 159.0 |
| | plank | $\mu_{x_{err,LCI}}$ | 33.4 | 45.0 | 20.6 | 86.9 | 88.0 | 7.14 | 24.4 | -4.82 | 76.8 | 87.3 | -21.3 | -3.97 | 13.9 | 55.1 | 79.1 |
| | | $\mu_{x_{err,0.9}}$ | 34.1 | 46.5 | 50.0 | 90.4 | 91.2 | 7.94 | 26.5 | 23.0 | 79.9 | 90.1 | -20.8 | -2.64 | 45.0 | 61.9 | 81.3 |
| | | $\mu_{x_{err,UCI}}$ | 34.8 | 48.0 | 79.3 | 93.9 | 94.3 | 8.74 | 28.5 | 50.9 | 82.9 | 92.8 | -20.3 | -1.3 | 76.1 | 68.6 | 83.6 |
| | vinyl | $\mu_{x_{err,LCI}}$ | 24.7 | 36.2 | 22.7 | 63.9 | 103.0 | 0.806 | 14.9 | -17.6 | 57.1 | 102.0 | -0.344 | 4.45 | 58.7 | 50.3 | 80.6 |
| | | $\mu_{x_{err,0.9}}$ | 25.3 | 37.5 | 53.1 | 70.4 | 105.0 | 1.26 | 16.4 | 7.47 | 61.3 | 105.0 | -0.911 | 1.7 | 33.6 | 48.7 | 77.3 |
| | | $\mu_{x_{err,UCI}}$ | 26.0 | 38.7 | 83.4 | 76.8 | 106.0 | 1.71 | 17.8 | 32.5 | 65.4 | 108.0 | -1.48 | -1.04 | 8.54 | 47.1 | 74.0 |
| Odometric error $y_{err}$ $[mm]$ | Carpet | $\mu_{y_{err,LCI}}$ | -7.45 | -81.0 | -97.6 | -137.0 | -101.0 | -6.49 | -83.7 | -156.0 | -157.0 | -128.0 | -7.17 | -82.8 | -179.0 | -182.0 | -162.0 |
| | | $\mu_{y_{err,0.9}}$ | -3.07 | -76.5 | -80.5 | -133.0 | -101.0 | -3.15 | -77.6 | -118.0 | -154.0 | -127.0 | -4.17 | -80.6 | -154.0 | -180.0 | -160.0 |
| | | $\mu_{y_{err,UCI}}$ | 1.31 | -72.1 | -63.4 | -129.0 | -99.6 | 0.195 | -71.5 | -80.4 | -151.0 | -125.0 | -1.17 | -78.5 | -128.0 | -179.0 | -158.0 |
| | plank | $\mu_{y_{err,LCI}}$ | -9.46 | -37.7 | -65.4 | -60.4 | -34.0 | -6.48 | -54.1 | -72.1 | -84.3 | -64.5 | -4.87 | -54.1 | -147.0 | -109.0 | -90.5 |
| | | $\mu_{y_{err,0.9}}$ | -7.03 | -34.2 | -35.2 | -59.0 | -33.1 | -4.94 | -49.2 | -44.2 | -82.8 | -63.9 | -0.992 | -51.3 | -113.0 | -105.0 | -89.7 |
| | | $\mu_{y_{err,UCI}}$ | -4.61 | -30.7 | -5.13 | -57.6 | -32.1 | -3.41 | -44.3 | -16.2 | -81.4 | -63.3 | 2.89 | -48.4 | -79.2 | -101.0 | -89.0 |
| | vinyl | $\mu_{y_{err,LCI}}$ | -0.0671 | -58.7 | -124.0 | -96.9 | -75.5 | -0.676 | -66.1 | -97.7 | -119.0 | -101.0 | 3.51 | -13.1 | -20.2 | -55.5 | -92.9 |
| | | $\mu_{y_{err,0.9}}$ | 4.18 | -56.0 | -90.5 | -94.1 | -74.5 | 5.14 | -62.7 | -72.3 | -117.0 | -99.5 | -3.34 | -18.8 | -47.3 | -57.0 | -93.5 |
| | | $\mu_{y_{err,UCI}}$ | 8.43 | -53.3 | -57.1 | -91.4 | -73.5 | 10.9 | -59.4 | -47.0 | -115.0 | -98.4 | -10.2 | -24.5 | -74.4 | -58.6 | -94.1 |
| Odometric error $\theta_{err}$ $[rad]$ | Carpet | $\mu_{\theta_{err,LCI}}$ | -0.00945 | -0.0017 | -0.0278 | 0.0186 | 0.0398 | -0.00717 | -0.00258 | -0.0222 | 0.0195 | 0.042 | -0.0055 | -0.00372 | -0.00968 | 0.0212 | 0.0481 |
| | | $\mu_{\theta_{err,0.9}}$ | -0.00769 | $-7.28\ 10^{-4}$ | -0.0114 | 0.0199 | 0.0414 | -0.00641 | $-8.28\ 10^{-4}$ | 0.00184 | 0.0212 | 0.0437 | -0.00474 | -0.00264 | 0.00999 | 0.0229 | 0.0518 |
| | | $\mu_{\theta_{err,UCI}}$ | -0.00593 | $2.48\ 10^{-4}$ | 0.00503 | 0.0212 | 0.0431 | -0.00564 | $9.2\ 10^{-4}$ | 0.0259 | 0.023 | 0.0454 | -0.00397 | -0.00156 | 0.0297 | 0.0246 | 0.0556 |
| | plank | $\mu_{\theta_{err,LCI}}$ | -0.00451 | 0.00115 | -0.0255 | 0.0206 | 0.0288 | -0.00321 | 0.00381 | -0.0321 | 0.022 | 0.0269 | -0.00532 | 0.00229 | -0.00727 | 0.0201 | 0.0241 |
| | | $\mu_{\theta_{err,0.9}}$ | -0.00353 | 0.00267 | -0.00351 | 0.0221 | 0.03 | -0.00257 | 0.00519 | -0.00979 | 0.0231 | 0.0279 | -0.0042 | 0.00351 | 0.0205 | 0.0225 | 0.0259 |
| | | $\mu_{\theta_{err,UCI}}$ | -0.00254 | 0.0042 | 0.0185 | 0.0236 | 0.0313 | -0.00194 | 0.00657 | 0.0125 | 0.0241 | 0.0289 | -0.00307 | 0.00473 | 0.0484 | 0.0249 | 0.0276 |
| | vinyl | $\mu_{\theta_{err,LCI}}$ | -0.00731 | -0.00127 | -0.0142 | 0.0185 | 0.0348 | -0.00804 | $6.03\ 10^{-5}$ | -0.0262 | 0.0215 | 0.0338 | -0.0103 | -0.00155 | -0.00888 | 0.0246 | 0.0405 |
| | | $\mu_{\theta_{err,0.9}}$ | -0.00593 | $-3.93\ 10^{-5}$ | 0.00826 | 0.0203 | 0.0358 | -0.00638 | 0.00112 | -0.00427 | 0.0228 | 0.0353 | -0.0079 | $-1.39\ 10^{-5}$ | 0.0129 | 0.0254 | 0.0422 |
| | | $\mu_{\theta_{err,UCI}}$ | -0.00454 | 0.00119 | 0.0307 | 0.022 | 0.0369 | -0.00472 | 0.00217 | 0.0177 | 0.0242 | 0.0368 | -0.00551 | 0.00152 | 0.0346 | 0.0262 | 0.044 |

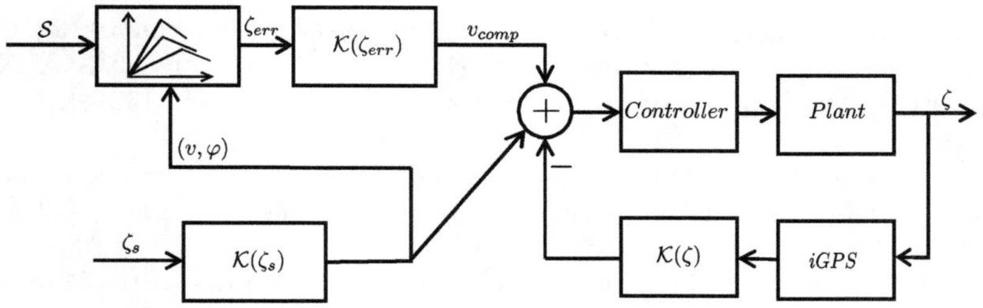


**Fig. 3.** Feed forward control scheme for odometric error compensation

A schematic plot of the control loop, involving feed forward compensation part, is presented in Fig. 3. In the block diagram of the Fig. 3, the vector $\boldsymbol{\zeta}$ introduces the set value of the pose vector $\boldsymbol{\zeta}$. The blocks "*Controller*" and "*Plant*" represent the "*pose controller*" and "*mobile robot*" respectively.

## 4 Conclusion

The statistical results of the indoor experiments aiming to investigate the odometric errors of s mecanum wheeled mobile robot have been presented in this paper. A feed forward pose control scheme has been introduced for reducing the odometric errors of the mobile robot. The presented motion control loop contributes to higher accuracy in position control and localization of the mobile robot.

**Acknowledgement.** The Authors would like to thank for the kind support of German Research Foundation DFG (Deutsche Forschungsgemeinschaft) under Germany's Excellence Strategy – EXC-2023 Internet of Production – 390621612.

## References

1. Belter, D., Nowicki, M., Skrzypczyński, P.: Improving accuracy of feature-based RGB-D SLAM by modeling spatial uncertainty of point features. In: 2016 IEEE International Conference on Robotics and Automation (ICRA), pp. 1279–1284. IEEE (2016)
2. Dellaert, F., Fox, D., Burgard, W., Thrun, S.: Monte Carlo localization for mobile robots. In: Proceedings 1999 IEEE International Conference on Robotics and Automation (Cat. No. 99CH36288C), vol. 2, pp. 1322–1328. IEEE (1999)
3. Depenthal, C.: Path tracking with IGPS. In: 2010 International Conference on Indoor Positioning and Indoor Navigation, pp. 1–6. IEEE (2010)
4. Engel, J., Schöps, T., Cremers, D.: LSD-SLAM: large-scale direct monocular SLAM. In: European Conference on Computer Vision, pp. 834–849. Springer (2014)
5. Ganganath, N., Leung, H.: Mobile robot localization using odometry and kinect sensor. In: 2012 IEEE International Conference on Emerging Signal Processing Applications, pp. 91–94. IEEE (2012)

6. Handa, A., Whelan, T., McDonald, J., Davison, A.J.: A benchmark for RGB-D visual odometry, 3D reconstruction and slam. In: 2014 IEEE INTERNATIONAL CONFERENCE on Robotics and Automation (ICRA), pp. 1524–1531. IEEE (2014)
7. Kiriy, E., Buehler, M.: Three-state extended Kalman filter for mobile robot localization. McGill University, Montreal, Canada, Technical report TR-CIM **5**, 23 (2002)
8. Krishnamoorthy, K.: Handbook of Statistical Distributions with Applications. CRC Press, Boca Raton (2016)
9. Lee, J.S., Chung, W.K.: Robust mobile robot localization in highly non-static environments. Auton. Robot. **29**(1), 1–16 (2010)
10. Martinelli, A.: Modeling and estimating the odometry error of a mobile robot. IFAC Proc. Vol. **34**(6), 407–412 (2001)
11. Martinelli, A.: The odometry error of a mobile robot with a synchronous drive system. IEEE Trans. Robot. Autom. **18**(3), 399–405 (2002)
12. Martinelli, A., Tomatis, N., Siegwart, R.: Simultaneous localization and odometry self calibration for mobile robot. Auton. Robot. **22**(1), 75–85 (2007)
13. Meng, Q., Bischoff, R.: Odometry based pose determination and errors measurement for a mobile robot with two steerable drive wheels, pp. 263–282. Springer (2005)
14. Nicolai, A., Skeele, R., Eriksen, C., Hollinger, G.A.: Deep learning for laser based odometry estimation. In: RSS Workshop Limits and Potentials of Deep Learning in Robotics, vol. 184 (2016)
15. Nikonmetrology: iGPS for dynamic tracking and alignment, 21 July 2020. www.nikonmetrology.com/images/brochures/ispace-dynamic-tracking-en.pdf
16. Palacin, J., Valganon, I., Pernia, R.: The optical mouse for indoor mobile robot odometry measurement. Sens. Actuators Phys. **126**(1), 141–147 (2006)
17. Santos, J.M., Portugal, D., Rocha, R.P.: An evaluation of 2D SLAM techniques available in robot operating system. In: 2013 IEEE International Symposium on Safety, Security, and Rescue Robotics (SSRR), pp. 1–6. IEEE (2013)
18. Smieszek, M., Dobrzanska, M., Dobrzanski, P.: The impact of load on the wheel rolling radius and slip in a small mobile platform. Auton. Robot. **43**(8), 2095–2109 (2019)
19. Team, R Core and others: A language and environment for statistical computing (2013)
20. Thrun, S.: Probabilistic Robotics, vol. 45. ACM, New York (2002)
21. Xue, Z., Schwartz, H.: A comparison of several nonlinear filters for mobile robot pose estimation. In: 2013 IEEE International Conference on Mechatronics and Automation, pp. 1087–1094. IEEE (2013)

# Usability of a Compliant Deltoid Mechanism as a Motion Transmission in Civil Engineering Machinery

Philip Johannes Steinbild[1(✉)], Peilin Zhang[1], Karl-Heinz Modler[2], Andreas Borowski[1], Marco Zichner[1], Niels Modler[1], and Martin Dannemann[1]

[1] Institute of Lightweight Engineering and Polymer Technology, Technische Universität Dresden, Dresden, Germany
philip_johannes.steinbild@tu-dresden.de
[2] Institute of Solid Mechanics, Technische Universität Dresden, Dresden, Germany

**Abstract.** Recent advances in battery and actuator technologies enable their use in various small- and mid-sized devices in civil engineering. The principle of operation of most devices in civil engineering is based on rotary motions originating from combustion engines that are converted to linear motions via slider cranks. These relatively high-revving combustion engines require the use of a transmission in order to achieve lower frequencies on the output which is a necessity in most applications, e.g. for ground compacting vibrating tampers, which need tamping frequencies of about 16 Hz. The use of linear actuators or plunger coils bear the potential of reducing the number of parts while enhancing the efficiency. For the use of linear actuators different mechanisms can be utilized. This paper describes the classical slider crank for converting rotary into linear motions and two mechanisms for use with linear actuators or plunger coils, the double slider mechanism and the deltoid mechanism. All three mechanisms are described and analyzed. The deltoid mechanism is selected to be used in a demonstrator. The mechanism's links and hinges are simplified using a compliant structure design. This compliant design and its construction and calculation process is described. The demonstrator is then built up and tested measuring its linear motion optically.

**Keywords:** Deltoid mechanism · Plunger coil · Linear actuator · Compliant structure

## 1 Introduction and State of the Art

In civil engineering, the used ground material needs to be compacted to enhance the mechanicals properties of subsoil or road surfaces. State of the art machinery used for compaction of such ground materials in small and narrow environments are vibrating tampers powered by combustion engines (e.g. Fig. 1). Due to future emission regulations and for health and safety reasons, the substitution of combustion engines with electric motors in civil engineering is a rational measure. The use of linear electric


E.-C. Lovasz et al. (Eds.): MTM&Robotics 2020, MMS 88, pp. 135–154, 2021.
https://doi.org/10.1007/978-3-030-60076-1_13

motors or plunger coils bear the potential to simplify the construction by using simpler rigid-body mechanisms or compliant mechanisms. Most vibrating tampers utilize a classical slider crank mechanism to convert the rotary motion of the combustion engine into a linear tamping motion. To prevent the engine from experiencing high forces and thus potential defects due to direct collision with the ground material, the engine is decoupled from the tampers foot using constructive measures involving springs or dampers. The aforementioned compliant mechanisms can also decouple or dampen depending on the used material and their shape.

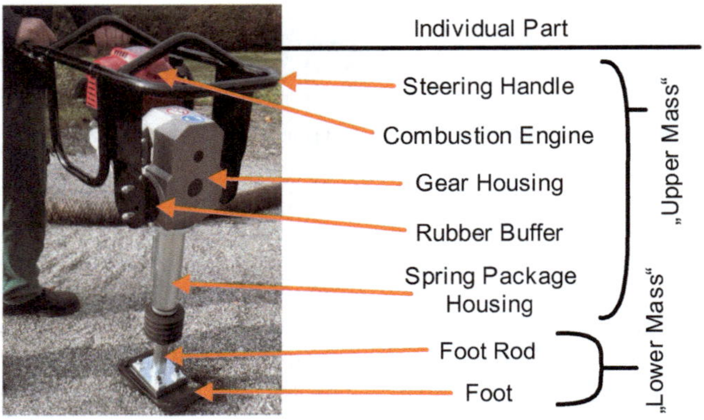


**Fig. 1.** Small vibrating tamper for use in small and narrow environments.

Rigid-body mechanisms, as shown in Sect. 2, consist of rigid links and movable joints. Its motion is composed of rigid body translations and rotations. Compliant mechanisms, however, generate some of their mobility from the deflection of flexible parts rather than from moveable joints only. Compliant mechanisms can be classified in two main categories (compare Fig. 2), mechanisms with distributed compliance and with concentrated compliance which are described and discussed in [1].

Other sources do not differ between distributed and concentrated compliance. Following [2], compliant mechanisms consist of special solid-state elements that are designed to feature high stiffness and high compliance, depending on the application and the functional aim. The properties of links and hinges become indistinct. Links are the parts of a compliant mechanism whose segments are firmly or rigidly bonded together, only allowing their flexibility through deformation [2]. In both cases, however, compliance comprises elastic and plastic as well as viscoelastic and viscoplastic deformability.

Detailed research on compliant mechanisms is a relatively young branch in engineering science. The base for most available approaches today are the works by Bisshop and Drucker from 1945 who described beams under high deformation. Selected validations, variations, and further development of the approach described by Bisshop and Drucker can be e.g. found in [3–11].

With the development of the finite element method (FEM) computer-assisted approaches for the synthesis of compliant mechanisms could be used. Most FEM-based approaches for compliant mechanisms employ parametric models optimizing positions of links and stress of the system iteratively. Several works on the use of FEM within the development of compliant mechanisms give a good overview over the subject [12–15, 17, 21–26]. The rising level of understanding since about 1970 made for analysis of potential applications in engineering as described in [2, 16–21].

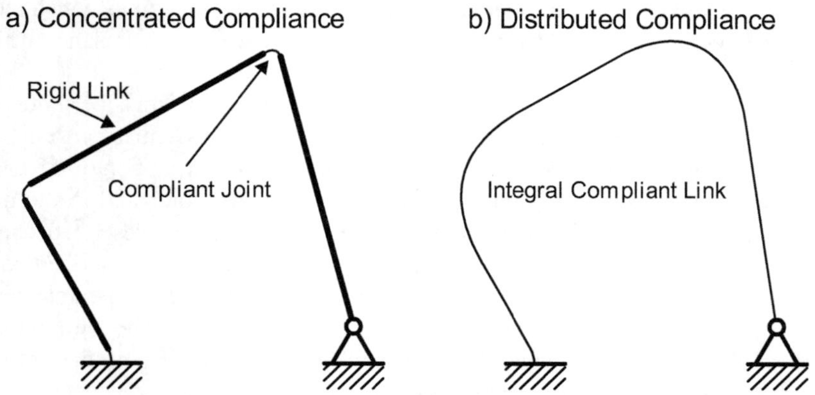


**Fig. 2.** Classification of compliant mechanisms: concentrated compliance (a) and distributed compliance (b) [2, 16, 27].

The field of analysis and precision synthesis of rigid body mechanisms developed parallel to compliant mechanisms [28–30], however these developments attracted interest in the field of compliant mechanisms only very slightly. In fact, iterative and uncertainty approaches are still state of the art regarding the analysis and synthesis of compliant mechanisms as described, for example, in [8, 19, 31, 32]. A new, much-noticed approach was then formulated by Howell and Midha [33], called Pseudo-Rigid-Body (PRB)-Model, which marked a new development in the field of compliant mechanisms. Approximating the compliant element using small length flexural pivots enabled the transfer of well-established methods of the synthesis of rigid body mechanisms to the synthesis of compliant mechanisms. In the following years Howell summarized and worked up the approach of PRB-Model in [34] and formulated it suitable for engineers. This practical approximation method leads to a nearly independent field of research which until today features more than 2,000 papers and works showing extensions, variations, and applications around the PRB-Method, e.g. [35–38]. The tremendous success of the PRB-Method shows the importance of simple approaches and formulations of the complex motion behavior of compliant mechanisms and the dynamics of the field of research on the interface between the theory of mechanisms, mathematics, and material science.

The application of topology optimization enables relatively simple, yet good and efficient design of compliant mechanisms [39–44]. The rapid development of additive manufacturing allows for fast validation of new development techniques for compliant mechanisms [45].

Compliant mechanisms gain their mobility through elastic deformation of flexible elements during deflection. Theoretically, their deformation pattern can be determined exactly by carefully choosing the shape and material of the elements. A big advantage compared to rigid body mechanisms is the reduction of parts in a mechanism and with it the reduction of the overall mass, also lowering the assembly effort and cost. Regarding applications in mechanical engineering, compliant mechanisms can be designed free of play, thus enabling exact motion. Additionally, compliant mechanisms do not require frictional contact elements or any form of movable joints which reduces wear in the system. These advantages also reduce the need for maintenance which is especially helpful for mechanisms that are hard to access.

The present study describes three classical rigid-body mechanisms, the slider crank, the double slider, and the deltoid mechanism. The deltoid mechanism is suitable for converting the linear motion of e.g. a plunger coil into the linear motion of the tampers foot while requiring relatively simple constructive measures compared to the double slider mechanism. A compliant deltoid mechanism is suggested and described, showing a concept and rough calculations as well as a FEM calculation. The described concept is then manufactured using additive manufacturing and polyamide as a flexible material for the deltoid. Optical measurements are taken to be able to describe the motion and deflection of the compliant deltoid mechanism. The applicability of the compliant deltoid mechanism in civil engineering machinery is discussed in the conclusion.

## 2 Classical Rigid-Body Mechanisms

Rigid-body mechanisms are the classical way to transfer or convert motions. In this section three rigid-body mechanisms are described and analyzed. The slider crank mechanism can be used to convert a rotational motion into a linear motion, whereas the other two examples, the double slider and the deltoid mechanism, can be used to transfer linear motions to linear motions of other directions.

### 2.1 Slider Crank

One of many classical mechanisms used to convert a rotational motion into a linear motion or vice versa is the slider crank mechanism (Fig. 3). A main advantage is the great potential for conveying high speeds and forces with relatively low constructional effort. The most popular example being the piston based combustion engine, where the piston (link 4, Fig. 3) moves up and down inside of the cylinder, transmitting the force ($f_{41}$) over the connection rod (link 3 with length $l_3$) to the crank shaft (link 2, $l_2$), producing a moment ($M_{21}$).

Equation (1) shows that the translational movement of link 4, $s$, follows no sinusoidal motion (with $\varphi$ being the rotation angle). Thus, the slider crank cannot deliver a perfect sinusoidal oscillation.

$$s = l_2 \cdot \cos\varphi + \sqrt{l_3^2 - l_2^2 . \sin^2\varphi} \tag{1}$$

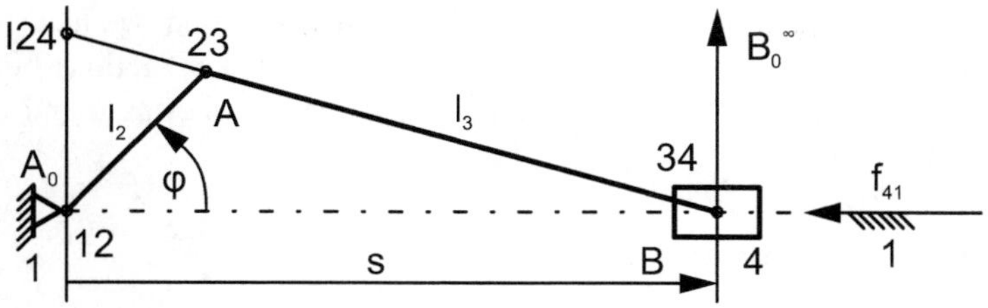


**Fig. 3.** Mechanical representation of a slider crank.

Through implicit differentiation of Eq. (1) with respect to the differentiation variable $\varphi$ and clever rearrangement, $s'$ and $s''$ can be formulated like in Eqs. (2) and (3).

$$s' = \frac{s \cdot l_2 \cdot \sin\varphi}{l_2 \cdot \cos\varphi - s} \tag{2}$$

$$s'' = \frac{(s')^2 + 2 \cdot s' \cdot l_2 \cdot \sin\varphi + s \cdot l_2 \cdot \cos\varphi}{l_2 \cdot \cos\varphi - s} \tag{3}$$

The velocity of the slider (link 4) relative to the frame (link 1), $v_{41}$, resembles to Eq. (4):

$$\dot{s} = v_{41} \tag{4}$$

and thus follows Eq. (5), where $\overline{12 - I24}$ is the distance between the joint 12 and the instantaneous center of rotation I24 (scalar):

$$v_{41} = \omega_{21} \cdot \overline{12 - I24}. \tag{5}$$

Thus, the relation of the angular velocity ($\dot{\varphi} = \omega_{21}$) to the velocity of the slider $v_{41}$ only depends on the position of the instantaneous center of rotation *I24*. Large coupling lengths $l_3$ provide small lengths for the distance $\overline{12 - I24}$. From the principle of virtual power with $f_{41}$ being the force acting on the slider (link 4) and $M_{21}$ being the moment acting on link 2, Eq. (6) can be derived:

$$0 = M_{21} \cdot \omega_{21} + f_{41} \cdot v_{41}. \tag{6}$$

With Eqs. (5) and (6), Eq. (7) can be formulated as

$$f_{41} = -M_{21} \cdot \frac{\omega_{21}}{v_{41}} = -\frac{M_{21}}{\overline{12 - I24}}. \tag{7}$$

In stretched position of $l_2$ and $l_3$ (maximum travel distance) point *12* equals *I24* and thus the slider crank shows a knee lever effect.

The slider crank mechanism converts a rotary motion into an oscillating translational motion or vice versa. With regard to the recent advances of battery and linear actuator technology mentioned in Sect. 1, in the following section two mechanisms are described which are able to transmit linear motions into linear motions but bear the potential of better damping or other functionalities between the input and the output side of the mechanism.

## 2.2 Double Slider Mechanism

The double slider mechanism (Fig. 4) is able to transmit a linear motion at the input (link 2 with velocity $v_{21}$ and force $f_{21}$) into a linear motion at the output (link 4, $v_{41}, f_{41}$).

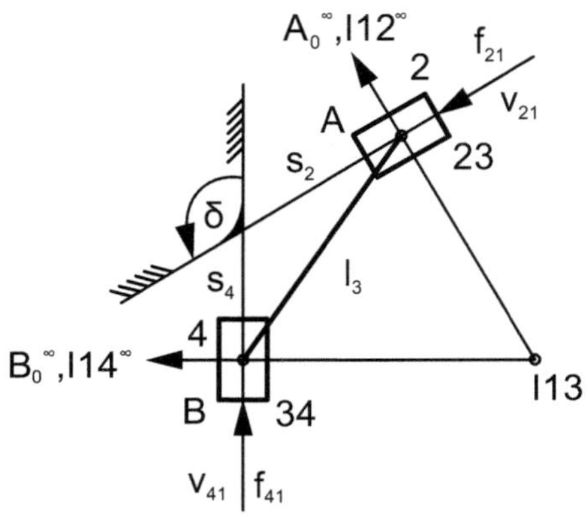


**Fig. 4.** Mechanical representation of a double slider mechanism.

The relation between the drive path on the input $s_2$ and the drive path on the output $s_4$ is given in Eq. (8) with respect to the coupling length $l_3$ and the angle between $s_2$ and $s_4$:

$$l_3^2 = s_2^2 + s_4^2 - 2 \cdot s_2 \cdot s_4 \cdot \cos\delta. \tag{8}$$

The instantaneous center of rotation *I13* and the joints *23* and *34* are forming the pole distances $\overline{I13-23}$ and $\overline{I13-34}$. The relation between the linear velocities ($v_{21}$ and $v_{41}$) and the rotational velocity of link 3 relative to the frame ($w_{31}$) can be described as in Eq. (9):

$$\frac{v_{21}}{w_{31}} = \overline{I13-23}, \frac{v_{41}}{w_{31}} = \overline{I13-34}. \tag{9}$$

Thus, the relation between the velocities $v_{41}$ and $v_{21}$ can be described as in Eq. (10):

$$\frac{v_{41}}{v_{21}} = \frac{\overline{I13-34}}{\overline{I13-23}}. \tag{10}$$

From the principle of virtual power Eq. (11) can be derived:

$$0 = f_{21} \cdot v_{21} + f_{41} \cdot v_{41}, \tag{11}$$

which describes the relation between the input force and the output force. The input force $f_{41}$:

$$f_{41} = f_{21} \cdot \frac{v_{21}}{v_{41}} = f_{21} \cdot \frac{\overline{I13-23}}{\overline{I13-34}}. \tag{12}$$

is large for small values of $\overline{I13-34}$ in relation to $\overline{I13-23}$. A knee lever effect ($f_{41} \rightarrow \infty$) can be achieved when the instant center of rotation are the same ($I13 = 34$).

## 2.3 Deltoid Mechanism

In a deltoid, the two diagonals are always perpendicular to each other. With respect to one diagonal, the deltoid is mirror-symmetrical. In one special case, all four sides are of the same length. For the mechanical application, the general deltoid consists of two mirror-symmetrical slider cranks (deltoid or kite mechanism, Fig. 5).

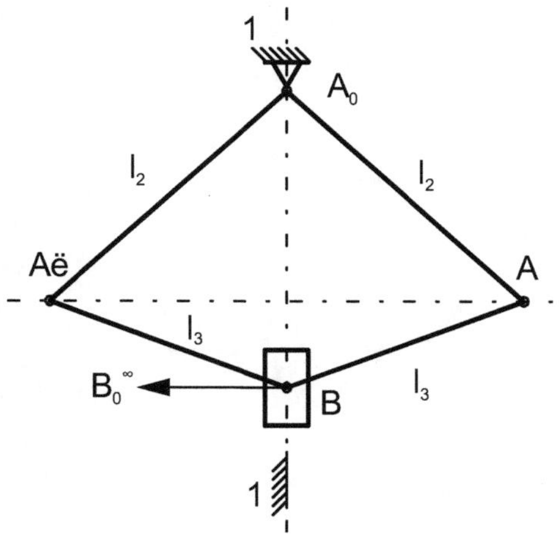


**Fig. 5.** Mechanical representation of the deltoid mechanism.

At points $A_0$ and $B$ double hinges are used, since three transmission links are joined in one point. From the point of view of mechanics, the deltoid mechanism is a two-stage mechanism consisting of six links. The degree of freedom $F$ equals 1. For the mathematical relations only the right side relating to the symmetrical axis $A_0B$ is observed (Fig. 6). The deltoid mechanism can be actuated by applying a force to the point $B$, $A$, or $A'$ respectively ($f_{21}$ or $f_{41}$ in Fig. 6). In the case that is discussed in this section, the actuation takes most likely place at point $B$ (link 4).

For the travel of link 4, $h$, Eq. (13) applies:

$$h = x_2 + x_3 \tag{13}$$

The drive path on the input is $2\Delta$ (compare Fig. 6). With the basic Eqs. (14) and (15):

$$l_2^2 = a^2 + b^2,\ l_3^2 = c^2 + b^2 \tag{14}$$

$$l_2^2 = (a + x_2)^2 + (b - \Delta)^2,\ l_3^2 = (c + x_3)^2 + (b - \Delta)^2, \tag{15}$$

and Eq. (13), the following relations can be described (Eq. (16)):

$$0 = x_2^2 + 2 \cdot a \cdot x_2 + \Delta^2 - 2 \cdot b \cdot \Delta$$

$$0 = x_2^2 - 2 \cdot (h + c) + 4 \cdot (h + 2 \cdot c) + \Delta^2 - 2 \cdot b \cdot \Delta. \tag{16}$$

The difference of the two equations in Eq. (16) resolves to Eq. (17):

$$0 = 2 \cdot (a + c + h) \cdot x_2 - 4 \cdot (h + 2 \cdot c). \tag{17}$$

Thus, $x_2$ can be calculated Eq. (18):

$$x_2 = \frac{2 \cdot (h + 2 \cdot c)}{a + c + h}. \tag{18}$$

Using $x_2$ from Eq. (18), $\Delta$ from Eq. (16) can be calculated. For the calculation of the output force $f_{41}$ Eq. (12) still applies.

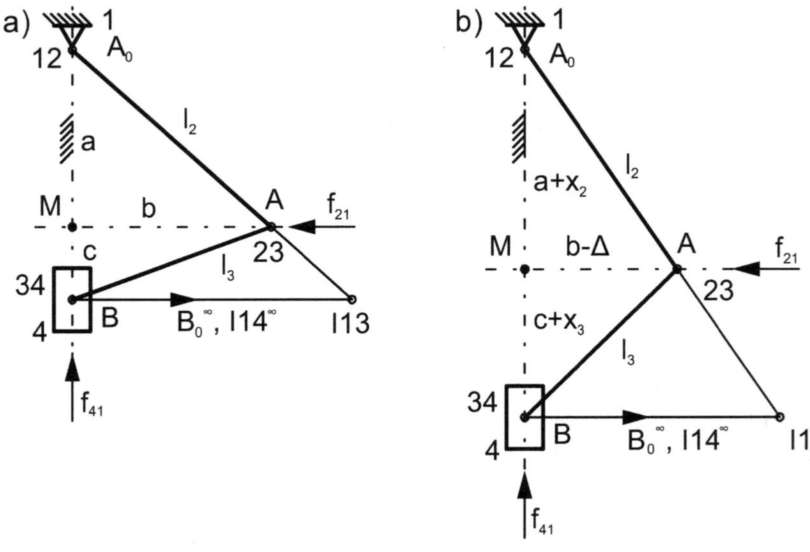


**Fig. 6.** Start (a) and end (b) position of one half of the deltoid mechanism for the mathematical description.

### 2.4 Discussion

For the application of a mechanism in a vibrating tamper (e.g. Fig. 1) an alternative to the use of a rotational motor might show various benefits. The approach in this paper is the substitution of the rotational motor employing a linear actor of any kind. In this chapter the mechanism 'slider crank' was described followed by the 'double slider mechanism' and a derivative, the 'deltoid mechanism'. Generally speaking, the double slider mechanism can convey the same motion and forces as the deltoid mechanism. From a construction point of view, a compliant variant of the deltoid mechanism is simpler to develop than a compliant variant of the double slider mechanism since it is generally more stable due to its symmetrical shape. Thus the deltoid mechanism is used in the following sections as the basis for the demonstrator that is the aim of this paper.

## 3 Compliant Deltoid Mechanism Concept for a Vibrating Tamper Demonstrator

This section describes the construction of a demonstrator for the deltoid mechanism described in Sect. 2.3 including the motivation, requirements, the concept and calculations as well as a simple Finite Element Model. The classical rigid-body deltoid mechanism from Sect. 2.3 is seen as the basis for a compliant version of the mechanism. Thus, rough calculations are made using geometric relations as in rigid-body mechanisms and the equations from Sect. 2.3. The calculation's results are used to design a concept for the compliant deltoid mechanism. A corresponding FE model is built up just to show the deformation of the compliant mechanism qualitatively without taking large deformations into account.

### 3.1 Motivation and Requirements

The aim of these investigations is to analyze the applicability of a compliant mechanism in civil engineering machinery. As the practical example for these investigations a vibrating tamper is used for comparison (Fig. 1). The tamper consists of the upper mass and lower mass, which are decoupled by the spring package inside the spring package housing. The combustion engine delivers a rotating motion which is transmitted over a gearbox (gearbox in gear housing, Fig. 1) to a slider crank mounted inside the spring package housing. The slider crank converts the rotational motion into a linear motion. This linear motion is delivered to the foot of the tamper, which is compacting the ground material with every hit to the floor.

An electrically driven actuator, like e.g. a linear motor or a simple plunger coil is an efficient option to replace the combustion engine. However, the linear motor, just like the combustion engine, should not be mounted directly to the tamper's foot, since the ground might trigger high peaks in tamping force, which can lead to damage at the motor. Thus, a decoupling still needs to be a feature in the design. Also, for construction reasons regarding installation space, the motor should be mounted horizontally. This requirement calls for a mechanism that is able to convert a horizontal into a vertical motion.

The double slider mechanism (Sect. 2.2) is capable of converting a horizontal into a vertical motion but the design of such mechanisms can get very complex, especially when high forces need to be transmitted. Therefore, the deltoid mechanism (Sect. 2.3) is chosen to be the base for the demonstrator in this paper. Using a rigid-body deltoid mechanism would still need special means in terms of decoupling the upper mass against the lower mass. The deltoid mechanism demonstrator is therefore designed as a compliant mechanism, combining the advantageous motion converting capabilities with the option of using the compliant links as springs and thus a decoupling. Furthermore, the high damping properties of the typical materials used for compliant mechanisms, as e.g. glass fiber-reinforced composites can be utilized [46].

### 3.2 Compliant Deltoid Mechanism Concept, Material, and Process

The conceptual work on the compliant deltoid mechanism starts with taking the rigid-body deltoid mechanism into account. In Fig. 7 the rigid-body deltoid mechanism is shown in the starting and end position when deflected by a linear actuator or a plunger coil.

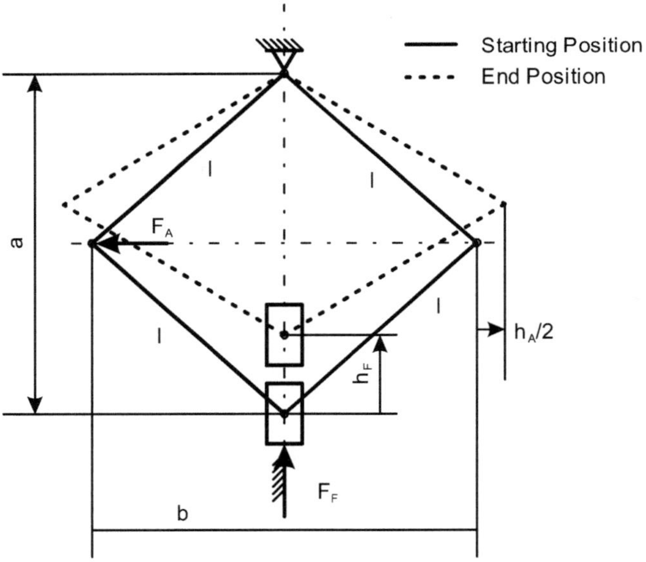


**Fig. 7.** Dimensions and main parameters of the concept for the draft of the compliant deltoid mechanism demonstrator at the rigid-body reference (compare Fig. 5 and Fig. 6).

The force $F_A$ is delivered by the actuator mounted between point A and A′ (compare Fig. 5). The displacement of the actuator is equal to the distance $h_A$. The motion of the actuator causes a travel of the foot $h_F$ (point B in Fig. 5). The transmitted force to the foot equals to $F_F$ ($F_{41}$ in Fig. 6). Using Eqs. (12) to (18) and given length l, distance a, and distance b, the parameters can be calculated as stated in Table 1.

**Table 1.** Parameters and calculated values for the base rigid-body deltoid mechanism (Fig. 7).

| Parameter | Value | Unit |
|---|---|---|
| l | 200 | mm |
| a | 346.42 | mm |
| b | 200 | mm |
| $h_A$ | 13 | mm |
| $F_A$ | 1000 | N |
| $h_F$ | 7.51 | mm |
| $F_F$ ($h_F$ = 0 mm to 8.61 mm) | 866 to 795.5 | N |

The transmission of the force delivered by the actuator depends on the current position of the mechanism and thus is not constant. The transmitted force $F_F$ in relation to the displacement of the foot $h_F$ is shown in Fig. 8. The graph shows that the maximum force is transmitted at the start of the deflection and has a minimum at the end position.

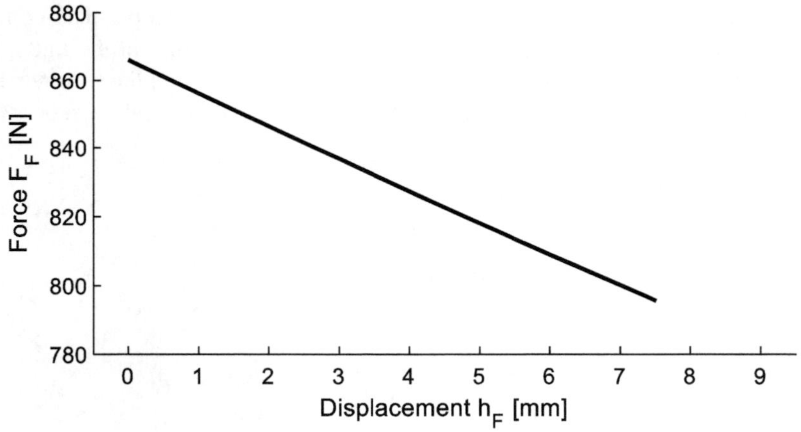


**Fig. 8.** Force at the deltoid's foot $F_F$ in relation to the foot displacement $h_F$.

The final concept of the compliant deltoid mechanism is shown in Fig. 9. The four rigid-body links of the classical deltoid mechanism are replaced and simplified through the use of two compliant elements. The deflection is still done by a linear actuator or plunger coil mounted in the middle of the structure. The two compliant elements for the demonstrator are identical in shape and dimension.

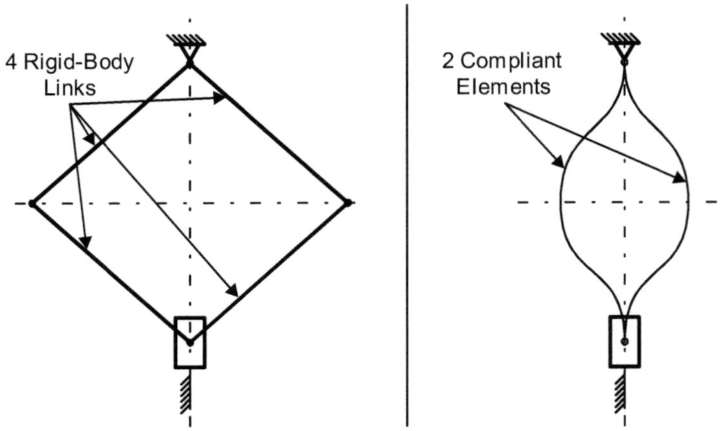


**Fig. 9.** Concept of the compliant deltoid mechanism demonstrator compared to the classical deltoid mechanism.

The design of the demonstrator is shown in Fig. 10. The top clamping is rigid while the bottom clamping allows translation in the up or down direction. The compliant elements are curved panels with a slot acting as a mounting point in the middle. The used actuator can be mounted in the middle between the two compliant elements. The height of the demonstrator is $a$ while the width is $b$ (Table 1). The wall thickness of one curved panel is $t$ = 2 mm.

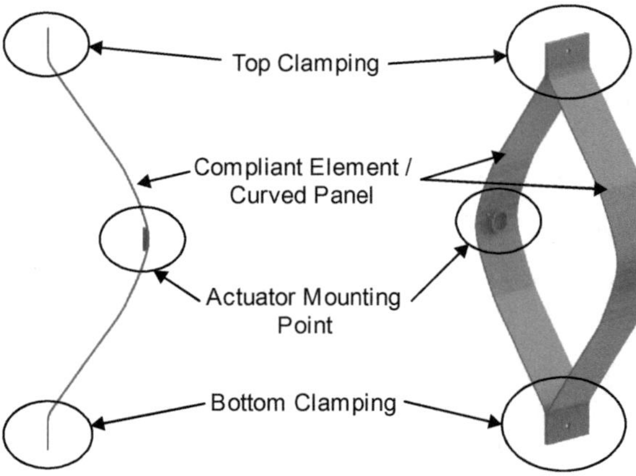


**Fig. 10.** Concept of the compliant deltoid mechanism demonstrator compared to the classical deltoid mechanism.

To realize a structure with a flexible behavior and a high resistance to fatigue, a polyamide material in a configuration featuring a high ultimate elongation was selected, the taulman Alloy 910 Nylon 6/69. The material has a good tensile strength value of around 56 MPa and an ultimate elongation of 32% [47]. For the manufacturing process a 3D printing process was selected.

The calculation of the deflection of the compliant elements is done using a finite element model described in the following section.

### 3.3 Finite Element Model of the Compliant Deltoid Mechanism

To approximate the deformation of the demonstrator a FE model was build up using Abaqus/CAE 2019 (Fig. 11 a).

The material properties used for the polyamide feature a Young's Modulus of E = 502.8 MPa and a Poisson's Ratio of $\nu = 0.34$. The displacement at the top is restricted while the translation in y-direction at the foot is free. At the mounting points a displacement of 6.5 mm per side is given as a loading. The mesh is made up of 3D linear tetrahedron (C3D4) elements. A static analysis taking large displacement into account ('Static, General') is calculated. The results of the FE model are shown in Fig. 11 b). A maximum displacement of 10.96 mm in y-direction at the foot was calculated.

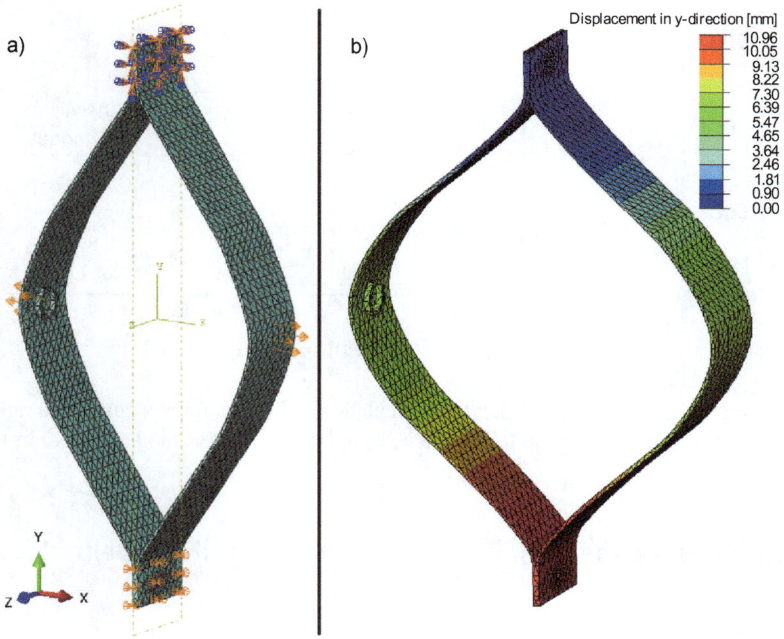


**Fig. 11.** Finite Element Model: a) Mesh, boundary conditions, and loading conditions and b) displacement in 2 (y)-direction [mm].

The overall deflection of the FE model of the demonstrator shows a higher displacement at the foot than the analytic calculations of the rigid-body mechanism in Sect. 3.2. This occurs due to the deflection of the entire structure.

To be able to compare the analytical calculation from Sects. 2.2, 2.3, or, 3.2 respectively, with the FE model, a second model was formulated, holding the structure in place by two boundary conditions and loading it with a force of 1000 N as depicted in Fig. 7 ($F_A$) for discrete foot displacements of 0 mm to 7.51 mm. The reaction force at the foot $F_F$ calculated using the FE model is shown in Fig. 12 alongside the already known values calculated using the analytical solution from Eq. (12).

Looking at Fig. 12 it becomes apparent that the forces calculated via the FE model are significantly lower than the theoretical values calculated using the analytical model (584.4 N at $h_F$ = 0 mm to 538.3 N at $h_F$ = 7.51 mm). This is mainly due to the flexural elements used in the FE formulation and the different shape of the two compliant elements compared to the simple rods used for the analytical model. It is to be noted that the values calculated using the FE model are purely theoretical since the force is not simply conveyed from the actuator mounting point to the foot (bottom clamping) but also deflects the whole structure. With further development a local enforcement of the compliant elements could increase the conveyed force to the foot.

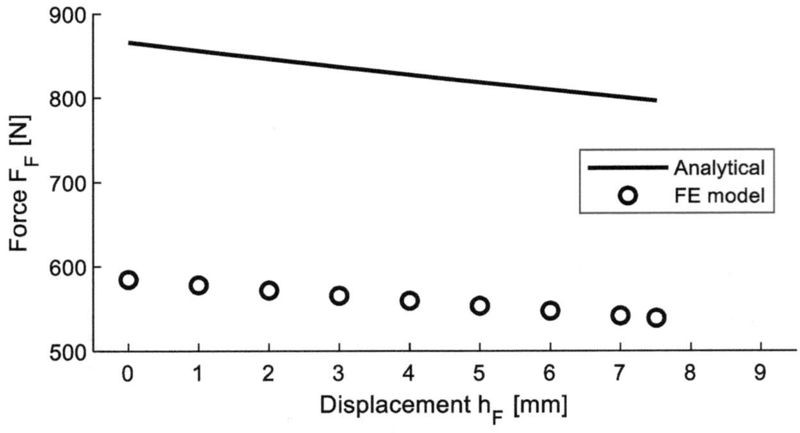


**Fig. 12.** Comparison between the analytical calculation of the foot force $F_F$ using the rigid-body mechanism from Sect. 3.2 and the foot force calculated using the FE model relative to the foot displacement $h_F$.

# 4 Validation of the Applicability of the Compliant Deltoid Mechanism

The aim of this paper is to determine the overall deflection of a compliant deltoid mechanism to show the applicability of such a mechanism for civil engineering machinery. For this purpose, the concept of a demonstrator was described and calculations were made in Sect. 3. The compliant deltoid mechanism described in the concept is build up and shown in the following section.

### 4.1 Manufactured Compliant Deltoid Mechanism Demonstrator

The compliant elements were manufactured on a 3D printer with enclosure at 260°C nozzle temperature, 60 mm/s printing speed and a build plate temperature of 60°C using the polyamide material described in Sect. 3.2 [47]. The print direction is defined along the Z-direction shown in Fig. 13. Using this printing direction, the material's potential regarding the mechanical properties can be fully utilized especially when the force path, perpendicular to the Z-direction, is taken into account.

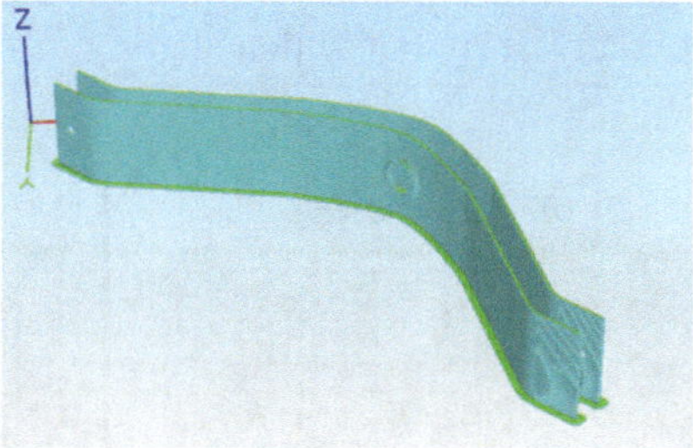


**Fig. 13.** Printing direction (Z-direction) and printing plane (XY-plane) as seen in Simplify 3D Software.

The compliant deltoid mechanism demonstrator consisting of two compliant elements mounted in a frame is shown in Fig. 14.

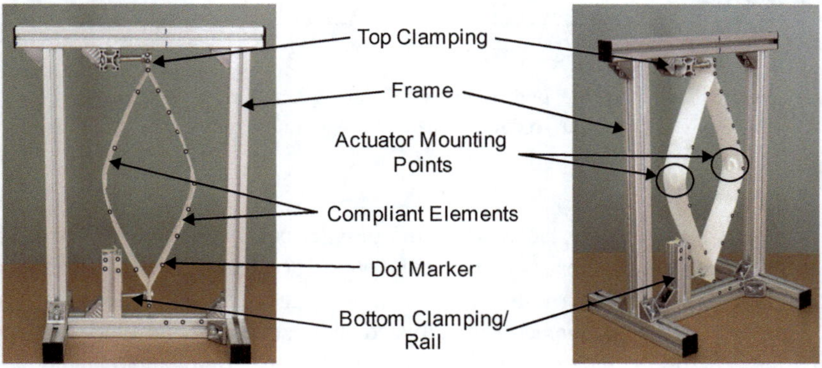


**Fig. 14.** Compliant deltoid mechanism demonstrator with frame.

The top is clamped to the frame using three screws which are also holding the compliant structure together. The bottom, or foot, of the compliant elements is also held together by a screw, guided in a rail, allowing movement of the foot in up and down direction. The mounting points were also formed in the 3D printing process.

### 4.2 Optical Measurement of Motion and Deflection

The optical measurements were done using the optical measurement system GOM ARAMIS SRX. The results are shown in Fig. 15 and Fig. 16. Figure 15 shows the compliant deltoid mechanism demonstrator with a plunger coil in between the mounting points of the compliant elements. Figure 15a) shows the reference state where the plunger coil is not loaded. Figure 15b) shows the state of maximum deflection while c) shows the structure after the plunger coil is deactivated. The arrows show the displacement of every dot marker in horizontal and vertical direction. The maximum displacement of 13.1 mm is carried out by the plunger coil horizontally. The foot follows this motion without any delay but with a slightly lower maximum of 11 mm.

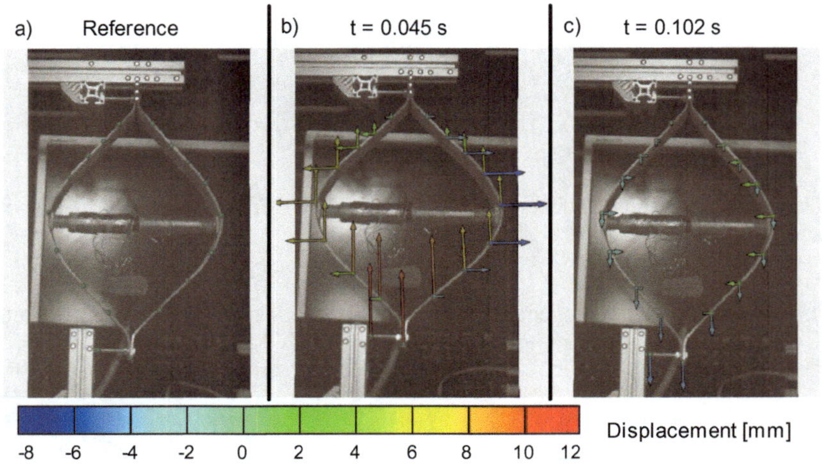


**Fig. 15.** Pictures taken by the optical measurement system GOM ARAMIS SRX at three different states of excitation: a) reference state, b) maximum deflection, c) rebound of the structure.

Figure 16 shows the displacement of the plunger coil and the foot over a time of 0.45 s. The plunger coil consists of two coils which accelerate two actuation magnets when supplied with high currents, one actuation magnet per mounting point. During Sect. 1 in Fig. 16 the plunger coil is activated, thus supplied with electrical current, causing the actuation magnets to accelerate and deflect the compliant elements. During Sect. 2 no electric current is supplied, thus the structure is free-floating, causing it to rebound and oscillate around the reference state (displacement of 0 mm).

The supply of the electric current is timed to be in phase with the oscillation while free-floating. This shows that with clever excitation of the structure, the energy can be minimized while the displacement is maximized.

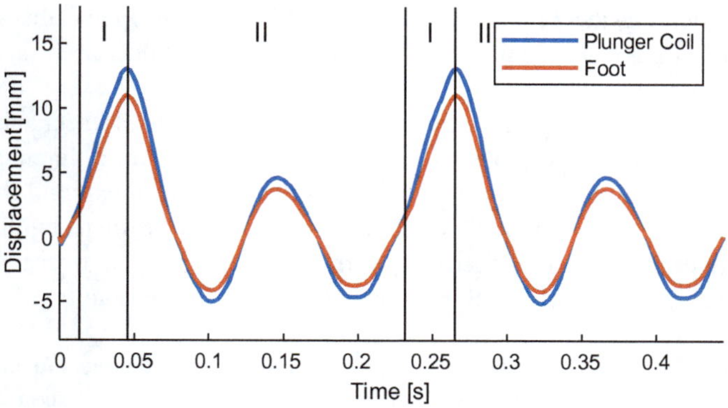


**Fig. 16.** Displacement carried out by the plunger coil (horizontal direction, Fig. 15) and by the foot (vertical direction) over time with Sects. 1 (activated plunger coil) and II (deactivated plunger coil).

### 4.3 Comparison of the Experimental and FE Model Results

The FE model's formulation and its results are depicted in Sect. 3.3 and show a maximum deflection in vertical direction of the compliant deltoid mechanism's foot of 10.96 mm with respect to a maximum horizontal displacement at the mounting points of 13 mm. Comparing this result to the experimental results shown in Sect. 4.2 (11 mm vertical displacement at foot with 13.1 mm displacement at mounting points), it becomes apparent, that the mechanical behavior of the static FE model and the dynamic deflection of the manufactured demonstrator are in good accordance with only slight differences at maximum deflection. Taking into account that the displacement caused by the plunger coil slightly differs from excitation cycle to excitation cycle, the displacement used in the FE model of 13 mm is a valid assumption and thus substantiate the validity of the FE model.

## 5 Discussion and Conclusion

In the paper at hand, a compliant deltoid mechanism was designed, built up, and tested. The design process involved the review of three classical mechanisms, from which one was chosen to be the base for a compliant mechanism demonstrator, the deltoid mechanism. This mechanism is able to convert a linear motion into a linear motion of a direction perpendicular to the input motion. The compliant design featured the substitution of the rigid-body links with compliant elements, reducing the number of links and making the use of joints obsolete. This design approach has the potential to decouple or dampen the movements done by the output side of the mechanism resulting from impact events e.g. when applied in civil engineering machinery, specifically in vibrating tampers.

When comparing the FE model from Sect. 3.3 with the optical measurements in Sect. 4.2, the numerical calculation seems to be in very good accordance with the experimental results.

The demonstrator shows that the application of compliant mechanisms in civil engineering machinery is possible and could be advantageous for different applications. The straightforward FEM approach showed good results for predicting the motion of such relatively simple compliant mechanisms. For a successful technology transfer, however, further investigations focusing on materials and possible fiber reinforcements are necessary, especially when taking conveyed forces into account.

**Acknowledgments.** The results shown in this paper originate from a project funded through the European Regional Development Fund (ERDF) and through means of taxation of the state of Saxony, Germany. This is here-by acknowledged.

## References

1. Albanesi, A.E., Fachinotti, V.D., Pucheta, M.A.: A review on design methods for compliant mechanisms. In: Dvorkin, E., Goldschmit, M., Storti, M. (eds.) Mecánica Computacional Vol XXIX 2010, pp. 59–72. Asiciación Argentina de Mecánica Computacional, Buenos Aires (2010)
2. Modler, N.: Nachgiebigkeitsmechanismen aus Textilverbunden mit intergrierten aktorischen Elementen. Dissertation. TU Dresden, Dresden, Germany (2008)
3. Chen, L.: And integral approach for large deflection cantilever beams. Int. J. Non Linear Mech. **45**(3), 301–305 (2010)
4. Kemper, J.D.: Large deflections of tapered cantilever beams. Int. J. Mech. Sci. **10**, 469–478 (1968)
5. Lee, K.: Large deflections of cantilever beams of non-linear elastic material under a combined loading. Int. J. Non Linear Mech. **37**(3), 439–443 (2002)
6. Lewis, G., Monasa, F.: Large deflections of cantilever beams of nonlinear materials. Comput. Struct. **14**(5–6), 357–360 (1981)
7. Nallathambi, A.K., Rao, C.L., Srinivasan, S.M.: Large deflection of constant curvature cantilever beam under follower load. Int. J. Mech. Sci. **52**, 440–445 (2010)
8. Sevak, N.M.: Optimal synthesis of flexible link mechanisms with large static deflections. Dissertation. The Ohio State University, Columbus, Ohio (1972)
9. Shvartsman, B.: Large deflections of a cantilever beam subjected to a follower force. J. Sound Vib. **304**, 969–973 (2007)
10. Wang, C.Y.: Large deflections of an inclined cantilever with an end load. Int. J. Non Linear Mech. **16**(2), 155–164 (1981)
11. Zhang, A., Chen, G.: A comprehensive elliptic integral solution to the large deflection problems of thin beams in compliant mechanisms. J. Mech. Robot. **5**(5), 1–10 (2013)
12. Banerjee, A., Bhattacharya, B., Mallik, A.: Large deflection of cantilever beams with geometric non-linearity: Analytical and numerical approaches. Int. J. Non Linear Mech. **43** (5), 366–376 (2008)
13. Dado, M., Al-Sadder, S.: A new technique for large deflection analysis of non-prismatic cantilever beams. Mech. Res. Commun. **32**(6), 692–703 (2005)
14. Modler, N., Hufenbach, W., Comsa, A., Maniu, I., Zichner, M., Friedrich, J.: Compliant structures in book handling applications. Appl. Mech. Mater. **162**(3), 543–548 (2012)

15. Wu, G., He, X., Pai, P.: Geometrically exact 3D beam element for arbitrary large rigid-elastic deformation analysis of aerospace structures. Finite Elem. Anal. Des. **47**(4), 402–412 (2011)
16. Christen, G., Pfefferkorn, H.: Zum Bewegungsverhalten nachgiebiger Mechanismen. Wissenschaftliche Zeitschrift der Technischen Universität Dresden **50**(3), 53–58 (2001)
17. Kota, S., Lu, K.-J., Kreiner, Z., Trease, B., Arenas, J., Geiger, J.: Design and application of compliant mechanisms for surgical tools. J. Biomech. Eng. **12**, 981–989 (2005)
18. Salomon, B.: Mechanical advantages aspects in compliant mechanisms design. Diploma Thesis, Purdue University, West Lafayette, Indiana (1989)
19. Sevak, N.M., McLarnan, C.W.: Optimal synthesis of flexible link mechanisms with large static deflections. J. Eng. Ind. **97**(2), 520–526 (1975)
20. Ehrig, T., Koschichow, R., Dannemann, M., Modler, N, Filippatos, A.: Design and development of an active wheelchair with improved lifting kinematics using CFRP-compliant elements. In: ECCM18 – 18th European Conference on Composite Materials, Athens, Greece, 24–28, June 2018
21. Zichner, M.: Mechanismenelemente mit lokal angepasster Nachgiebigkeit. Dissertation. TU Dresden, Dresden, Germany (2018)
22. Lin, J., Luo, Z., Tong, L.: A new multi-objective programming scheme for topology optimization of compliant mechanisms. Struct. Multidiscip. Optim. **40**(1), 241–255 (2010)
23. Lu, K.-J., Kota, S.: Topology and dimensional synthesis of compliant mechanisms using discrete optimization. J. Mech. Des. **128**(5), 1080–1091 (2006)
24. Pandey, A., Datta, R., Bhattacharya, B.: Topology optimization of compliant structures and mechanisms using constructive solid geometry for 2-d and 3-d applications. Soft. Comput. **21**(5), 1157–1179 (2017)
25. Panganiban, H., Jang, G.-W., Chung, T.-J.: Topology optimization of pressure-actuated compliant mechanisms. Finite Elem. Anal. Des. **46**(3), 238–246 (2010)
26. Wang, M.Y., Chen, S., Wang, X., Mei, Y.: Design of multimaterial compliant mechanisms using level-set methods. J. Mech. Des. **127**(5), 941–956 (2005)
27. Zentner, L., Böhm, V.: Zur Anwendung nachgiebiger Mechanismen. Konstruktion: Zeitschrift für Produktentwicklung und Ingenieur-Werkstoffe **57**(11–12), 49–50 (2005)
28. Hears, W.B.: Rigid Body Mechanics. WILEY-VCH Verlag GmbH & Co. KGaA, Weinheim (2006)
29. Lohse, P.: Getriebesynthese: Bewegungsabläuft ebener Koppelmechanismen. Springer, Heidelberg (1986)
30. Modler, K.-H., Luck, K.: Getriebetechnik – Analyse Synthese Optimierung, 2nd edn. Springer, Heidelber (1995)
31. Limaye, P., Ramu, G., Pamulapati, S., Ananthasuresh, G.K.: A compliant mechanism kit with flexible beams and connectors along with analysis and optimal synthesis procedures. Mech. Mach. Theory **49**, 21–39 (2012)
32. Venanzi, S., Giesen, P., Parenti-Castelli, V.: A novel technique for position analysis of planar compliant mechanisms. Mech. Mach. Theory **40**(2), 230–241 (2005)
33. Howell, L.L., Midha, A.: A method for the design of compliant mechanisms with small-length flexural pivots. J. Mech. Des. **116**(1), 280–290 (1994)
34. Howell, L.L.: Compliant Mechanisms. Wiley, New York (2001)
35. Howell, L.L., Midha, A.: A loop-closure theory for the analysis and synthesis of compliant mechanisms. J. Mech. Des. **118**, 121–125 (1996)
36. Pucheta, M.A., Cardona, A.: Design of bistable compliant mechanisms using precision-position and rigid-body replacement methods. Mech. Mach. Theory **45**(2), 304–326 (2010)
37. Tian, Y., Shirinzadeh, B., Zhang, D.: Design and dynamics of a 3-DOF flexure-based parallel mechanism for micro/nano manipulation. Microelectron. Eng. **87**(2), 230–241 (2010)

38. Venkiteswaran, V.K., Su, H.-J.: A parameter optimization framework for determining the pseudo-rigid-body model of cantilever-beams. Precis. Eng. **40**, 46–54 (2015)
39. Sigmund, O.: On the design of compliant mechanisms using topology optimization. J. Struct. Mech. **25**(4), 493–524 (1997)
40. Bendsoe, M.P., Kikuchi, N.: Generating optimal topologies in structural design using a homogenization method. Comput. Methods Appl. Mech. Eng. **71**, 197–224 (1988)
41. Bendsoe, M.P.: Optimization of Structural Topology, Shape and Material. Springer, Heidelberg (1995)
42. Ananthasuresh, G.K., Kota, S., Gianchandani, Y.: A methodical approach to the design of compliant micromechanisms. In: Solid-State Sensor and Actuator Workshop, pp. 189–192 (1994)
43. Sigmund, O.: Some inverse problems in topology design of materials and mechanisms. In: Bestle, D., Schielen, W. (eds.) Symposium on optimization of mechanical systems, pp. 277–284. Kluwer, Netherlands (1996)
44. Larsen, U.D., Sigmund, O., Bouwsta, S.: Design and Fabrication of Compliant Micromechanisms and Structures with Negative Poisson's Ratio. J. Microelectromech. Syst. **6**(2), 99–106 (1997)
45. Gaynor, A.T., Meisel, N.A., Williams, C.B., Guest, J.K.: Multiple-material topology optimization of compliant mechanisms created via PolyJet three-dimensional printing. J. Manuf. Sci. Eng. **136**(6), 061015 (2014)
46. Dannemann, M., Holeczek, K., Leimert, J., Friebe, S., Modler, N.: Adapted measuring sequence for the determination of directional-dependent dynamic material properties using a bending resonance method. Polym. Test. **79**, 1–10 (2019)
47. taulman3D, LLC: Datasheet taulman Alloy 910. https://taulman3d.com/alloy-910-spec.html. Accessed 29 Mar 2020

# A New General Methodology for the Topological Structure Analysis of Multiloop Mechanisms with Multiple Joints and Crossing Links

Vladimir Pozhbelko and Ekaterina Kuts(✉)

South Ural State University (National Research University), Chelyabinsk, Russia
ekaterina.n.kuts@gmail.com

**Abstract.** Complex multiloop mechanisms are widely used in mechanical engineering and robotics, ones can contain diverse types and number of multiple joints and crossing links for the best performance. This paper proposes a novel approach to topological structure analysis of multiloop mechanisms based on new set of general topological and degrees of freedom equations. Various examples of topological structure analysis of multiloop mechanical systems as well as multibody deployable lifting mechanism with numerous links and independent loops, lifting-turning manipulator-loader with numerous multiple joints and multiloop mechanism of folding reflector antenna with crossing links are conducted to show effectiveness, convenience and wide possibilities of proposed approach. The results of this work can be useful for structural synthesis as well as topological structure analysis of complex multibody multiloop mechanisms with different numbers of degrees of freedom (*F*-DOF), any possible multiple joints, crossing links and independent closed loops for creative design of mechanisms without redundant constrains.

**Keywords:** Topology · Mobility · Analysis · Multiple joint · Crossing link

## 1 Introduction

Structural analysis is one of the important stages of conceptual creative mechanical design and is the study of the nature of relationship between the elements of a mechanism and its mobility as well as problem of isomorphism identification [1–5]. It is concerned with the fundamental relationships between the number of degrees of freedom, the number of independent closed loops, the number and types of joints, the number and types of bodies used in the mechanism [6–8].

Hence mechanical task of establishment of its topology and mobility for creative design of complex multiloop mechanisms with optimal structure (without harmful redundant constraints or uncontrolled motions) is difficult problem in mechanisms and machine theory [6–15].

Over the past century, according to various researches, different structural analysis methods in form of computational analysis [6], Assur-group decomposition [8], algorithm SAM [10] and modular approach on the basis of adjacency matrices were


E.-C. Lovasz et al. (Eds.): MTM&Robotics 2020, MMS 88, pp. 155–163, 2021.
https://doi.org/10.1007/978-3-030-60076-1_14

proposed. However the using existing methods and formulas including classical Chebychev-Grübler-Kutzbach criterion [3, 13, 14] does not allow to solve the problem. These methods are limited because it is difficult to apply it for complex multiloop mechanisms with multiple joints and crossing links for determining of the full set of topological characteristics such as the number and assortments of independent closed loops, the number and assortments of types of multiple joints and the number of redundant constraints in the structure. On contrary, the last Pozhbelko's works [15, 16] describe the common structural properties for any possible kinematic chains and contain structural mathematical models for the solution of a wide range of tasks related to the complex kinematic structures of multibody mechanical systems. In this way it can be used for our task.

The aim of this paper is to propose a new general methodology for the topological structure analysis of diverse multiloop mechanisms with different numbers of degrees of freedom (*F*-DOF), any possible types and number of multiple joints, crossing links and independent closed loops, for example, in complex parallel mechanisms, robots and manipulators.

## 2 Basic Conceptual Methodology

We propose a new systematic algorithm for topology and mobility analysis of diverse planar kinematic structures for creative design of complex mechanisms. It includes the following successive steps:

**Step 1.** Establish the link assortment [*LA*] of multiple joints $(n_2, n_3, n_4, n_5, \ldots, n_{i\max})$ for given closed kinematic chain (KC):

$$[LA] = [n_2 n_3 n_4 n_5 \ldots n_{i\max}], \tag{1}$$

where $n_2, n_3, n_4, n_5, \ldots, n_{i\max}$ denote the number of binary ($n_2$), ternary ($n_3$), quaternary ($n_4$), pentagonal ($n_5$) links etc.; $\tilde{n} = \Sigma n_i$ – the total number of links in the mechanism (including fixed link).

**Step 2.** Establish the multiple joint assortment [*MJA*], including the number of various types of multiple joints ($v_j$):

$$[MJA] = [v_2 \cdot v_3 \cdot v_4 \cdot \ldots \cdot v_{j\max}], \tag{2}$$

and total multiple joint factor (*V*) of the KC (case $j \geq 2$):

$$V = v_2 + 2v_3 + 3v_4 + \ldots + (j_{max} - 1)v_{j\max}, \tag{3}$$

where $j = n' - 1$ – the multiplicity of joints which number is $v_j$; $n'$ is the number of adjacent links forming multiple joints ($n' \geq 3, j \geq 2$); $v_2$ – the number of two-pin joints (denoted by "$j_2$"), $v_3$ – the number of three-pin joints (denoted by "$j_3$"), $v_4$ – the number of four-pin joints (denoted by "$j_4$"), etc.

*Remark.* The value varies in between 1 and $V_{max}(1 \leq V \leq V_{max})$ for multiple joint kinematic chains (MJKCs).

**Step 3.** Calculate the number of independent closed loops ($K$) in the kinematic chain:

$$K = 1 + \frac{1}{2}[V + n_3 + 2n_4 + 3n_5 + \ldots + (i - 2)n_i], \tag{4}$$

each of $K$ loops differs in at least one link or kinematic pair.

**Step 4.** Establish independent loop assortment $[L_K]$:

$$[L_K] = [\alpha_4 - \alpha_5 - \alpha_6 - \ldots - \alpha_{k\max}], \tag{5}$$

where $\alpha_k$ denotes the number of $k$-sides loops in the kinematic chain.

**Step 5.** Calculate "Assembly verification criterion" of the valid KC ($AVC = 0$) in general form:

$$AVC = 2(K - 1) - V - [n_3 + 2n_4 + 3n_5 + \ldots + (i - 2)n_i] = 0, \tag{6}$$

where if $AVC \neq 0$ then the kinematic chain cannot be used as closed structure.

**Step 6.** Calculate the number of degrees of freedom ($F$-DOF) of the mechanism ($F \geq 1$) using the following three equivalent mobility equations. These equations are more universal than well-known Gruebler $F$-DOF criterion [2] and present the following set of simple formulas containing the number of $V$ and $c^x$:

$$\text{(a) } F = (n_2 + p_2) - (n_4 + 2n_5 + \ldots) - V - 3; \tag{7}$$

$$\text{(b) } F = (\tilde{n} - 1) - 2K + p_2; \tag{8}$$

$$\text{(c) } F = (v_\Sigma - c^x - 3) + p_2, \tag{9}$$

where $c^x$ – the total number of crossing links within the peripheral closed loop which contains the total number of vertices $v_\Sigma (v_\Sigma = v_1 + v_2 + v_3 + \ldots + v_k)$; $v_1$ – the number of simple joints; $p_2$ – the number of 2-dof of higher kinematic pairs in the KC.

**Step 7.** Calculate the total number of redundant constraints ($r_c$) in the closed kinematic chain in the following way:

$$r_c = (N + V + 3) - n_2 + [n_4 + 2n_5 + \ldots + (i - 3)n_i] - p_2, \tag{10}$$

where $N$ – the number of given activators in the KC, the value $r_c = 0$ denotes the optimal multiloop structure (without both redundant constraints) or on contrary case $r_c \geq 1$ and case $r_c < 1$ (uncontrolled motions).

**Step 8.** The concept of isomorphic identification: the two kinematic chains are non-isomorphic structures if its multiple joint assortments $[MJA] = [v_2 \cdot v_3 \cdot v_4 \cdot \ldots \cdot v_j]$, for example, $[MJA]_1$ and $[MJA]_2$ are different:

$$[MJA]_1 \neq [MJA]_2. \tag{11}$$

# 3 General Topology and Mobility Analysis of Multiloop Mechanism with Numerous Links and Independent Loops

For multibody deployable lifting mechanism shown in Fig. 1, where its corresponding closed kinematic chain has the following structural parameters: $\tilde{n} = 27$(the total number of links), $n_2$ = 8 (the number of binary links), $n_3$ = 18 (the number of ternary links), $n_4$ = 1 (the number of quaternary links), $v_2$= 2 (the number of two-pin joints is denoted by "$j_2$"), the Eqs. (1)–(10) yield:

**Step 1.** (a) Link assortment $[LA] = [n_2 = 8, n_3 = 18, n_4 = 1]$;
(b) The total number of links $\tilde{n} = n_2 + n_3 + n_4 = 8 + 18 + 1 = 27$.

**Step 2.** (a) Multiple joint assortment $[MJA] = [v_2] = [2]$;
(b) Total multiple joint factor $V = v_2 = 2$.

**Step 3.** The number of independent loops (shown in the Fig. 1 as $L_i$)

$$K = 1 + \frac{1}{2}[V + n_3 + 2n_4] = 1 + \frac{1}{2}[2 + 18 + 2 \cdot 1] = 12.$$

**Step 4.** Independent loop assortment $[L_K]$

$$[L_K] = [\alpha_4 - \alpha_6 - \alpha_4 - \alpha_4 - \alpha_4 - \alpha_4 - \alpha_4 - \alpha_4 - \alpha_4 - \alpha_4 - \alpha_4 - \alpha_4].$$

**Step 5.** Assembly verification criterion ($AVC$ = 0)

$$AVC = 2(K - 1) - V - [n_3 + 2n_4] = 2(12 - 1) - 2 - [18 + 2 \cdot 1] = 22 - 22 = 0.$$

**Step 6.** The number of degrees of freedom of the mechanism ($p_2$ = 0)

(a) $F = (n_2 - 3) - V - n_4 = (8 - 3) - 2 - 1 = 2$;
(b) $F = (\tilde{n} - 1) - 2K = (27 - 1) - 2 \cdot 12 = 2 \Rightarrow N = F = 2$.

**Step 7.** The total number of redundant constraints (with given $N$ = $F$ = 2)

$$r_c = (N + V + 3) - (n_2 - n_4) = (2 + 2 + 3) - (8 - 1) = 7 - 7 = 0.$$

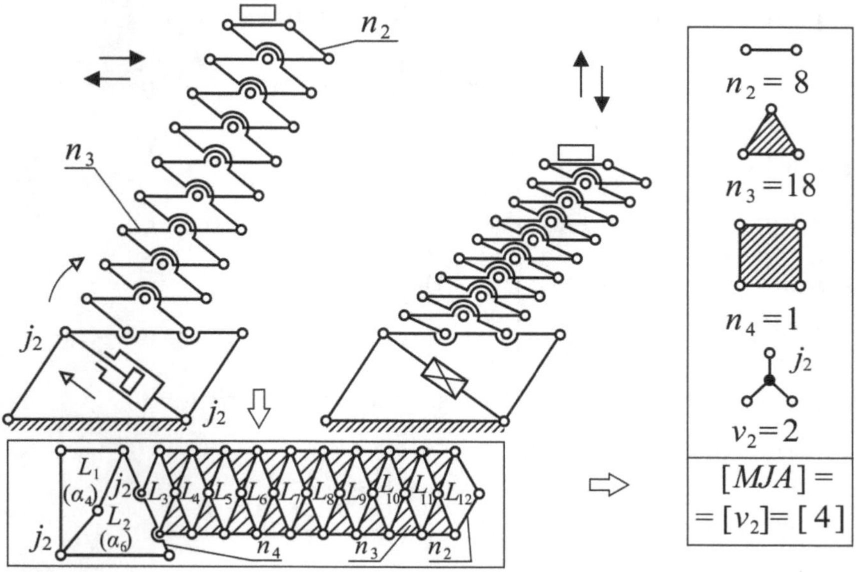


**Fig. 1.** Deployable lifting mechanism of horizontal platform ($F = 2$, $V = 2$, $K = 12$).

Hence, this multiloop mechanical system presents 2-DOF ($F = 2$) optimal structure of mechanism without redundant constraints ($r_c = 0$). Therefore, it can be used in mechanical engineering and construction industry.

*Remark.* We can derive the same result ($W = F = 2$) using the classical Grübler mobility formula [2], but it requires greater calculation effort as we need to calculate the all numerous $\tilde{n} = 27$ links (in contrast, only a calculation of $n_2 = 8, n_4 = 1$) and all the numerous 1-DOF kinematic pairs $p_1 = 38$ (in contrast, only a calculation of two multiple joints): $W = 3(\tilde{n} - 1) - 2p_1 - p_2 = 3(27 - 1) - 2 \cdot 38 - 0 = 2$.

## 4 General Topology and Mobility Analysis of Multiloop Mechanism with Numerous Multiple Joints

For multibody lifting-turning manipulator-loader shown in Fig. 2, where its corresponding closed kinematic chain has the following structural parameters: $\tilde{n} = 23$ (the total number of links), $n_2 = 18$ (the number of binary links), $n_3 = 3$ (the number of ternary links), $n_4 = 2$ (the number of quaternary links), $v_2 = 6$ (the number of two-pin joints is denoted by "$j_2$"), $v_4 = 1$ (the number of four-pin joints is denoted by "$j_4$"), the Eqs. (1)–(10) yield:

**Step 1.** (a) Link assortment $[LA] = [n_2 = 18, n_3 = 3, n_4 = 2]$;
(b) The total number of links $\tilde{n} = n_2 + n_3 + n_4 = 18 + 3 + 2 = 23$.

**Step 2.** (a) Multiple joint assortment $[MJA] = [v_2 \cdot v_3 \cdot v_4] = [6 \cdot 0 \cdot 1]$;
(b) Total multiple joint factor $V = v_2 + 2v_3 + 3v_4 = 6 + 2 \cdot 0 + 3 \cdot 1 = 9$.

**Step 3.** The number of independent loops (shown in the Fig. 2 $L_i$)

$$K = 1 + \frac{1}{2}[V + n_3 + 2n_4] = 1 + \frac{1}{2}[9 + 3 + 2 \cdot 2] = 9.$$

**Step 4.** Independent loop assortment $[L_K]$:

$$[L_K] = [\alpha_4 - \alpha_4 - \alpha_4 - \alpha_4 - \alpha_5 - \alpha_4 - \alpha_6 - \alpha_4 - \alpha_5].$$

**Step 5.** Assembly verification criterion ($AVC = 0$)

$$AVC = 2(K - 1) - V - [n_3 + 2n_4] = 2(9 - 1) - 9 - [3 + 2 \cdot 2] = 16 - 16 = 0.$$

**Step 6.** The number of degrees of freedom of the mechanism ($p_2 = 0$)

(a)$F = (n_2 - 3) - V - n_4 = (18 - 3) - 9 - 2 = 4;$
(b) $F = (\tilde{n} - 1) - 2K = (23 - 1) - 2 \cdot 9 = 4 \Rightarrow N = F = 4.$

**Step 7.** The total number of redundant constraints (with given $N = F = 4$)

$$r_c = (N + V + 3) - (n_2 - n_4) = (4 + 9 + 3) - (18 - 2) = 16 - 16 = 0.$$

Hence, considered multiloop mechanical system presents 4-DOF ($F = 4$) optimal structure of mechanism without redundant constraints ($r_c = 0$). Therefore, it can be applied for design of industrial robot manipulators.

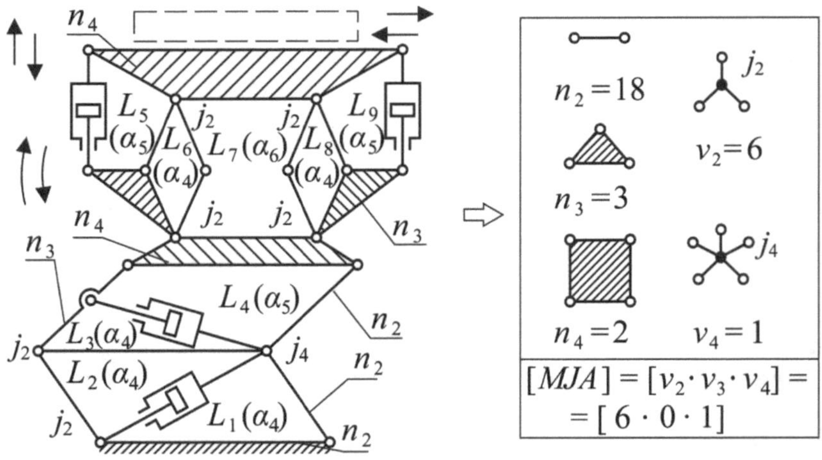


**Fig. 2.** Lifting-turning manipulator-loader with multiple joints ($F = 4$, $V = 9$, $K = 9$).

## 5 Topological Structure Analysis of Multiloop Mechanism with Numerous Crossing Links

The complex mechanical system shown in Fig. 3 consists of two different parts. First part is the mechanism *A* (presented in two types of mechanisms $A_1$ and $A_2$), second is the mechanisms *B*:

(1) For multiloop mechanism *A* of folding reflector antenna with crossing links ($c^x = 8$), where its corresponding closed kinematic chain has the following structural parameters: $\tilde{n} = 20$ (the total number of links), $n_2 = 20$ (the number of binary links), $p_2 = 0$, $v_2 = 8$ (the number of two-pin joints is denoted by "$j_2$"), $v_3 = 4$ (the number of three-pin joints is denoted by "$j_3$"), the Eqs. (1)–(11) yield:

**Step 1.** (a) Link assortment $[LA] = [n_2 = 20]$;
(b) The total number of links $\tilde{n} = n_2 = 20$.

**Step 2.** (a) Multiple joint assortment $[MJA] = [v_2 \cdot v_3] = [8 \cdot 4]$;
(b) Total multiple joint factor for mechanism $A_1$ $V = v_2 + 2v_3 = 8 + 2 \cdot 4 = 16$; total multiple joint factor for mechanism $A_2$ $V = v_2 + 3v_4 = 10 + 3 \cdot 2 = 16$.

**Step 3.** The number of independent loops $(L_i)K = 1 + V/2 = 1 + 16/2 = 9$.

**Step 4.** The number of vertices in a peripheral loop $v_\Sigma = v_2 + v_3 = 8 + 4 = 12$.

**Step 5.** Assembly verification criterion ($AVC = 0$)

$$AVC = 2(K - 1) - V = 2(9 - 1) - 16 = 16 - 16 = 0.$$

**Step 6.** The number of degrees of freedom of the mechanism

(a) $F_1 = v_\Sigma - c^x - 3 = 12 - 8 - 3 = 1 \Rightarrow N = F_1 = 1$;
(b) $F_1 = (n_2 - 3) - V = (20 - 3) - 16 = 1$;
(c) $F_1 = (\tilde{n} - 1) - 2K = (20 - 1) - 2 \cdot 9 = 1$.

**Step 7.** The total number of redundant constraints (with given $N = F_1 = 1$)

$$r_c = (N + V + 3) - n_2 = (1 + 16 + 3) - 20 = 20 - 20 = 0.$$

**Step 8.** On the basis of established in Fig. 3 inequality $[MJA]_1 \neq [MJA]_2$ we can derive that the both of multiloop mechanisms $A_1$ and $A_2$ are non-isomorphic structures with common number of independent loops ($K = 9$), multiple joints ($V = 16$) and crossing links ($c^x = 8$).

(2) For spatial spherical gear mechanism *B* for turning of reflector antenna, where its corresponding closed kinematic chain has the following structural parameters: $n_2 = 3$, $p_2 = 1$ (the number of higher gearing kinematic pairs), the Eqs. (1)–(10) yield:

1) $[LA] = [n_2 = 3] \Rightarrow \tilde{n} = 3$; 2) $[MJA] = [0] \Rightarrow V = 0$; 3) $K = 1$; 4) $AVC = 2\,(K - 1)\ - V = 0$; 5) $F_2 = (n_2 + p_2) - 3 = 1$; 6) $r_c = (N + 3) - n_2 - p_2 = (1 + 3) - 3 - 1 = 4 - 4 = 0$.

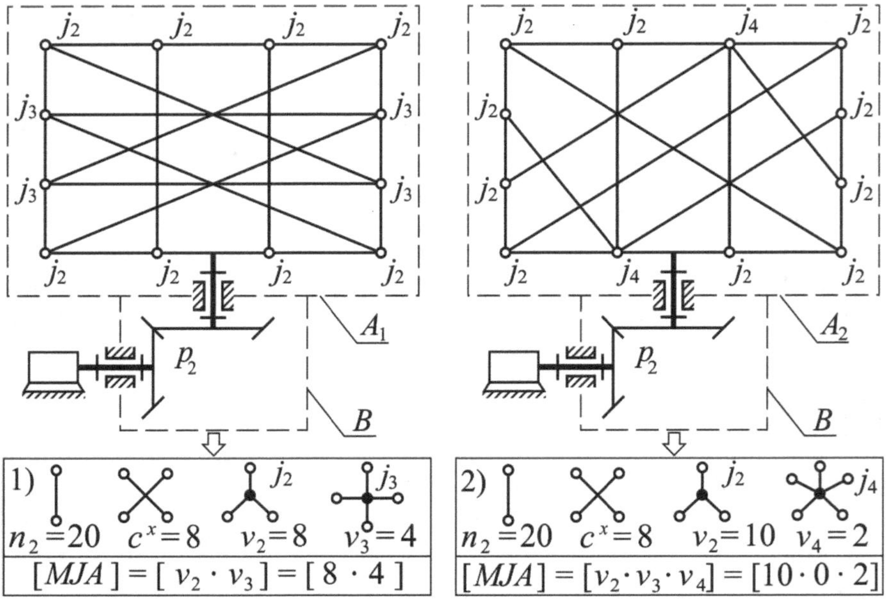


**Fig. 3.** Multiloop mechanism of folding reflector antenna with crossing links ($F$ = 2, $V$ = 16, $c^x$ = 8).

Hence, this multiloop mechanical system consisting of mechanism $A$ and $B$ presents 2-DOF ($F = F_1 + F_2 = 2$) optimal structure of mechanism without redundant constraints ($r_c = 0$). Therefore, it can be applied for mobile antennas.

## 6 Conclusions

In this paper, we have presented an original approach to general topology and mobility analysis of various complex multiloop structures. Then, various examples of structure analysis of novel mechanisms (provided by the authors) are included to demonstrate the effectiveness of proposed approach. Note that presented mobility formulas allow to do a quick and concise mobility calculation and mobility property analysis for various kinds of mechanisms, especially, for complex multiloop mechanisms. Moreover, these formulas (Eqs. (7)–(9)) have low calculation complexity that is essential for the design of mechanisms with numerous number of links, independent closed loops and multiple joints. The results of this paper can be applied to structural synthesis and analysis of complex mechanical systems for creative mechanical design.

**Acknowledgments.** The work was supported by Act 211 Government of the Russian Federation, contract № 02.A03.21.0011.

## References


1. Ding, H., Yang, W., Huang, P., Kecskeméthy, A.: Automatic structural synthesis of planar multiple joint kinematic chains. J. Mech. Des. **135**(9), 091007 (2013)
2. Erdman, A.G., Sandor, G.: Mechanism Design: Analysis and synthesis, vol. 1. Prentice-Hall, Englewood Cliffs (1984)
3. Gogu, G.: Mobility of mechanisms: a critical review. Mech. Mach. Theory **40**(9), 1068–1097 (2005)
4. Gogu, G.: Chebychev- Grübler-Kutzbach's criterion for mobility calculation of multi-loop mechanisms revisited via theory of linear transformations. Eur. J. Mech. A/Solids **24**, 427–441 (2005)
5. Phillips, J.: Freedom in Machinery. Cambridge University Press, Cambridge (1984)
6. Saura, M., Celdran, A., Dopico, D., Cuadrado, J.: Computational structural analysis of planar multibody systems with lower and higher kinematic pairs. Mech. Mach. Theory **71**(1), 79–92 (2014)
7. Ceccarelli, M.: Fundamentals of Mechanics of Robotic Manipulation. Kluwer Academic Publishers, Dordrecht (2004)
8. Uicker, J.J., Pennock, G.R., Shigley, J.E.: Theory of Machines and Mechanisms, 3rd edn. Oxford University Press, New York (2003)
9. Mueller, A.: Kinematic topology and constraints of multi-loop linkages. Robotica **36**(11), 1–23 (2018)
10. Buśkiewicz, J.: A method of optimization of solving a kinematic problem with the use of structural analysis, algorithm (SAM). Mech. Mach. Theory **41**(7), 823–837 (2006)
11. Manolescu, N.I.: The history of the original methods used in the synthesis of the planar kinematic chains with different degrees of liberty. In: Proceeding of V Conference on the Theory of Machines and Mechanisms, Liberec, pp. 145–157 (1988)
12. Servatius, B., Shai, O., Whiteley, W.J.: Combinatorial characterization of the Assur graphs from engineering. Eur. J. Comb. **31**, 1091–1104 (2010)
13. Rico, J.M., Ravani, B.: On mobility analysis of linkages using group theory. Mech. Des. **125** (1), 70–80 (2003)
14. Rico, J.M., Gallardo, J., Ravani, B.: Lie algebra and the mobility of kinematic chains. Robot. Syst. **20**(8), 477–479 (2003)
15. Pozhbelko, V., Ermoshina, E.: Number structural synthesis and enumeration process of all possible sets of multiple joints for 1-DOF up to 5- loop 12-link mechanisms on base of new mobility equation. Mech. Mach. Theory **90**(8), 108–127 (2015)
16. Pozhbelko, V.: A unified structure theory of multibody open, closed loop and mixed mechanical systems with simple and multiple joint kinematic chains. Mech. Mach. Theory **100**(6), 1–16 (2016)

# A New Modelling Approach to Determine the DOF of Compliant Mechanisms

Marco Zichner[1(✉)], Niels Modler[1], Joanna Katarzyna Wollmann[1], Karl-Heinz Modler[2], Philip Johannes Steinbild[1], and Martin Dannemann[1]

[1] Institute of Lightweight Engineering and Polymer Technology, Technische Universität Dresden, Dresden, Germany
marco.zichner@tu-dresden.de

[2] Institute of Solid Mechanics, Technische Universität Dresden, Dresden, Germany

**Abstract.** The focus of the work was the evaluation of the real degree of freedom of compliant mechanisms or their correct prediction by means of standard an a new adapted modelling. For this purpose, the widely used Pseudo Rigid Body Model (PRBM) was initially applied, whereby contradictions arise between modelling and real motion behavior. Based on this approach an extension was developed. This allows the structural analysis of flexible elements to be performed realistically. The development engineer is thus provided with an essential aid for the selection of suitable support and coupling conditions for the mechanics synthesis. On the basis of this new replacement model, the fuzzy motion behavior of the compliant mechanisms can also be clarified and solutions for completely constrained structures can be identified.

**Keywords:** Compliant mechanisms · Degree of freedom · Modelling · Analysis

## 1 Preliminary Considerations and State of Science

Compliant elements are predestined for the development of lightweight mechanisms, because they allow the integration of additional functionalities and could be developed following a given solution scheme. At the same time, the use of compliant mechanisms helps to reduce the number of mechanism elements, thus minimizing overall mass, maintenance requirements and noise emissions.

However, the obvious advantages of compliant mechanisms are counterbalanced by some considerable disadvantages, which generally prevent their wider technical application. For example, the analysis and synthesis of those mechanisms require a detailed description of the complex, nonlinear motion and deformation behavior of the flexible components [1, 2, 6, 7]. Nevertheless, the challenges in the development of compliant mechanisms are not only in its synthesis. Engineers already have various iterative approximation methods at their hands [2, 3], which often provide sufficiently accurate starting solutions for a following FEM analysis. However, this alone will hardly promote the use of compliant structures. Rather, it is the unbound motion


E.-C. Lovasz et al. (Eds.): MTM&Robotics 2020, MMS 88, pp. 164–171, 2021.
https://doi.org/10.1007/978-3-030-60076-1_15

behavior of the flexible elements itself that often impede their use. Therefore, an investigation of the kinematic properties of the compliant mechanism, which is detached from the statics and dynamics, is necessary.

### 1.1 Brief Introduction to the Analysis of the DOF of Planar Mechanisms

For the following considerations, the concept of the degree of freedom of a general rigid body mechanism is briefly introduced. According to [4] this degree of freedom is derived directly from the lack of freedom $u$ of the used joints. The following applies

$$u = b - f, \tag{1}$$

where $f$ corresponds to the possible relative movements of two links to each other (degree of freedom of joint) and $b$ describes the general degree of freedom of the bound movement of the links in the reference system. The degree of freedom $F$ is given by TSCHEBYSCHEW or GRÜBLER (cf. [4]) to

$$F = b \cdot (n - 1) - \sum_{j=1}^{e} u_j, \tag{2}$$

Where $e$ is the number of joints and $n$ is the number of links. This equation applies to movements in the plane ($b$ = 3), on a spherical surface ($b$ = 3) or any movement in space ($b$ = 6). For the plane mechanisms considered here, in which only swivel joints with $u$ = 2 are used, this equation goes into the so-called *1. Constraint condition*:

$$F = 3(n - 1) - 2e, \tag{3}$$

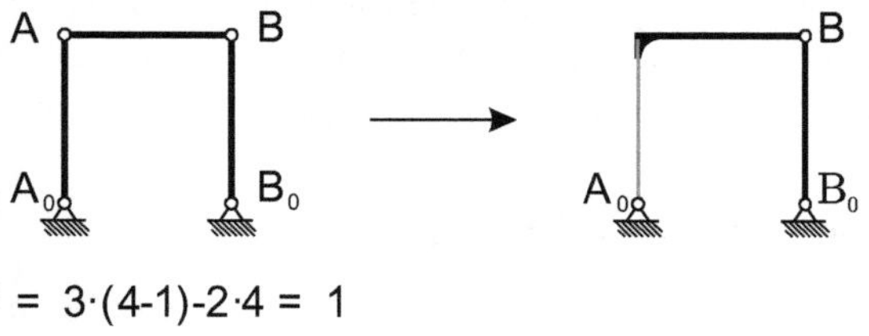


**Fig. 1.** DOF of a planar parallel-crank RRRR mechanism (left) and conversion into a compliant mechanism (right)

Any arrangement of links and rotation joints can have following degrees of freedom:

- $F > 1$: The arrangement is indefinite. The mechanism needs exactly $F$ drives to be used as a guide or transmission gear.
- $F = 1$: The arrangement is constrained. The mechanism can be used with only one drive as a guide or transfer gear.

- $F = 0$: The arrangement is static determined. No transmission of motion can take place.
- $F < 0$: The arrangement is static over-determined. No motion transfer can take place.

A classic planar, four bar mechanism analogous to Fig. 1 has a DOF $F = 1$. This means that only one drive is required to define the position of all links uniquely.

## 1.2 Nomenclature of Compliant Mechanisms

By using the standard nomenclature of rigid-body mechanisms [4], which only includes movable couplings with the classification (selection):

**R** Rotational joint with $F = 1$,
**P** Prismatic joint with $F = 1$,
**C** Cylindrical joint with $F = 2$,

mechanisms with compliant elements cannot be described sufficiently.

To characterize compliant mechanisms more detailed, N. Modler introduced in [5] a new nomenclature by combining the bearing and coupling conditions of the members as well as their properties into a kind of structural formula. Building on the fundamental work of Cristian [6] and Zentner [7] on the classification of compliant mechanisms, he defines:

**g** jointed bearing or coupling,
**f** fixed bearing or coupling,
**s** rigid link,
**n** compliant link.

The hybrid compliant mechanism according to Fig. 1 (left) is of a *BRR* structure or according to N. Modler as *g-n-f-s-g-n-g* type. Depending on the purpose of the description, both variants are useful and common in our work today.

## 1.3 Short Summary on the PRBM-Approach

Figure 2 shows a flexible beam element which is deformed from its initial position $B_1$ to the position $B_2$, while the tip moves on a cochleoid (dashed line). In this position $B_2$, which is defined by the angle $\vartheta_{12}$, the element is in equilibrium with the load $F$. For small angles, this path can be described approximately as a segment of a circle, whereby the mapping error increases as the angle increases (see Fig. 3).

This is used by Howell, who has made a significant contribution to the analysis and synthesis of compliant structures with the so-called Pseudo-Rigid-Body-Model (PRBM). His method is based on the substitution of the flexible element, which has to be calculated in a complex way, by two rigid links connected with a rotating joint. In addition, he adds a torsion spring so that the force-deformation ratios are approximately maintained. The key to creating a sufficiently accurate PRBM depends on the joint position and the spring constants based on the formula introduced in [2]. This position is defined in relation to the beam length $L$, where the length fraction is called distance

$\gamma L$ (so-called characteristic radius). In addition, the angle of rise $\vartheta_{12}$ at the end of the beam and the load-deformation relationships can be defined as a function of $\Theta$, the angle between the undeformed and the deformed position of the PRB model (Fig. 3). Howell now uses this approach to compute compliant mechanisms with easy-to-use geometric relationships, making them accessible not only to engineers.

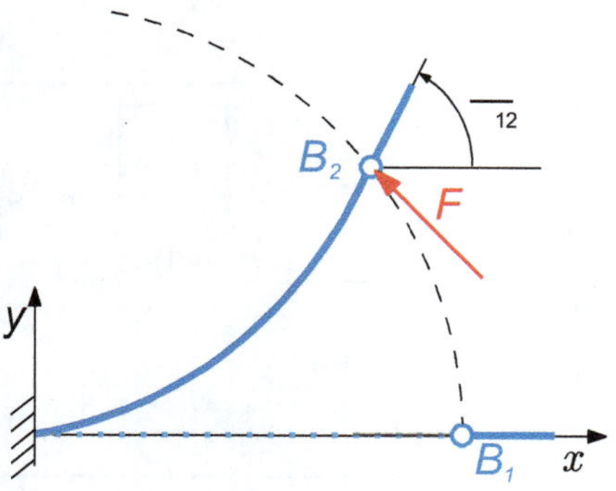


**Fig. 2.** Deformed and undeformed (dotted) compliant element

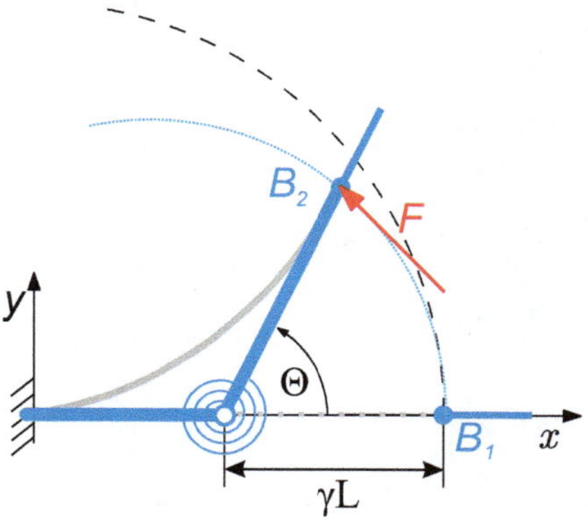


**Fig. 3.** Derived PRBM

## 2 Calculation the Mechanism DOF Using PRBM

For the evaluation of the DOF prediction of the PRBM-modelling, a planar 4-bar mechanism according to Fig. 1 is to be used in the following. Figure 4 shows eight derived, different configurations of this mechanism as well as their translation to the mechanism model $A_0ABB_0$ or $A_0AB_0$ using the PRBM approach introduced previously. The examples in Fig. 4. (1)–(3) show structures with only one compliant element (blue line). Due to the PRB modelling, structures with a DOF $F = 1$ are created. The change of the fixed restraint as well as elements clamped on both sides – each with its own modeling step – lead to a fully constrained mechanism.

However, applying PRB modeling to structures as shown in Fig. 4 (7)–(8) leads to challenging results. It is to be noted that the change of the joint from the *f-n-f-s-g-s-g* structure (Fig. 4 (3)) to a *f-n-g-s-f-s-g* structure (Fig. 4 (7)) alone is sufficient to design a static determined mechanism with theoretically no movability ($F = 0$). Also the mechanism in Fig. 4. (8) would not be functional. Nevertheless, it is known that such structures of an *f-n-f-s-g-s-g* and *f-n-f-s-f-s-g* type are indeed moveable, so there is a contradiction in applying the PRBM to those.

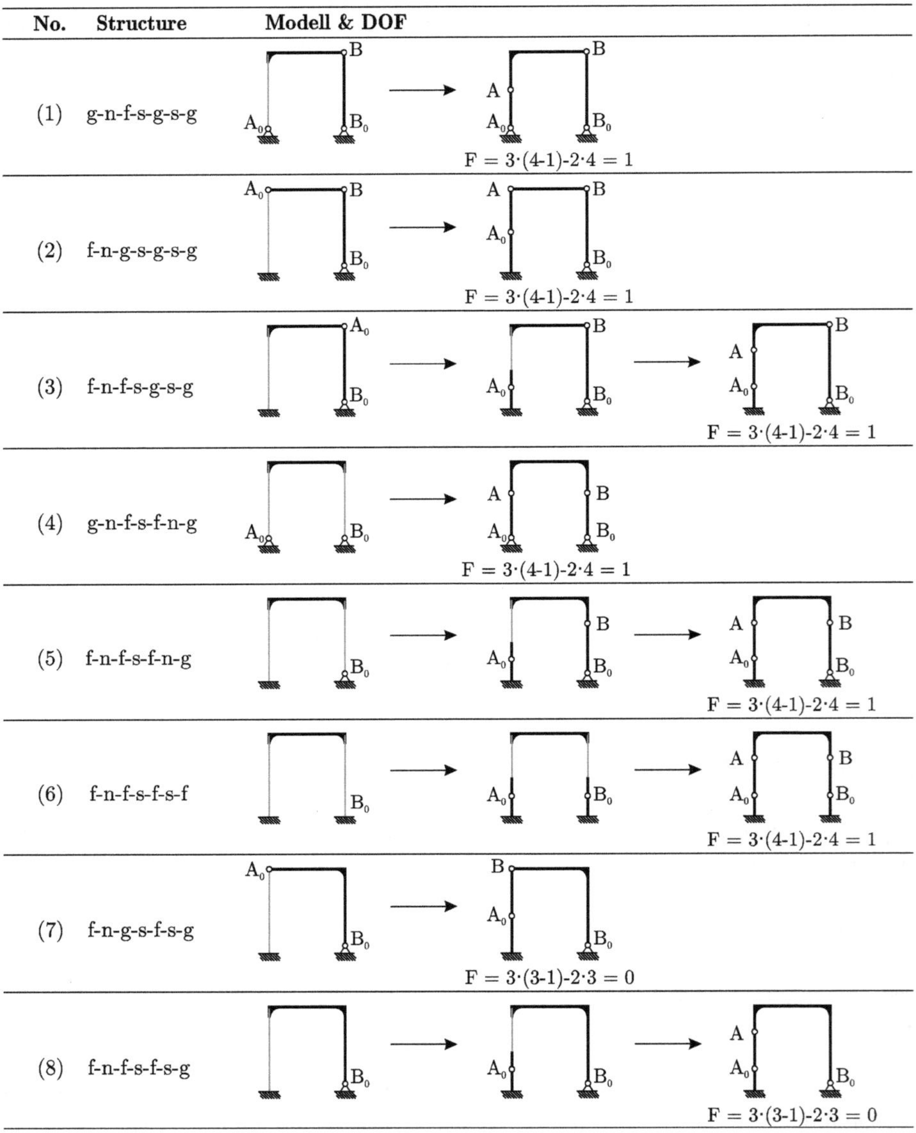


**Fig. 4.** Compliant mechanisms and their transformation into rigid body mechanisms using the PRBM approach

# 3 Extended Substitute Model

The goal was to find a substitute model that describes the actual freedom of movement of the mechanism without contradiction. Figure 5 shows an extension of the PRBM previously used by applying a three-bar chain. This allows the possible deformation states and DOF of compliant structures to be reproduced more accurately. Especially the structures deformed in the so-called *Mode-2* – where the force vector intersects the compliant element (see [8]) – can only be represented accurately with this extended approach. When modelling two-sided clamped compliant element, two substitute models with correspondingly superimposed joints are to be used – analogous to the method previously introduced when considering the PRBM. Identical freedoms are eliminated according to the standard procedure [4].

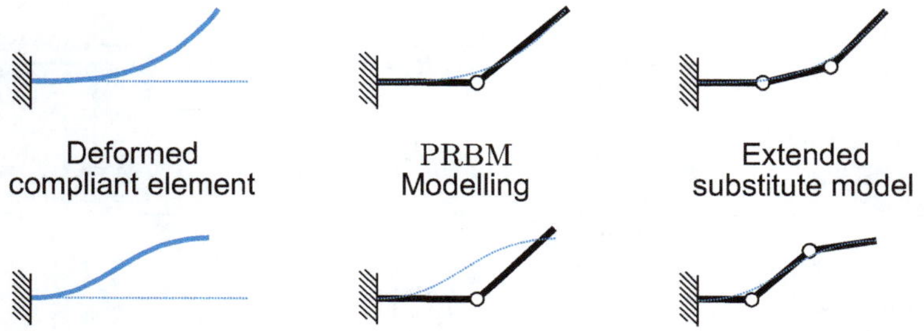


**Fig. 5.** PRBM and extended substitute model for deformed compliant elements in different modes

With approximation of the compliant element by three bars and two rotational joints the contradictions for the structures *f-n-g-s-f-s-g* and *f-n-f-s-f-s-g* can be resolved.

Figure 6 (7) and (8) shows the equivalent models, which now form constrained structures with a DOF of $F = 1$. The motion behavior of the compliant structures (exemplarily shown in Fig. 6 (8)) indicates that the deformed element corresponds to mode 2 and can ultimately only deform in this way. When examining all other structures shown, it becomes clear that numerous indefinite or undetermined structures are now forming. The selected examples according to Fig. 6 (1)–(3) lead to structures with $F = 2$. Mechanisms with two yielding elements even result in three unbound motions in translation (see Fig. 6 (4)–(6)). The developed model thus corresponds to the practical experience of a fuzzy motion transmission of the compliant mechanisms, which so far allow a use as guide and transmission gear only to a limited extent. Rather, depending on the magnitude and orientation of the loads, very different deformation states are formed, so that a specific prediction of the position of the individual mechanism elements is hardly possible without experimental validation.

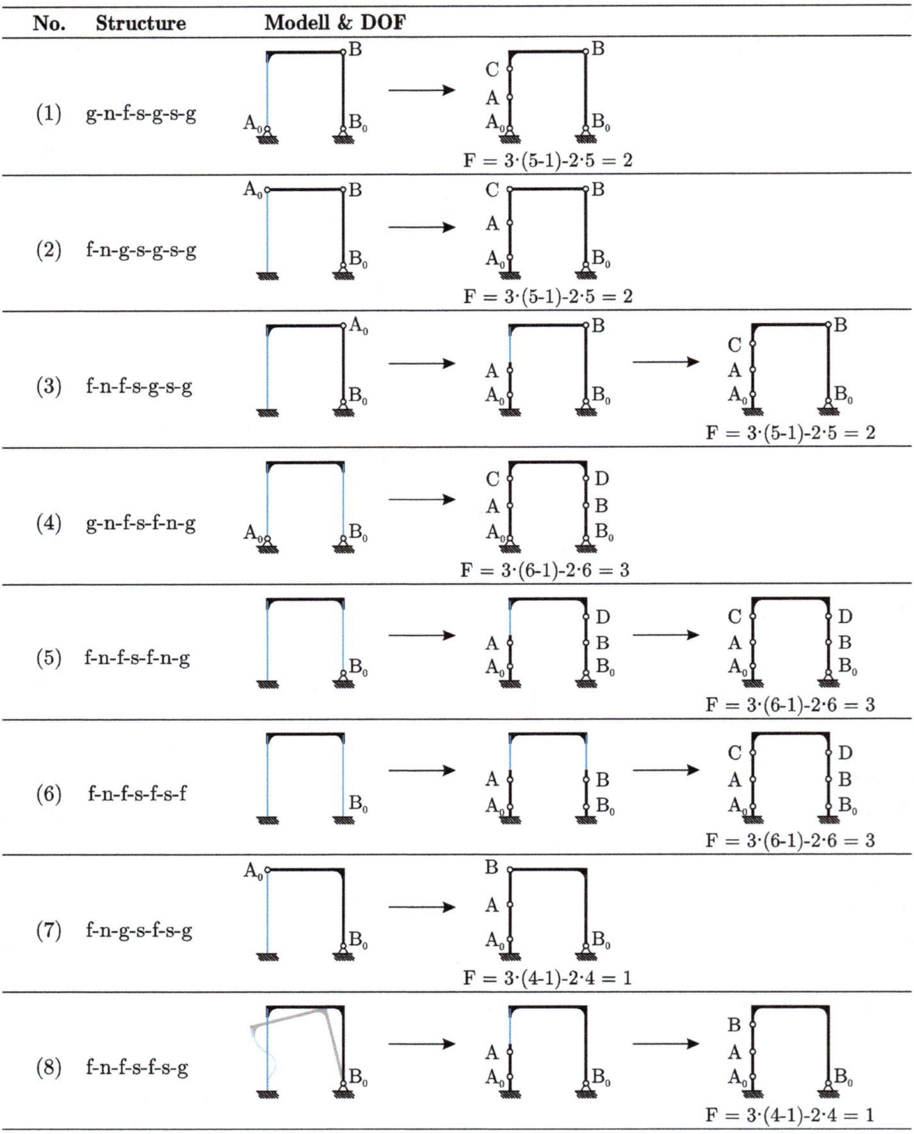


**Fig. 6.** Compliant mechanisms and their transformation using the extended substitute model

# 4 Conclusion

Compliant elements form their own class of joints, which ultimately correspond to *spring elements* (i.e. mechanism organs). This is because they are in an equilibrium with the applied load exactly only when the internal deformation energy corresponds to the external potential. Typical "4-bar" compliant mechanisms (given in Fig. 6) therefore seem to be defined sufficiently precisely with only one drive. However, due to their free deformability, they are able to assume a new state of equilibrium because of altered external potentials. This also means a change in the position of the input, output or coupler element. If the flexible elements are virtually removed when looking at Fig. 6 open chains of mechanisms with exactly the same DOF are formed. Therefore, the use of compliant structures in transmissions with time-varying loads requires either an iterative development that discretizes the transient change of the load and thus enables the optimization of the overall system or the use of individual links, which ultimately act like simple joints (see Fig. 6 (7)–(8)).

With the extended substitute model, a tool is available that allows the modelling of real constrained compliant mechanism and thus allows their use in guidance and, if the above mentioned conditions are taken into account, also as transmission applications.

**Acknowledgement.** The authors would like to express their gratitude towards the Deutsche Forschungsgemeinschaft (DFG) for funding this research within the scope of Research Training Group "Interactive Fiber-Rubber-Composites" - GRK 2430/380321452.

## References

1. Hanke, U., Zichner, M., Tudorache, A., Modler, N., Ashir, M.: Design of monolithic compliant mechanisms with beam elements of distributed stiffness. Materialwissenschaft und Werkstofftechnik **47**(11), 1132–1139 (2016). https://doi.org/10.1002/mawe.201600638
2. Howell, L.L.: Compliant Mechanisms. Wiley, New York (2001)
3. Zichner, M., Hanke, U., Modler, K.-H., Hornig, A., Modler, N.: Compliant mechanisms with distibuted stiffness - synthesis and application. In: The 14th IFToMM World Congress, Taipei, Taiwan, pp. 1–7 (2015)
4. Modler, K.-H.; Luck, K.: Getriebetechnik – Analyse Synthese Optimierung. Springer, Heidelberg (1995). https://doi.org/10.1007/978-3-642-78383-8
5. Modler, N.: Nachgiebigkeitsmechanismen aus Textilverbunden mit integrierten aktorischen Elementen. TU Dresden, Dissertation (2008)
6. Christen, G., Pfefferkorn, H.: Zum Bewegungsverhalten nachgiebiger Mechanismen. Wissenschaftliche Zeitschrift der Technischen Universität Dresden **50**(3), pp. 53–58 (2001)
7. Zentner, L., Böhm, V.: Zur Anwendung nachgiebiger Mechanismen. Konstruktion: Zeitschrift für Produktentwicklung und Ingenieur-Werkstoffe **57**(11/12), pp. 49–50 (2005). ISSN 07205953
8. Zichner, M.: Mechanismenelemente mit lokal angepasster Nachgiebigkeit. TU Dresden, Dissertation (2018)

# Design and Development of a Search and Rescue Robot with TriSTAR Locomotion Units

Marton Gyarmati, Mihai Olimpiu Tătar, and Călin Rusu(✉)

Department of Mechatronics and Machine Dynamics,
Technical University of Cluj-Napoca, Cluj-Napoca, Romania
calin.rusu@mdm.utcluj.ro

**Abstract.** In this paper the authors present a search and rescue robot with hybrid locomotion system. The robot uses a locomotion system with two TriSTAR units and two conventional wheels. Details are also presented regarding the design of the robot and the TriSTAR locomotion unit, the robot locomotion and the robot actuation and control. In order to provide the possibility to manipulate small objects the robot will be equipped with a manipulator presented in detail.

**Keywords:** Search and rescue robot · Hybrid locomotion system

## 1 Introduction

The field of search and rescue robots is a relatively new domain being first presented in the 1980s [1]. The application and usage of search and rescue robots increased in the emergency situations. This domain remains a challenging field, with many open problems such as: mobility, communication, control, sensors, autonomy and human-robot interaction [2].

Searching and rescuing in collapsed buildings is a difficult and very dangerous activity for rescuers.

Moreover, some debris of the collapse building are inaccessible to the rescuers. Under these conditions, urban search and rescue mobile robot (USAR - Urban Search and Rescue Robot) can be sent to the collapsed buildings to help rescuers complete the search and rescue task.

Experience from search and rescue has shown that after 48 h, the mortality of victims increases drastically due to radiation exposure, lack of food, water and lack of medical treatment [3]. Therefore, rescuing victims after disasters is under extreme time pressure.

In this sense, one of the biggest challenges is that rescue robot operations to be more efficient in finding more survivors or in providing faster and clearer information to the rescuers [3].

The search and rescue robots are different from the domestic or industrial robots. Its operating environment is uncertain and dangerous. The robot can thus replace the


E.-C. Lovasz et al. (Eds.): MTM&Robotics 2020, MMS 88, pp. 172–186, 2021.
https://doi.org/10.1007/978-3-030-60076-1_16

human operator in the dangerous environment, then implement the rescue work and transport the necessary means to the victims.

In principle, the Search and Rescue Robots (SRR) have the following advantages [1]:

- They have a high adaptability in various operating environments, environments which can reach high temperatures and pressures and in which rescuers cannot arrive.
- It can travel on a rough train (among the ruins) to detect signs of life.
- They should be small and flexible.
- Being equipped with functional sensors and detection equipment, such as the infrared detector and video cameras to search for survivors.
- Using batteries with low power consumption and long operating time.
- It should be easy to transport.

In [4] a classification of the search and rescue field is presented (Fig. 1) and it is considered necessary to standardize all phases of USAR operations (deployment, search, location, extrication, on-site medical support) and increase the speed of rescue efforts.

**Fig. 1.** The different environments of SAR operations

In the specialized literature, several researches in the field of search and rescue robots are presented [5, 6]. A classification of these robots according to the environment in which they operate can be done as follows: robots for the surface of the ground [7], aerial [8], underwater and water surface. The type of robot has a significant impact on the design and capabilities of the robot [9–11].

The locomotion systems used in this field are especially important in carrying out specific tasks of search and rescue.

For the singular locomotion systems (with feet, wheels and tracks) that move on the surface of the earth can be mentioned that:

- The walking robot has high mobility, but its stability and weight limit it in urban search and rescue applications. It is easy for this robot to move among the ruins in the case of collapsed buildings.
- The wheeled robot has high velocity when traveling on a flat surface but cannot climb over an obstacle greater than the radius of the wheels.
- The crawler-based robot has high propulsion and in general it is used for outdoor environment. It is difficult to enter in a collapsed building.

The thin-body snake robot can have many advantages in USAR applications. This robot can enter the collapsed building through cracks and crevices by performing a proper walk. In addition, articulated robots can be a solution in this sense.

As described in [12] the search and rescue robot with hybrid locomotion system would have great possibilities to successfully reach hard-to-reach places.

From the category of hybrid systems, the design of the TriSTAR unit is aimed at a more comprehensive solution that makes robots more manageable, reliable and robust.

Taking into consideration the fundamental problems of the search and rescue robots listed above and the importance of the field in helping and rescuing victims of various disasters, research in this field is fully justified.

A great advantage of the TriSTAR locomotion unit proposed by the authors is that it can move on the stairs. Even if the proposed robot has limitations in various operating environments, the authors consider that it can be a solution in the locomotion systems of search and rescue robots.

The paper is organized as following. In the next chapter we will discuss about the proposed search and rescue robot, the TriSTAR locomotion unit and design relationships in the third chapter, the details on the actuation and control of the robot in the fourth chapter, in the fifth chapter are presented the research directions and finally the conclusions of the paper.

## 2 Description of the Search and Rescue Mobile Robot

We propose a search and rescue mobile robot made up of a central platform, with a locomotion system that includes classic wheels and TriSTAR locomotion units (Fig. 2).

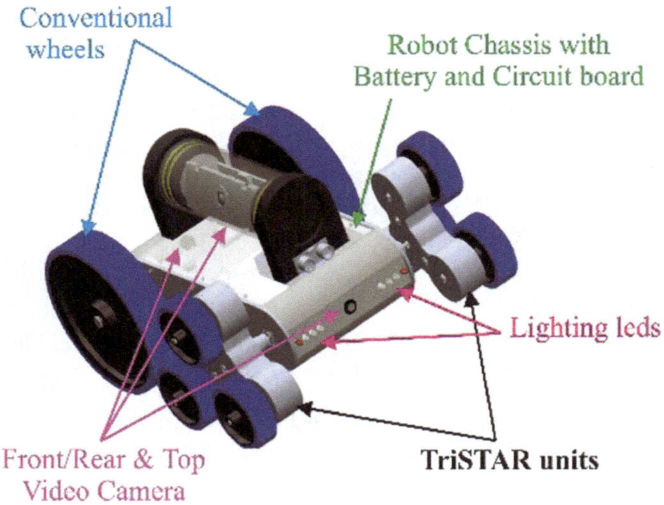


**Fig. 2.** The proposed search and rescue robot

In this paper, the authors have introduced in addition to previous works, a camera module, which can be changed if needed, with a serial robotic arm module. The basic concept of operation and control of the proposed search and rescue robot was also outlined.

In the front and rear side of robot platform (Fig. 2) are included several video cameras and on the top of platform, there is a module with a video camera at 180°. Also included in the platform are lighting LEDs, battery, control board and four DC motors with GHM-01 reducer [13].

The locomotion system can be easily changed according to needs, thus other configurations of mobile robots can be obtained as shown in Fig. 3.

The robots with the three types of locomotion systems that can be obtained are with conventional wheels (Fig. 3a); with TriSTAR locomotion units (Fig. 3b) and with a hybrid system, that combines classic wheels with TriSTAR locomotion units (Fig. 3c). The mobile robots obtained with these locomotion systems are shown in Fig. 3.

a) b) c)

**Fig. 3.** The 3D model of the mobile robot with the proposed locomotion systems

An important feature of the search and rescue robots is the size of the robot. The proposed robot was designed so that it can be easily transported in a backpack. The overall dimensions are 400 × 380 × 180 mm, so there is no need for special transportation to reach the intervention site.

## 3 TriSTAR Locomotion Unit

The 3D model of the TriSTAR locomotion unit and the classic wheel used for proposed robot are presented in Fig. 4. The diameter of the classic wheel is 200 mm, and the overall height of the TriSTAR unit is also 200 mm. The maximum height of the obstacles that can be exceeded with this type of wheels is 100 mm.

The proposed locomotion unit is shown in Fig. 5 [14].

**Fig. 4.** 3D model of TriSTAR locomotion unit a) and conventional wheels b)

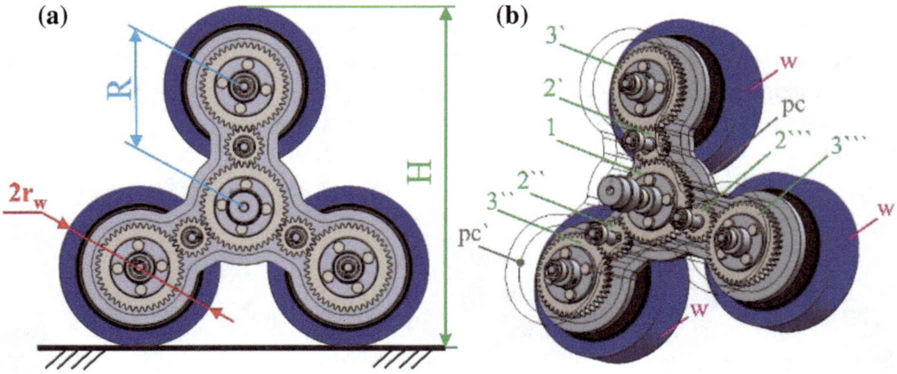


**Fig. 5.** 3D model of the TriSTAR locomotion unit a) design parameters b) component elements

The components of the locomotion unit are the following [14]: three conventional wheels noted with w, a solar gear with the number of teeth $z_1$ noted with 1, three primary satellite gears with the number of teeth $z_2$ noted with 2′, 2″, 2‴, three secondary satellite gears with the number of teeth $z_3$ noted with 3′, 3″, 3‴, pc - planet carrier, pc′ - cover of planet carrier.

The geometry of a TriSTAR type locomotion unit allows switching between wheel locomotion (advancing mode) and walking locomotion (automatic climbing mode) [14–16].

This type of design as can be observed in Fig. 5a, b uses a classic wheel with the radius $r_w$ attached at the end of each leg. The legs are arranged at an angle of 120° and contain gear wheel transmissions.

The great advantage of these TriSTAR locomotion units is that they use a single motor, a motor that is fixed to their central axis.

Figure 6 shows the movement of the TriSTAR unit as follows: movement on flat surface (Fig. 6a), on rough terrain (Fig. 6b) and climbing of a stair (Fig. 6c) [14, 17].

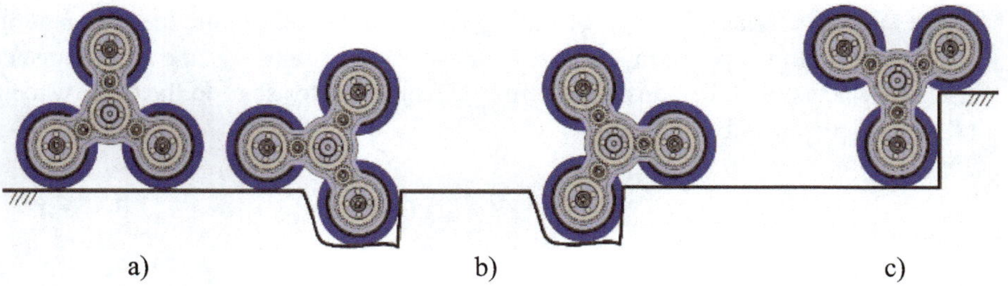


**Fig. 6.** Movement of the TriSTAR locomotion unit a) flat surface b) rough terrain c) on stairs

### 3.1 Design Relationships

The radius of the locomotion unit R is depending on the *m* module of the gear and the teeth numbers $z_1$, $z_2$, $z_3$ there of $R = f(m, z_1, z_2, z_3)$. This radius is also the axial distance between the axes of the gears 1 and 3 and can be determined with the Eq. (1) [16]:

$$R = m\left(\frac{z_1}{2} + z_2 + \frac{z_3}{2}\right) \tag{1}$$

And the height of the locomotion unit H is:

$$H = 3\frac{R}{2} + 2r_w \tag{2}$$

If the TriSTAR locomotion unit is designed according to the height *h* and the width *w* of a step, the Eqs. (1), (2) and (3) proposed by [18, 19] can be used. In this case, $r_{w\ min}$ and $r_{w\ max}$ represent the minimum and maximum limit between which the radius of the outer wheels must fit; *w, t* - is the foot width of the locomotion unit [18].

$$r_{w\ min} = \frac{6Rt + h\left(3w - \sqrt{3}h\right)}{\left(3 - \sqrt{3}\right)h + \left(3 + \sqrt{3}\right)w} \tag{3}$$

$$r_{w\ max} = \sqrt{\frac{w^2 + h^2}{2}} \tag{4}$$

In addition, *R* can be determined with the equation:

$$R = \sqrt{\frac{w^2 + h^2}{3}} \tag{5}$$

The width *t* of the foot of the locomotion unit must be designed in such a way to avoid a collision with the step. In this sense in [18, 19] there are presented the relations for $t_{max}$. Different forms of design of the legs of the locomotion unit are presented in [15, 17]. The authors propose in [16] that in the determination of $t_{min}$, the head diameter of the gear wheels used for transmission from the drive motor shaft to the outer wheels $w$ to be also considered.

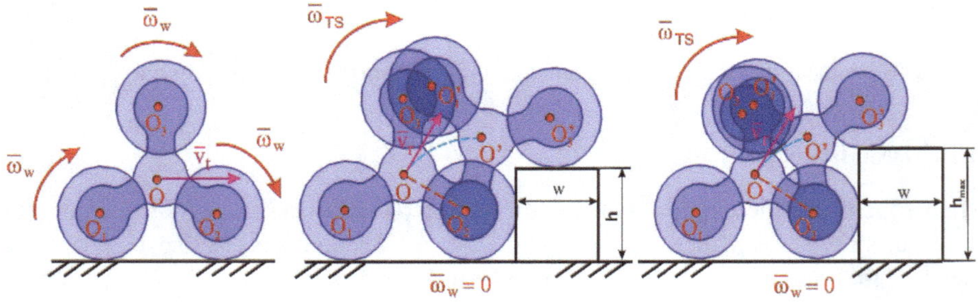


a) b) c)

**Fig. 7.** Locomotion unit's velocity a), b), c) and step parameters b), c)

In Fig. 7 it is noted with: $\overline{\omega}_w$ - the angular velocity of the outer wheels, $\overline{v}_t$ - the linear speed of the locomotion unit ($\overline{v}_t = \overline{v}_O$), $\overline{\omega}_{TS}$ - the angular velocity of the TriSTAR unit, $\overline{v}_r$ - the tangent linear velocity at the trajectory center O of the translation unit, h - the height of the step ($h_{max}$ is the maximum height of the step), w - the width of a step.

The maximum height $h_{max}$ of an obstacle that can be overpassed with this locomotion unit having $z_1 = 42$, $z_2 = 19$, $z_3 = 40$ teeth and the gear module m = 1 [mm], is 100 mm. The radius of the TriSTAR locomotion unit is R = 60 [mm] and the radius of the outer wheels $r_w = 40$ [mm].

If the robot moves on a flat surface, the two wheels of the TriSTAR locomotion units will be in contact with the locomotion surface, and the robot will move with a higher speed (Fig. 8a) than in the case of overpassing obstacles.

When the wheels cannot be used in movement due to obstacles or terrain irregularity, the TriSTAR locomotion units will stop and will rotate around the wheel axis, which is in contact with the obstacle/denivelation. For example, in Fig. 7b, 7c the TriSTAR unit rotates around the $O_2$ axis of the wheel 2.

The disadvantage of such a locomotion system (TriSTAR) is that the wheels at the end of the legs require a high torque, which means that the drive motors must be carefully chosen.

In the following figures (Fig. 8) the robot is presented in some of the situations which can be found in the operating field. One of the great advantages of using this type of hybrid locomotion system is the ability to switch from wheel to legs. To highlight this, in the following figures, we presented the robot moving in several situations: on flat surfaces where it can reach high speed (Fig. 8a), then overcoming obstacles such as stairs (Fig. 8b) and finally moving on uneven terrain (Fig. 8c, d).

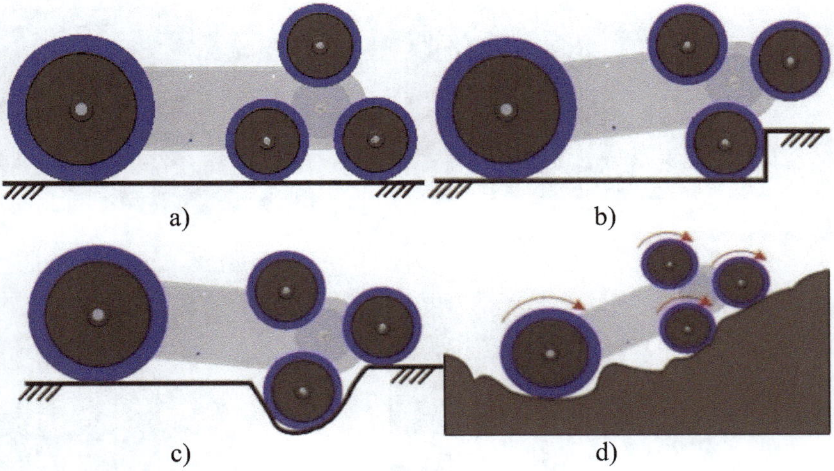


**Fig. 8.** The robot in different situations in the field of operation

Uneven terrain is encountered in various cases in the field of search and rescue and can be considered as a combination of different types of surfaces. Taking this into account, in the following figures, the robot is presented in such situations: in ascent or descent on a longitudinal slope (Fig. 9b, c) when climbing or descending stairs (Fig. 9d, e) or passing over pits (Fig. 9f).

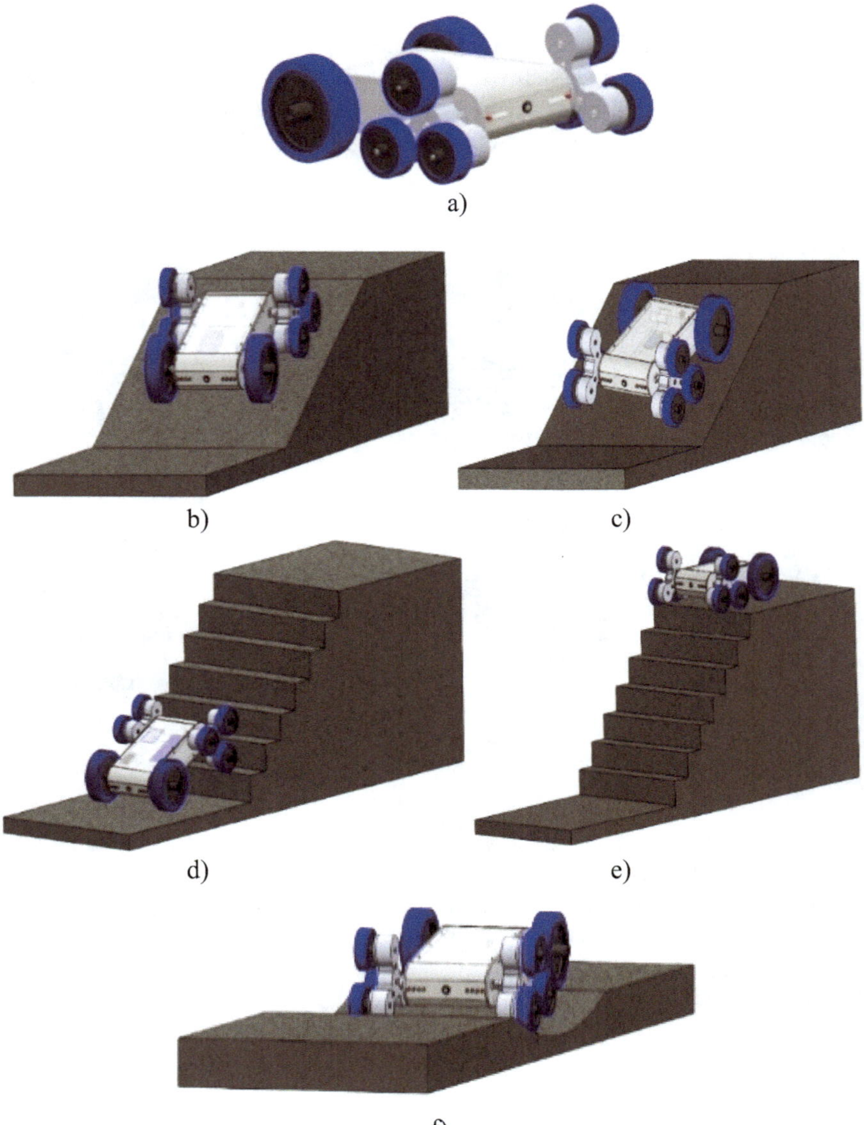

a) b) c) d) e) f)

**Fig. 9.** The robot on fundamental terrains to maneuver on uneven terrain a) basic posture b) ascension c) descending d) climbing stairs e) descending stairs f) uneven surface

# 4 Operation and Control of the Robot

The TriSTAR locomotion units and wheels will be driven by a 12-V DC motor GHM-01 with 30:1 gear ratio reduction and 200 RPM rated speed [13].

In order to be able to move the robot in different directions with variable speeds, each DC motor must be controlled with a driver. It usually has an H-bridge configuration as shown in Fig. 10 and is driven with PWM signals.

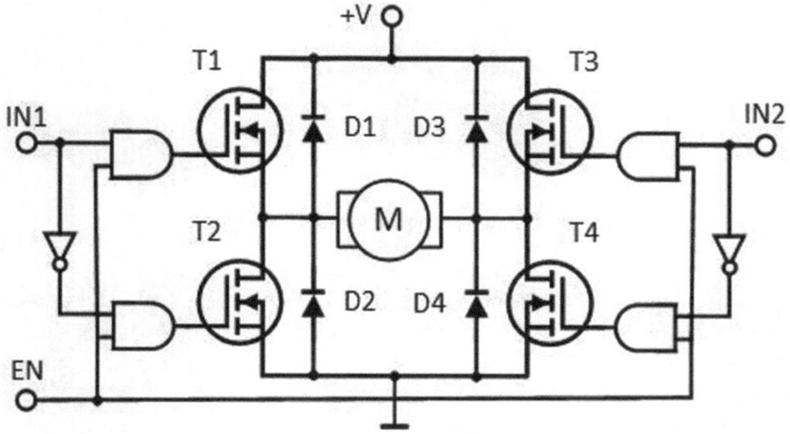


**Fig. 10.** H-bridge diagram

The logic table for actuating a motor is shown in Table 1.

**Table 1.** Logic table for motor actuation

| EN | IN1 | IN2 | MOTOR STATE |
|---|---|---|---|
| 0 | – | – | STOP |
| 1 | 0 | 0 | BRAKE |
| 1 | 0 | 1 | CLOCKWISE ROTATION |
| 1 | 1 | 0 | COUNTER CLOCKWISE ROTATION |
| 1 | 1 | 1 | BRAKE |

Modules equipped with L298 driver were used to control the motors of the robot proposed in the paper. One module with L298N driver is presented in Fig. 11. This module operates at voltages between 5 and 35 V, the logic control is between 4.5 and 7 V, supports a maximum current of 3 A per channel and it can drive two DC motors.

**Fig. 11.** Image of the L298N Driver [20]

The basic diagram of the robot operation is shown in Fig. 12. The power supply ensures all the voltages for components and peripherals. The Arduino Uno development board, which will be used, contains a 5 V regulator that supplies the microcontroller.

For the control of peripherals that include video cameras, camera actuators, lights, the serial robotic arm and other elements/components that will equip the robot, other electronic modules or components will be used. Also included in the peripheral category is the Bluetooth module for data transfer between the robot and the operator.

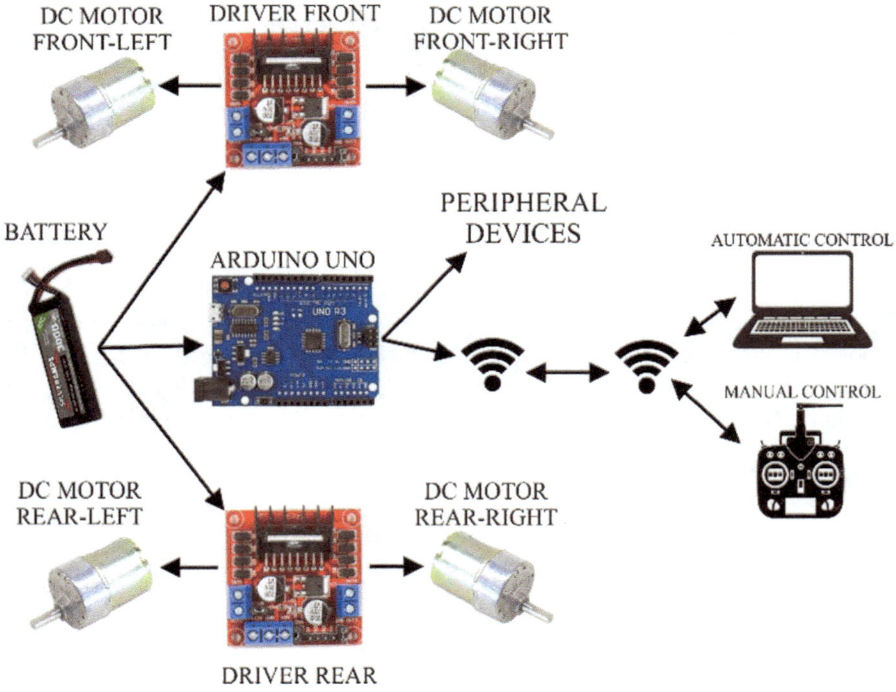


**Fig. 12.** Actuation diagram of the mobile robot

As stated, before in case of the encountered situations in field if there is necessary to remove objects which are blocking the way of the robot or it necessary to observe small places where the robot cannot enter, the mobile robot can be equipped with a serial manipulator with 5 DOF shown in Fig. 13.

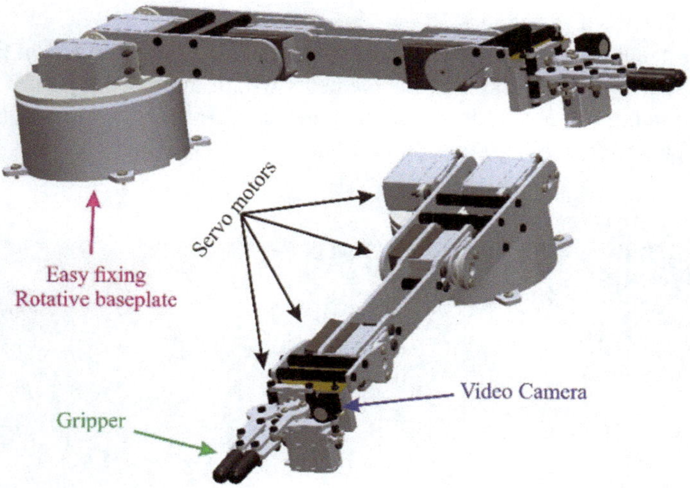


**Fig. 13.** The proposed serial robotic arm

The action diagram of the robotic arm containing the 7 servomotors (denoted M1, M2, M2′… M6) is shown in Fig. 14. The servomotors M2 and M2′ from the base of the robot are connected and actuated simultaneously. The M6 servomotor will be used to operate the gripper.

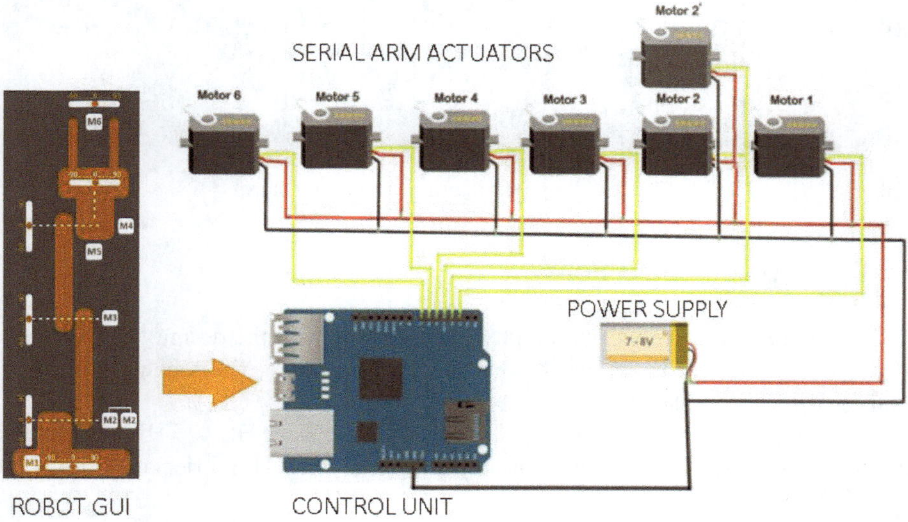


**Fig. 14.** The operation of the serial robotic arm

For the *control of the robot*, a *graphical interface* developed in Processing will be used, which will display the image received from the video cameras, the values of the sensors placed on the robot and the control elements of the robot including its peripherals such as the orientation of the camera, the lights, the actuation of the robotic arm, and so on.

The proposed graphical interface for controlling the robot and cameras and sensor monitoring is shown in Fig. 15.

Control panel for the serial robotic arm is arranged in the interface on the left side. To modify the angular positions of the elements within the robot structure, the control bars corresponding to each rotation coupling are used. At the bottom right, the interface contains a panel for moving the mobile robot.

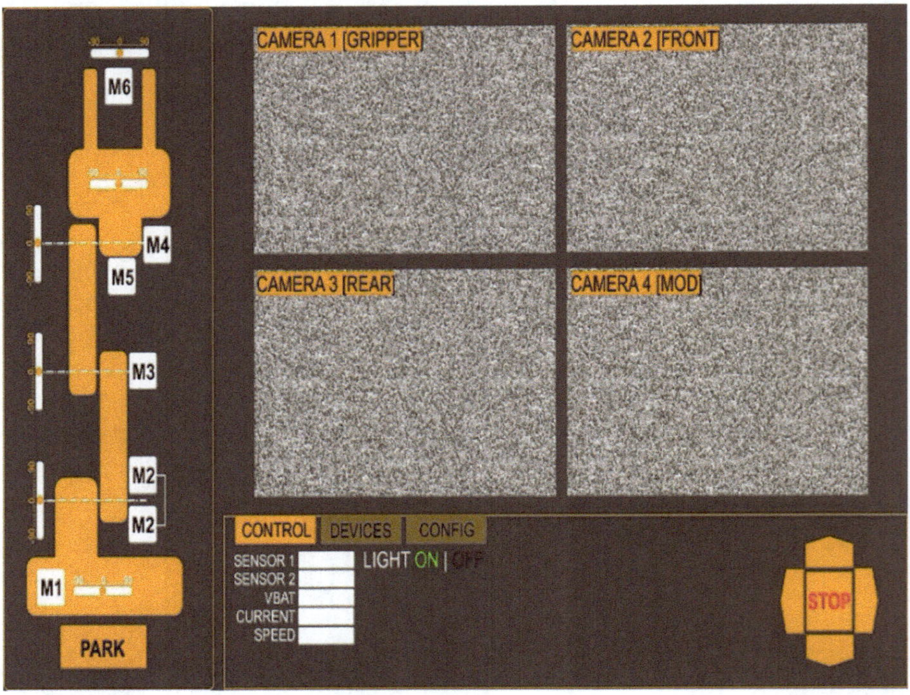


**Fig. 15.** Proposed interface for robot control

## 5 Development Directions

In the next period the robot will be physically realized. As the design has been done modularly, the front camera can be disassembled and the serial robotic arm can be connected for manipulation.

The manipulator will have 5 DOF. To operate, three Hitec HS-475 HB servomotors and one HS-422 servomotor will be used for the rotary base. For orientation, it will be used a 2 DOF mechanism that will be operated by the HS-475 HB and HS-85 servomotors and the gripper will be operated by an HS-81 servomotor.

The model of the proposed robot with the manipulator arm is shown in Fig. 16.

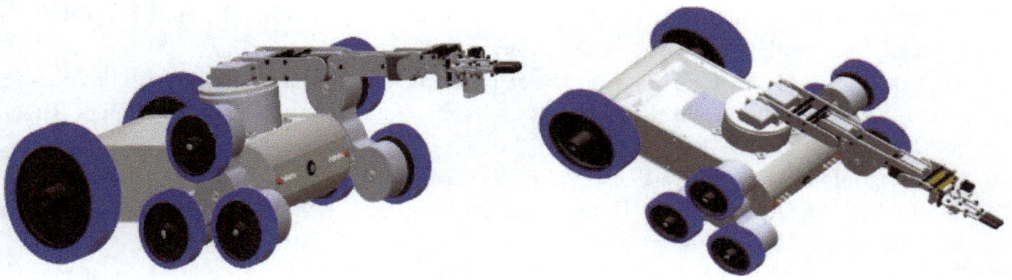

a) b)

**Fig. 16.** The proposed robot with the manipulator device

Besides mobility, there is also a need for a good communication method between the robot and the rescue team. In this sense the robot will contain sensors (e.g. ultrasonic, temperature, humidity), front-rear cameras, infrared camera, microphone, speaker and a GPS system which will help rescuers to obtain information from the operating environment. All of these will be connected to a wireless transmission-reception module so that the robot is in constant contact with the rescue team regardless of its positioning.

In addition to wireless communication and energy autonomy (on battery), the robot will also have a safety cable that also has the role of auxiliary power supply.

After the physical realization the robot will be tested in the field in order to bring the corresponding continuous improvements.

## 6 Conclusions

The authors propose in this paper a search and rescue robot with hybrid locomotion system that combines TriSTAR units and conventional wheels, which are an improvement for the mobility chapter of the search and rescue field. The locomotion units have the great advantage of moving on stairs and using only one motor for driving, which is an improvement for the autonomy chapter from the search and rescue field. To can transport easily in a backpack the robot designed it's a small size robot, which is a fulfillment of the search and rescue field most important requirements. Also, the modularly designed camera system which can be exchanged with the serial robotic arm is a useful concept in this field where human-machine interaction, mobility and communication are an important point.

## References


1. Dai, F., Kang, Q., Yue, Y., Xie, P.: Research on visualized search and rescue robot. J. Robot. Network. Artif. Life **3**(3), 201–204 (2016)
2. Burke, J., Murphy, R., Coovert, M., Riddle, D.: Moonlight in Miami: an ethnographic study of human-robot interaction in USAR. Hum. Comput. Interact. **19**(1–2), pp. 85–116 (2004). (Special Issue on Human-Robot Interaction)
3. Ye, C.L., Ma, S.G., Li, B.: Design and basic experiments of a shape-shifting mobile robot for urban search and rescue. In: Proceedings of International Conference on Intelligent Robots and Systems, pp. 3994–3999 (2006)
4. Statheropoulos, M., Agapiou, A., Pallis, G.C., Mikedi, K., Karma, S., Vamvakari, J., Dandoulaki, M., Andritsos, F., Thomas, C.P.: Factors that affect rescue time in urban search and rescue (USAR) operations. Nat. Hazards **75**(1), 57–69 (2015)
5. Delmerico, J., Mintchev, S., Giusti, A., Gromov, B., Melo, K., Horvat, T., Cadena, C., Hutter, M., Ijspeert, A., Floreano, D., Gambardella, L., Siegwart, R., Scaramuzza, D.: The current state and future outlook of rescue robotics. J. Field Robot. **36**(7), 1171–1191 (2019)
6. Choi, B., Lee, W., Park, G., Lee, Y., Min, J., Hong, S.: Development and control of a military rescue robot for casualty extraction task. J. Field Robot. **36**(4), 656–676 (2019)
7. Wang, W., Dong, W., Su, Y.U., Wu, D., Du, Z.: Development of search-and-rescue robots for underground coal mine applications. J. Field Robot. **31**(3), 386–407 (2014)
8. Qi, J., Song, D., Shang, H., Wang, N., Hua, C., Wu, C., Qi, X., Han, J.: Search and rescue rotary-wing UAV and its application to the Lushan Ms 7.0 earthquake. J. Field Robot. **33**(3), 290–321 (2016)
9. Bogue, R.: Search and rescue and disaster relief robots: has their time finally come? Ind. Robot Int. J. **43**(2), 138–143 (2016)
10. Murphy, R., et al.: Search and rescue robotics. In: Siciliano, B., Khatib, O. (eds.) Springer Handbook of Robotics, pp. 1151–1173. Springer, Heidelberg (2008)
11. Faruk, Kececi E.: Design and prototype of mobile robots for rescue operations. Robotica **27** (5), 729–737 (2009)
12. Bruzzone, L., Quaglia, G.: Review article: locomotion systems for ground mobile robots in unstructured environments. Mech. Sci. **3**, 49–62 (2012)
13. Data sheet for GHM-01. https://www.electrokit.com/uploads/productfile/41002/ghm01.pdf. Accessed 24 Apr 2020
14. Gyarmati, M., Tătar, M.O.: Locomotion systems for search and rescue robots. Robotica Manage. **24**(1), 8–13 (2019)
15. Quaglia, G., Franco, W., Oderio, R.: Wheelchair.q, a motorized wheelchair with stair climbing ability. Mech. Mach. Theory **46**(11), 1601–1609 (2011)
16. Tătar, M.O., Gyarmati, M.: Locomotion unit for mobile robots. In: IOP Conference Series: Materials Science and Engineering, p. 052024. IOP (2018)
17. Quaglia, G., Bruzzone, L., Bozzini, G., Oderio, R., Razzoli, R.P.: Epi.q - TG: mobile robot for surveillance. Ind. Robot Int. J. **38**(3), 282–291 (2011)
18. Dalvand, M.M., Moghadam, M.M.: Stair climber smart mobile robot (MSRox). Auton. Robots **20**, 3–14 (2006)
19. Moghadam, M.M., Ahmadi, M.: Climbing robots. In: Habib, M.K. (ed.) Bioinspiration and Robotics: Walking and Climbing Robots. I-Tech, Vienna (2007)
20. Image of the L298N Driver. https://www.makerfabs.com/l298n-motor-driver-board.html. Accessed 04 Feb 2020

# Design and Simulation of Gait Rehabilitation Parallel Robotic System

Bogdan Gherman[1], Iuliu Nadas[1,2], Paul Tucan[1], Giuseppe Carbone[1,3], and Doina Pisla[1(✉)]

[1] CESTER-Research Center for Industrial Simulation and Testing, Technical University of Cluj-Napoca, Cluj-Napoca, Romania
doina.pisla@mep.utcluj.ro
[2] National Institute for Research and Development of Isotopic and Molecular Technologies, Cluj-Napoca, Romania
iuliu23@yahoo.com
[3] Department of Mechanical, Energy, and Management Engineering, University of Calabria, Rende, Italy
Giuseppe.Carbone@unical.it

**Abstract.** The aim of this paper is to present design considerations regarding a parallel robotic system for gait rehabilitation of patients who have suffered a stroke. The paper focuses on how the design of the robot meets anthropometric and motion amplitude requirements. The robotic system is analyzed using commercially available software, thus proving the workability in medical gait recovery. CAD based simulations are made to show the performance of the robot, using as inputs data Optitrack Motion Capture tracking. A set of testing protocols for the robotic system are described in order to clearly show the operation features of the robot.

## 1 Introduction

The number of people that are affected by stroke all over the world are in continuous growth also the number of patients affected by vascular accident were increased in the last decade [1]. Increasing the life span of the population the stroke is the major cause of disabilities leads to temporary or permanent motor disfunction affecting the life quality and activity daily living [2]. A problem that will occur in the next decades will be the growing number of elderly people and the need to care for them, involving the medical recovery because most of them will survive after a stroke but will remain with motor disabilities [3]. The human resources ready to provide medical assistance to people in the process of post-stroke medical recovery will be insufficient, therefore a technological alternative must be prepared so that medical needs regarding post-stroke medical recovery will be covered [4]. Robotic rehabilitation system can compensate this shortcoming by being precise [5, 6], not signs of fatigue, they can be set to operate at different intensities, are programable and can be substitute physiotherapist in the training necessary for stroke rehabilitation [7]. The robotic systems solutions for gait rehabilitation available in the market have an average cost between $60000 and $300000, this makes them difficult to access for the medical units an alternative


E.-C. Lovasz et al. (Eds.): MTM&Robotics 2020, MMS 88, pp. 187–200, 2021.
https://doi.org/10.1007/978-3-030-60076-1_17

solution represents low-cost robotic systems that may be used in the post-stroke gait rehabilitation [8]. Some examples of rehabilitation systems are quickly presented in this paper. A gait rehabilitation system that have to be mentioned is ANdROS [9] (active knee rehabilitation) is a portable robotic device used for gait rehabilitation and for monitoring, the exoskeleton consisting of two orthoses an actuated one placed on the impaired link and an orthosis equipped with sensors placed on healthy leg. *The disadvantage* of this portable orthosis consists of the fact that the patients that are in the acute or post-acute phase after stroke cannot use this device because of the impossibility to adopt an upright position without an external sustaining. Lokomat [10] is one of the well-known robotic systems for rehabilitation after stroke used in gait rehabilitation. The system is based on a treadmill and it is developed for the patients who suffered diseases or injuries of the spine or for patients who suffered a stroke. *The drawback* of this robotic system is that the patient has to maintain a minimum balance even though the body is hanged in a harness, so in this case, the therapy cannot start in the stage when the patient is mobilized in the bed. The WalkTrainer [11] is composed of a mobile frame, a pelvic orthosis, a leg orthosis actuated by three motorized joints hip, knee, and ankle. Trainer active body weight support and electrical muscle stimulator that is also an active element necessary in the mobilization of lower limbs. *The disadvantage* of this rehabilitation device consists of the low ability of acute and post-acute patients to hold the mobile frame and to move at the same time.

To fulfill and enlarge the specter of research regarding of robotic system designed to perform gait rehabilitation, this paper comes with a solution consisting of two parallel structures linked together, in order to realized motion for hip, knee, and ankle.

## 2 A Brief Classification of a Gait Rehabilitation Device

Researches suggest that rehabilitation intervention should begin as early as possible and the intervention intensive, having a specific task and multisensory simulation favoring brain plasticity [12]. Neuroplasticity can lead to rehabilitation mechanisms and functional adapting that yields from global modification of neuronal organization.

It is difficult to make a complete analysis for all the devices used worldwide in rehabilitation after stroke because of the big number of prototypes [13].

However, a general classification of the robotic rehabilitation systems can be made according to the type of robots, and the motions that they are applied to the patients, for instance [14]:

i. Exoskeleton type, able to move joints as hip, knee, ankle and they can to control the force, torque, speed and acceleration.
ii. End-effector type moves only the feet, usually a footplate working as support for foot is moved thus the lower limb achieved gait trajectories of gait.
iii. Parallel robotic systems having kinematic parallel chains actuated by active joints, the kinematic chains can move the lower limb like an exoskeleton (as is the system presented in this paper) or as an end-effector, making the mobilization of the lower limb gripped from a certain point.

Robotic systems can be also classified as:

i. Robotic systems that can be static devices, when the weight of the patient is not sustained by the system, i.e. robotic systems that move the lower limb of a patient who is laying on the bed or sitting on chair [15].
ii. The robotic systems that are able to sustain patient's weight and are able to move the patient in the environment such as hospital rooms, domestic places, even outsides environments. An example could be portable exoskeletons, or hybrid robotic systems parallel structure combined with an exoskeleton.

## 3 Design of Hip-Knee Module of RECOVER System

### 3.1 The Design of Hip-Knee (HK) Module

Based on kinematic analyses [16, 17] the new design of the parallel robot shown in Fig. 1, has been developed. This robot addresses the rehabilitation of the lower limb, targeting of bedridden patients, who are unable to stand an upright position. The possible rehabilitation motions achieved with RECOVER are gait training which involves the following motions as hip flexion and extension, knee flexion, plantar flexion, and dorsiflexion the ankle eversion and inversion. The design takes into account the human upper limb's anthropometric proportions, the design starts from the knee-hip module taking into consideration the stroke of active prismatic joints q1 and q2 shown in Fig. 2, these lengths give the total length of the table where the active joint is mounted. The 4 DOF of the robot allows increasing the range of motion for each joint, by adding a very good possibility of customization of the physical characteristics of patients.

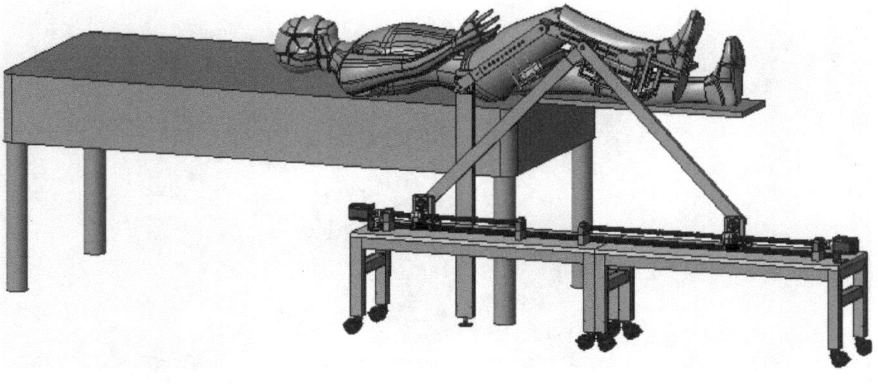

**Fig. 1.** RECOVER the CAD of experimental model

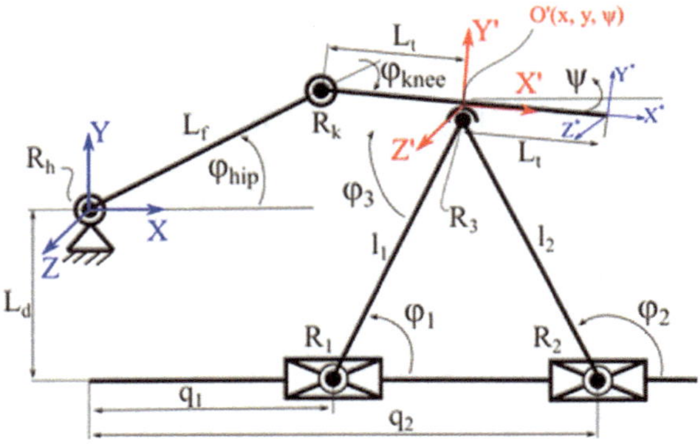


**Fig. 2.** RECOVER Hip-Knee module kinematic scheme [16]

The robot must be in accordance with anthropometric variations of the human body through the $L_f$ and $L_t$ parameters. These values have been chosen based on human male with a height of 1800 mm having approx. 430 mm thigh length and 480 mm leg length [18], imposing the constrain $(q_1 - q_2)/2 \geq 200$ mm. $L_d = 435$ mm taking into consideration the bed height, the equal links are $l_1 = l2 = 850$ mm. The module knee-hip is a 2DOF parallel robot with the use of two prismatic joints. As can be seen in Fig. 3, the linear motions of two prismatic joints are performed by two stepper motors, whose rotational motion is converted by the two ball screws into translational motion. The prismatic joints are guided by two aligned linear rail guides. Each joint slide on a guide block and each of them is connected with a ball nut. The two kinematic chains have one prismatic joint that slides along X axes and two revolute joints situated on guide blocks and two links $L_1$and $L_2$ both intersect in one rotation axes $R_3$ in the origin of moving frame $O'X'Y'Z'$.

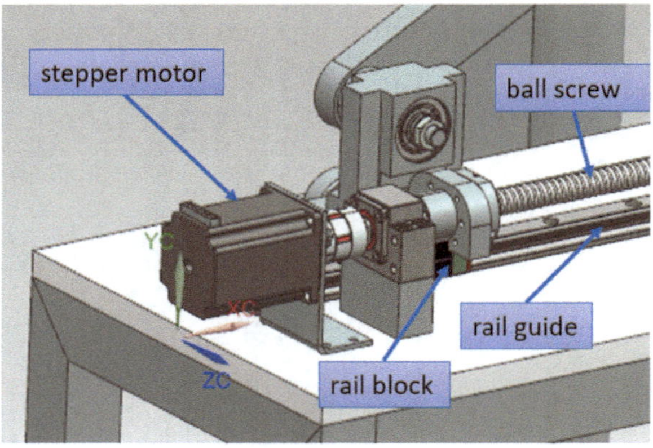


**Fig. 3.** Detailed view representing the linear mechanism for prismatic joints q1 and q2

To assure a good rotation in revolute joints several sizes of radial bearings were used. The third kinematic chain has two revolute joints $R_h$ and $R_k$, and the links $L_f$ and $L_t$: first revolute joint is placed on a fix point Rh where the link $L_f$ starts and intersects link $L_t$ in revolute joint $R_k$; the thigh support is located on the $L_f$ link and the ankle module is placed on the $L_t$ link. The entire translational mechanism is placed on the long table with the 1800 mm length being a solid structure made from welded steel bars. In Fig. 4 can see an overview image where is shown the mechanical links who are designed to be built mainly from Al alloy square pipes, thus having a good material strength and lightweight.

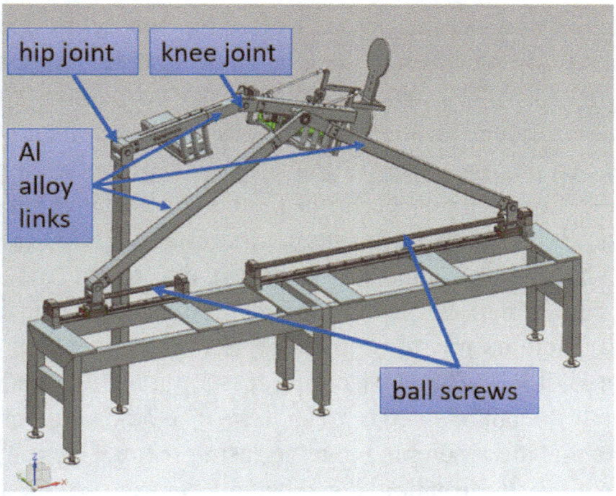


**Fig. 4.** RECOVER experimental model with the main components

The ball screws with 16 mm diameter and 5 mm thread pitch were chosen, as well the stepper motors were chosen to be able to reach high speed and good torque, to have a better sliding beside of using ball screws transmission, the ball nut mechanism slides on

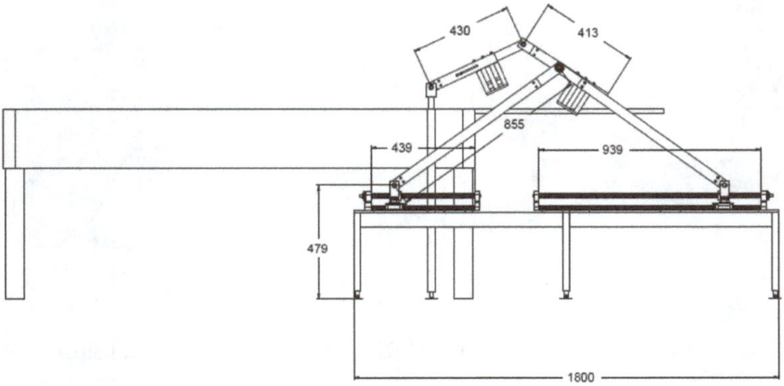


**Fig. 5.** Recover hip-knee module showing main dimensions

the linear guiding rails HGR20. All revolute joints are equipped with double radial bearings. The knee-hip module is able to perform hip flexion and extension and knee flexion and extension. In Fig. 5 is presented the assembly with main dimension of robot.

## 3.2 The Design of Ankle Module

With respect to the ankle module, the ankle joint must be aligned with the revolution axes $R_{a1}$, $R_{a2}$ so that there is no need to adjust the lengths of $L_a$, $l_a$, $L_O$. The only anthropometric adjustment is the distance ($l_S$) between the patient's sole and the intersection point of the $R_{a1}$ and $R_{a2}$ rotation axes. The lengths for the links are: $L_a$ = 140 mm, $L_O$ = 35 mm, $l_a$ = 175 mm.

The ankle module is a parallel structure having two active prismatic joints; the ankle module is fixed on the lower leg support and is made preponderantly from Al alloy combined with ABS. The ankle module is able to perform flexion/dorsiflexion motion through the two identical kinematic chains $\underline{P}SS$, each chain having one active prismatic joint and two passive spherical joints. The linear motion is performed using a set of ball screws. The sole support performs the flexion/dorsiflexion of the ankle when the prismatic joints q3, q4 execute translation motion in the same direction and having the same speed. The revolute joint $R_{a1}$ is driven due to the motion of q3 and q4 described above, thus it will do flexion/dorsiflexion motion. The eversion/inversion motion is achieved when the prismatic joints perform movements in opposite direction actuate the rotation joint $R_{a2}$. The two existing rotation joints $R_{a1}$ and $R_{a2}$, see Fig. 6b), must be aligned with the ankle's center of motion. The ankle support structure is built using Al alloy parts that are assembled together using metric screws, the revolute joints uses radial bearings for better rotation, the spherical joins are achieved using commerce ball joints. The two active joints are actuated by properly sized stepper motor each of them, the sliding motion is achieved with ball screws and a complementary set of linear guiding rails MGN 9H are used to stiffen transmission and reduce friction.

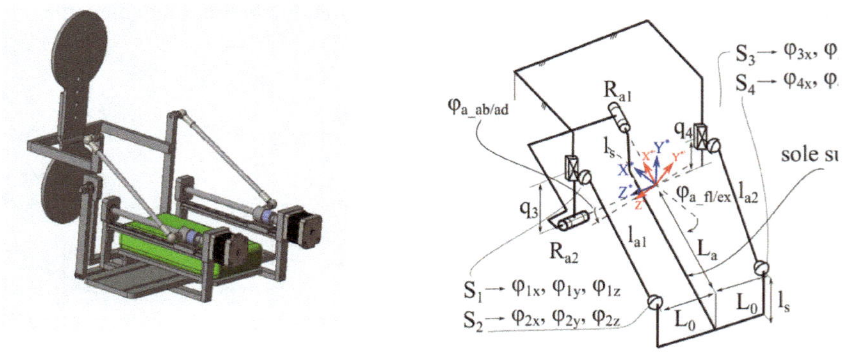


a) b) [16]

**Fig. 6.** a) RECOVER experimental model; b) RECOVER ankle module kinematic scheme [16]

## 4 Performance Simulation of RECOVER Rehabilitation System

The RECOVER robotic rehabilitation system design was subjected to simulation tests in NX. In Figs. 7 and 8 are shown the trace of a point placed on the sole support of ankle module, Fig. 7a) shows the trace simulation performed by HK module without the moving of ankle module, in Fig. 7b) shows the trace of the entire robot system, the robot performs hip and knee flexion/extension and ankle performs plantar flexion. In Fig. 8a) and b) the traces made by a point situated on sole support where is described the trajectory of point in plantar flexion and in inversion/eversion motions are shown.

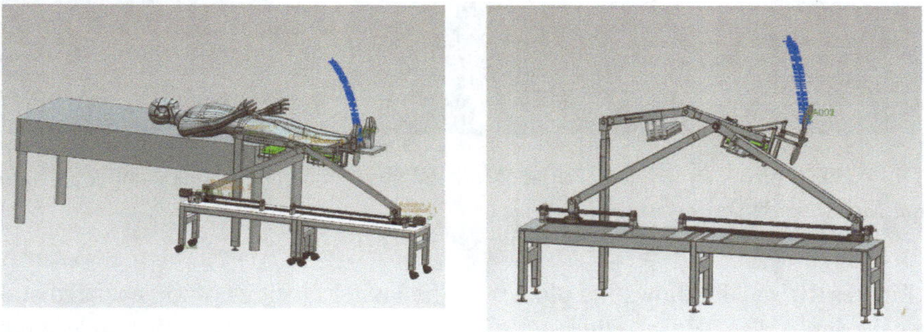

a) b)

**Fig. 7.** RECOVER trace during exercise a) a trace performed by HK module only; b) a trace performed by both modules

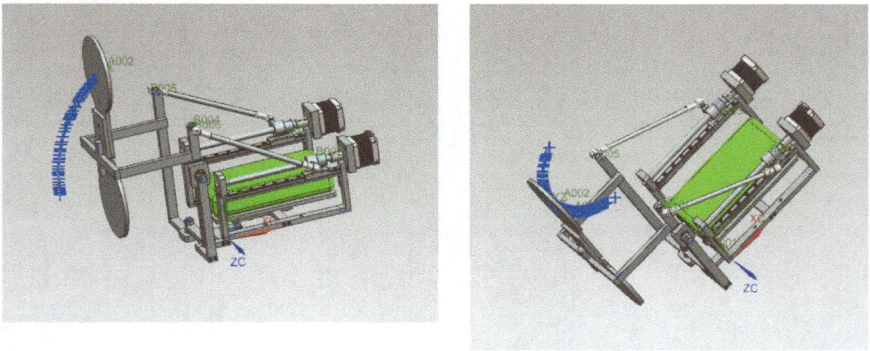

a) b)

**Fig. 8.** RECOVER ankle module showing a) plantar flexion trace; b) inversion/eversion trace

To gather the input data for the Siemens NX simulation, the authors have carried out a gait study experiment using the Optitrack Motion Capture tracking system [19]. A set of 8 highly reflective markers attached to a subject's leg during normal walking placed as in Fig. 9 were tracked. The vectors describe the angles between lower limb segments as follows: the vectors $\boldsymbol{v}_1$ and $\boldsymbol{v}_2$ describe the ankle flexion-dorsiflexion angle, vectors $\boldsymbol{v}_2$ and $\boldsymbol{v}_3$ describe the knee flexion and $\boldsymbol{v}_3$ and $\boldsymbol{v}_4$ the hip flexion. The results of this study in terms of position, speed and acceleration have been used as input data for the NX simulation as follows:

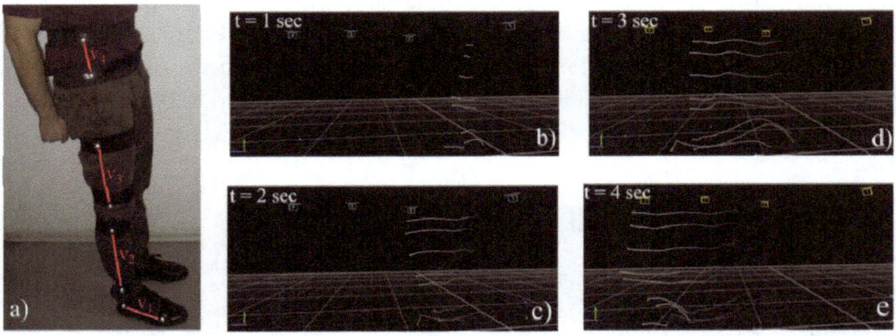


**Fig. 9.** [19] Optitrack experimental setup. Markers' position on the subject's body (a). Markers within Motive software (b–d).

Figures 10 and 11 shows the plots with the input velocities of the actuated sliding joints q1 and q 2 for HK module and q3 and q4 actuated joints of the ankle module.

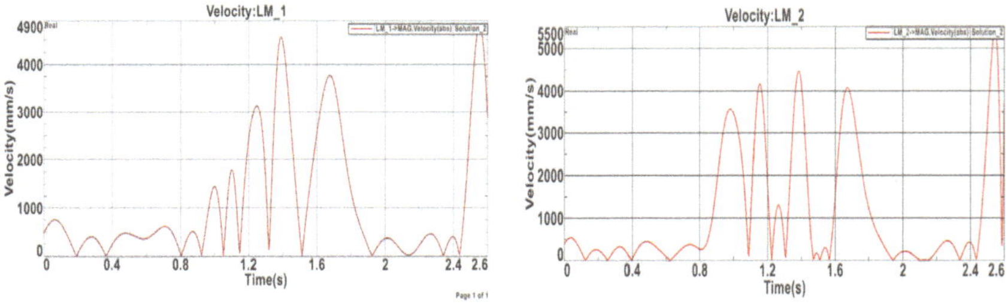


a) b)

**Fig. 10.** The input velocity of active joints: a) q1 and b) q2 of HK module

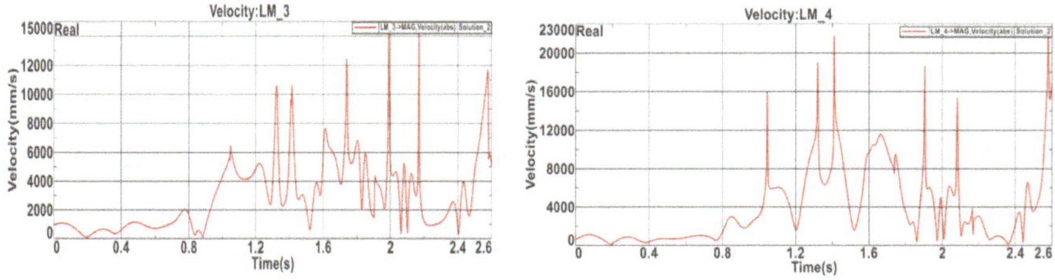


a) b)

**Fig. 11.** The input velocity of active joints: a) q3 and b) q4 of ankle module

In Fig. 12 the values of forces necessary to move the sliding elements, the force range of q1 and q2 are around 400N with a peak that reaches 940N in q1 joint, necessary to move the entire weight of the robot are shown. Also, Fig. 13 shows the torque of prismatic joints.

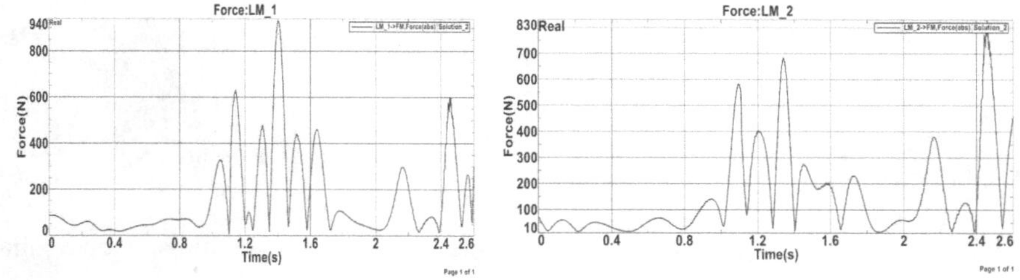


a) b)

**Fig. 12.** Plot representing the forces of prismatic joints: a) q1 and b) q2

Figure 13 represent the torque necessary in the q1and q2 motor's axels to rotate the ball screw thus obtaining a sliding motion. The range value is around 6–7 Nm.

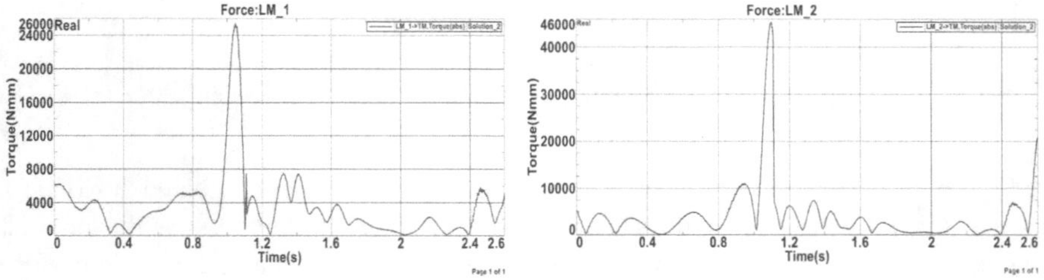


a) b)

**Fig. 13.** Plot representing the torque of prismatic joints: a) q1 and b) q2

The force developed in hip revolute joint Rh shows in Fig. 14a), have values between 250N and maximum 1500N. The force developed in knee revolute joint Rk shows in Fig. 14b), have values around 250N and a peak that reach 1500N when the motion direction changes.

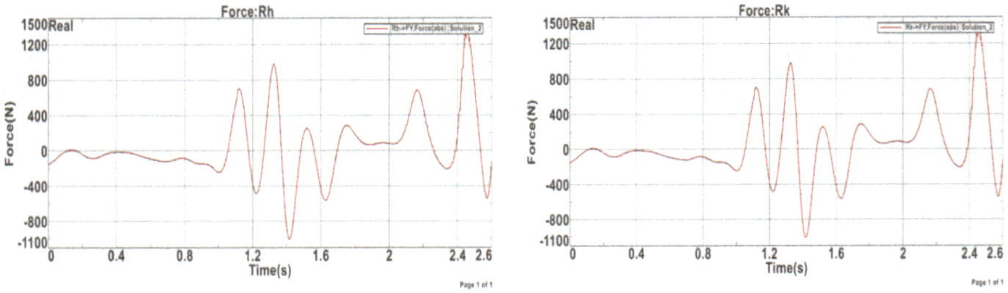


a) b)

**Fig. 14.** Plot representing the forces of hip and knee joints

Figures 15a) and b) presents the torque values developed in the hip and knee joints.

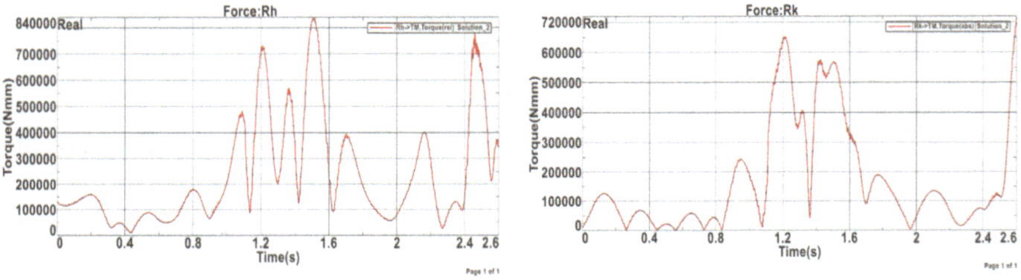


a) b)

**Fig. 15.** Plot representing the torque of hip and knee joints

Figure 16a) and b) shows the forces in the prismatic joints q3 and q4 of ankle module, the force values of these two joints have same values, when the prismatic joints are moving in the same direction having same speed, the ankle module performs dorsiflexion and plantar flexion of the ankle, the moving of prismatic joints in opposite direction impose to ankle module a motion that helps in rehabilitation of eversion/inversion motion of the ankle.

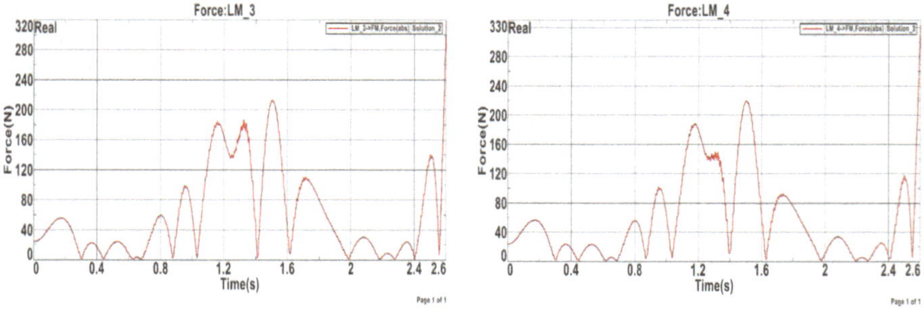


a) b)

**Fig. 16.** Plot representing the force of prismatic joints q3 and q4 of ankle module

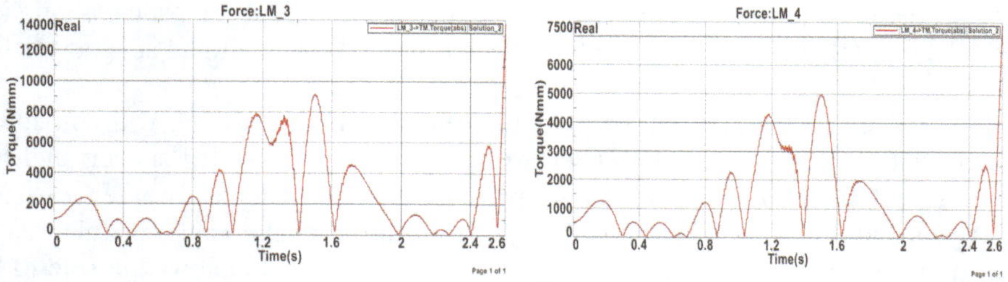


a) b)

**Fig. 17.** Plot representing the torque of prismatic joints q3 and q4

Figures 17a) and b) represent the torque necessary in the q3 and q4 motor's axels to rotate the ball screw thus obtaining a sliding motion. The range value is around 2 Nm.

## 5 Testing Procedures for RECOVER Rehabilitation Robot

To have a clear image of how the robotic system has to be tested, some procedures for usage are necessary to be known. The human subject must lie on the edge of the bed, with the pelvis positioned at the end of the bed so that the limb movement can be achieved beyond the upper side of the bed.

The rotational axes of human hip and rotational axes of robot $R_h$ must be collinear. The length of femoral link $L_f$ of the robot must be adjusted to fit with the anthropometric length of the thigh so that the rotational knee joint axes of the human lower limb and the axes of $R_k$ having to be collinear. The lower limb to which the rehabilitation training is not applied will be placed on support located in the extension of the bed so that it stays perfectly horizontal. As is shown in Fig. 18 the robotic system will be positioned, so that the angle between femoral link and bed must be around 6° and the angle between tibial link and bed must be around 17° (special markers are placed on the

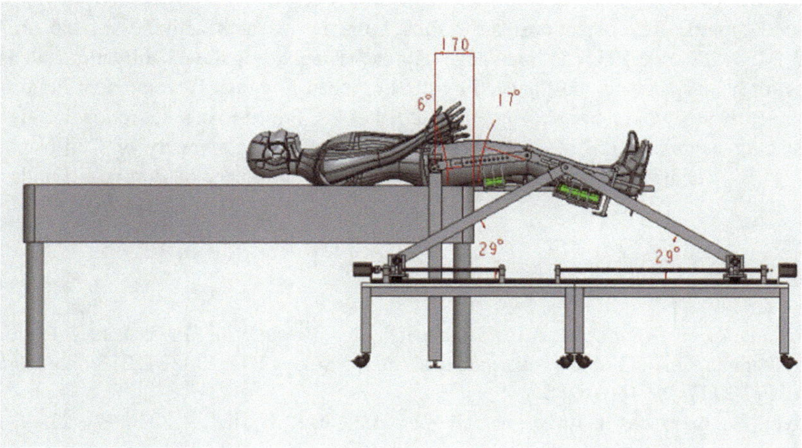


**Fig. 18.** Recover showing the start potion of robotic system

robot to help perform this procedure), so that the lower limb at which is applied rehabilitation training can be easily fixed on the support for thigh and support for the lower leg.

The trained lower limb will be placed on thigh support, the thigh will be tied on support with straps, also the lower leg will be placed on lower leg support and the lower leg will be tied on support with straps. Once the thigh and the lower leg are fixed on their support, the foot is automatically placed inside of the ankle module that is placed on lower limb support, the adjustment length of link $L_t$ allows the sole to be fixed on sole support and the foot will be tied with straps to sole support. The rehabilitation procedure is initiated based on physiotherapist recommendations. The human subject performs motions with the amplitude and speed of the robot set within the safety limit. When the testing procedure is over, the robot will be positioned back on start potion the lower limb will be unstrapped from sole, leg and thigh support and the subject lower link will be unattached from robot.

## 6 Conclusions

This paper has been addressing the design and performance estimation for a parallel robotic rehabilitation system. In particular, the paper highlights the feasible motions for achieving a gait rehabilitation system. CAD models and simulations demonstrate the capability of the robot to perform the required range of motions for gait rehabilitation. Moreover, joint forces and torques are numerically simulated to estimate the performance and feasibility of the proposed robotic system. Moreover, the paper proposes specific operation modes for using the proposed robotic device. Future work will consist of building a prototype with the purpose to demonstrate the rehabilitation feasibility and confirming the estimated performances. The prototype will be evaluated in CESTER (Research Center for Industrial Robots Simulation and Testing) lab to see that the design of the inputs will fulfil the requirements of post-stroke therapy. After the checks, both ergonomic and potential safety issues must be enhanced in order to actually provide a robotic rehabilitation device that is healthy and comfortable.

**Acknowledgment.** The paper presents results from the research activities of the project ID 37_215, MySMIS code 103415 "Innovative approaches regarding the rehabilitation and assistive robotics for healthy ageing" co-financed by the European Regional Development Fund through the Competitiveness Operational Programme 2014-2020, Priority Axis 1, Action 1.1.4, through the financing contract 20/01.09.2016, between the Technical University of Cluj-Napoca and ANCSI as Intermediary Organism in the name and for the Ministry of European Funds.

## References

1. Amanda, G.T., Dominique, A.C., Tharshanah, T., Howard, G., Howard, J.V., Rothwell, P. M., Donnan, G.A.: Global stroke statistics. Int. J. Stroke **12**(1), 13–32 (2017). https://doi.org/10.1177/1747493016676285
2. Silver, B.: Advances in stroke over the past decade. R. I. Med. J. **2014**(97), 27–30 (2013)

3. Erol, D., Sarkar, N.: Intelligent control for robotic rehabilitation after stroke. J. Intell. Robot. Syst. **50**, 341–360 (2007). https://doi.org/10.1007/s10846-007-9169-2
4. Abdullah, H.A., Tarry, C., Lambert, C., et al.: Results of clinicians using a therapeutic robotic system in an inpatient stroke rehabilitation unit. J. NeuroEng. Rehabil. **8**, 50 (2011). https://doi.org/10.1186/1743-0003-8-50
5. Nedezki, C.M., Julean, D., Kerekes, G.: Study of the workspace for parallel manipulator 3RUU. In: Annals of DAAAM for 2009 & Proceedings of 20th DAAAM International Symposium, vol. 20, no. 1, pp. 1445–1446 (2009). ISBN 978-3-901509-70-4, ISSN 1726-9679
6. Nedezki, C.M.: The maximal workspace with constant orientation of the 3 DOF RPR parallel manipulator. In: 3nd International Conference Advanced Engineering in Mechanical Systems 2013, Index Copernicus Journals Master List ID 4755, Cluj-Napoca, Romania, November 2013, vol. 56, no. IV, pp. 725–728 (2013). ISSN 1221-5872
7. Geonea, I.D., Tarnita, D.: Design and evaluation of a new exoskeleton for gait rehabilitation. Mech. Sci. **8**, 307–321 (2017). https://doi.org/10.5194/ms-8-307-2017
8. Nadas, I.A., Pisla, D., Vaida, C., Gherman, B.G., Carbone, G.: Towards cost-oriented user-friendly robotic systems for post-stroke rehabilitation. In: Handbook of Research on Biomimetics and Biomedical Robotics, pp. 99–141 (2018)
9. Kawasaki, H., Cox, D., Jeon, D., Saint-Bauzel, L., Mouri, T.: Rehabilitation robotics. J. Robot. **2011**, 1–3 (2011). Article ID 937875, Hindawi
10. Nam, K.Y., Kim, H.J., Kwon, B.S., et al.: Robot-assisted gait training (Lokomat) improves walking function and activity in people with spinal cord injury: a systematic review. J NeuroEng. Rehabil. **14**, 24 (2017). https://doi.org/10.1186/s12984-017-0232-3
11. Yves, S., Bouri, M., Clavel, R., Yves, A., Brodard R.: A Novel verticalized reeducation device for spinal cord injuries: the WalkTrainer, from design to clinical trials. In: Abdellatif, H. (ed.) Robotics 2010 Current and Future Challenges (2010). ISBN 978–953-7619-78-7
12. Major, K., Carbone, G., Major, Z., Vaida, C., Pisla, D.: Predefined assessment tools in robot assisted physical therapy after stroke. Bull. Transylvania Univ. Brasov Ser. I Eng. Sci. **10** (59) No. 1, pp. 47–52 (2017). In 3rd International Conference for Doctoral Students – IPC 2017
13. Vaida, C., Birlescu, I., Pisla, A., Carbone, G., Plitea, N., Ulinici, I., Gherman, B., Puskas, F., Tucan, P., Pisla, D.: RAISE - an innovative parallel robotic system for lower limb rehabilitation. In: Carbone, G., Ceccarelli,, M., Pisla, D. (eds.) New Trends in Medical and Service Robotics. Mechanisms and Machine Science, vol. 65, pp. 293–302. Springer, Cham (2019). https://doi.org/10.1007/978-3-030-00329-6_33
14. Morone, G., Paolucci, S., Cherubin, A., et al.: Robot-assisted gait training for stroke patients: current state of the art and perspectives of robotics. Neuropsychiatr. Dis. Treat. **13**, 1303–1311 (2017)
15. Cafolla, D., Russo, M., Carbone, G.: CUBE, a cable-driven device for limb rehabilitation. J. Bionic Eng. **16**(3), 492–502 (2019)
16. Gherman, B., Birlescu I., Tucan, P., Vaida, C., Pisla, A., Pisla, D.: Modelling and simulation of a robotic system for lower limb rehabilitation. In: ASME. International Design Engineering Technical Conferences and Computers and Information in Engineering Conference, Volume 5B: 42nd Mechanisms and Robotics Conference ():V05BT07A083. https://doi.org/10.1115/detc2018-85872
17. Nadas, I.A., Gherman, B., Bîrlescu, I., Banica, A., Carbone, G., Pisla, D.: Analysis of the design and dynamic balancing of the RECOVER robotic system. In: ACME 2020, Iasi, status: pending decision

18. Hora, M., Soumar, L., Pontzer, H., Sládek, V.: Body size and lower limb posture during walking in humans. PLoS ONE **12**(2), e0172112 (2017). https://doi.org/10.1371/journal.pone.0172112
19. Gherman, B., Birlescu, I., Plitea, N., Carbone, G., Tarnita, D., Pisla, D.: On the Singularity-free workspace of a parallel robot for lower-limb rehabilitation. Proc. Rom. Acad. Ser. A **20**, 383–391 (2019)

# Study on High Productivity Manufacturing Line for Deo Roll Balls

Dan Florin Teuşdea[1(✉)], Erwin-Christian Lovasz[1], Mircea Vodă[1], Nicuşor-Alin Sîrbu[2], Inocenţiu Maniu[1], and Jean Philippe Bizet[3]

[1] Universitatea Politehnica Timisoara, Bld. Mihai Viteazu nr. 1, Timisoara, Romania
dan-florin.teusdea@student.upt.ro
[2] National R&D Institute for Welding and Material Testing–ISIM, Timisoara, Romania
[3] Institute ISEN, 41 Boulevard Vauban, 59800 Lille, France

**Abstract.** The paper proposes the design of a high productivity line with alternative technology for manufacturing deodorant roll-on balls made of polypropylene. The current manufacturing technology of these balls involve high raw material and energy consumption and very low productivity. The alternate method of manufacturing these balls aims to form the ball from two ball halves joined together. This process will result in a significant saving of raw material and energy at significantly higher productivity. Experimental results and similar applications on assembly technologies enable the authors to propose potential solutions for automation and industrialization phase. Some quality issues have been reported along the experimental phase of alternative technologies by joining the ball halves. Several remedial solutions are available but need to be validated by larger amount of tests and measurements. This study generates a novel concept for a high productivity manufacturing line for mass production.

**Keywords:** Deo roll · Balls · Hot plate · Plastic welding · Cam mechanism · Assembly

## 1 Introduction

The increasing use of plastic in various fields has brought many benefits and significant cost savings. Thus, plastic materials have successfully replaced other materials that require higher energy consumption and costs [1]. Unfortunately, unwanted side effects occur mainly because of human behavior and the local authorities' weak involvement in selective collection and waste recovery. The increased pollution of the environment with plastic waste led to wide debates in the European and world forums which triggered the Directive EU 2019/04 of the European Parliament [2].

Currently, there is great emphasis on removing or replacing certain parts of plastic materials and also on reducing specific consumption from the very design phase [2].


E.-C. Lovasz et al. (Eds.): MTM&Robotics 2020, MMS 88, pp. 201–211, 2021.
https://doi.org/10.1007/978-3-030-60076-1_18

This article refers to the cosmetic industry and as a case study we chose as a general term the roll-on deodorant. An important landmark of this product is the deo roll ball (see Fig. 1).

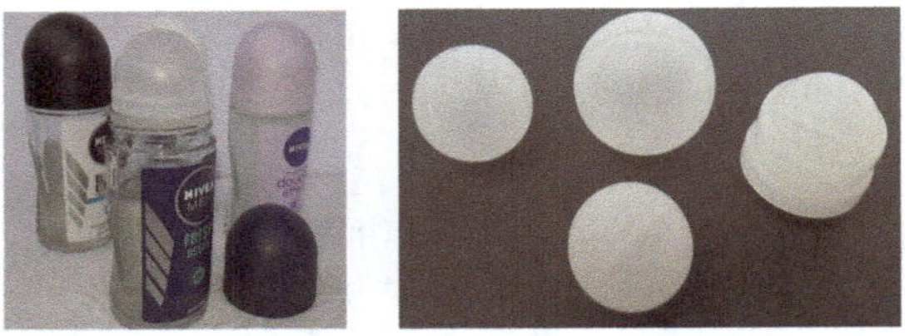

**Fig. 1.** Examples of deo roll components.

According the definition, the deo roll ball made of plastic is a spherical shaped object with thin walls, which rotates in the dosing device and provides the transfer of the liquid (deodorant) from the container to the human body [3, 4].

The classical methods of obtaining deo roll balls consist in an extrusion-blowing process that forms the balls as a single piece (see Fig. 2a).

According to the current technology, the manufacturing of these balls results in an uneven wall thicknesses and a large amount of technological waste of material (see Fig. 2b), which needs to be grinded and reintroduced into the injection equipment. In addition, the productivity of this technology is very low [5].

**Fig. 2a.** Deo roll ball by classic method.

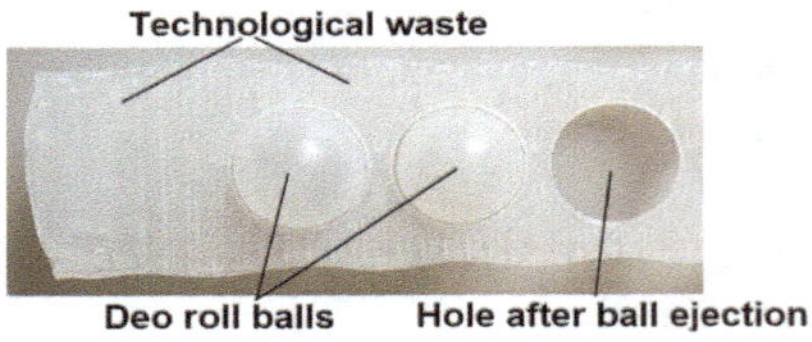


**Fig. 2b.** Technological waste by classic method.

This article is based on an *alternative method* of manufacturing these balls, which creates the ball from *two ball halves* joined together by hot plate welding (see Fig. 3).

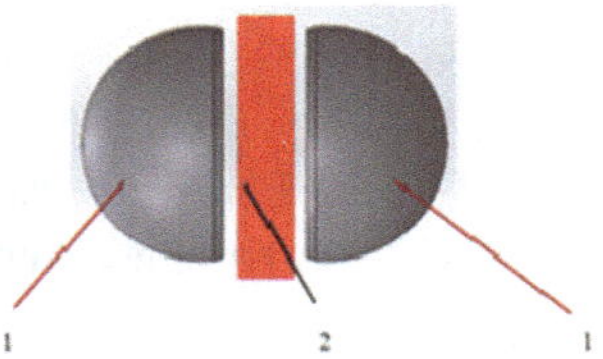


**Fig. 3.** Deo roll ball welding by hot plate. 1 - Half ball (hemisphere) 2 - Hot plate

Using this technology, the wall thickness is better controlled and the weight of the deo roll ball can be reduced [6]. The technological waste related to classic technology is eliminated [8, 9].

The manufacturing method we have chosen consists in two ball halves joined together by hot plate welding [7]. The welding operations are outlined below (see Fig. 4).

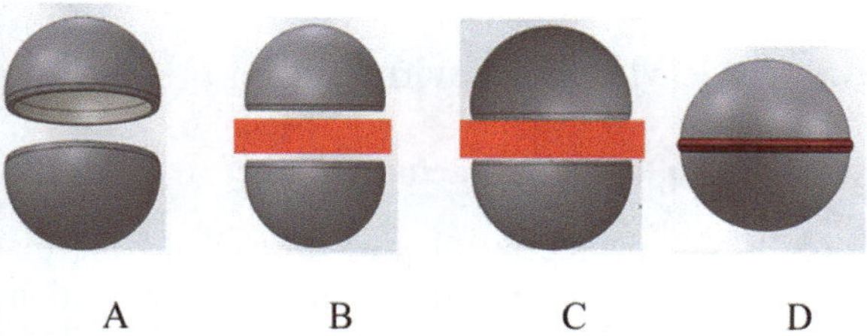


**Fig. 4.** Schematic of deo roll ball welding operation with hot plate (A - Ball halves before welding; B - Alignment; C - Heating; D - Welded ball)

## 2 Manufacturing Process

The technical system performing the welding of the ball halves by hot plate (see Fig. 5) is very similar with the stand used for Infrared Radiation welding [9]. The two half balls (1) are heated in the joining area by the hot plate (2).

The loading of the polypropylene ball halves on the carrier devices (3) is done manually, and the retention in the carrier devices is performed by the vacuum generator (4). Carrier devices are also equipped with temperature sensors (8) with digital temperature controller (5) for each ball half. Upon reaching the joining temperature of 200–250 °C, the pneumatic cylinder (7) acts to withdraw the hot plate (2), and the ball halves are pressed together by the linear pneumatic actuators (6). Weld bead is generated. After welding, the vacuum in the carrier devices is interrupted in order to remove the welded piece and retract the actuators to the ball halves loading position.

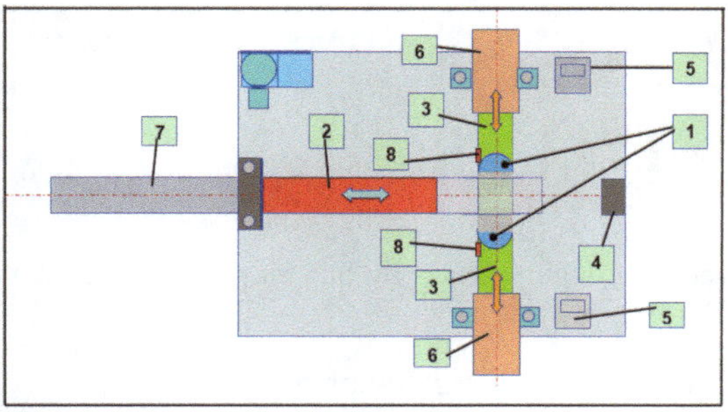


**Fig. 5.** Experimental stand for deo roll balls welding by hot plate.

The technological parameters for our experiment are:

- Welding temperature on the welding bead area: 200–250 °C
- Heating time: 2–3 s
- Welding time: 1.5–2.3 s
- Heating pressure: 0.1 $N/mm^2$
- Welding pressure: 0.2 $N/mm^2$.

## 3 High Productivity Manufacturing Line

### 3.1 Concept of the Complete Manufacturing Line

The experimental results and the recent study of other similar machines on the market [7] allow the configuration of an integrated system for the production and assembly of polypropylene deo roll balls for the cosmetic industry at a rate of 150–200 welded balls/minute (see Fig. 6).

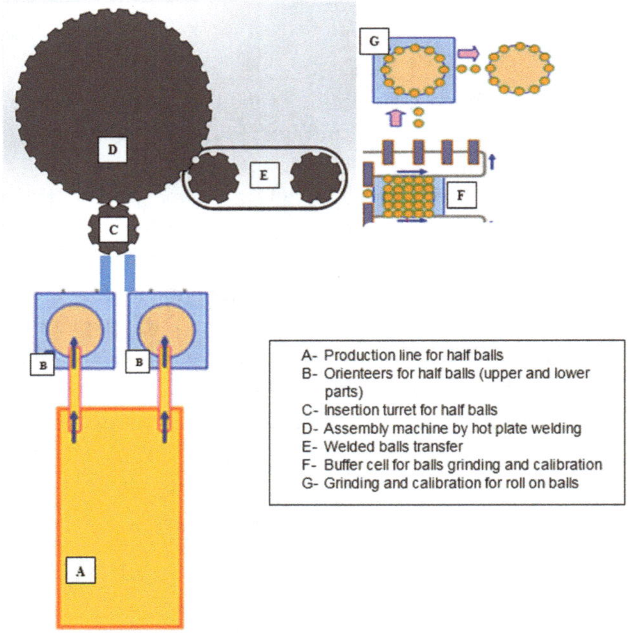


**Fig. 6.** Concept of high productivity line for deo roll balls manufacturing.

This line consists of the following equipment:

**A. Production equipment** for deo roll hemispheres (see Fig. 6A). The equipment comprises of an injection machine and a mold with 24 or 32 cavities. The mold that produces hemispheres of several sizes can be installed on the injection machine.

**B. Orienting and feeding equipment** (see Fig. 6B). The best performing system used for such systems consists of a centrifugal orienteer [10]. The turntable has a (10… 15) degree inclination to the vertical axis and has peripheral "pockets" that allow the entry of the hemispheres only if they are in the correct position. The hemispheres that do not enter the pocket due to incorrect orientation are eliminated by compressed air as they pass through the blowing hole.

**C. Insertion turret** (see Fig. 6C) will provide the separation of the hemispheres in order to be taken over by the assembly device with an exact defined rate. The rotational movement of the insertion unit syncs the rotational movement of the assembly equipment (turret), through mechanical transmissions with gear wheels. The insertion unit will provide the supply of the assembly equipment through two streams of ball halves corresponding to upper and lower assembly tools (see Fig. 7).

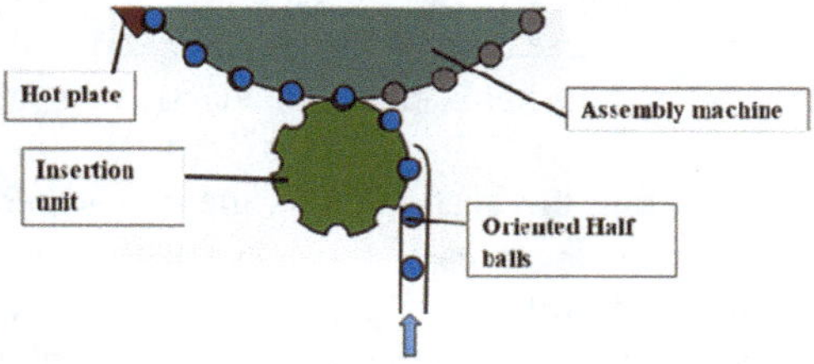


**Fig. 7.** Insertion of the half balls.

**D. The assembly equipment** (see Fig. 6D) by hot plate welding of the hemispheres, is rotational. The main components of the active parts (tooling) of the assembly equipment are shown in the 3D image below (see Fig. 8).

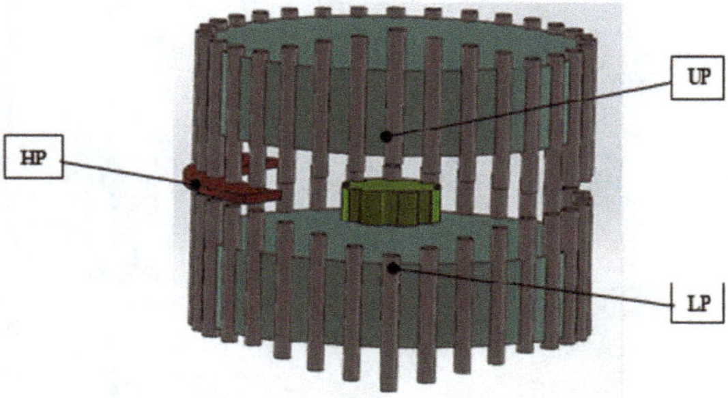


**Fig. 8.** 3D image for deo roll rotational assembly equipment. UP - upper tool; LP - Lower tool; HP - hot plate

The position of the upper and lower tools is given by the cam follower of end cylindrical cam (see Fig. 12). The welding operations described in Fig. 4A, B, C, D are also found in the assembly equipment (see Fig. 9). These operations can be outlined in the following phases:

1. *First phase* (see Fig. 9.1). The 32 pairs of upper (UP) and lower (LP) tools along the carousel grasp the hemispheres. Then the hemispheres come into contact with the alignment system which provides the parallel plane orientation of the hemisphere middle surfaces. The main purpose is to provide a uniform heating of the surfaces to be joined in the second phase.

2. *Second phase*. Uniform heating is maintained until the melting temperature of the adjacent surfaces to be joined is reached (see Fig. 9.2). The heating system is equipped with a ceramic plate (HP) heated up to 280–350 °C and is set along the diameter of the carousel between the upper and lower tools. The hemispheres placed in the synchronized individual carrier units (LP and UP) are brought into contact with the hot plate by precise movements generated by the cam profile. When exiting the heated area, the temperature of the thermoplastic material will reach a value that will allow welding.

The tools in which the hemispheres have been placed are cooled with water at 15–20 °C in order to prevent the overheating of the assembly equipment in time, and also to prevent the "softening" of the hemispheres made of plastic during the welding process.

3. *Third phase*. After exiting the heating zone, the hemispheres enter the welding area (see Fig. 9.3) where they are pressed against each other over a very precise distance determined by the cam profile until the parts interpenetrate in the molten area of the welding bead. A uniform controlled thermoplastic material is flowing inward and outward the joint to obtain a sealed welding, with a mechanical resistance joint according to the required specifications.

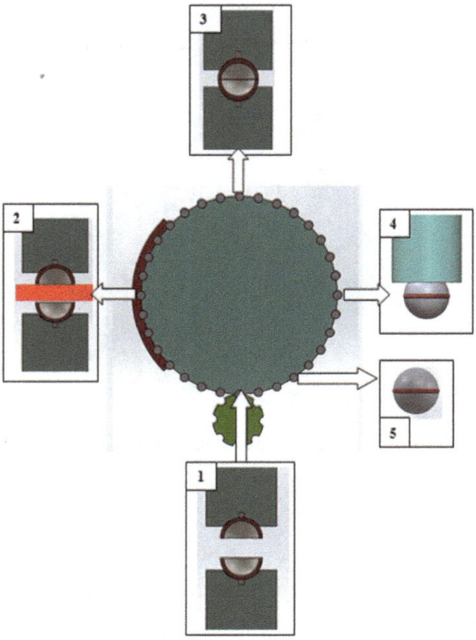


**Fig. 9.** Outlining of the welding operations.

4. *Fourth phase.* The pre-ejecting phase of the welded balls is the next step (Fig. 9.4). This is a phase where the welded ball remains in the upper punch (UP) through the vacuum technique, while the lower punch (LP) drops on the cam.

5. *Fifth phase.* In this phase, the welded ball reaches the evacuation area, the vacuum ceases and the ball is released from the upper punch (see Fig. 9.5).

**E. Transfer equipment** for deo roll welded balls (see Fig. 6E). After ejection, the welded balls are took by a synchronized system that consists of a toothed belt which catch the welded balls from the assembly turret in order to transfer them to the next technological phase (see Fig. 14) (Fig. 10).

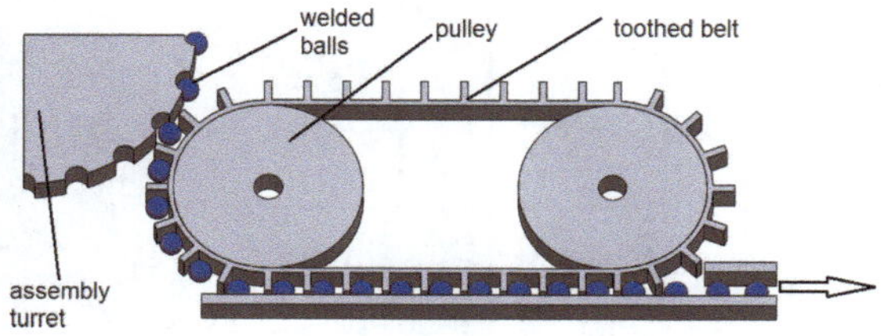


**Fig. 10.** Synchronized system with toothed belt.

**F. The buffer system** (see Fig. 6F). In order to provide an autonomous and continuous operation of the manufacturing line, a buffer zone of a given number of balls is necessary to transfer them to the final grinding and calibration.

**G. Grinding and calibration** equipment (see Fig. 6G) performs the mechanical processing of the semi-manufactured deo roll balls in order to remove all the burrs and flashes resulted from the welding process. It also performs dimensional calibration and seeks to obtain a predefined roughness. The outline of the grinding and final calibration equipment of the deo roll ball is depicted in Fig. 11.

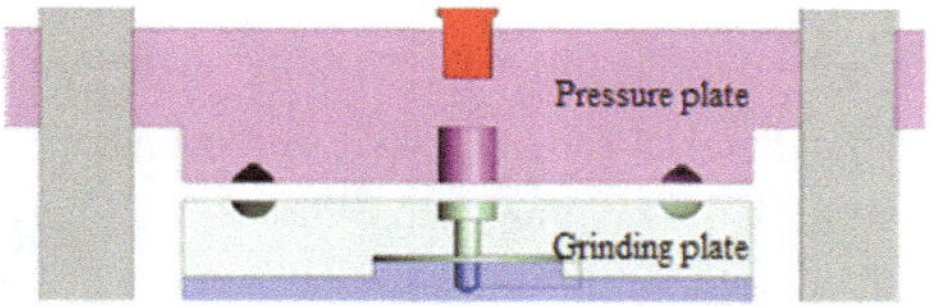


**Fig. 11.** Schematic of the grinding and final calibration equipment of the deo roll balls.

## 3.2 Design of the End Cylindrical Cams

The controlled movement of the tools for each welding assembly operation is achieved by using of a cylindrical cams mechanism [11]. The assembly equipment is provided with two fixed conjugate end cylindrical cams (upper and lower). The cam follower will be in permanent contact with the end cylindrical cam profiles through a compression spring and will ensure a constrained motion of the end cylindrical cam mechanisms. The followers support the devices in which the hemispheres are inserted before assembly, respectively the welded balls after the welding operation (see Fig. 12).

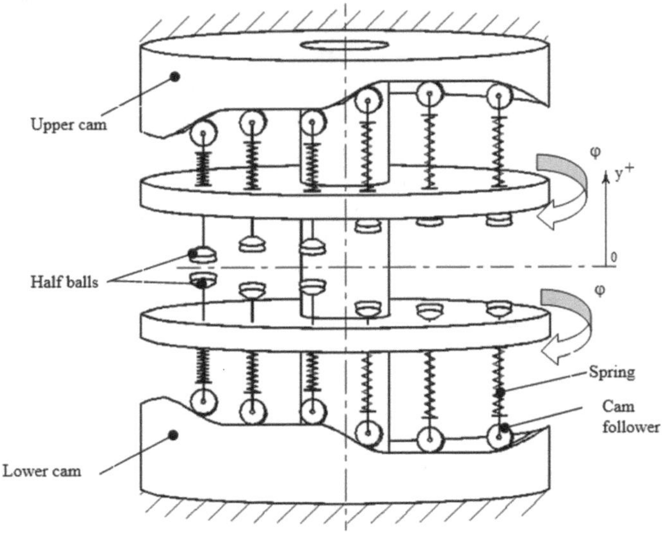


**Fig. 12.** Assembly machine schema by using end cylindrical cam mechanisms.

The displacement diagram according the motion phases described before is shown in Fig. 13 using for the rise and return motions 3-4-5 polynomial curves. The 3-4-5 polynomial curves were selected in order to avoid high accelerations and high tendency to vibrations, because the lower acceleration and jerk characteristic values are $c_a = 5.78$, respectively $c_j = 60$.

**Table 1.** Geometrical parameters of the motion phases.

| Upper end cylindrical cam | | | | Lower end cylindrical cam | | | |
|---|---|---|---|---|---|---|---|
| No. | Phase | φ [°] | y [mm] | No. | Phase | φ [°] | y [mm] |
| 1 | dwell 1 | 22.5° | 21 | 1 | dwell 1 | 22.5° | −21 |
| 2 | return 1 | 22.5° | 14 | 2 | rise 1 | 22.5° | −14 |
| 3 | dwell 2 | 22.5° | 14 | 3 | dwell 2 | 22.5° | −14 |
| 4 | return 2 | 22.5° | 5 | 4 | rise 2 | 22.5° | -5 |
| 5 | dwell 3 | 90° | 5 | 5 | dwell 3 | 90° | −5 |
| 6 | return 3 | 22.5° | −0.15 | 6 | rise 3 | 22.5° | 0.15 |
| 7 | dwell 4 | 45° | −0.15 | 7 | dwell 4 | 45° | 0.15 |
| 8 | rise 1 | 45° | 14 | 8 | return 1 | 45° | −14 |
| 9 | dwell 5 | 22.5° | 14 | 9 | dwell 5 | 22.5° | −14 |
| 10 | rise 2 | 22.5° | 21 | 10 | return 2 | 22.5° | −21 |
| 11 | dwell 6 | 22.5° | 21 | 11 | dwell 6 | 22.5° | −21 |

The corresponding geometrical parameters of the motion phases are given in Table 1 in accordance with the notations in Fig. 12. For all motion phases it is imposed the same minimal transmission angle, $\mu_{min}$.

The displacement diagram of the upper and lower cam profile is presented in Fig. 13.

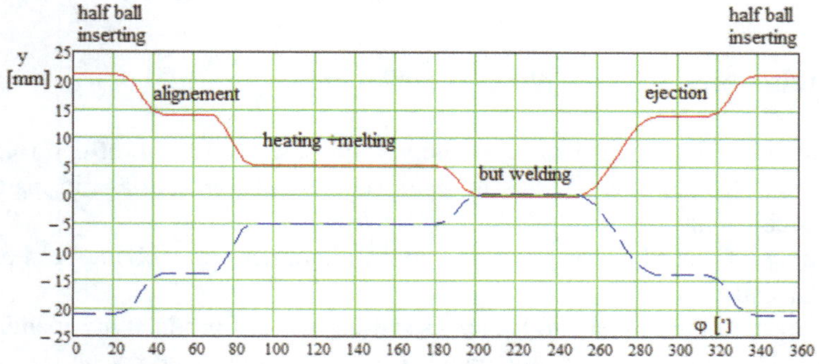


**Fig. 13.** Displacement diagram for upper and lower cam profile.

In order to design the upper and lower end cylindrical cams is necessary to compute the base radius of the cylindrical cams, by using the relationship in Eq. (1):

$$r_b \geq \max\{r_1 \ldots r_5\}, \tag{1}$$

where for each corresponding rise or return motion, the minimum base radius is described in Eq. (2):

$$r_j \geq \left| \frac{s_{j+1} - s_j}{\varphi_{j+1} - \varphi_j} \cdot c_v \cdot \tan \mu_{\min} \right|, \quad \text{j} = 1\ldots5, \tag{2}$$

with: $c_v = 1.88$ - velocity characteristic values for 3-4-5 polynomial curve;
$\mu_{\min} = 30°$ - minimum transmission angle.

The base radius of the end cylindrical cams results $r_b \geq 24.87$ mm. The cylinder radius, which fulfills the base radius condition, is chosen R = 800 mm, taking into account especially the technological conditions.

The developed lower and upper end cylindrical cams profile is shown in Fig. 14.

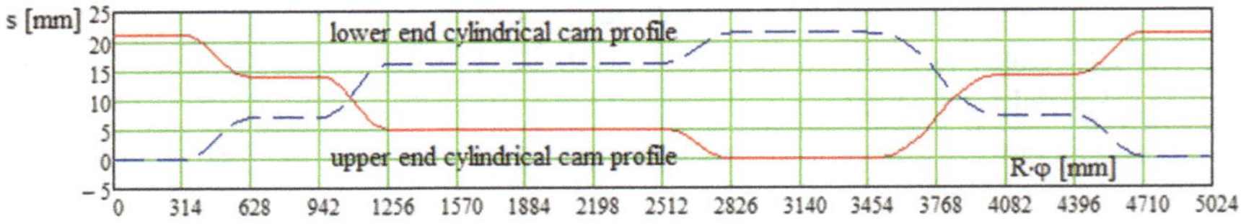


**Fig. 14.** Lower and upper end cylindrical cams profile.

# 4 Conclusions

1. Alternative technologies can be employed for mass production of deo roll balls.
2. This novel technology can lead to significant savings up to 20%, in material and energy consumption.
3. Further studies need to concentrate on better alignment and welding bead control of the half balls.
4. The experimental model can lead to designing of high productivity manufacturing line of deo roll balls with different ball dimensions (diameters of 1″/1.14″/1.3″/1.4″).
5. This concept can be extended for hollow plastic balls manufacturing.

**Aknowledgements.** This work was supported by Obrist Eastern Europe Romania and Universitatea Politehnica Timisoara. Research and experiments were supported by National R&D Institute for Welding and Material Testing ISIM Timisoara. The authors would like to thank Mr. Octavian Oanca and Radu Cojocaru from National R&D Institute for Welding and Material Testing ISIM Timisoara, for their contribution in editing this article.

# References

1. DuPont.: General design principles for DuPont engineering polymers, E.I.du Pont de Nemours and Company, Switzerland (2000)
2. Directive (EU) 2019/904 of the European Parliament and of the Council. http://data.europa.eu/eli/dir/2019/904/oj. Accessed 20 Mar 2020
3. Schmidt, P., Hindle, D., Goede, R., Stoelben, P.: Improvements in or relating to deodorant balls, U.S. patent 20170079403A1 (2015)
4. Stoelben, P., Schmidt, P., Schawo, R.: Roll on dispenser balls, U.S. patent 20170013938A1 (2013)
5. Fully automatic high speed plastic balls blow molding machine, 06 September 2016. https://www.youtube.com/watch?v=Baq-EfbJafI. Accessed 20 Mar 2020
6. Stan, D.V.: Aplicaţii ale ultrasunetelor la injectarea şi extrudare materialelor polimerice, Ed. Politehnica, Timişoara (2003)
7. MMC Packaging, Roll-On Ball Sphere Welding. https://www.youtube.com/watch?v=aEDXSnwk0jo. Accessed 15 Mar 2020
8. Teuşdea, D., Oancă, O., Vodă, M., Sîrbu, N-A., Lovasz, E-C.: TIMA 2019. In: Sîrbu, N.-A. (eds.) Researches on Manufacturing Deo-Roll Balls by Ultrasonic Welding, vol. 1557, pp. 58–72. Advanced Materials Research, Switzerland (2020)

9. Teuşdea, D., Sîrbu, N-A., Vodă, M., Oancă, O., Lovasz, E-C.: TIMA 2019. In: Sîrbu, N.-A. (ed.) Researches on Manufacturing Deo-Roll Balls by Infrared Radiation Welding, vol. 1557, pp. 21–30. Advanced Materials Research, Switzerland (2020)
10. AGIR. http://www.agir.ro/carte/mecanismele-robotilor-industriali-volii-mecanisme-de-prehensiune-mecanisme-pentru-ordonare-orientare-transfer-110602.html. Accessed 21 Mar 2020
11. Luck, K., Modler, K.-H.: Getriebetechnik. Springer, New York (1990)

# An Overview of Grippers in Agriculture Robotic Systems

Mihai Şerdean[1(✉)], Florina Şerdean[2], and Dan Mândru[1]

[1] Faculty of Automotive, Mechatronics and Mechanical Engineering, Technical University of Cluj-Napoca, 103-105 Muncii Blvd., 400641 Cluj-Napoca, Romania
Mihai.Serdean@mdm.utcluj.ro

[2] Faculty of Machine Building, Technical University of Cluj-Napoca, 103-105 Muncii Blvd., 400641 Cluj-Napoca, Romania

**Abstract.** An area of interest for researchers in the recent decades is to reduce production costs by using independent robotic systems in agriculture, especially in farms and crops. The conducted research encompassed in the open literature and the technological developments have been undertaken in the three major areas of smart farming: Management Information System, Precision Agriculture and Agricultural Automation and Robotics. This paper is focused on presenting a comprehensive overview of the different types of agricultural grippers, which are an integrating part of the robotics systems used in smart farming technologies. First, the need for harvesting and phenotyping grippers is briefly presented, followed by a table-summary of their techno-functional properties. Then, the main attention is paid to a detailed analysis which includes several classifications of the grippers presented in the open literature based on their main characteristics, such as: overall dimensions, type of fingers, number of fingers, actuation solution and so on.

**Keywords:** Gripper · Smart farming · Agriculture robotic systems · Harvesting automation

## 1 Introduction

Smart framing can be described as the utilization of modern Information and Communication Technologies in conventional agriculture, this being the start point of what is known as the third green revolution. The 1st Green Revolution (1960–1970) had focused on ingenious plant propagation methods to increase production significantly (especially rice and wheat production at a time when the world was struggling to feed a population half that of today's size). The 2nd Green Revolution, which is still underway, has considered a rapid expansion of knowledge on plant and animal genetics and the ability to optimize genes to improve important features.

Smart Farming has a real potential to provide a more productive and sustainable agriculture based on a more accurate and cost-effective approach. However, while in the United States around 80% of farmers currently use certain types of intelligent


E.-C. Lovasz et al. (Eds.): MTM&Robotics 2020, MMS 88, pp. 212–225, 2021.
https://doi.org/10.1007/978-3-030-60076-1_19

farming technologies, the same Smart Farming technologies are used in Europe at around 24% [1, 2].

Agriculture will be highly challenged due to the fact that it will have to provide food for the 9.6 billion people living on our planet by 2050 according to the Food and Agricultural Organization (FAO) predictions. In this respect, food production must increase by up to 70% by then and all this must be done despite the limited availability of arable land, the growing need for drinking water (agriculture accounts for about 70% of the total drinking water) and other less predictable factors such as the impact of global climate change, which could, among other things, contribute to major changes in seasons that interfere with the life cycle of plants and animals.

One way to address this issue is to increase the quantity and quality of agricultural production by using detection technologies to operate the farms in a smarter way and much more interconnected by so-called precision farming or smart farming [3].

It is already happening that large corporations invest in collecting, storing and analyzing information to increase the quantity and quality of agricultural products they produce: crop yield, land mapping, soil fertilizer, weather data, machinery and animal health. Depending on what is measured and when it is measured, the farmer can be notified very quickly (via sms, e-mail, etc.) and like so, the reaction time is considerably reduced, and the monitored system (harvest, livestock) can be saved [3].

From the point of view of farmers, smart farming technologies (Fig. 1) should provide the user with added value in the form of much better decisions based on new knowledge and information and more efficient use of available resources. In this way, smart farming is strongly linked to three major areas of technology:

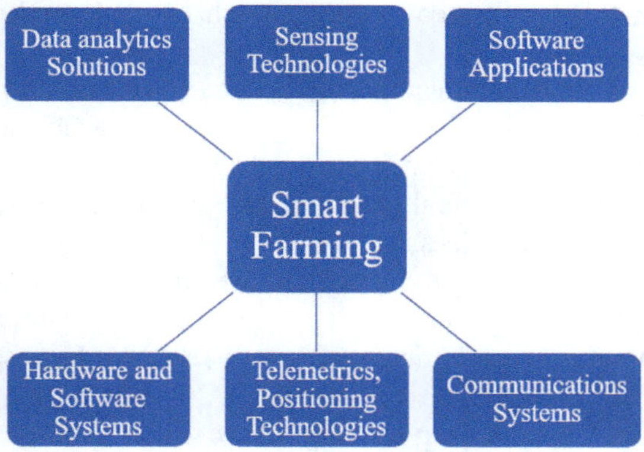


**Fig. 1.** The main elements of a Smart Farming technologies [2].

- **Management Information System**

Scheduled system for collecting, processing, storing and analyzing data in the form that is needed for the user to be able to properly conduct farm activities and operations [1].

- **Precision Agriculture**

Managing spatial and temporal variability to improve economic returns through the use of inputs and reducing environmental impact. It also includes decision support systems (DDS) for the entire farm to optimize returns with as little effort as possible and fewer resources by using systems such as GPS, GNSS, aerial images obtained with the help of the drones and the last generation of satellite hyperspectral images, thus allowing the generation of detailed spatial variability maps with as many variables as they can be measured (e.g. crop yield, land features/topography, organic matter content, moisture levels, nitrogen levels, etc.) [1].

- **Agricultural Automation and Robotics**

The process of applying robotics, automated control techniques and artificial intelligence at all levels of agricultural production, including farm bots and farm drones [1].

## 2 Robotic Systems for Smart Farming

In the past years, field of agriculture has seen a fast expansion especially in terms of perception systems, such as computer-vision and global positioning systems. With the help of these new improvements, tasks can now be more easily automated, in areas like weed control where high accuracy is needed in order to maintain efficient plant treatment at a low cost. A big advantage of this new technologies is that they are modular and can be easily used even on old agricultural machines. What is now an outdated tractor can be easily upgraded or retrofitted and therefore becoming a semi-automatic or even fully automated machine which can independently execute agricultural tasks at a low cost [4, 5].

The autonomous agricultural equipment integrates perception systems, for gathering information from the field decision making systems for analyzing all the collected information and helping the operator in tacking the best decision and actuation systems which are executing the agricultural operation. In Fig. 2 are presented some examples of automated agricultural machines.

Another example in which all the systems above are used together combined with a gripper and a complex stereo vision camera system in closed environments is presented in Fig. 3. Here an autonomous agricultural robot is used for harvesting tomatoes and sweet pepper [6–8].

Computer vision based on deep learning, image acquisition and image processing are very important for agriculture. Recent years have shown a big advancement in these fields. Complex stereovision systems are used not only in harvesting, for detecting the exact location of fruit and guiding the gripper to grasp it but also in counting the fruit on a plant or tree. Yield estimation is extremely important for farmers who need to prepare for what is ahead and take care of all the logistics that is required with harvesting, storing and transporting the fruits. In Fig. 4 a counting system based on deep learning [9] is presented which can make such important estimations for farmers which they need to know about their plantations.

**Fig. 2.** Examples of smart agricultural vehicles [4, 5].

**Fig. 3.** Fruit harvesting in closed and controlled environments equipped with stereovision camera and custom-made grippers [6–8].

Another important aspect, especially for harvesting but not only, the separation of fruit from the plant stem when the image is processed. In [10] such a system is presented using an Xbox Kinect.

A performant stereo vision system can be used also for plant phenotyping. Using autonomous mobile platforms to constantly monitor the crops in real time and based on such detailed inspections the farmer can act immediately on each individual plant or group of plants, saving costs by not allowing the diseases to spread and damage the whole plantation. For even more detailed plant phenotyping a 3D scan of each sick plant can be made (Fig. 5) and after that a specialist can examine the scan and offer a more complete solution for the problem and, more importantly, stopping the disease from spreading even more [11].

**Fig. 4.** Crop monitoring system with yield estimation capabilities [9].

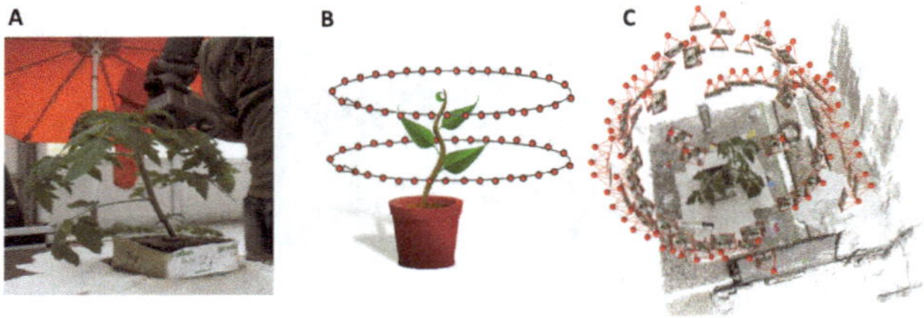


**Fig. 5.** Plant phenotyping system [11].

A more exhaustive approach is presented in Fig. 6 where a small-scale complete food production system is shown. This type of agricultural system can be programmed from the seeding stage and all the way up to the harvesting stage which is done separately. The success of a crop is influenced by multiple factors including spacing between the plants and therefore the seeds, fact taken into consideration by the farmbot [13] in the seeding stage. This robot also takes care of plant nurturing by giving it the right amount of water and nutrients at the right time and this enables growing different types of plants on a small surface area.

Personal food computer [12] goes to the next level by enabling small scale climate control. It can control the temperature, humidity, C02 levels, light intensity and so based on what is planted inside and what the plant requires at the exact moment. There are multiple advantages in using fully automated climate-controlled food production systems and one of them is that, through IoT, it enables the creation and sharing of climate recipes. Almost anywhere in the world one can now grow food with the exact characteristics of different areas of the world. By adopting the exact temperature, humidity C02, fruits grown in one side of the planet can have the same taste and characteristics as plant or fruits from any other part of the world.

**Fig. 6.** Complete food production systems with temperature, climate control and automated nutrient delivery system [12, 13].

# 3 Robotic Systems for Smart Farming

## 3.1 Current Situation with Harvesting

Harvesting fruits is a very difficult task for independent robotic systems, due to the several aspects that it involves such as high precision in fruit identification, real time decision making about harvesting and storing, handling without deteriorating both the plant and the fruit and so on. There is an increasing demand for efficient and cost-effective harvesting operators due to competition from low cost country sourcing. Some studies show that in order to remain competitive the costs related to harvesting have to be diminished by half [14].

At this moment more than half of the total costs regarding production is determined by the labor cost for harvesting. In this context when one also considers the existing competition, a good option is given by minimizing production costs through harvesting automatization [15]. Due to the particularities of shape, dimension, weight and so on of the harvested products, there cannot be used a unique gripper and hence, a lot of research is conducted in order to design customized grippers. Depending on the crops, mechanical shakers are useful in some situations. The biggest advantage of the mechanical shakers is that it will harvest more in less time and with almost no sophisticated sensing technologies after the right branches are identified [14].

## 3.2 Summary of Techno-Functional Gripper Properties

See Table 1.

**Table 1.** Summary of gripper characteristics encompassed in the open literature.

| Field of use | Gripper structure | Product dimensions | Drive system | Types of sensors used |
|---|---|---|---|---|
| Horticulture (trees shrubs and vines) in open field [16] | Gripping with fingers | Small to medium size | Pneumatic with levers | Pressure sensor |
| Horticulture, especially for fruit-bearing shrubs [17] | Clamps with rubber protection for the bark of the tree or shrub | Small size | Hydraulic | Ultrasonic and magnetic sensors |
| Harvesting in a controlled environment for suspended plants [18] | Electrically operated gripper | Small size | Electric | Proximity sensor, stereo vision, track position |
| Horticulture, fruit trees [19] | Articulated pliers type gripper with final gripping elements in the shape of a spoon | Small to medium size | Electric and pneumatic | Proximity, pressure, position sensor vision system - camera |
| Wood harvesting for industry - forest work environment [20] | Pliers type gripper | Very large size | Hydraulic and electric | Proximity, pressure, force sensor |
| Harvesting apples and pears in laboratory conditions [21] | Finger type gripper | Small to medium size | Electric | Stereo vision system |
| Tomato harvesting in greenhouses mounted on a mobile platform with a predefined running track [8] | Gripper with clamp mechanism and collecting container | Small to medium size | Electric and pneumatic | Proximity sensor, image acquisition system, pressure sensor |
| Tomato harvesting in greenhouses [6] | Two arms with cylindrical grippers | Small to medium size | Electric and pneumatic | Image acquisition system, CCD, vacuum, pressure and flow meter sensors |

(*continued*)

**Table 1.** (*continued*)

| Field of use | Gripper structure | Product dimensions | Drive system | Types of sensors used |
|---|---|---|---|---|
| Harvesting sweet peppers in greenhouses [22] | 1. Parallel finger gripper,<br>2. Gripper with suction cup and knife for detaching the stem | Small to medium size | Electric and pneumatic | RGB and 3D time of flight cameras, CCD, vacuum, pressure and flow meter sensors |
| Grape harvesting in vineyard trellis system [23] | Two finger clamp grippers | Medium size | Electric | Vision system and force sensor |
| Can harvest multiple types of crops [24] | Customized to the harvested agricultural product | Small to medium size | Electric | Position sensor, proximity sensor, torque sensing |
| Harvesting strawberries in greenhouse environments [25] | Parallel gripper with two fingers | Small size | Pneumatic and electric | Vision system and force sensor |
| Manipulator arm operated from the air for open and closed environments [26, 27] | Customized to the harvested agricultural product | Small to medium size | Electric | Position and force sensors |
| Harvesting cherry tomatoes in closed environments [28] | Parallel gripper | Small size | Electric | Video detection system, force sensor, infrared sensor |
| Phenotyping in sorghum [29] | Parallel gripper mounted on a manipulator with 3DOF | Small size | Electric | Vision system, position sensor, force sensor |
| Harvesting the pedicle from Gerbera Jamesonii flowers [30] | Parallel gripper | Small size | Electric, pneumatic | Multispectral vision system, proximity sensor |
| Cucumber harvesting [31] | Parallel gripper | Small to medium size | Electric | Position, force, proximity sensor |
| Harvesting Saffron flowers [32] | Pliers parallel gripper | Small and thin | Electric | Position, distance, and proximity sensors |

(*continued*)

**Table 1.** (*continued*)

| Field of use | Gripper structure | Product dimensions | Drive system | Types of sensors used |
|---|---|---|---|---|
| Melon harvesting [33] | Parallel gripper, with spoon type gripping elements | Medium to large size | Electric | Force and position sensors |
| Testing the level of ripeness in eggplants by palpation [34] | Parallel gripper | Medium size | Pneumatic, electric | Position, force, pressure sensor, |
| Citrus harvesting [35, 36] | Three-fingered gripper (at 120 degrees) | Small to medium size | Electric, | Position, force and proximity sensors |
| Agriculture and mechanical assembly [37] | Parallel gripper | Small size | Electric | Position sensor, accelerometer |
| Agriculture, horticulture [38] | Three-fingered gripper (at 120 degrees) | Small to medium size | Electric | Force sensor |
| Eggplant harvesting [39] | Gripper with flexible fingers | Medium size | Pneumatic, electric | CCD camera, photoelectric sensor |
| Radicchio harvesting [40] | Gripper with 2 bucket-fingers | Medium | Pneumatic, electric | CCD camera |
| Sweet pepper harvesting [41] | Vacuum tube gripper with a cutting element | Small to medium | Pneumatic, electric | CCD camera, vacuum and position sensors |

### 3.3 Classifications Based on Gripper Properties

**Overall Dimensions.** Multiple dimensions have been considered when designing grippers for harvesting agricultural products. One of the most common dimensions is the medium one which refers to apples, pears and medium size tomatoes. For example, in [8, 16] and [6] the authors have approached the specific problems for tomato horticulture, proposing prototype grippers which are slightly adjustable from small to medium size tomatoes. Another application for medium size grippers is apple harvesting, see for example the pneumatic actuated gripper presented in [19] or pear harvesting see for example the automatic fruit harvester [21]. In the same category is the sweet-pepper harvesting robot as the one presented in [22], the cucumber harvesting gripper mounted on a mobile platform shown in [31], the grape harvesting agricultural robot from [23] or the robust servo-controlled manipulator for citrus harvesting [35, 36]. Other researchers have developed ground multipurpose agricultural manipulators [24] and aerial multipurpose manipulators [26, 27].

For small agricultural products such as strawberries the harvesting process is difficult and labor intensive. Due to this aspect the hanging culture is preferred, and special small size gripping manipulators have been developed, such the one in [18] and [25]. In the same category, for small agricultural products, are the robots with grippers for small cherry tomatoes [28].

For large agricultural product handling an illustrative example is the Tree gripper used in wood harvesting machines [20].

For some agricultural cultures the harvesting is not made by picking up each individual fruit but through a type of shaker gripper which grabs the stem and shakes the whole tree, see for example the inertial limb shaker in [17].

In other areas of agriculture, different types of mobile platforms equipped with grippers are used for phenotyping. An example can be seen in [29] where a ground based mobile robot equipped with a parallel gripper and different types of non-contact sensors can perform a thorough inspection of the plants. In [30] with the help of image analytic algorithms, grippers are used for harvesting the flower pedicels while in [32] a special type of gripper was developed for harvesting saffron flowers.

Some applications require different types of sorting capabilities. For example, in [34] a special type of pneumatic gripper was developed which allows to sort the eggplant products according to their firmness and therefore determine if they are ripe or not. However, when it comes to eggplant harvesting different gripper designs are required [39].

**Constructive Solutions.** In general, depending on the culture type and the harvested product, there can be several types of constructive solutions. A more common approach is the one presented in [16] and consists in a two-finger gripper pneumatically actuated. A similar approach where the gripper is mounted on a mobile platform used for strawberry harvesting can be found in [18, 25, 29, 31] or for grapes harvesting in [23]. A similar robot can be found in [28] which is used for cherry tomato harvesting. A more complex approach is presented in [37] where a variable aperture gripper can be used for a wide variety of agricultural products.

A different type of gripper, namely a two-finger gripper with spoon shaped gripping elements incorporated on a 5 DOF manipulator was designed for harvesting apples [19].

An innovative approach is proposed in [8] where the gripper mainly consists of a fruit stalk clamping mechanism, a fruit clamping mechanism, a separating mechanism and a control unit.

A different constructive solution is proposed in [6] where the dual arm tomato harvesting robot uses vacuum to grasp the fruits and a small circular saw for separating them from the stem. In contrast to this solution, in [24, 26] and [27] multiple grippers can be mounted on the manipulator arm depending on the harvest type. A multi arm harvesting solution is presented in [33] where the gripper is used for melon harvesting.

Some researchers have designed and tested two types of grippers: the classical two finger parallel gripper and the lip-type gripper which uses vacuum for grasping [22].

In [17] is presented a solution for the pistachio harvesting using a hydraulically powered shacking gripper.

There are constructive solutions which combine the robotic arm with a separate module for acquiring visual information (stereo vision camera) [21].

A big and heavy hydraulically driven agricultural gripper is used in wood harvesting with side rollers to hold and guide the wood while it is cut to pieces [20].

**Number of Fingers/Gripping Elements.** Considering parameters like size, weight or hardness of the products there are different requirements regarding the number of gripping elements. The minimum number of fingers reported in the open literature is two and grippers with two fingers have been usually used for small or medium agricultural products such as tomatoes [16], strawberries [18, 25] apples [19], pears [21], sweet peppers [22], grapes [23], cherry tomatoes [28].

Sometimes, depending on the handled product, the number of fingers is increased. For example, in [20] the manipulator has 4 gripping elements divided in in two pairs. However, there are approaches in which the gripper has no fingers [6, 8].

A three gripping element solution is presented in [38], where the developed gripper is used for grasping different horticultural products.

**Type of Fingers/Gripping Elements.** Based on the harvested product the gripper must have a certain type of gripping elements. Commonly used in agriculture is the flexible articulated type of griping elements. A basic approach is presented in [16], where it is shown only an articulated type of gripper with soft tips for not damaging the horticultural products. A similar type of fingers with soft protection is proposed in [21] or with soft structure (Fin-Ray fingers) is presented in [22]. The same basic gripper but without the soft protection can be found in [28] where it is used for cherry tomatoes harvesting. Some researchers also opted for spoon-shaped griping elements [19].

For some applications the gripping fingers have a second function as knifes for cutting down the tree branches faster such as the solution proposed in [20] or for cutting grapes rachis as presented in [23]. Another relevant example is the gripper with two bucket like fingers that can cut the radicchio 10 mm underground [40].

There are also mechanical structures that encompass only a clamping mechanism for grasping the agricultural products [8] or a vacuum gripper suction device [6]. Nevertheless, there are design solutions which combine a grasping device with a cutting tool. For example, the particular kind of vacuum gripper equipped with a semicircular cutting element presented in [41] for sweet pepper harvesting.

Not all grippers in agriculture are used for fruit grasping and harvesting. Some type of grippers are developed only for plant phenotyping such as the one presented in [11]. In [29] the electric actuated gripper is used for inspection of sorghum stalk.

**Comparative Appreciations Regarding Actuation.** There is a wide variety of actuating solutions in the field of agriculture. In [6, 16, 34] the authors propose pneumatic actuated grippers for harvesting tomatoes while in [8, 18, 21, 25, 28–32, 35] the electrical actuation is chosen for the strawberry, cherry tomatoes, cucumber, melon, citrus and pears harvesting robot. Some researchers have chosen to combine the two actuation solutions [19, 22].

**Comparative Appreciations Regarding the Types of Sensors Used.** Most grippers are able to detect the agricultural product based on the feedback given by the proximity sensors [16, 18]. Some have complex cameras instead (see [22]) or visual sensors [23] and [25], while, some of them also have pressure sensors [16, 25, 34], or force sensors [28, 38].

Compared to other structures there are grippers which have incorporated collision sensors, pressure sensor, position sensors, and vision sensors [8, 19], other have chosen to incorporate accelerometer and internal sensors [34]. Several researchers choose to use a separate stereo vision system separate to the gripper as it can be seen in [21, 35].

Others have different sensors depending on the chosen constructive solution, for example, in [6] is was used a vacuum sensor alongside other sensors such as proximity and pressure sensors.

A more complete approach of sensors array can be seen in [29] where, alongside the above-mentioned types of sensors, the mobile platform is equipped with planar laser scanners and GPS for better understanding the surrounding and so, the gripper can precisely be directed towards the plant, or group of plants and the task required to be done.

## 4 Conclusions

In this paper, a broad overview of robotic systems for smart farming and grippers for agriculture is given.

As other sectors directly correlated to food supply for world's population, the agricultural one is constantly adapting by increasing the level of automation due to new challenges that occur. It can be noticed that the level of automation of robotic systems for smart farming varies from simply upgrading old machinery to developing new fully automated and independent food production systems.

One of the main challenges addressed by the researches published in open literature is the high cost of human labor associated with harvesting of agricultural products. A major contribution in reducing these costs is achieved by implementing harvesting automatization with grippers. For this purpose, a variety of grippers with different overall dimensions and constructive solutions have been developed, including the basic standard two finger parallel gripper, as well as the multi flexible fingers grippers. Additionally, depending on the culture type and the domain there can be customized grippers such as the ones with flexible fingers with soft tips for protection of the harvested product or spoon shaped fingers for a better handling. Other aspects directly linked to the harvested culture are the used sensors and actuation which lead to the development of specialized grippers, for instance the electrical actuated gripper with force sensing tips used in harvesting of the fragile agricultural products.

Moreover, based on the terrain and plantation type, grippers can be mounted on independent mobile platforms or even unmanned aerial vehicles.

Overall, a lot of effort has been invested in the advancement of current gripping technologies and the development of new robotic harvesting systems paving the road for cutting-edge technologies that will revolutionize harvesting as we know it.

## References

1. Smart AKIS, Smart Farming Thematic Network. https://www.smart-akis.com/index.php/network/what-is-smart-farming/. Accessed 28 May 2020

2. Beechan Research - Towards Smart Farming Agriculture: Embracing the IoT Vision. http://www.beechamresearch.com/files/BRL%20Smart%20Farming%20Executive%20Summary.pdf. Accessed 28 May 2020
3. Guerrini, F., The Future Of Agriculture? Smart Farming. https://www.forbes.com/sites/federicoguerrini/2015/02/18/the-future-of-agriculture-smart-farming/#4035a5623c42. Accessed 28 May 2020
4. Emmi, L., Gonzalez-de-Soto, M., Pajares, G., Gonzalez-de-Santos, P.: Integrating sensory/actuation systems in agricultural vehicles. Sensors **14**(3), 4014–4049 (2014)
5. Penisi, O., Bocca, J., Aguilar, H., Bocca, P.: One DOF mechanism for the mechanical harvest of vines in an arbor structure and the validation of the acceleration of grape berry harvesting. Front. Mech. Eng. **10**, 221–232 (2015)
6. Zhao, Y., Gong, L., Liu, C., Huang, Y.: Dual-arm robot design and testing for harvesting tomato in greenhouse. IFAC-PapersOnLine **49**(16), 161–165 (2016)
7. Hemming, J., Ruizendaal, J., Hofstee, J.W., van Henten, E.J.: Fruit detectability analysis for different camera positions in sweet-pepper. Sensors **14**(4), 6032–6044 (2014)
8. Wang, G., Yu, Y., Feng, Q.: Design of end-effector for tomato robotic harvesting. IFAC-PapersOnLine **49**(16), 190–193 (2016)
9. Rahnemoonfar, M., Sheppard, C.: Deep count: fruit counting based on deep simulated learning. Sensors **17**(4), 905 (2017)
10. Li, D., Xu, L., Tan, C., Goodman, E.D., Fu, D., Xin, L.: Digitization and visualization of greenhouse tomato plants in indoor environments. Sensors **15**(2), 4019–4051 (2015)
11. Rose, J.C., Paulus, S., Kuhlmann, H.: Accuracy analysis of a multi-view stereo approach for phenotyping of tomato plants at the organ level. Sensors **15**(5), 9651–9665 (2015)
12. Ferrer, E.C., Rye, J., Brander, G., Savas, T., Chambers, D., England, H., Harper, C.: Personal food computer: a new device for controlled-environment agriculture. In: Arai, K., Bhatia, R., Kapoor, S. (eds.) Future Technologies Conference (FTC) 2018, vol. 2, pp. 1077–1096. Springer, Cham, Switzerland (2018)
13. Cruz, J., Herrington, S., Rodriguez, B.: Farmbot final design report, 2014. https://www.researchgate.net/publication/304162022_Farmbot. Accessed 28 May 2020
14. Vu Q., Kuzov M., Ronzhin A.: Hierarchical classification of robotic grippers applied for agricultural object manipulations. In: MATEC Web of Conferences, vol. 161, p. 03015 (2018)
15. Broun, G.K.: New mechanical harvester for the Florida Citrus juice industry. HortTechnology **15**(1), 69–72 (2005)
16. Ceccarelli, M., Figliolini, G., Ottaviano, E., Mata, A.S., Criado, E.J.: Designing a robotic gripper for harvesting horticulture products. Robotica **18**(1), 105–111 (2000)
17. Polat, R., Gezer, I., Güner, M., Dursun, E., Erdogan, D., Bilim, H.C.: Mechanical harvesting of pistachio nuts. J. Food Eng. **79**(4), 1131–1135 (2007)
18. Yamamoto, S., Hayashi, S., Saito, S., Ochiai, Y., Yamashita, T., Sugano, S.: Development of robotic strawberry harvester to approach target fruit from hanging bench side. IFAC Proc. Vol. **43**(26), 95–100 (2010)
19. De-An, Z., Jidong, L., Wei, J., Ying, Z., Yu, C.: Design and control of an apple harvesting robot. Biosyst. Eng. **110**(2), 112–122 (2011)
20. Goubet, D., Fauroux, J.-C., Gogu, G.: Gripping mechanisms in current wood harvesting machines. Front. Mech. Eng. **8**(1), 42–61 (2013)
21. Font, D., Pallejà, T., Tresanchez, M., Runcan, D., Moreno, J.J., Martínez, D., Teixidó, M., Palacín, J.: A proposal for automatic fruit harvesting by combining a low cost stereovision camera and a robotic arm. Sensors **14**(7), 11557–11579 (2014)
22. Hemming, J., Bac, C.W., van Tuijl, B.A.J., Ruud, B., Bontsema, J., Pekkeriet, E.: A robot for harvesting sweet-pepper in greenhouses. In: International Conference of Agricultural Engineering, Zurich (2014)

23. Monta, M., Kondo, N., Shibano, Y.: Agricultural robot in grape production system. In: Proceedings of 1995 IEEE International Conference on Robotics and Automation, Nagoya, Japan, pp. 2504–2509 (1995)
24. Baur, J., Pfaff, J., Ulbrich, H., Villgrattner, T.: Design and development of a redundant modular multipurpose agricultural manipulator. In: 2012 IEEE/ASME International Conference on Advanced Intelligent Mechatronics (AIM), Kachsiung, pp. 823–830. IEEE (2012)
25. Qingchun, F., Wengang, Z., Quan, Q., Kai, J., Rui, G.: Study on strawberry robotic harvesting system. In: 2012 IEEE International Conference on Computer Science and Automation Engineering, Zhangjiajie, China, pp. 320–324. IEEE (2012)
26. Korpela, C., Orsag, M., Jun, Y., Brahmbhatt, P., Oh, P.: A hardware-in-the-loop test rig for aerial manipulation. In: 2013 International Conference on Unmanned Aircraft Systems (ICUAS), Atlanta, GA, pp. 982–987. IEEE (2013)
27. Korpela, C., Orsag, M., Oh, P.: Towards the realization of mobile manipulating unmanned aerial vehicles (MM-UAV): peg-in-hole insertion tasks. In: 2013 IEEE Conference on Technologies for Practical Robot Applications (TePRA), Woburn, MA, pp. 1–6. IEEE (2013)
28. Taqi, F., Al-Langawi, F., Abdulraheem, H., El-Abd, M.: A cherry-tomato harvesting robot. In: 2017 18th International Conference on Advanced Robotics (ICAR), Hong Kong, pp. 463–468. IEEE (2017)
29. Mueller-Sim, T., Jenkins, M., Abel, J., Kantor G.: The robotanist: a ground-based agricultural robot for high-throughput crop phenotyping. In: 2017 IEEE International Conference on Robotics and Automation (ICRA), Singapore, pp. 3634–3639. IEEE (2017)
30. Rath, T., Kawollek, M.: Robotic harvesting of Gerbera Jamesonii based on detection and three-dimensional modeling. Comput. Electron. Agric. **66**(1), 85–92 (2009)
31. Van Henten, E.J., Schenk, E.J., van Willigenburg, L.G., Meuleman, J., Barreiro, P.: Collision-free inverse kinematics of the redundant seven-link manipulator used in a cucumber picking robot. Biosyst. Eng. **106**(2), 112–124 (2010)
32. Asimopoulos, N., Parisses, C., Smyrnaios, A., Germanidis, N.: Autonomous vehicle for saffron harvesting. Procedia Technol. **8**, 175–182 (2013)
33. Zion, B., Mann, M., Levin, D., Shilo, A., Rubinstein, D., Shmulevich, I.: Harvest-order planning for a multiarm robotic harvester. Comput. Electron. Agric. **103**, 75–81 (2014)
34. Blanes, C., Ortiz, C., Mellado, M., Beltrán, P.: Assessment of eggplant firmness with accelerometers on a pneumatic robot gripper. Comput. Electron. Agric. **113**, 44–50 (2015)
35. Mehta, S.S., MacKunis, W., Burks, T.F.: Robust visual servo control in the presence of fruit motion for robotic citrus harvesting. Comput. Electron. Agric. **123**, 362–376 (2016)
36. Mehta, S.S., MacKunis, W., Burks, T.F.: Nonlinear robust visual servo control for robotic citrus harvesting. IFAC Proc. Vol. **47**(3), 8110–8115 (2014)
37. Rosati, G., Minto, S., Oscari, F.: Design and construction of a variable-aperture gripper for flexible automated assembly. Robot. Comput. Integr. Manuf. **48**, 157–166 (2017)
38. Russo, M., Ceccarelli, M., Corves, B., Hüsing, M., Lorenz, M., Cafolla, D., Carbone, G.: Design and test of a gripper prototype for horticulture products. Robot. Comput. Integr. Manuf. **44**, 266–275 (2017)
39. Hayashi, S., Ganno, K., Ishii, Y., Tanaka, I.: Robotic harvesting system for eggplants. Jpn. Agric. Res. Q. JARQ **36**(3), 164–168 (2002)
40. Foglia, M.M., Reina, G.: Agricultural robot for radicchio harvesting. J. Field Robot. **23**(6–7), 363–377 (2006)
41. Bac, C.W., Roorda, T., Reshef, R., Berman, S., Hemming, J., van Henten, E.J.: Analysis of a motion planning problem for sweet pepper harvesting in a dense obstacle environment. Biosys. Eng. **146**, 85–97 (2016)

# Evaluation by Simulation of Reaction Forces that Occur in Spherical Joints of Parallel Topology Robots

Calin-Octavian Miclosina(✉), Zoltan-Iosif Korka, and Vasile Cojocaru

Faculty of Engineering and Management, "Eftimie Murgu" University of Resita, Resita, Romania
c.miclosina@uem.ro

**Abstract.** The paper shows a method of evaluation by simulation of reaction forces that occur in the spherical joints of a parallel topology robot. The geometrical model of the parallel topology robot, operated as a flight simulator, is presented. The structure of the model is of $FP_3 + 6{\cdot}SPS + MP_3$ (or MSSM) type. Based on the input parameters, which are the relative displacements of the driving kinematical joints links versus time, certain motions of the mobile platform are studied. The modeling and simulation processes were accomplished by using the SolidWorks software.

**Keywords:** Parallel topology robot · Reaction forces · Modeling · Simulation · SolidWorks

## 1 Introduction

In terms of the topology of the guiding device mechanism, the robots can be classified into serial topology robots, parallel topology robots, and mixed topology robots [1].

The guiding device mechanism of serial topology robots is based on an open kinematical chain, with one extremity as fixed link, and the other extremity link carrying the end effector.

Parallel topology robots guiding device mechanism is based on closed kinematical chains. It usually contains a mobile platform linked to a fixed one by open kinematical chains. The structure of various parallel topology guiding device mechanisms is studied in literature [2–5].

Mixed topology robots guiding device mechanism is based on both open and closed kinematical chains.

Some applications of parallel guiding device mechanisms, without being exhaustive, may be mentioned: tool positioners for NC machine tools [6], experimental positioners [7], precision medical devices in surgery [8, 9], material handling devices [10], satellite antenna positioners [10], and flight or driving simulators [10–12].

Determination of forces that occur on the guiding device mechanism links while operating is important for fatigue and life time estimation calculus [13–15].


E.-C. Lovasz et al. (Eds.): MTM&Robotics 2020, MMS 88, pp. 226–234, 2021.
https://doi.org/10.1007/978-3-030-60076-1_20

The paper is based on previous work [11, 12], and it proposes the evaluation of reaction forces variation that occur in spherical kinematical joints of a parallel topology robot used as a flight simulator, as a result of certain motions simulation in dynamic conditions. The modeling and simulation processes were performed by using SolidWorks software, and the SolidWorks Motion module.

## 2 Geometrical Model of the Parallel Topology Robot

The guiding device mechanism of the parallel topology robot has the structure $FP_3$ + 6·SPS + $MP_3$ [11, 12], equivalent to MSSM [1]. It contains a fixed platform $FP_3$, and a mobile one, $MP_3$, linked together by 6 open kinematical chains of SPS structure. Every SPS kinematical chain is composed of 2 binary links, 1 driving prismatic joint (P) linking them, and 2 spherical joints (S) to extremities. Both at fixed platform level and at mobile platform level, the centers of the spherical joints belonging to different kinematical chains, are coincident 2 by 2, thus forming double spherical joints [11, 12].

The minimum distance between the centers of the 2 spherical joints of the same kinematical chain SPS is 800 [mm], the maximum distance is 1300 [mm], thus the maximum stroke of the driving prismatic joint is 500 [mm] [11, 12].

The geometrical model of the parallel topology robot, obtained by using the SolidWorks software, is presented in Fig. 1. The geometrical model of a seat is locked on the mobile platform, together with the geometrical model of a human operator [11, 12].

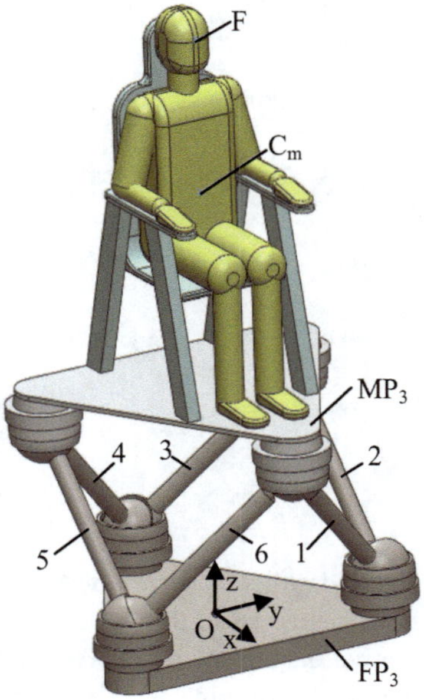


**Fig. 1.** The geometrical model of the parallel robot, where [11, 12]: $FP_3$ - the fixed platform; $MP_3$ - the mobile platform; 1 ÷ 6 - SPS kinematical chains; Oxyz - fixed reference system; $C_m$ - the center of mass of the human operator model; F - frontal point of the human operator model.

The materials of the components are chosen from the SolidWorks materials library. The geometrical model of the human operator is meant to be an average one, with the height of 1.75 [m] and the mass of 75 [kg]. In order to obtain the weight of this model, a custom material with the mass density of 1007.27 [kg/m$^3$] is defined [11, 12].

## 3 Simulation of Parallel Topology Robot Motion

The parallel topology robot simulation is accomplished in the SolidWorks Motion module [11, 12]. In order to accomplish a kinematics calculus in this module, no discretization is needed. It is possible to perform a FEM analysis while motion, to obtain stresses and deformations, but with discretization limitations. For complex FEM analysis, SolidWorks Simulation module is recommended.

In order to be as close as possible to the real case, the gravitational acceleration of 9806.65 [mm/s$^2$] is defined.

For every SPS kinematical chain, a linear motor is defined. It establishes the length between the centers of the two spherical joints of the same kinematical chain, as shown in Table 1.

**Table 1.** Variation of lengths $L_1 \div L_6$, between the centers of the spherical joints.

| Length | Motion 1 | | Motion 2 | |
|---|---|---|---|---|
| | Initial position ($t_0$ = *0* s) | Final position ($t_f$ = *0.5* s) | Initial position ($t_0$ = *0* s) | Final position ($t_f$ = *0.5* s) |
| $L_1$ [mm] | 1300 | 800 | 1050 | 1050 |
| $L_2$ [mm] | 1300 | 800 | 1300 | 800 |
| $L_3$ [mm] | 1300 | 800 | 1300 | 800 |
| $L_4$ [mm] | 1300 | 800 | 800 | 1300 |
| $L_5$ [mm] | 1300 | 800 | 800 | 1300 |
| $L_6$ [mm] | 1300 | 800 | 1050 | 1050 |

The motors are considered hydraulically driven, with a maximum speed limit set to 1 [m/s]. The friction forces are neglected in the simulation process [11, 12].

For every motor, the relative displacement between the piston and the hydraulic cylinder versus time is imposed, determining the variations of velocity and acceleration. The input parameters of the hydraulic motors, corresponding to motions 1 and 2, are presented in Fig. 2 and in Fig. 3, respectively.

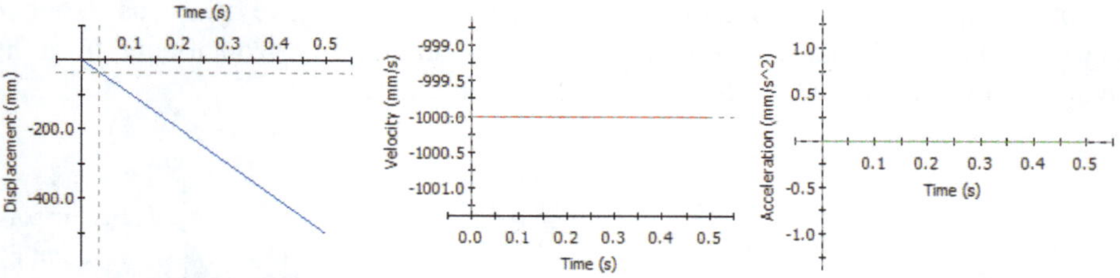


a) Displacement for motors 1÷6;    b) Velocity for motors 1÷6;    c) Acceleration for motors 1÷6.

**Fig. 2.** The variations of input parameters of the hydraulic motors for the motion 1.

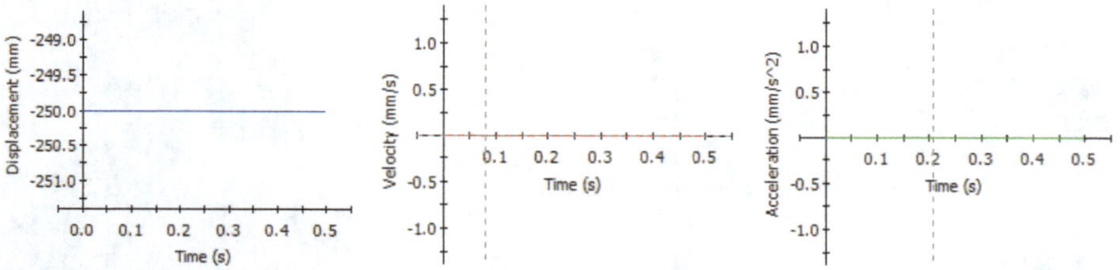


a) Displacement for motors 1 and 6;    b) Velocity for motors for motors 1 and 6;    c) Acceleration for motors 1 and 6;

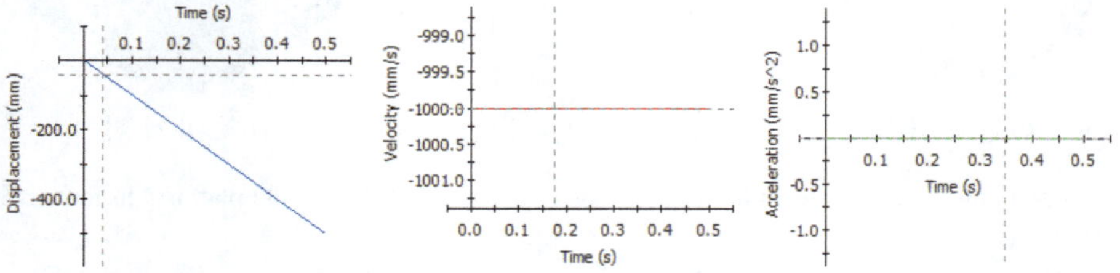


d) Displacement for motors 2 and 3;    e) Velocity for motors 2 and 3;    f) Acceleration for motors 2 and 3;

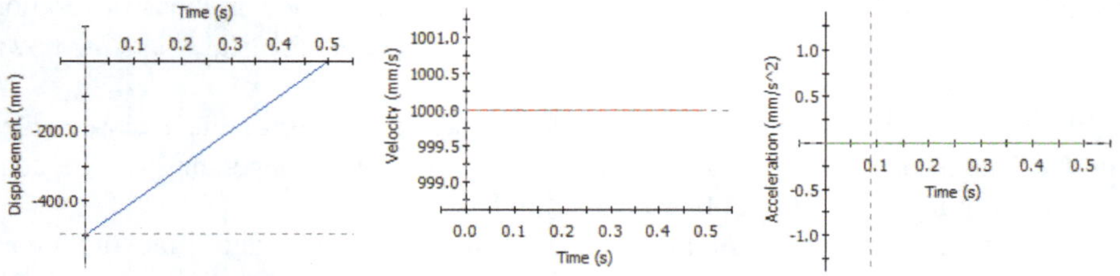


g) Displacement for motors 4 and 5;    h) Velocity for motors 4 and 5;    i) Acceleration for motors 4 and 5.

**Fig. 3.** The variations of input parameters of the hydraulic motors for the motion 2.

In motion 1, a descent of the mobile platform along z axis takes place, as shown in Fig. 4. In motion 2, the mobile platform balances around a horizontal axis, from the right side of human operator to the left side, as presented in Fig. 5.

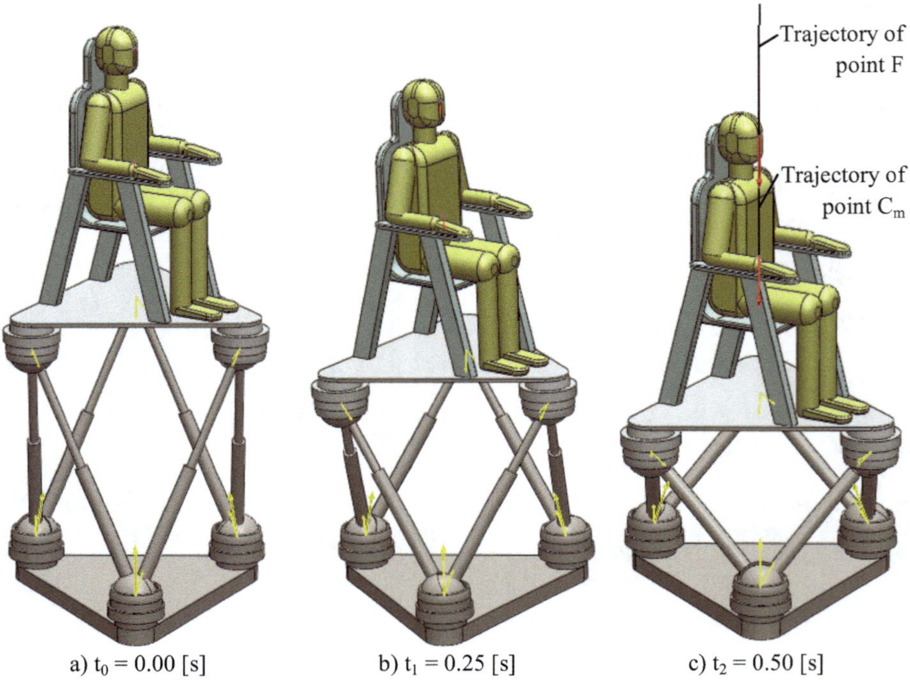


**Fig. 4.** Initial, intermediate and final positions-orientations of the mobile platform in motion 1.

The variations of reaction forces modulus which occur in the lower and upper levels spherical joints during motions 1 and 2, computed in Solidworks Motion module, are shown in Fig. 6, 7, and Fig. 8, 9, respectively. The variations of reaction forces plots are denoted with number of the corresponding kinematical chain, as shown in Fig. 1.

In motion 1 (descent of the mobile platform), the reaction forces have close values for the kinematical chains 1, 3, 5, and 2, 4, 6, respectively, depending on the constructive variant (parts of double spherical joints).

In motion 2 (balancing of the mobile platform from the right side of human operator to the left side), the values of reaction forces varies between around 100 [N] (upper level spherical joint of the kinematical chain no. 1) and 2000 [N] (lower level spherical joint of the kinematical chain no. 3).

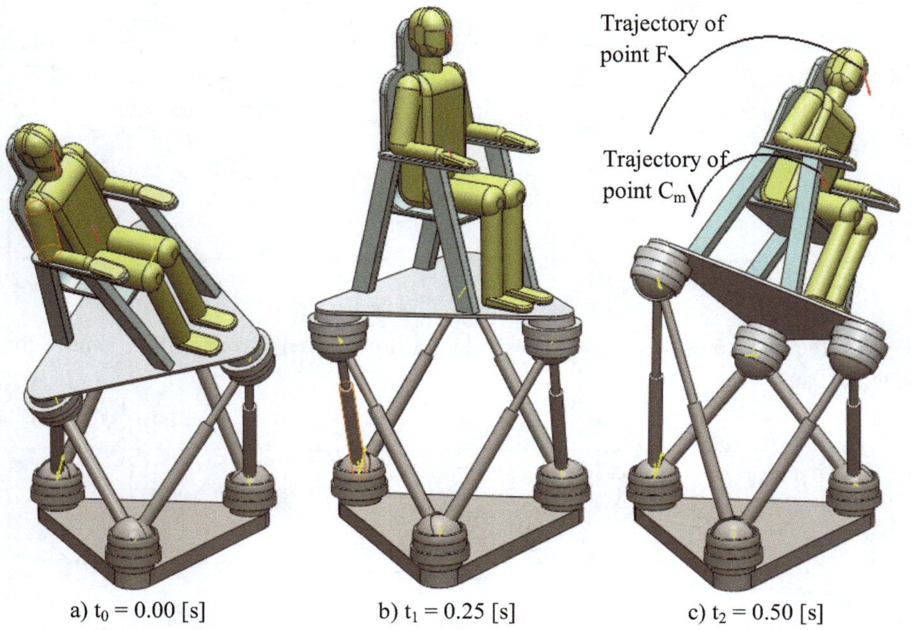


**Fig. 5.** Initial, intermediate and final positions-orientations of the mobile platform in motion 2.

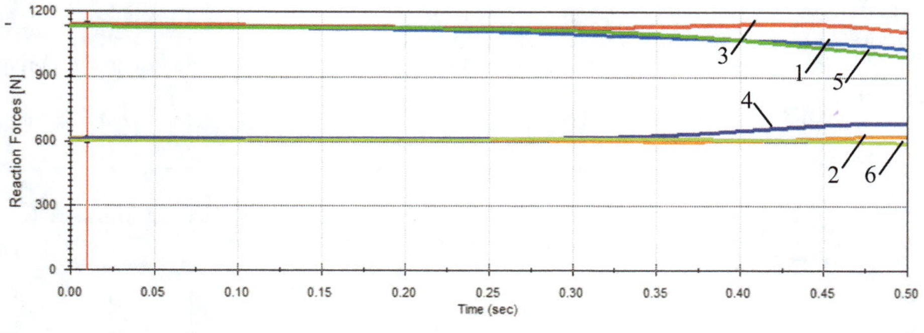


**Fig. 6.** The variations of reaction forces modulus in the lower level spherical joints during motion 1.

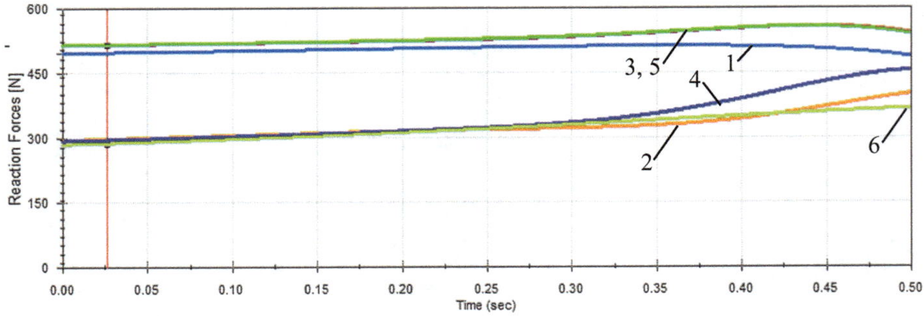


**Fig. 7.** The variations of reaction forces modulus in the upper level spherical joints during motion 1.

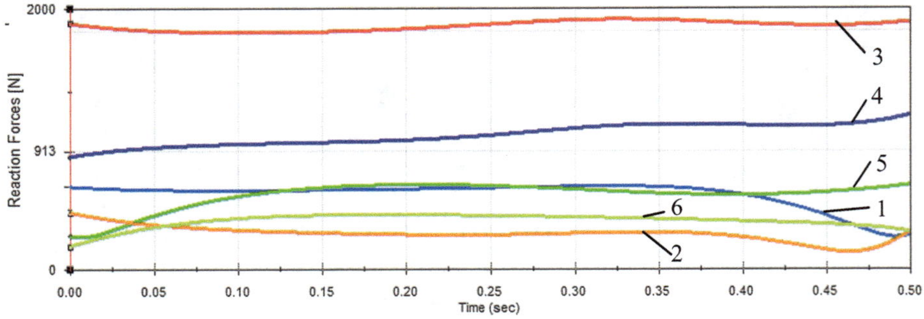


**Fig. 8.** The variations of reaction forces modulus in the lower level spherical joints during motion 2.

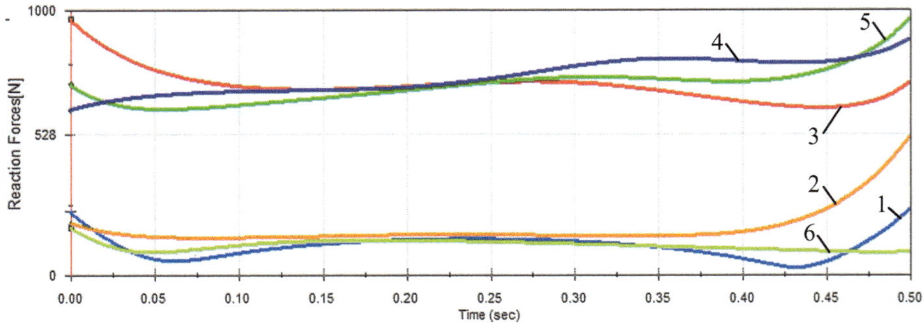


**Fig. 9.** The variations of reaction forces modulus in the upper level spherical joints during motion 2.

## 4 Conclusions

After the modeling and simulation processes, the following conclusions may be drawn:

- for both motions studied, the variations of reaction forces in the spherical joints are determined, depending on the mobile platform behavior during simulation;
- for the case of motion 1, the reaction forces have close values, depending on the disposal of kinematical chain, and on the constructive variant of the spherical joint;
- the highest value of the reaction forces occurs in the lower level spherical joint of the kinematical chain no. 3.

This method of reaction forces evaluation can be used for other types of forces and for various mechanical systems, including different parallel robots structures.

As further research, the inclusion of the friction forces and of the clearances in the motion study can be considered, by setting the contact properties between the considered components, and by modifying the components dimensions, respectively.

## References


1. Kovacs, F.V., Pau, C.V.: Introducere in Robotica/Introduction to Robotics. Printech Publishing House, Bucharest (2000)
2. Merlet, J.P.: Parallel Robots, 2nd edn. Springer, Dordrecht (2006)
3. Gogu, G.: Structural Synthesis of Parallel Robots: Part 1: Methodology. Springer, Dordrecht (2008)
4. Acevedo, M., Ceccarelli, M., Carbone, G.: Application of counter-rotary counterweights to the dynamic balancing of a spatial parallel manipulator. In: Gogu, G., Maniu, I., Lovasz, E. C., Fauroux, J.C., Ciupe, V. (eds.) 11th International Conference on Mechanisms and Mechanical Transmissions/International Conference on Robotics, Applied Mechanics and Materials, vol. 162, pp. 224–233. Trans Tech Publications Ltd., Durnten-Zurich (2012)
5. Lovasz, E.C., Grigorescu, S.M., Margineanu, D.T., Gruescu, C.M., Pop, C., Ciupe, V., Maniu, I.: Geared linkages with linear actuation used as kinematic chains of a planar parallel manipulator. In: Corves, B., Lovasz, E.C., Husing, M. (eds.) 3rd Conference of the Mechanisms Transmissions and Applications, Mechanisms and Machine Science - Mechanisms, Transmissions And Applications, vol. 31, pp 21–31. Springer, Heidelberg (2015)
6. Martini, A.: Gravity compensation of a 6-UPS parallel kinematics machine tool through elastically balanced constant-force generators. FME Trans. **46**(1), 10–16 (2018)
7. Miclosina, C.O.: Six inextensible wire device for determination of position-orientation matrix of a robot's mobile reference system. In: Katalinic, B. (ed.) Annals of DAAAM for 2009 & Proceedings of the 20th International DAAAM Symposium, Annals of Daaam and Proceedings, vol. 20, pp. 841–842. DAAAM International, Vienna (2009)
8. Miller, K.: Design and applications of parallel robots. In: Jarvis R.A., Zelinsky A. (eds.) Robotics Research. Springer Tracts in Advanced Robotics, vol. 6. Springer, Heidelberg (2003)
9. Anderson, P.L., Mahoney, A.W., Webster III, R.J.: Continuum reconfigurable parallel robots for surgery: shape sensing and state estimation with uncertainty. IEEE Robot. Autom. Lett. **2** (3), 1617–1624 (2017)

10. Patel, Y.D., George, P.M.: Parallel manipulators applications—a survey. Mod. Mech. Eng. **2** (3), 57–64 (2012)
11. Miclosina, C.O., Cojocaru, V., Korka, Z.I.: Dynamic simulation of a parallel topology robot operation. In: Tabara, I., Vladareanu, L., Doicin, C., Ionescu, N., Savu, T., Bruja, A., Ocnarescu, C., Opran, C.G. (eds.) The Forth International Conference on Robotics - Robotics 2014, Applied Mechanics and Materials, vol. 762, pp. 107–112, Trans Tech Publications Ltd., Durnten-Zurich (2015)
12. Miclosina, C.-O., Cojocaru, V., Korka, Z.:. Analysis of a 6-DOF parallel robot motion simulation. J. Phys. Conf. Ser. **1426** (2020). International Conference on Applied Sciences 2019. IOP Publishing, Bristol
13. Vela, D.G., Madaras, L., Miclosina, C.O., Vela, I.: Calculation of forces and reactions from kinematical joints of double harmonic transmission with wave generator dephased with π/2 used in industrial robots structure. In: Brisan, C., Maties, V., Stan, S.D., Brad, S. (eds.) 5th International Conference on Robotics, Solid State Phenomena - Robotics and Automation Systems, vols. 166–167, pp. 451–456. Trans Tech Publications Ltd., Durnten-Zurich (2012)
14. Miclosina, C.O., Campian, C.V.: Fatigue analysis of low level links of a parallel topology robot guiding device mechanism. In: Gogu, G., Maniu, I., Lovasz, E.C., Fauroux, J. C., Ciupe, V., (eds.) 11th International Conference on Mechanisms and Mechanical Transmissions/International Conference on Robotics, Applied Mechanics and Materials, vol. 162, pp. 98–105. Trans Tech Publications Ltd., Durnten-Zurich (2012)
15. Budai, A.M., Campian, V.C., Frunzaverde, D., Miclosina, C., Pepa, D.: Lifetime estimations of the operating mechanism of Kaplan turbine runner blades. Tehnicki Vjesnik-Technical Gazette **24**(Suppl. 2), 271–277 (2017)

# Dynamic Modelling of an Isoglide T3 Type Parallel Robot

Nadia Cretescu and Mircea Neagoe(✉)

Renewable Energy Systems and Recycling R&D Center,
Transilvania University of Brasov, Brașov, Romania
{ncretescu,mneagoe}@unitbv.ro

**Abstract.** The dynamic modelling of parallel robots rise difficult development issues due to the structural, kinematic and dynamic complexity of parallel mechanisms. The paper presents the application of a dynamic approach aiming at reducing the modelling complexity by replacing the parallel robotic structure with simpler open kinematic chains derived from the parallel structure. This method was successfully applied to obtain the closed-form dynamic model of an Isoglide T3 parallel robot and it was validated by numerical simulations using the ADAMS software.

**Keywords:** Parallel robots · Dynamic modelling · Closed-form dynamic model · Simulation · ADAMS software

## 1 Introduction

The parallel robots are closed kinematic chain type mechanisms, composed by a mobile platform connected to the base by two or more kinematic arms (legs or limbs). For each arm a simple open or a complex kinematic chain can be associated [1].

The parallel robots have the advantages of higher speeds and precision, higher loads and stiffness comparing with serial manipulators. The most important drawback is related their reduced workspace.

Different approaches can be applied to obtain the dynamic model of parallel robots, e.g. Euler-Lagrange method, Lagrange equations with multipliers, Newton-Euler method, were applied in literature aiming at identifying their dynamic behaviour [2–7].

A dynamic modelling method for hybrid robotic structures is presented in [8]. The direct and inverse dynamic models of the robotic arms and the mobile platform were obtained by using the Newton Euler equations, where Jacobian matrices are also involved.

The influence of the load-rigidity correlation on the dynamic response of a medical Triglide parallel robot was identified using ADAMS simulations in the hypothesis of rigid and elastic connecting rods [9]. The same approach was applied for a 4 DOF parallel robot of type T3R1with decoupled motions [10]. The Lagrange with multipliers method was used in [11] to obtain the analytical dynamic models of two parallel robots with two and three degrees of coupling, respectively, followed by a virtual simulation in SimMechanics environment.


E.-C. Lovasz et al. (Eds.): MTM&Robotics 2020, MMS 88, pp. 235–248, 2021.
https://doi.org/10.1007/978-3-030-60076-1_21

The Lagrange equations of second kind is used in this paper to derive the closed-form dynamic model of an Isoglide T3 parallel robot, considering the rigid link hypothesis, based on an approach presented by Ibrahim [12]. The structural and kinematic analysis is presented in Sect. 3, followed by the analytical dynamic modelling of the three open arms and the end-effector platform and the projection of the dynamic equations in the space of active joints (Sect. 4), developed using the Maple software. The close-form dynamic model is numerically simulated and validated in the ADAMS environment software (Sect. 5). The conclusions of the study are drawn in the Sect. 6.

## 2 Problem Formulation

The dynamic modelling of parallel type robotic structures, used in this paper, is based on an algorithm proposed by Ibrahim [12] and composed of the following steps:

1) breaking the kinematic joints between the mobile platform (end-effector) and the component robotic arms;
2) inverse dynamic modelling of each open robotic arm with serial kinematic chain configuration, considering the joint variables as independent generalized coordinates;
3) inverse dynamic modelling of the end-effector platform based on the operational variables associated to the parallel robot, considering known the end-effector motion;
4) projection of the dynamic equations obtained for the open robotic arms and the mobile platform into the space of the independent joint variables (in the active joints) of the parallel robot through Jacobian matrices.

The dynamic modelling of the robot arms, considered after removing their connections with the end-effector platform, can be done using known approaches applied to serial robots, such as the Newton-Euler method or Lagrange equations of the second kind. The inverse dynamic model of the end-effector platform can be described through the Newton-Euler equations, which allow to obtain the forces and moments applied on a rigid body with known motion.

The inverse dynamic modelling algorithm of each robotic arm, isolated from the parallel robot, has the following steps:

- establishing the kinetic energy of the component kinematic links;
- establishing the potential energy of the component kinematic links;
- expressing the Lagrangian of the robotic arm;
- deriving the inverse dynamic equations for each arm.

The generalized forces (forces and torques) resulting in the inverse dynamic models of the end-effector platform and the robotic arms, considered independently, are reduced in the space of active joints and thus the actuating generalized forces are obtained by applying the superposition principle.

## 3 Geometric and Kinematic Modelling

The Isoglide T3 robot has a three degrees of freedom (DOF) parallel structure of 3TRRR type (3 Translation, **R**otation, **R**otation, **R**otation) with three translational inputs and three translational outputs (Fig. 1). It is composed by a mobile end-effector platform (4) connected to the base by three arms *a, b c*, each one composed by three links ($1_i$, $2_i$, $3_i$, where $i = a, b, c$) and four kinematic joints: one active prismatic joint disposed along the axis of the global coordinate system $O_0X_0Y_0Z_0$ (denoted by $q_i$, $i = a, b, c$) and 3 passive revolute joints characterized by the relative angular displacements $\varphi_{1i}$, $\varphi_{2i}$, $\varphi_{3i}$, where $i = a, b, c$. The four joints have parallel axes to each other.

The three output motions of the end-effector along the axes $X_0$, $Y_0$ and $Z_0$ are decoupled. Therefore, each output motion is commanded from a single motor: the output motion along the $X_0$ axis is controlled by the active joint $q_a$, the one along the $Y_0$ axis by the active joint $q_b$, and the one along $Z_0$ axis by the active joint $q_c$. In addition, the input motions are transmitted unchanged to the outputs [13]. Therefore, this parallel robot type is also called "maximally regular" [1, 14] and refers to the particular case of Jacobian matrix that equals the identity matrix.

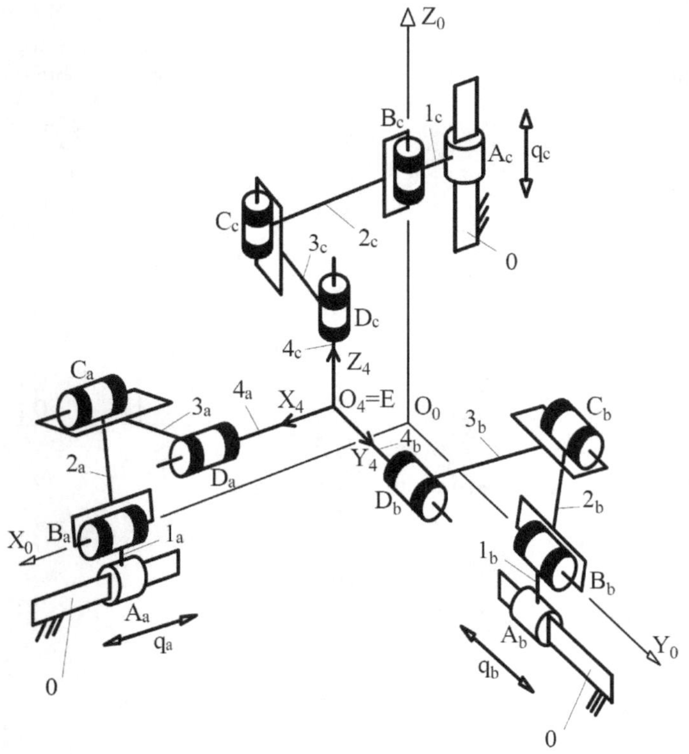


**Fig. 1.** Kinematic diagram of an Isoglide T3 parallel robot

The traveling coordinate system method [15] was used to obtain the direct and inverse geometric and kinematic models of the Isoglide T3 parallel robot. The results of the direct geometric model is the absolute translational displacement function, i.e. the displacements of the end-effector characteristic point $E(x_E, y_E, z_E) \equiv O_4$ in relation with the relative displacements in the active joints ($q_a$, $q_b$, $q_c$):

$$\begin{cases} x_E = q_a - l_{4a} \\ y_E = q_b - l_{4b} \\ z_E = q_c - l_{4c} \end{cases} \tag{1}$$

where $l_{4a}$, $l_{4b}$, $l_{4c}$ represent the lengths of the mobile platform 4 in the direction of the axes of the local coordinate system $O_4X_4Y_4Z_4$ as depicted in Fig. 1.

The inverse geometric model allows to establish the dependence of the relative displacements in the active joints ($q_a$, $q_b$, $q_c$), in relation with the end-effector Cartesian coordinates ($x_E$, $y_E$, $z_E$):

$$\begin{cases} q_a = x_E + l_{4a} \\ q_b = y_E + l_{4b} \\ q_c = z_E + l_{4c} \end{cases} \tag{2}$$

The direct kinematic model expresses the dependences of the end-effector speeds ($\dot{x}_E$, $\dot{y}_E$, $\dot{z}_E$) and acceleration ($\ddot{x}_E$, $\ddot{y}_E$, $\ddot{z}_E$) on the relative linear speeds ($\dot{q}_a$, $\dot{q}_b$, $\dot{q}_c$) and accelerations ($\ddot{q}_a$, $\ddot{q}_b$, $\ddot{q}_c$) in the active joints, through the robot Jacobian matrix $J_e$:

$$\begin{bmatrix} \dot{x}_E \\ \dot{y}_E \\ \dot{z}_E \end{bmatrix} = J_e \cdot \begin{bmatrix} \dot{q}_a \\ \dot{q}_b \\ \dot{q}_c \end{bmatrix} \text{ and } \begin{bmatrix} \ddot{x}_E \\ \ddot{y}_E \\ \ddot{z}_E \end{bmatrix} = \dot{J}_e \cdot \begin{bmatrix} \dot{q}_a \\ \dot{q}_b \\ \dot{q}_c \end{bmatrix} + J_e \cdot \begin{bmatrix} \ddot{q}_a \\ \ddot{q}_b \\ \ddot{q}_c \end{bmatrix} \tag{3}$$

where $J_e$ and $\dot{J}_e$ are define by relations:

$$J_e = \begin{bmatrix} \frac{\partial x_E}{\partial q_a} & \frac{\partial x_E}{\partial q_b} & \frac{\partial x_E}{\partial q_c} \\ \frac{\partial y_E}{\partial q_a} & \frac{\partial y_E}{\partial q_b} & \frac{\partial y_E}{\partial q_c} \\ \frac{\partial z_E}{\partial q_a} & \frac{\partial z_E}{\partial q_b} & \frac{\partial z_E}{\partial q_c} \end{bmatrix} = \begin{bmatrix} 1 & 0 & 0 \\ 0 & 1 & 0 \\ 0 & 0 & 1 \end{bmatrix} \text{ and } \dot{J}_e = \begin{bmatrix} 0 & 0 & 0 \\ 0 & 0 & 0 \\ 0 & 0 & 0 \end{bmatrix}. \tag{4}$$

As a result, the kinematical model is expressed by:

$$\begin{bmatrix} \dot{x}_E \\ \dot{y}_E \\ \dot{z}_E \end{bmatrix} = \begin{bmatrix} \dot{q}_a \\ \dot{q}_b \\ \dot{q}_c \end{bmatrix} \text{ and } \begin{bmatrix} \ddot{x}_E \\ \ddot{y}_E \\ \ddot{z}_E \end{bmatrix} = \begin{bmatrix} \ddot{q}_a \\ \ddot{q}_b \\ \ddot{q}_c \end{bmatrix} \tag{5}$$

# 4 Dynamic Modelling

The inverse dynamic modelling begins for the Isoglide T3 parallel robot with the preliminary stage of disconnecting the three arms from the mobile platform, i.e. disassembling the passive revolute joints $D_a$, $D_b$ and $D_c$ as represented in Fig. 2. Thus, three separate kinematic open chains of T-R-R type (the arms $1_j$-$2_j$-$3_j$, $j = a, b, c$) and the end-effector 4 are obtained.

The three robotic arms $a$, $b$ and $c$ have a serial structure, formed by the active prismatic joints $A_j$ and two passive revolute joints $B_j$ and $C_j$, $j = a, b, c$, Fig. 2. The three kinematic joints $A_j$, $B_j$ and $C_j$ have axes parallel to the axis $O_0x_0$ for the robotic arm $a$, with axis $O_0y_0$ for the robotic arm $b$ and to the axis $O_0z_0$ for the robotic arm $c$, respectively. The independent joint variables (the linear relative displacement in the active joints $A_j$) are denoted by $q_j$, and the dependent ones by $\varphi_{1j}$ for the joints $B_j$ and by $\varphi_{2j}$ for the joints $C_j$, $j = a, b, c$.

The links of the robotic arms are rigid bodies and characterized by the length $l_k$ (distance between the axes of the two adjacent joints), the gravity centre $G_k$, the mass $m_k$ and the matrix of the moments of inertia $J_k$ established in a local coordinate system with origin point in the gravity centre $G_k$, $k = 1_a, 2_a, 3_a, 1_b, 2_b, 3_b, 1_c, 2_c, 3_c$. As a simplifying hypothesis, the gravity centres $G_k$ are located at middle of the length $l_k$ and the centrifugal moments of inertia are zero (i.e. the axes of the local coordinate system coincide with the axes of inertia of the link).

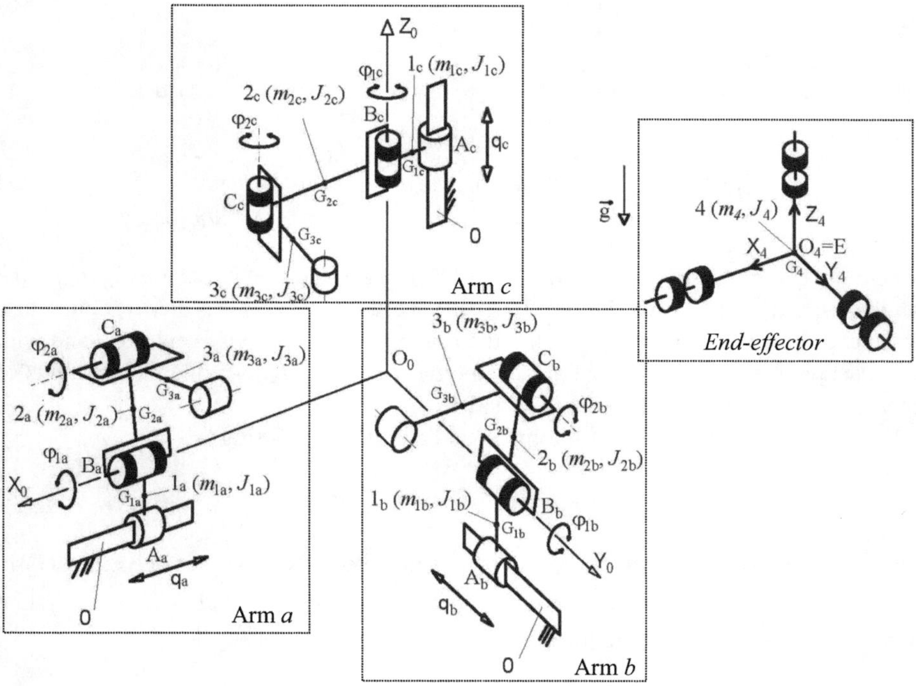


**Fig. 2.** Dynamic scheme of the Isoglide T3 robot considering the robotic arms detached from the end-effector platform.

The end-effector platform 4 performs only translational motions, it has the mass $m_4$ and the gravity centre $G_4$ in the origin $O_4$ of the local coordinate system $O_4X_4Y_4Z_4$, as well as the matrix of moments of inertia $J_4$. The length of the element 4 in the $X_4$ direction is denoted by $l_{4a}$, in the $Y_4$ direction by $l_{4b}$ and in the $Z_4$ direction by $l_{4c}$.

The inverse dynamic model of the Isoglide T3 parallel robot is derived based on the following algorithm:

1. Stage I: inverse dynamic modelling of the 3 robotic arms *a*, *b* and *c*, considered as independent serial kinematic chains. The Lagrange equations of second kind are used in this study.
2. Stage II: inverse dynamic modelling of the end-effector platform 4, considered as a rigid body with independent translational motions along the three axes of the global coordinate system $O_0X_0Y_0Z_0$.
3. Stage III: projection of the dynamic equations of the robotic arms and of the end-effector platform in the force space of the active joints $A_j$, $j = a, b, c$, Fig. 2.

## 4.1 Inverse Dynamic Models of the Robotic Arms

The inverse dynamic modelling based on the Lagrange equations of second kind, in the case of a robotic arm $j = a, b, c$, involves determining the total kinetic energy $K_j$ and the total potential energy $P_j$, necessary to establish the system Lagrangian $L_j$:

$$L_j = K_j - P_j = \sum_{i=1}^{3} K_{ij} - \sum_{i=1}^{3} P_{ij}, \tag{6}$$

where $K_{ij}$ represent the kinematical energy of the $i^{th}$ link from the arm $j$:

$$K_{ij} = \frac{1}{2} m_{ij} v_{Gij}^2 + \frac{1}{2} \omega_{ij}^T J_{Gij} \omega_{ij}, \; J_{Gij} = \begin{bmatrix} J_{ijxx} & 0 & 0 \\ 0 & J_{ijyy} & 0 \\ 0 & 0 & J_{ijzz} \end{bmatrix}, \tag{7}$$

where $m_{ij}$ is the mass of the link *ij*, $v_{Gij}$ – the speed of the gravity centre $G_{ij}$, $\omega_{ij}$ – the angular speed vector of the link *ij*, $J_{Gij}$ – the inertial matrix of the link *ij*, define in a local coordinate system $O_{ij}x_{ij}y_{ij}z_{ij}$ with the origin $O_{ij}$ in the gravity centre $G_{ij}$ and with the axes parallel to the principal axis of inertia, $J_{ijxx}$, $J_{ijyy}$, $J_{ijzz}$ – principal moments of inertia in relation with the axis $x_{ij}$, $y_{ij}$ and $z_{ij}$.

The potential energy $P_{ij}$ of the link *ij* is determined by using:

$$P_{ij} = -m_{ij}\vec{g} \cdot \vec{r}_{Gij} = m_{ij} g \cdot z_{Gij}, \tag{8}$$

where $\vec{r}_{Gij} = x_{Gij}\vec{i}_0 + y_{Gij}\vec{j}_0 + z_{Gij}\vec{k}_0$ is the position vector of the gravity centre $G_{ij}$ in global coordinate system, $\vec{i}_0$, $\vec{j}_0$, $\vec{k}_0$ – the unit vectors of the global coordinate system axes, $\vec{g} = -g\vec{k}_0$ is the gravitational acceleration, Fig. 2.

Using the geometric and kinematic models, the Langrangian $L_j$ of the robotic arm $j$ is described by:

$$L_j = \frac{1}{2}A_{0j}\ddot{q}_j^2 + \frac{1}{2}A_{1j}\dot{\varphi}_{1j}^2 + \frac{1}{2}A_{2j}\dot{\varphi}_{2j}^2 + A_{3j}\dot{\varphi}_{1j}\dot{\varphi}_{2j} + A_{4j}g, j = a,b,c \tag{9}$$

where

$$\begin{cases} A_{0j} = m_{1j} + m_{2j} + m_{3j} \\ A_{1j} = \ J_{2j} + J_{3j} + m_{2j}c_{2j}^2 + m_{3j}\left(l_{2j}^2 + c_{3j}^2\right) + 2m_{3j}l_{2j}c_{3j}\cos\varphi_{2j} \\ A_{2j} = J_{3j} + m_{3j}c_{3j}^2 \\ A_{3j} = J_{3j} + m_{3j}c_{3j}^2 + m_{3j}l_{2j}c_{3j}\cos\varphi_{2j} \\ A_{4a} = m_{1a}c_{1a} - m_{2a}c_{2a}\sin\varphi_{1a} - m_{3a}l_{2a}\sin\varphi_{1a} - m_{3a}c_{3a}\sin(\varphi_{1a} + \varphi_{2a}) \\ A_{4b} = m_{1b}c_{1b} - m_{2b}c_{2b}\cos\varphi_{1b} - m_{3b}l_{2b}\cos\varphi_{1a} - m_{3b}c_{3b}\cos(\varphi_{1b} + \varphi_{2b}) \\ A_{4c} = -(m_{1c} + m_{2c} + m_{3c})\, q_c \end{cases} \tag{10}$$

with notation $J_{2a} = J_{2axx}$, $J_{3a} = J_{3axx}$, $J_{2b} = J_{2byy}$, $J_{3b} = J_{3byy}$, $J_{2c} = J_{2czz}$, $J_{3c} = J_{3czz}$, and $c_{1j} = B_jG_{1j}$, $c_{2j} = B_jG_{2j}$ and $c_{3j} = C_jG_{3j}$ (Fig. 2).

After performing the calculus according to Lagrange equations, the inverse dynamic model of a robotic arm $j$ is expressed by:

$$\begin{bmatrix} F_j \\ \tau_{1j} \\ \tau_{2j} \end{bmatrix} = M_j \cdot \begin{bmatrix} \ddot{q}_j \\ \ddot{\varphi}_{1j} \\ \ddot{\varphi}_{2j} \end{bmatrix} + V_j \begin{bmatrix} \dot{q}_j \\ \dot{\varphi}_{1j} \\ \dot{\varphi}_{2j} \end{bmatrix} + G_j, j = a,b,c \tag{11}$$

where: $F_j$ represents the force on the active prismatic joint $j$, $\tau_{ij}$ - the torque on the revolute joint $i$ from arm $j$, $M_j$ - the matrix of inertial coefficients, $V_j$ - the matrix of centrifugal and Coriolis coefficients, and $G_j$ - the vector of gravitational terms:

$$M_j = \begin{bmatrix} A_{0j} & 0 & 0 \\ 0 & A_{1j} & A_{3j} \\ 0 & A_{3j} & A_{2j} \end{bmatrix} \tag{12}$$

$$V_j = \begin{bmatrix} 0 & 0 & 0 \\ 0 & -2m_{3j}l_{2j}c_{3j}\sin\varphi_{2j}\dot{\varphi}_{2j} & -m_{3j}l_{2j}c_{3j}\sin\varphi_{2j}\dot{\varphi}_{2j} \\ 0 & m_{3j}l_{2j}c_{3j}\sin\varphi_{2j}\dot{\varphi}_{1j} & 0 \end{bmatrix} \tag{13}$$

$$G_a = \begin{bmatrix} 0 \\ m_{2a}c_{2a}\cos\varphi_{1a} + m_{3a}(l_{2a}\cos\varphi_{1a} + c_{3a}\cos(\varphi_{1a} + \varphi_{2a})) \\ m_{3a}c_{3a}\cos(\varphi_{1a} + \varphi_{2a}) \end{bmatrix} \tag{14a}$$

$$G_b = \begin{bmatrix} 0 \\ -m_{2b}c_{2b}\sin\varphi_{1b} - m_{3b}(l_{2b}\sin\varphi_{1b} + c_{3b}\sin(\varphi_{1b} + \varphi_{2b})) \\ -m_{3b}c_{3b}\sin(\varphi_{1b} + \varphi_{2b}) \end{bmatrix} \tag{14b}$$

$$G_c = \begin{bmatrix} m_{1c} + m_{2c} + m_{3c} \\ 0 \\ 0 \end{bmatrix} \tag{14c}$$

### 4.2 Inverse Dynamic Model of the End-Effector Platform

As the mobile platform can only perform translational motions, and considering the Eq. (5), the inverse dynamic model is expressed by:

$$F_4 = \begin{bmatrix} F_{4x} \\ F_{4y} \\ F_{4z} \end{bmatrix} = \begin{bmatrix} m_4 \ddot{q}_a \\ m_4 \ddot{q}_b \\ m_4(\ddot{q}_c - g) \end{bmatrix}. \tag{15}$$

### 4.3 Projection of the Dynamic Equations in the Force Space of the Active Joints

The dynamic model of a robotic arm $j$ can be projected in the force space of the active joints by applying the principle of virtual power:

$$\begin{bmatrix} F_j \\ \tau_{1j} \\ \tau_{2j} \end{bmatrix}^T \cdot \begin{bmatrix} \dot{q}_j \\ \dot{\varphi}_{1j} \\ \dot{\varphi}_{2j} \end{bmatrix} = \begin{bmatrix} F_{a.j} \\ F_{b.j} \\ F_{c.j} \end{bmatrix}^T \cdot \begin{bmatrix} \dot{q}_a \\ \dot{q}_b \\ \dot{q}_c \end{bmatrix}, j = a, b, c \tag{16}$$

where $F_{a.j}$, $F_{b.j}$ and $F_{c.j}$ represent the components of the forces in the active joints generated by the dynamic effects of the robotic arm $j$.

It is known that:

$$\begin{bmatrix} \dot{q}_j \\ \dot{\varphi}_{1j} \\ \dot{\varphi}_{2j} \end{bmatrix} = J_{qj} \cdot \begin{bmatrix} \dot{q}_a \\ \dot{q}_b \\ \dot{q}_c \end{bmatrix} \rightarrow \begin{bmatrix} F_{a.j} \\ F_{b.j} \\ F_{c.j} \end{bmatrix} = J_{qj}^T \begin{bmatrix} F_j \\ \tau_{1j} \\ \tau_{2j} \end{bmatrix} \tag{17}$$

where

$$J_{qj} = \begin{bmatrix} \frac{\partial q_j}{\partial q_a} & \frac{\partial q_j}{\partial q_b} & \frac{\partial q_j}{\partial q_c} \\ \frac{\partial \varphi_{1j}}{\partial q_a} & \frac{\partial \varphi_{1j}}{\partial q_b} & \frac{\partial \varphi_{1j}}{\partial q_c} \\ \frac{\partial \varphi_{2j}}{\partial q_a} & \frac{\partial \varphi_{2j}}{\partial q_b} & \frac{\partial \varphi_{2j}}{\partial q_c} \end{bmatrix}. \tag{18}$$

The absolute positions of the central points of the passive joint $D_j$, $j = a, b, c$ can be expressed depending on both the operational variables ($x_E$, $y_E$, $z_E$) and on the joint variables ($q_j$, $\varphi_{1j}$, $\varphi_{2j}$). The relationships thus obtained make possible to obtain the partial derivatives in the matrix (18):

$$J_{qa} = \begin{bmatrix} 1 & 0 & 0 \\ 0 & \frac{\cos(\varphi_{1a}+\varphi_{2a})}{l_{2a}\sin\varphi_{2a}} & \frac{\sin(\varphi_{1a}+\varphi_{2a})}{l_{2a}\sin\varphi_{2a}} \\ 0 & \frac{-l_{2a}\cos\varphi_{1a}-l_{3a}\cos(\varphi_{1a}+\varphi_{2a})}{l_{2a}l_{3a}\sin\varphi_{2a}} & \frac{-l_{2a}\sin\varphi_{1a}-l_{3a}\sin(\varphi_{1a}+\varphi_{2a})}{l_{2a}l_{3a}\sin\varphi_{2a}} \end{bmatrix} \tag{19a}$$

$$J_{qb} = \begin{bmatrix} 1 & 0 & 0 \\ 0 & \frac{\sin(\varphi_{1b}+\varphi_{2b})}{l_{2b}\sin\varphi_{2b}} & \frac{\cos(\varphi_{1b}+\varphi_{2b})}{l_{2b}\sin\varphi_{2b}} \\ 0 & \frac{-l_{2b}\sin\varphi_{1b}-l_{3b}\sin(\varphi_{1b}+\varphi_{2b})}{l_{2b}l_{3b}\sin\varphi_{2b}} & \frac{-l_{2b}\cos\varphi_{1b}-l_{3b}\cos(\varphi_{1b}+\varphi_{2b})}{l_{2b}l_{3b}\sin\varphi_{2b}} \end{bmatrix} \tag{19b}$$

$$J_{qc} = \begin{bmatrix} 1 & 0 & 0 \\ 0 & \frac{\cos(\varphi_{1c}+\varphi_{2c})}{l_{2c}\sin\varphi_{2c}} & \frac{\sin(\varphi_{1c}+\varphi_{2c})}{l_{2c}\sin\varphi_{2c}} \\ 0 & \frac{-l_{2c}\cos\varphi_{1c}-l_{3c}\cos(\varphi_{1c}+\varphi_{2c})}{l_{2c}l_{3c}\sin\varphi_{2c}} & \frac{-l_{2c}\sin\varphi_{1c}-l_{3c}\sin(\varphi_{1c}+\varphi_{2c})}{l_{2c}l_{3c}\sin\varphi_{2c}} \end{bmatrix} \tag{19c}$$

Similarly, the dynamic equations of the end-effector platform are projected into the force space of the active joints using:

$$\begin{bmatrix} F_{a.4} \\ F_{b.4} \\ F_{c.4} \end{bmatrix} = J_e^T \begin{bmatrix} F_{4x} \\ F_{4y} \\ F_{4z} \end{bmatrix} \tag{20}$$

where $J_e$ is the identity matrix according to Eq. (4).

By applying the superposition principle for the projected dynamic equations, the total driving forces in the active joints $A_j$, $j = a, b, c$ are:

$$\begin{bmatrix} F_{A_a} \\ F_{A_b} \\ F_{A_c} \end{bmatrix} = \begin{bmatrix} F_{a.a} \\ F_{b.a} \\ F_{c.a} \end{bmatrix} + \begin{bmatrix} F_{a.b} \\ F_{b.b} \\ F_{c.b} \end{bmatrix} + \begin{bmatrix} F_{a.c} \\ F_{b.c} \\ F_{c.c} \end{bmatrix} + \begin{bmatrix} F_{a.4} \\ F_{b.4} \\ F_{c.4} \end{bmatrix}. \tag{21}$$

## 5 Simulation Results

The closed-form dynamic model described by Eq. (21), obtained by using the Maple software, was simulated on a test trajectory implemented both on Maple and ADAMS software, that allows the modelling and simulation of mechanisms as multibody systems (MBS) [16]. The CAD model of the Isoglide T3 parallel robot (Fig. 3) includes bodies of simple cylindrical shape; this ADAMS model can be used for validation of the closed-form model and to deliver the necessary input data in numerical simulations related to the inertial link properties, according to Table 1.

**Table 1.** Geometric and inertial properties of the robot links

| Link/body | Length [m] | Mass [kg] | Principal moments of inertia $[\text{kg} \cdot \text{m}^2]$ | | |
|---|---|---|---|---|---|
| | | | Jxx | Jyy | Jzz |
| $1j$ | 0.1 | 0.98 | $9.14 \cdot 10^{-4}$ | $9.14 \cdot 10^{-4}$ | $1.96 \cdot 10^{-4}$ |
| $2j$ | 0.63 | 6.17 | 0.204 | $1.235 \cdot 10^{-3}$ | 0.204 |
| $3j$ | 0.63 | 6.17 | 0.204 | $1.235 \cdot 10^{-3}$ | 0.204 |
| 4 | 0.20 | 5.66 | $5.302 \cdot 10^{-2}$ | $5.302 \cdot 10^{-2}$ | $5.302 \cdot 10^{-2}$ |
| $j = a, b, c$ | | | | | |

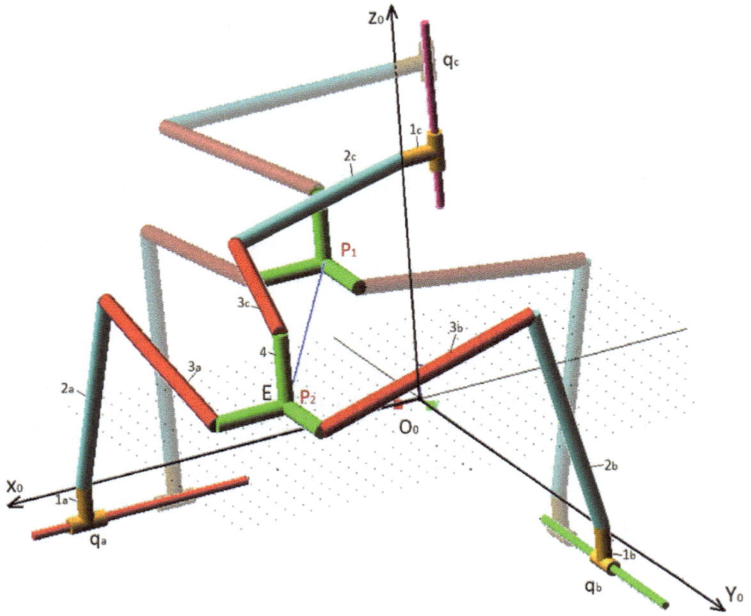


**Fig. 3.** ADAMS model of the Isoglide T3 parallel robot and its positions at the trajectory ends

The numerical simulations were performed on a trajectory defined in the joint space, characterized by:

- motion duration t = 10 s;
- strokes in active joints $\Delta q_a = \Delta q_b = 0.25$ m and $\Delta q_c = -0.25$;
- linear trajectory between the points $P_1(0.7, 0.8, 0.9)$ and $P_2(0.95, 1.05, 0.65)$;
- time interpolation functions: fifth degree polynomials, as represented in Fig. 4 for the active prismatic joint $A_a$.

The simulations on the selected test trajectory (Fig. 3) allowed obtaining the driving forces as depicted in Fig. 5, 6 and 7.

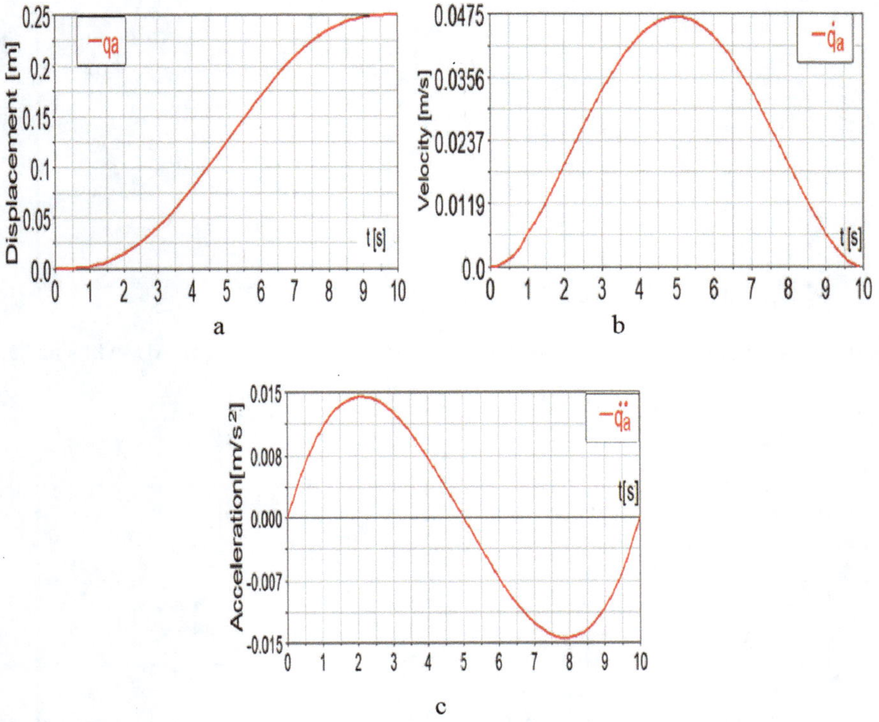


**Fig. 4.** Simulated motion for the active joint $A_a$: linear displacement $q_a$ (a); linear speed $\dot{q}_a$ (b); linear acceleration $\ddot{q}_a$(c).

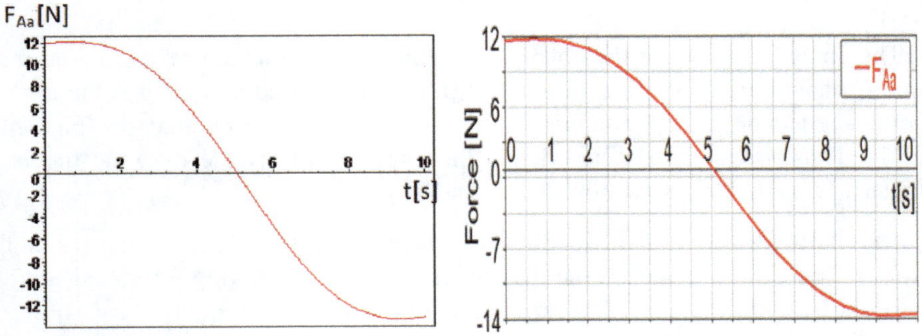


**Fig. 5.** Driving force $F_{Aa}$ obtained by: numerical simulation in Maple (a); ADAMS simulation (b).

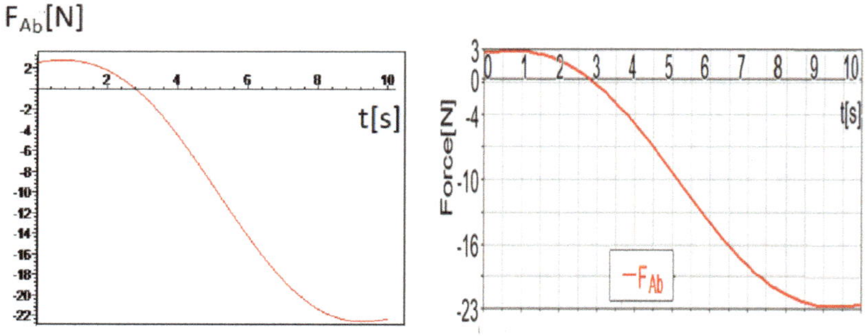


**Fig. 6.** Driving force $F_{Ab}$ obtained by: numerical simulation in Maple (a); ADAMS simulation (b).

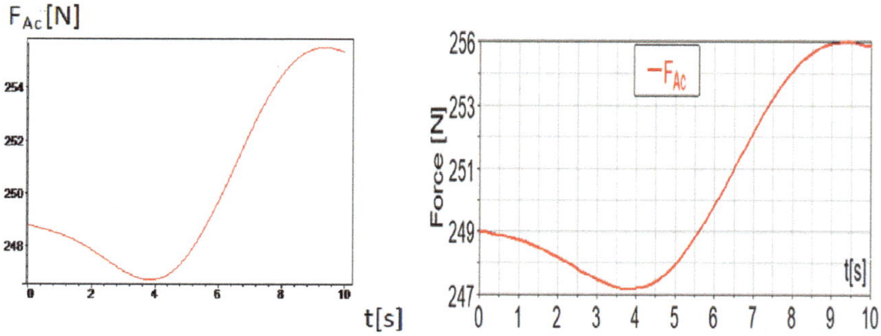


**Fig. 7.** Driving force $F_{Ac}$ obtained by: numerical simulation in Maple (a); ADAMS simulation (b).

The analysis of the results obtained by numerical simulation of the closed-form model (Maple) and of the CAD model (ADAMS), it can be noticed that the driving forces have identical time variations for the same active joint in both cases (Figs. 5, 6 and 7). Thus, it can be concluded that the closed-form inverse dynamic model is validated by ADAMS simulation. It can also be observed that:

- the driving forces $F_{Aa}$ and $F_{Ab}$ follow the shape as the displacements imposed in the active joints (Fig. 5 and 6); the maximum values of the driving forces are $\sim$12 N for the $F_{Aa}$ force and $\sim$22 N for the $F_{Ab}$ force, respectively;
- the force $F_{Ac}$ has higher values in relation to the other two driving forces (ranges between $\sim$249 N up to $\sim$255 N), due to the gravitational forces directed in the negative direction of the $O_0z_0$ axis;
- the maximum deviations in the calculus of the driving forces by using Maple and ADAMS software are of $10^{-8}$ N, that is the theoretical dynamic model is validated by numerical simulation.

## 6 Conclusion

The paper presents the inverse dynamic modelling of an Isoglide T3 parallel robot, using the approach of preliminary decomposing the parallel structure into open kinematic chains by disassembling the revolute joints between the robotic arms and the end-effector platform.

The application of the investigated inverse dynamic modelling for the Isoglide T3 parallel robot allowed to highlight the following aspects:

- By dismantling the revolute joints with the mobile platform, the parallel structure was decomposed into three serial robotic arms of TRR type and the end-effector link with translational motion;
- The inverse dynamic models of the three robotic arms, considered as independent mechanisms, were derived using the Lagrange equations of the second kind. The dynamic model of the mobile platform resulted based on the Newton's second law of motion;
- The driving forces were obtained by projecting the dynamic equations of the robotic arms and end-effector platform on the force space of the active joints, based on Jacobian matrices determined in the kinematic analysis stage;
- The closed-form of the inverse dynamic model was developed using the Maple software, that also allowed numerical simulation;
- The inverse dynamic model was validated by developing the CAD model of Isoglide T3 parallel robot and by simulating its dynamic behaviour using the ADAMS software.

The main advantages of the applied dynamic modelling method refer to the simplicity of its application: the complex modelling of a parallel structure is reduced to simpler modelling of serial structures (the robotic arms) and the end-effector platform, followed by the projection of their dynamic equations in the force space of the active joint through Jacobian matrices.

## References

1. Gogu, G.: Structural Synthesis of Parallel Robots. Part 1: Methodology. Springer, Dordrecht (2008)
2. Lee, K.M., Shah, D.K.: Dynamic analysis of a three-degrees-of-freedom in-parallel actuated manipulator. IEEE J. Robot. Autom. **4**, 361–368 (1988)
3. Bhattacharya, S., Nenchev, D.N., Uchiyama, M.: A recursive formula for the inverse of the inertia matrix of a parallel manipulator. Mech. Mach. Theory **33**(7), 957–964 (1998)
4. Liu, M.-J., Li, C.-X., Li, C.-N.: Dynamics analysis of the Gough-Stewart platform manipulator. IEEE Trans. Robot. Autom. **16**(1), 94–98 (2000)
5. Dasgupta, B., Mruthyunjaya, T.S.: A Newton-Euler formulation for the inverse dynamics of the Stewart platform manipulator. Mech. Mach. Theory **33**(8), 1135–1152 (1998)
6. Dasgupta, B., Choudhury, P.: A general strategy based on the Newton-Euler approach for the dynamic formulation of parallel manipulators. Mech. Mach. Theory **34**(6), 801–824 (1999)

7. Khalil, W., Guegan, S.: Inverse and direct dynamic modeling of Gough-Stewart robots. IEEE Trans. Robot. Autom. **20**(4), 754–762 (2004)
8. Ibrahim, O., Khalil, W.: Inverse and direct dynamic models of hybrid robots. In: Mechanism and Machine Theory, vol. 45, pp. 627–640 (2010)
9. Rat, N.R., Neagoe, M., Diaconescu, D., Stan, S.D.: Dynamic simulations regarding the influence of the load-rigidity correlation on the working accuracy of a medical Triglide parallel robot. MECHANIKA **17**(2), 178–181 (2011). ISSN 1392–1207
10. Rat, N., Rizk, R., Gogu, G., Neagoe, M., Comportement des robots parallèles à mouvements découplés et corps déformables, Congrès Français de Mécanique (2005)
11. Raţ, N.R., Neagoe, M., Stan, S.D.: Comparative dynamic analysis of two parallel robots. In: The 2010 International Conference on Robotics (ROBOTICS 2010), vol. 166–167, pp. 345-356 (2010)
12. Ibrahim, O.: Contribution a la modelisation dynamique des robots parallels et des robots hybrids, Ph.D. thesis, L'Ecole Centrale de Nantes et l'Universite de Nantes (2006)
13. Gogu, G.: Structural synthesis of fully-isotropic translational parallel robots via theory of linear transformations. Eur. J. Mech. A/Solids **23**, 1021–1039 (2004)
14. Gogu, G.: Structural Synthesis of Parallel Robots, Part 3: Topologies with Planar Motion of the Moving Platform. Springer, Dordrecht (2010)
15. Gogu, G.: Optimization on the Modeling of Industrial Robots. Ph.D. thesis, Transilvania University of Brasov (1995). (in Romanian)
16. Costa, A., Jones, R.P.: Automotive Vehicle Chassis Simulation for Motion Control Studies using Multibody Systems (MBS) Modelling Techniques, SAE Paper No. 921443 (1992)

# Kinematics of the Planar Parallel Manipulator Using Geared Slider-Crank with Linear Actuation as Legs 3-R(PRRGR)RR

Antonio-M.-F. Luputi(✉), Sanda M. Grigorescu(✉), Erwin-Christian Lovasz(✉), Carmen Sticlaru(✉), and Iosif Cărăbaş(✉)

Politehnica University of Timişoara, Timişoara, Romania
antonio.luputi@student.upt.ro, {sanda.grigorescu, erwin.lovasz, carmen.sticlaru, iosif.carabas}@upt.ro

**Abstract.** Geared linkage with linear actuation shows several advantages as large rotation angle with proper transmission angle, a sturdy design and a roughly linear transmission ration in a large range. The paper shows the positional kinematics of a planar parallel manipulator actuated with a geared linkage with linear actuation, which use as base mechanism a slider-crank. The paper shows the stroke of the linear actuators of the geared slider-crank, in order to perform a translation along the x-axis, along y-axis and oblique line. Also, in two examples is considered the rotation of the mobile platform.

**Keywords:** Parallel planar manipulator · Geared linkage with linear actuator · Kinematic analysis · Inverse transmission function

## 1 Introduction

In recent decades the robotic technology has evolved exponential, new branches appeared in which the advantages of the parallel manipulator are being used. Domain such as manufacturing [1], shipping [2] make the most of the speed of the manipulator whereas health service, surgery [3] and military [4] benefit also from the precision of the manipulator.

Manipulator can be divided into two major types: serial and parallel [5]. The two types of manipulators have different areas of applications. The serial manipulators are being used extensively in industry for assembly, painting and palletizing [6]. Parallel manipulators are used in areas such as material handling, outer space applications, flights and automobile simulators.

An important characteristic of the parallel manipulator is the possibility to reconfigure the manipulator depending to the task [7–9] and [10]. Many studies are made to optimize [11] and increase the singularity-free workspace [12–17].

The motor actuators of manipulators can be hydraulic, pneumatic, piezoelectric or electromechanical. Linear actuators transform rotational movement into translation movement. A linear actuator with lesser elements gives the manipulator a robust structure, lowering the cost for building it [18–20].


E.-C. Lovasz et al. (Eds.): MTM&Robotics 2020, MMS 88, pp. 249–261, 2021.
https://doi.org/10.1007/978-3-030-60076-1_22

The present paper focuses on direct and invers kinematics for a class of planar parallel manipulator with a new base structure. The positional kinematic approach for the planar parallel manipulator using a slider crank as base structure is based on similar approach used for geared linkages with linear actuation with inverted slider-crank as base structure [21].

## 2 Design of the Novel Planar Parallel Manipulator

Studies illustrate that planar parallel manipulators having linear actuation with geared linkages give the mechanism a large rotation angle, an approximately constant transmission function in a large range and a proper transmission angle. Due to these characteristics, the manipulator structure allows a wider range for which the velocity vector is non-zero, therefore avoiding the singularity.

In Fig. 1 is presented the planar parallel manipulator having geared linkages with linear actuation as chain connections. The kinematic chain abridgement for the planar parallel manipulator is 3-R($\underline{P}$RRGR)RR, where R stands for revolute, P stands for prismatic and G stands for gear kinematic pairs. Inside the brackets displays the order of the joints of actuating chain and underlined is the active joint.

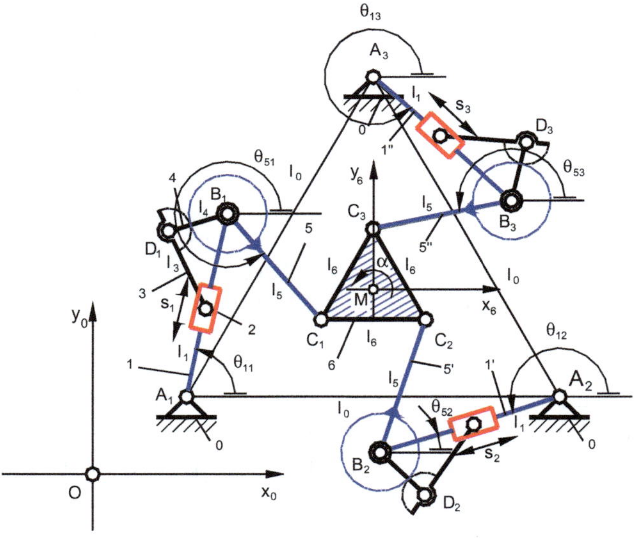


**Fig. 1.** Kinematic model of the novel planar parallel manipulator

Due to the linear actuator, the self-locking condition is satisfied and great transmission ratio is obtained through converting the movement. The geared linkages allow wide range for the rotation angle of the driven element (5) with almost constant transmission ratio. Many other references are available in [22, 23] and [24].

## 3 Kinematics Analysis of 3-R(PRRGR)RR Chain

The mobile platform of the planar parallel manipulator 3-R(PRRGR)RR allows 3 degrees of freedom: rotation around z-axis, translation along the x-axis and translation along the y-axis. The simplified kinematic schema, shown in Fig. 2 is used for the forward and invers positional kinematic of the manipulator.

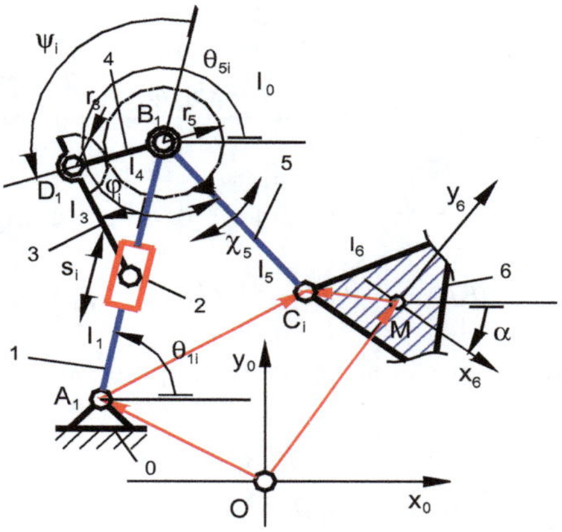


**Fig. 2.** The simplified scheme for a kinematic chain of the planar parallel manipulator

The matricial form of the Chasles vector equation used for the positional-kinematics is:

$$\begin{bmatrix} x_{A_i} \\ y_{A_i} \end{bmatrix} + \begin{bmatrix} l_1 \cos\theta_{1i} + l_5 \cos\theta_{5i}(s_i) \\ l_1 \sin\theta_{1i} + l_5 \sin\theta_{5i}(s_i) \end{bmatrix} = \begin{bmatrix} x_M \\ y_M \end{bmatrix} + \begin{bmatrix} \cos\alpha & -\sin\alpha \\ \sin\alpha & \cos\alpha \end{bmatrix} \cdot \begin{bmatrix} x_{C_i}^{(6)} \\ y_{C_i}^{(6)} \end{bmatrix}, i = \overline{1,3} \tag{1}$$

### 3.1 Forward Positional Kinematic

In the forward positional kinematic analysis for the planar parallel manipulator 3-R (PRRGR)RR the centroid coordinates of the mobile platform ($x_M$, $y_M$, $\alpha$) are computed. This is achieved by knowing the length side of the mobile platform and the frame, $l_0$ and $l_6$, the sliding parameters of the linear actuators $s_i$, coordinates corresponding to the 3 sets of joints present in the fixed platform ($x_{Ai}$, $y_{Ai}$), gear ratio $\rho$ and the lengths of the chains $l_1$, $l_5$. The geometry of the mobile platform is an equilateral triangle, all side are equal to $l_6$.

The rotation angle of the mechanism $\chi_i(s_i)$ can be obtained by:

$$\chi_i(s_i) = (1-\rho) \cdot \psi_i(s_i) + \rho \cdot \varphi_i(s_i) \tag{2}$$

where:

$$\rho = \pm r_4/r_6 \tag{3}$$

$$\varphi_i(s_i) = 2 \cdot arctan\frac{\sqrt{[2 \cdot l_6 \cdot (s_0+s)]^2 - [(s_0+s)^2 + l_6^2 - l_4^2]^2}}{2 \cdot l_6 \cdot (s_0+s) - [(s_0+s)^2 + l_6^2 - l_4^2]} \tag{4}$$

and

$$\psi_i(s_i) = arcsin\frac{l_6 \cdot sin(\phi_i(s_i))}{l_4} \tag{5}$$

If we consider the variation $\Delta\chi_i$ constant we can then compute:

$$\theta_{5i}(s_i) - \theta_{1i}(s_i) = \chi_i(s_i) + \varDelta\chi_i \tag{6}$$

Using Eq. (1) where $C_i$ is referenced to the $M$ point, we obtain the matrix:

$$\begin{bmatrix} x_M \\ y_M \end{bmatrix} = \begin{bmatrix} x_{A_i} \\ y_{A_i} \end{bmatrix} + \begin{bmatrix} l_1 \cos\theta_{1i} + l_5 \cos\theta_{5i}(s_i) \\ l_1 \sin\theta_{1i} + l_5 \sin\theta_{5i}(s_i) \end{bmatrix} - \begin{bmatrix} \cos\alpha & -\sin\alpha \\ \sin\alpha & \cos\alpha \end{bmatrix} \cdot \begin{bmatrix} x_{C_i}^{(6)} \\ y_{C_i}^{(6)} \end{bmatrix} \tag{7}$$

with $i = \overline{1,3}$, where:

$$x_{C_i}^{(6)} = l_6 \cos(\frac{\pi}{2} + i\frac{\pi}{3}), \quad y_{C_i}^{(6)} = l_6 \sin(\frac{\pi}{2} + i\frac{\pi}{3}). \tag{8}$$

The coordinates of the characteristic point M are unique therefor we can write:

$$\begin{cases} x_M = x_{A_i} + l_1 \cos\theta_{1i} + l_5 \cos\theta_{5i}(s_i) - x_{C_i}^{(6)} \cos\alpha + y_{C_i}^{(6)} \sin\alpha \\ y_M = y_{A_i} + l_1 \sin\theta_{1i} + l_5 \sin\theta_{5i}(s_i) - y_{C_i}^{(6)} \cos\alpha - y_{C_i}^{(6)} \sin\alpha \end{cases} \tag{9}$$

The results obtained are the 3 coordinates of the mobile platform ( $x_M$, $y_M$,$\alpha$) and the 3 angle coordinates $\theta_{1i}$, $i = \overline{1,3}$.

### 3.2 Inverse Positional Kinematic

The task of the inverse positional kinematic analysis is to compute the sliding parameters of the linear actuators $s_i$ by knowing the characteristic point and orientation angle of the mobile platform ($x_M(t)$, $y_M(t)$, $\alpha(t)$) where $t \in [0, 1]$, the lengths of the mobile and fixed platform $l_0$ and $l_6$, the frame joints coordinates ($x_{Ai}$, $y_{Ai}$), the gear ratio $\rho$ and the link lengths of the chains $l_1$, $l_5$.

Rewriting Eq. (1) two solutions are obtained for the generalized coordinates $\theta_{1i}$ and $\theta_{5i}$, $i = \overline{1,3}$:

$$\theta_{1i}(t) = 2 \cdot arctan \frac{B_{1i}(t) \mp \sqrt{A_{1i}^2(t) + B_{1i}^2(t) - C_{1i}^2(t)}}{A_{1i}(t) - C_{1i}(t)}, \tag{10}$$

where:

$$A_{1i}(t) = 2 \cdot l_1 \cdot \left(x_M(t) + x_{C_i}^{(6)} \cdot cos\,\alpha(t) - y_{C_i}^{(6)} \cdot sin\,\alpha(t) - x_{A_i}\right),$$

$$B_{1i}(t) = 2 \cdot l_1 \cdot \left(y_M(t) + x_{C_i}^{(6)} \cdot sin\,\alpha(t) + y_{C_i}^{(6)} \cdot cos\,\alpha(t) - y_{A_i}\right), \tag{11}$$

$$C_{1i}(t) = l_5^2 - l_1^2 - (x_M(t) + x_{C_i}^{(6)} \cdot cos\,\alpha(t) - y_{C_i}^{(6)} \cdot sin\,\alpha(t) - x_{A_i})^2$$
$$- (y_M(t) + x_{C_i}^{(6)} \cdot \sin\alpha(t) + y_{C_i}^{(6)} \cdot \cos\alpha(t) - y_{A_i})^2 .$$

and

$$\theta_{5i}(t) = 2\, arctan \frac{B_{2i}(t) \mp \sqrt{A_{2i}^2(t) + B_{2i}^2(t) - C_{2i}^2(t)}}{A_{2i}(t) - C_{2i}(t)}, \tag{12}$$

where:

$$A_{2i}(t) = 2 \cdot l_5 \cdot \left(x_M(t) + x_{C_i}^{(6)} \cdot cos\,\alpha(t) - y_{C_i}^{(6)} \cdot sin\,\alpha(t) - x_{A_i}\right),$$

$$B_{2i}(t) = 2 \cdot l_5 \cdot \left(y_M(t) + x_{C_i}^{(6)} \cdot sin\,\alpha(t) + y_{C_i}^{(6)} \cdot cos\,\alpha(t) - y_{A_i}\right), \tag{13}$$

$$C_{2i}(t) = l_1^2 - l_5^2 - (x_M(t + x_{C_i}^{(6)} \cdot cos\,\alpha(t) - y_{C_i}^{(6)} \cdot sin\,\alpha(t) - x_{A_i})^2$$
$$- (y_M(t) + x_{C_i}^{(6)} \cdot sin\,\alpha(t) + y_{C_i}^{(6)} \cdot cos\,\alpha(t) - y_{A_i})^2.$$

The invers transmission function is needed for obtaining the actuation parameters $s_i$. Due to the mechanism construct of the 3-R(PRRGR)RR manipulator the transmission function can be obtained from Eq. (6)

$$\chi_i(t) = \theta_{5i}(t) - \theta_{1i}(t) + \Delta\chi_i \cdot i = \overline{1,3} \tag{14}$$

The variable $\Delta\chi_i$ has the role of a constructive angle. The value of $\Delta\chi_i$ was chosen empirical to compensate the mechanism construct parameters. Based on Eq. (3) the inverse transmission function can be obtained:

$$\psi_i(t) + \rho \cdot (\varphi_i(t) - \psi_i(t)) - \chi_i(t) = 0 \tag{15}$$

Knowing that the geared linkages with linear actuation have the transmission function approximately linear it is possible to obtain the inverse transmission function with the Lagrange function:

$$s_i(\chi_i) = \sum_{k=1}^{4} \left( (s_i)_k \cdot \prod_{\substack{j=1 \\ j \neq k}}^{4} \frac{(\chi_i) - (\chi_i)_j}{(\chi_i)_k - (\chi_i)_j} \right) \tag{16}$$

The inverse transmission function is interpolated as a polynom of 3rd order.

## 4 Measurements and Simulations

Simulations are made on the planar parallel manipulator based on the following geometric parameters show in Table 1. The geometric parameters reduce the singularities that may appear in the planar parallel manipulator.

**Table 1.** The geometric parameters used for computation

| Geometric parameters | Notation and value |
|---|---|
| Frame platform length (0) | $l_0$ = 240 mm |
| Mobile platform length (6) | $l_6$ = 40 mm |
| Chain link length (1, 1′, 1″) | $l_1$ = 120 mm |
| Chain link length (5, 5′, 5″) | $l_5$ = 100 mm |
| Carrier length (4) | $l_4$ = 120 mm |
| Gear ratio | $\rho$ = 2 |

Using Eqs. (3)–(6) presented in Sect. 3.1 the transmission function of the geared linkage was computed and plotted:

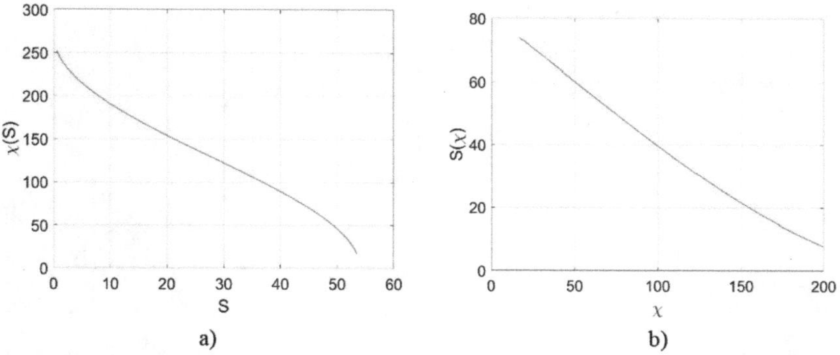


**Fig. 3.** The transmission function $\chi(s)$ (a) and the interpolated inverse transmission function $s(\chi)$ (b) of the geared linkage with linear actuation

In hindsight, the graph of the transmission function of the geared linkage with linear actuation in Fig. 3a is similar to a first-degree function for the stroke varying across the range [10, 40].

The inverse transmission function of the geared linkage with linear actuation was computed in Fig. 3b using the Lagrange function (16), detailed reference [21]. For creating the 3nd order polynom of the Lagrange function, 4 point from the direct function were carefully picked. The considered values are given in Table 2:

**Table 2.** Values of the points used for the Lagrange interpolation

| χ(°) | s(mm) |
|---|---|
| 238.4590 | 1.6364 |
| 221.6003 | 3.8182 |
| 206.2401 | 6.5455 |
| 193.6046 | 9.2727 |

## 4.1 Translation Along the X-Axis

To produce a nonuniform translation along the x-axis, the initial and final x coordinate of the mobile platform is linear interpolation in computing the $x_M(t)$:

$$x_M(t) = x_i + (x_f - x_i) \cdot t, \quad y_M(t) = k, \quad \alpha(t) = 0 \tag{17}$$

The $y_M(t)$ coordinate and the rotation $\alpha$ remain constant during the displacement. The motion is aptly illustrated in Fig. 4. The parameters of the first chain are shown by continuous red line, the dashed blue line for the second chain and the third chain is illustrated by the color green. The values considered for the coordinates are show in Table 3.

**Table 3.** Coordinates for motion along x-axis

| Initial coordinates | Final coordinates |
|---|---|
| $x_i = 120.433$ mm | $x_f = 58.290$ mm |
| $y_i = 69.503$ mm | $y_f = 69.503$ mm |
| $\alpha_i = 0.000°$ | $\alpha_f = 0.000°$ |

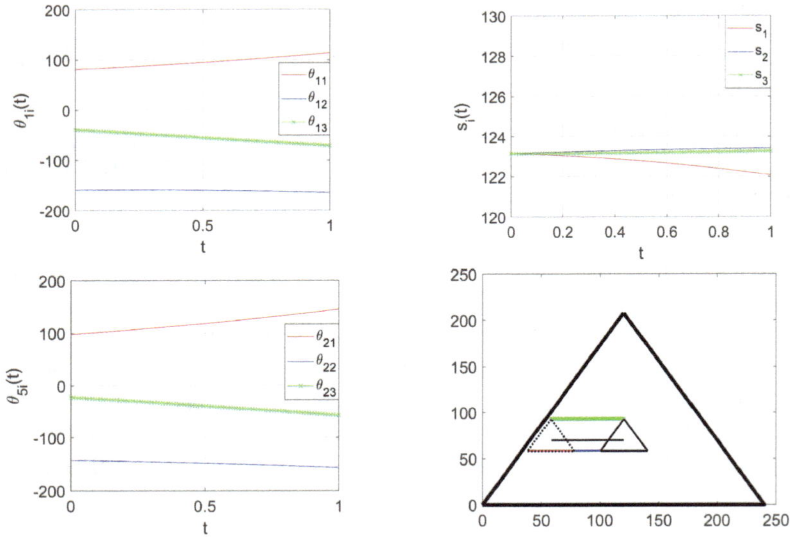


**Fig. 4.** The corresponding graphs for angles in chain links, stroke of the geared linkage with linear actuation and the trajectory for motion along the x-axis

## 4.2 Translation Along the Y-Axis

The equation for translation along the y-axis from the initial position ($x_i$, $y_i$, $\alpha_i$) to the final position ($x_f$, $y_f$, $\alpha_f$) is:

$$x_M(t) = k, \quad y_M(t) = y_i + (y_f - y_i) \cdot t, \quad \alpha(t) = 0 \tag{18}$$

The variation over time of the gear linkage angles $\theta_{1i}(t)$ and $\theta_{5i}(t)$ and the strokes $s_i$ are shown in Fig. 5 with the corresponding motion parameters in Table 4.

**Table 4.** Coordinates for motion along y-axis

| Initial coordinates | Final coordinates |
|---|---|
| $x_i = 120.000$ mm | $x_f = 120.000$ mm |
| $y_i = 80.000$ mm | $y_f = 100.000$ mm |
| $\alpha_i = 0.000°$ | $\alpha_f = 0.000°$ |

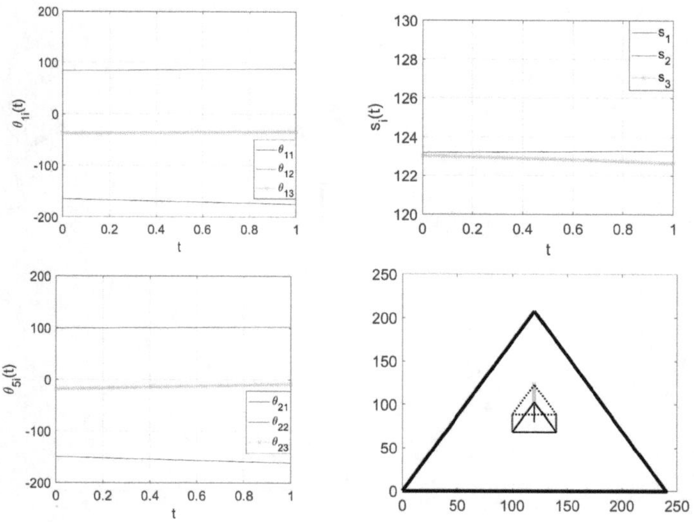


**Fig. 5.** The corresponding graphs for angles in chain links, stroke of the geared linkage with linear actuation and the trajectory for motion along the y-axis

### 4.3 Rotation Around Z-Axis

Considering the rotation around the z-axis between the initial position ($x_i$, $y_i$, $\alpha_i$) and the final position ($x_f$, $y_f$, $\alpha_f$) the equations are as following:

$$x_M(t) = p,\ \ y_M(t) = q,\ \ \alpha(t) = \alpha_i + (\alpha_f - \alpha_i) \cdot t \tag{19}$$

The resulting graph for the mobile platform is displayed in Fig. 6 with the considered values shown in Table 5.

**Table 5.** Coordinates for rotation around the z-axis

| Initial coordinates | Final coordinates |
|---|---|
| $x_i$ = 120.000 mm | $x_f$ = 120.000 mm |
| $y_i$ = 69.503 mm | $y_f$ = 69.503 mm |
| $\alpha_i$ = 0.000° | $\alpha_f$ = 50.000° |

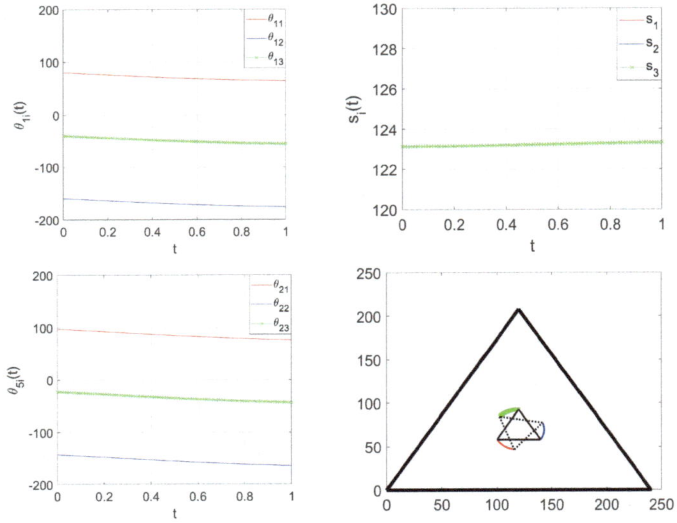


**Fig. 6.** The corresponding graphs for angles in chain links, stroke of the geared linkage with linear actuation and the trajectory for rotation around z-axis

## 4.4 Combined Motion

For motion that translates along the x-axis and y-axis with rotation along the z-axis all 3 coordinates are linear interpolated in obtaining the following equations:

$$x_M(t) = x_i + (x_f - x_i) \cdot t, \; y_M(t) = x_i + (x_f - x_i) \cdot t, \;\; \alpha(t) = \alpha_i + (\alpha_f - \alpha_i) \cdot t \quad (20)$$

with linear actuation and the trajectory for the combined motion

In Fig. 7 are represented the graphs for the generalized coordinates $\theta_{1i}(t)$ and $\theta_{5i}(t)$ of the gear linkage and the strokes $s_i$. Table 6 contains the considered values for this example.

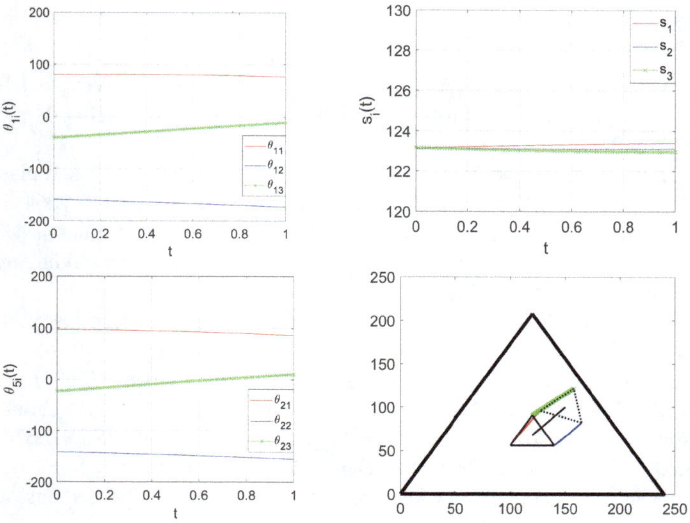


**Fig. 7.** The corresponding graphs for angles in chain links, stroke of the geared linkage

**Table 6.** Coordinates for the combined motion

| Initial coordinates | Final coordinates |
|---|---|
| $x_i$ = 120.000 mm | $x_f$ = 150.000 mm |
| $y_i$ = 67.6948 mm | $y_f$ = 100.000 mm |
| $\alpha_i$ = 0.000° | $\alpha_f$ = −20.000° |

## 5 Conclusions

The resulting data from the kinematic analysis program were compared with values calculated in AutoCAD.

The main focus was on the singularity linkages position. In our study we found two possible solution for positioning each of the three chains. These solutions were taken in account when computing the generalized coordinates $\theta_{1i}(t)$ and $\theta_{5i}(t)$ and the stroke $s_i$ for different positionings of the mobile platform.

When comparing the output data of the kinematic analysis with the values from AutoCAD remarks were made about transmission functions and the angles of the gear linkages with linear actuation. The values obtained in our paper and the values from AutoCAD were the same. Motion parameters were also compute using Mathcad for different translations and rotations of the mobile platform, the data was the same.

## References


1. Pott, A., Mutherich, H., Kraus, W., Schmidt, V., Miermeister, P., Verl, A.: IPAnema – a family of cable-driven parallel robots for industrial applications (2013). https://doi.org/10.1007/978-3-642-31988-4_8
2. Firthouse Begum, N., Vignesh, P.: Design and implementation of pick and place robot with wireless charging application. Int. J. Sci. Res. **4**(2) (2015). ISSN 2319-7064
3. Chalongwongse, S., Chumnanvej, S., Suthakorn, J.: Analysis of Endonasal Endoscopic Transsphenoidal (EET) surgery pathway and workspace for path guiding robot design. Asian J. Surg. **42**, 814–822 (2019)
4. Zhu, C., Zhang, H.: Parallel robotics control strategy study based on fuzzy-PID, pp. 664–669. Springer (2011)
5. Merlet, J.-P.: Parallel Robots. Kluwer Academic Publishers, Boston (2006)
6. Pandilov, Z., Dukovski, V.: Comparison of the characteristics between serial and parallel robots. Acta Tehnica Corviniensis – Bull. Eng. Tome **VII** (2014). ISSN2067-3809
7. Ibarreche, J.I., Altuzarra, O., Petuya, V., Hernandez, A., Pinto, C.: Structural synthesis of the families of parallel manipulators with 3 degrees of freedom. In: Proceedings of 19th CSIM IFToMM Symposium, pp. 35–42. Springer (2013)
8. Hernandez, A., Ignacio Ibarreche, J.I., Petuya, V., Altuzarra, O.: Structural synthesis of 3-DoF spatial fully parallel manipulators. Int. J. Adv. Robot. Syst. **11**, 101, 1–8 (2014)
9. Pisla, D., Vidrean, A., Prodan, B., Gherman, B., Lese, D.: Kinematics and design of two variants of a reconfigurable parallel robot. In: Proceedings of ReMAR, London, pp. 624–631 (2009)
10. Plitea, N., Lese, D., Pisla, D., Vaida, C.: Structural design and kinematics of a new parallel reconfigurable robot. Robot. Comput. Integr. Manuf. **29**(1), 219–235 (2013)
11. Huang, M.Z.: Design of a planar parallel robot for optimal workspace and dexterity. Int. J. Adv. Robot. Syst. **8**(4), 176–183 (2011)
12. Duca, C., Buium, F.: Singularities classification for structural groups of dyad type. Appl. Mech. Mater. **658**, 47–54 (2014)
13. Kotlarski, J., Heimann, B., Ortmaier, T.: Influence of kinematic redundancy on the singularity-free workspace of parallel kinematic machines. Front. Mech. Eng. **7**(2), 120–134 (2012)
14. Kotlarski, J., de Nijs, R., Abdellatif, H., Heimann, B.: New interval-based approach to determine the guaranteed singularity-free workspace of parallel robots. In: ICRA 2009, Proceedings of the IEEE International Conference on Robotics and Automation, Kobe, Japan, pp. 1256–1261 (2009)
15. Buium, F., Leonchi, D., Doroftei, I.: A workspace characterization of the 0/3/3R parallel mechanism. Appl. Mech. Mater. **658**, 563–568 (2014)
16. Huang, M.Z., Thebert, L.L.: A study of workspace and singularity characteristics for design of 3-DOF planar parallel robots. Int. J. Adv. Manuf. Technol. **51**(5–8), 789–797 (2010)
17. Kotlarski, J. Abdellatif, H., Ortmaier, T., Heimann, B.: Enlarging the useable workspace of planar parallel robots using mechanisms of variable geometry. In: Proceedings of ReMAR 2009, London, pp. 63–72 (2009)
18. Briot, S., Bonev, I., Chablat, D., Wenger, P., Vigen, A.: Self-motions of general 3-RPR planar parallel robots. Int. J. Robot. Res. **27**(7), 855–866 (2008)
19. Chablat, D., Wenger, P.: Self motions of a special 3-RPR planar parallel robot. In: Advances in Robot Kinematics, pp. 221–228. Springer (2006)
20. Duca, C., Buium, F.: Transmission indices adoption at 0/3/0 structural group. Appl. Mech. Mater. **658**, 55–58 (2014)

21. Lovasz, E.-C., Pop, C., Grigorescu, S.M., Gruescu, C.M., Mărgineanu, D.T., Maniu, I.: Kinematics of the planar parallel manipulator using geared linkages with linear actuation as kinematic chains 3-R(RPRGR)RR. In: 14th IFToMM World Congress Proceedings, pp. 493–498 (2015)
22. Lovasz, E.-C., Grigorescu, S.M., Mărgineanu, D.T, Gruescu, C.M., Pop, C., CiupeV., Maniu, I.: Geared linkages with linear actuation used as kinematic chains of a planar parallel manipulator. In: Proceedings of the Third MeTrApp Conference 2015. Series: Mechanism and Machine Science, vol. 31, pp. 21–31. Springer (2015)
23. Modler, K.-H., Hollmann, C., Lovasz, E.-C., Perju, D.: Geared linkages with linear displacement actuator used as function generating mechanisms. In: Proceedings of the 11-th World Congress on TMM, Tian Jin, 01 April–05 April 2004, (2005), pp. 1254–1259
24. Lovasz, E.-C., Ciupe, V, Modler, K-H, Gruescu, C.M., Hanke, U, Maniu, I, Mărgineanu, D.: Experimental design and control approach of an active knee prosthesis with geared linkage. In: New Advances in Mechanisms, Transmissions and Applications. Mechanisms and Machine Science, vol. 17, pp. 149–156. Springer (2013)

# Model-Free Continuous to Discrete Workspace Transformation and Path Planning of a 2DOF Serial Arm for Visual Obstacle Avoidance

Cristian Moldovan(✉), Valentin Ciupe, Hannelore Filipescu, Robert Kristof, and Valer Dolga

Universitatea Politehnica Timisoara, 300006 Timisoara, Romania
cristian.moldovan@upt.ro

**Abstract.** In this paper we present our approach on path planning for a 2DOF serial robotic arm without constructing a kinematic model of it. Our approach is different from the standard approach, where you have to know the robot's kinematic parameters such as length of elements, through the fact that we do not use a kinematic model to position or execute path planning for the arm, but rather rely on visual information from a camera to complete the task. We consider only the positions of the end-effector when planning the path, velocities are ignored. The problem to be solved in this paper is formulated as find a path that positions the robot's end-effector at the target position and avoids any obstacles. Our approach on solving this problem is to sample the workspace (randomly or by sweeping it incrementally) in order to get a correspondence between the end-effector position to the corresponding joint angle values (equivalent to deriving the kinematic model of the robot, but not through equations) and then use the subset of points as a look-up table to compute a path between the points that take us closest to the target.

To prove our concept we designed a system that consists of a 3D printed 2 DOF arm, driven by Dynamixel servos, Videocamera, OpenCV and Python. The precise positioning of the end-effector at the target position is achieved using algorithms presented in the literature and in our previous work.

**Keywords:** Serial robot · Path planning · Workspace sampling · Machine vision

## 1 Introduction

In the period 2018–2020 our group has been focusing on alternative methods to solve various problems in the field of robotics manipulation. The starting idea is that in the field of Artificial Intelligence researchers try to model the way humans perceive, think and act on their environment with the aim to implement these models in machines. Upon closer analysis on robotic compared with human manipulation, it seems that robotic manipulators rely heavily on very precise kinematic models in order to achieve the imposed task. On the other side, humans seem not to have a very precise model or their arm, yet achieve spectacular results in manipulation, hence our investigations on alternative methods to be applied in robotics. An example kinematic parameter is the


E.-C. Lovasz et al. (Eds.): MTM&Robotics 2020, MMS 88, pp. 262–271, 2021.
https://doi.org/10.1007/978-3-030-60076-1_23

length of elements comprised in the arm. Standard robotics uses the length of elements along with angle values to determine the position of the end-effector. Humans seem not to have a crisp value of their arm length and don't use complex mathematic operations (sinus, cosines, derivatives), yet they can reach any position in the arm's workspace effortlessly.

When thinking of classic applications such as cutting, precise assembly, welding etc. the classic model-based approach on robotics will remain standing, but the environments for this classic robots are very clearly defined, with strict tolerances and do not change uncontrolled over time. The performance of such classic robots greatly surpasses that of a human. For example, a classic industrial robot is capable to draw an extremely precise line or circle, while a human, even with great effort (without tools) fails at this task. On the other hand, humans have no problem picking up an object from a table thrown randomly, plan a hand motion to avoid a collision and still grab an object, grab an object through a hole etc.

Taking this into consideration, our conclusion regarding robot manipulation is that classic approaches are one possibility and work well especially for well-structured environments [1, 2], but humans have a different approach suited especially for unstructured environments [3–5].

When studying robot motion, we can define level of tasks for the robot such as

- Path planning problem: figure out the points in space through which the end-effector will pass in order to reach the target. There are two ways in which a path can be defined, first one, through a sequence of points which takes the end-effector to the target position, and second one as an equation which relates the position of the end-effector with the joint angles.
- Trajectory generation problem: figure out the velocity components of the end-effector motion along the path. Trajectory includes both speed and orientation of the end-effector.

In this paper of concern to us is the path planning problem for a serial robot in order to avoid collisions. The experiments were conducted on a platform consisting of a 2DOF arm and overlooking video camera, details of the system design being presented in Sect. 4 of the current paper.

Our previous research on the subject was concerned on solving the precise positioning task of the end-effector without knowing the kinematic model. The problem was that our method is based on sampling of the workspace, obtaining only a subset of points, which do not cover the entire workspace. This problem is broadly discussed in Sect. 2. What we did is develop algorithms that allow us to precisely position the end-effector at any point in the workspace tough only some are sampled [3, 4].

Similar research was conducted by [6] for MATLAB Research, where an Adaptive Neuro-Fuzzy Inference Systems (ANFIS) was tuned to infer arm axis angles to solve the direct and inverse kinematics problem as an example application of the ANFIS Toolbox. The main reference used to develop this system is [7].

Workspace sampling techniques are also implemented in [5] when planning the path of a robot. The resulting algorithms are optimized and implemented in an FPGA chip to drastically reduce planning time, resulting in a path reconfiguration in under 1 ms. Their work is broadly based on the work of Leven and Hutchinson [8].

Alternative control methods for the control of a robotic arm are studied in [9]. Here they propose that the robot learns the effects of various actions of actuators. They coin this concept as motion primitives and use it to reach the target position of the end-effector without knowing the kinematics of the robot, a very interesting approach. Classic approaches on path planning are presented in [10, 11] using visual servoing techniques; these rely on a mathematical model of the robot that maps control inputs to outputs.

Our paper is structured as following, Sect. 2 presents our method for sampling the workspace and obtaining a Discrete subspace, in Sect. 3 we present our path planning algorithm for two situations without and with obstacle in the path, Sect. 4 presents the System Design that allows the robotic arm to move, Sect. 5 Results and Conclusion.

## 2 Method for Obtaining a Discrete Subspace from the Continuous Workspace

Since the links length of the robotic arm is not available to the controller, for our 2DOF arm we propose a sampling method to determine the correlation between

$$\text{End–Effector Position}\,(X, Y) \longleftrightarrow \text{Axis Angles}(\theta_1, \theta_2)$$

that we can use interchangeably having in mind that our objective for this paper is to control the arm without a kinematic model. One also has to note that the system is comprised of a 2DOF arm and an overlooking video camera. Details on the system design are presented in Sect. 4. A schematic view of the system is presented in Fig. 1, where we denote with g – the direction of the gravitational acceleration.

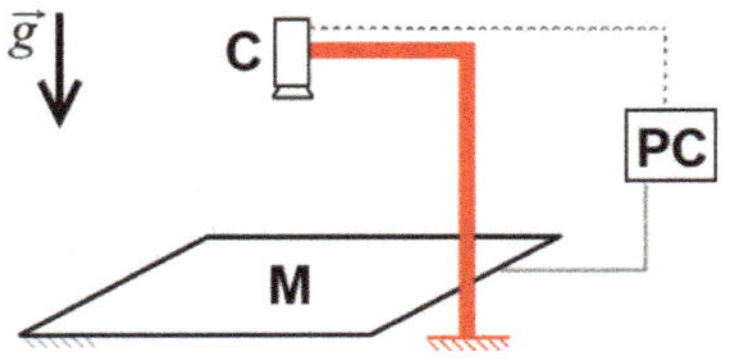


**Fig. 1.** Schematic system diagram (M – 2DOF arm, PC – Personal Computer, C – Camera)

The sampling phase can be executed in two ways, both with similar results:

**Method 1:** The arm is programmed to move randomly for a period of time during which the end-effector's (X, Y) coordinates are "photographed" by the camera. Simultaneously, for every "photograph", the joint coordinates are recorded. One has to note that the recorded joint values take the end-effector to the recorded (X, Y) coordinates, and that the (X, Y) coordinates are with respect to the camera's coordinate system and represent pixel values.

**Method 2:** The arm is programmed to sweep the workspace in specified steps for every joint. At each step, the end-effector's (X, Y) position is recorded in the image (from the camera), together with the joint angle values. Using this method, there can be obtained a sampled version of the workspace as shown in Fig. 2.

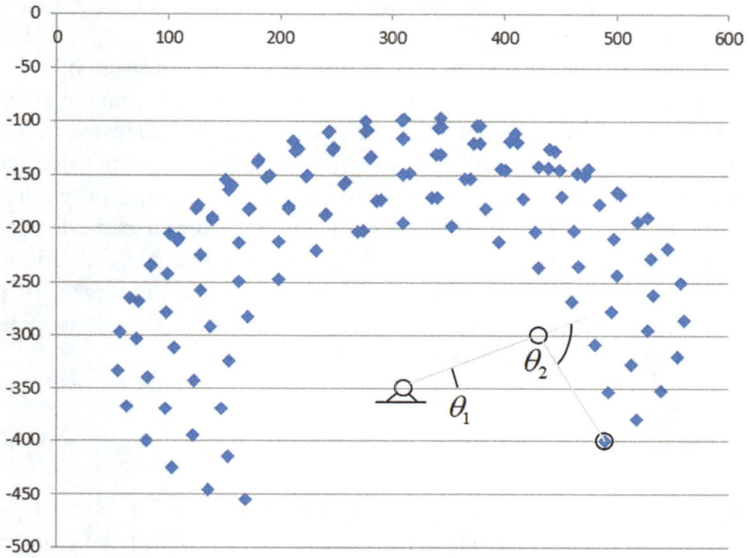


**Fig. 2.** Sampled workspace (with Method 2 – sweeping). 2DOF arm and joint angles

After applying any of these methods, results a table of positions together with joint angles that take the end-effector to that (X, Y) position. A sample recording from this table is presented in Table 1, (X, Y) represent the end-effector position in the image frame and ($\theta_1$, $\theta_2$) represent the joint angle values (in encoder impulses). What this means, is that if we want to position the end-effector at coordinates (144, 342) in the image frame, we need to command joint 1 to move at $\theta_1$ = 284 and $\theta_2$ = 782.

**Table 1.** Sample recording from the workspace

| X | Y | $\theta_1$ | $\theta_2$ |
|---|---|---|---|
| 144 | 342 | 284 | 782 |

At the end of the sampling step, the system has obtained a table of recordings which make the correspondence between a coordinate system (in the image) to the robot's joint space. This works similarly to the direct and inverse kinematics in the standard approach. This is similar to the situation where a toddler learns to walk, when he tries various configurations for his limbs and observes the effects of his actions. Later in his life he will use these motor skills to solve different tasks, such as path planning.

When analyzing the workspace of any robot, one might be tempted to think at it as a continuous manifold, in reality this is rather not true since we are limited by the encoders on every joint and their limitations with respect to positioning in space.

One example could be that for a joint, we cannot position the robot at ½ impulse, it is either 0 or 1. We consider this sampling method similar to the case of using a rather inaccurate equivalent robot, with very low resolution encoders. To obtain better positioning accuracy, that uses the full capacity of the encoders, we recommend the algorithms presented in [3, 4, 6, 7].

The way we are using this system is the following, we construct a HMI (Human Machine Interface) where we are able to click on the live image from the video camera designating the target position. Now the system determines the nearest sampled point to the current position of the end-effector and, if it is occupable commands the arm to move to that position, else finds the next nearest point and so on. In the next step, a search algorithm tries to find a route between the now current end-effector position toward a sampled point that is nearest to the target position. If the algorithm finds a solution, the end effector follows the route points until near the target. To precisely position the end-effector at the target position we use one of algorithms presented in [3, 4, 6, 7]. The search algorithm is presented in Sect. 3.

## 3 Path Planning and Object Detection in the Workspace

In the last section, through workspace sampling we obtained a table of values for position (X, Y) and the corresponding joint angles ($\theta_1$, $\theta_2$) of the end-effector. In this section we want to determine a path from the current position to the target position of the end-effector.

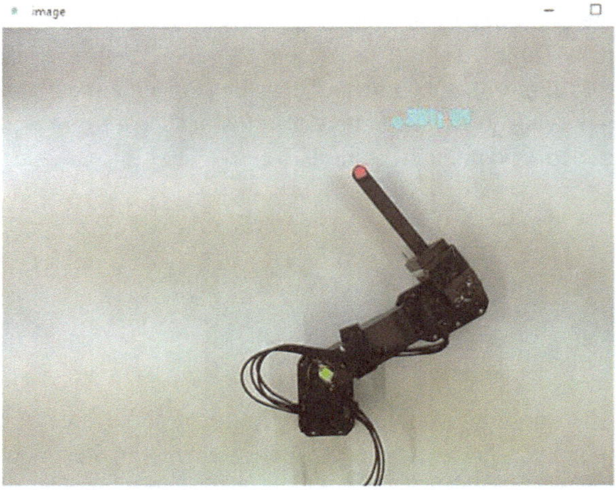


**Fig. 3.** HMI for commands input (one can see the target coordinates in the image frame)

To communicate with the system we use an HMI (Fig. 3), where by clicking onto a live image of the arm from the camera we are able to designate the target position of the end-effector. We call this position $P_{target}\left(X_{target}, Y_{target}\right)$. There is a great possibility that

$P_{target}$ does not overlap any point from the sampled workspace. In this case we need to determine the closest unoccupied sampled point from the workspace and move the end-effector there. We designate this point $P_{ts}$ (ts-target sampled). We proceed similarly for the starting position, we take the current position of the and-effector and call it $P_{start}$, find the nearest position that is sampled and call it $P_{ss}$ (ss-start sampled). We apply

### 3.1 Path Planning Algorithm

To find a path between the two points $P_{start}$ and $P_{target}$ we can follow the next algorithm noted with A1:

01: START
02: MOVE_FROM $P_{start}$ TO $P_{ss}$
03: FIND_PATH ($P_{ss}$, $P_{ts}$)
04: MOVE_PRECISE $P_{ts}$ TO $P_{target}$

It can be observed that in A1, the $P_{ss}$ and $P_{ts}$ points are sampled and we can connect them using a search technique. The proposed search algorithm is A* with a distance measure function d $(P_1, P_2)$ of Euclidian type – Eq. (1)

$$d(P_1, P_2) = \sqrt{(x_1 - x_2)^2 + (y_1 - y_2)^2} \quad (1)$$

In literature it is shown that A* is guaranteed to be complete and optimal. An example path is presented in Fig. 4, as an expanded and augmented region of Fig. 1

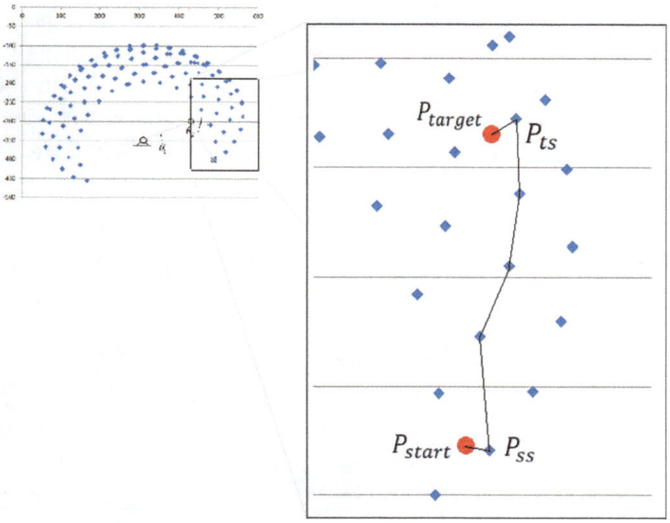


**Fig. 4.** Path obtained when applying A* between $P_{ss}$ and $P_{ts}$

If an obstacle is obstructing the path, A* is able to recalculate the route between $P_{start}$ and $P_{target}$, this case is presented in Fig. 5.

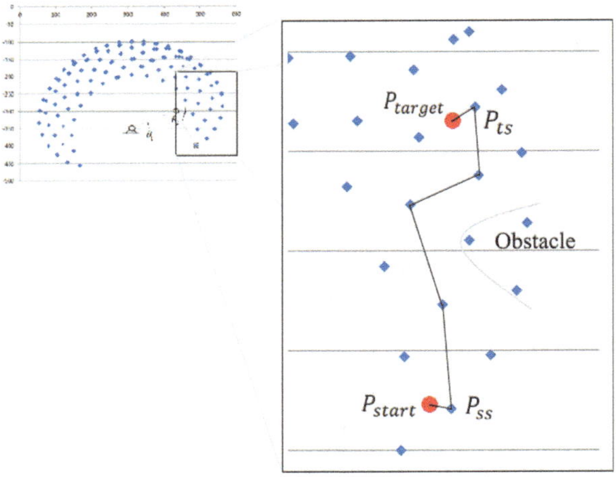


**Fig. 5.** Recalculated path when applying A* between $P_{ss}$ and $P_{ts}$ for obstacle avoidance

## 3.2 Object Detection in Workspace

For object detection in the workspace of the 2DOF arm we used the Background subtraction algorithms provided by OpenCV [12]. The OpenCV algorithm produces a mask, a binary image with pixel values 0 for background and 1 for detected object. We use this mask to filter out the occupied points from our learned workspace, obtaining a valid positions workspace. This process is presented in Fig. 6. To note is that Step 1 was presented in Sect. 2 of this paper. The process is functional since the sampled points are the coordinates of the end-effector in the image frame. The valid workspace is presented in Fig. 7 and is permanently updated repeating Steps 2 and 3.

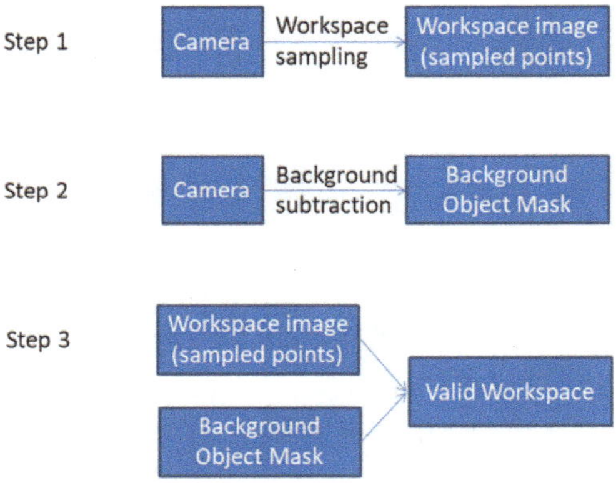


**Fig. 6.** Process flow for obtaining a viable workspace for path planning

For the internal representation of the valid workspace we have added an extra column to the sampled workspace table (Table 1) with 0 or 1 values, if the respective position is available or not. The search algorithm takes also in consideration the valid points as candidate positions

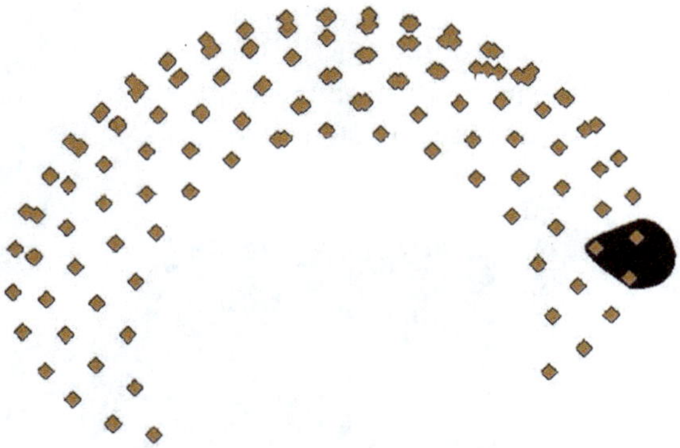

**Fig. 7.** Workspace overlapped with mask (mask obtained from object background subtraction process)

## 4 System Design

The general architecture diagram for the system designed as proof of concept demonstrator for human-like algorithms is presented in Fig. 1. The system is comprised of the following main components: 2 DOF arm, Video Camera, Controller PC

For the structure of the arm we have chosen a simple serial link that respects the kinematic diagram in Fig. 8a. Using 3D printed components we integrated two Dynamixel RX24F smart servos, the CAD assembled model is presented in Fig. 8b. The video-camera was mounted on top of the arm as shown in Fig. 8c

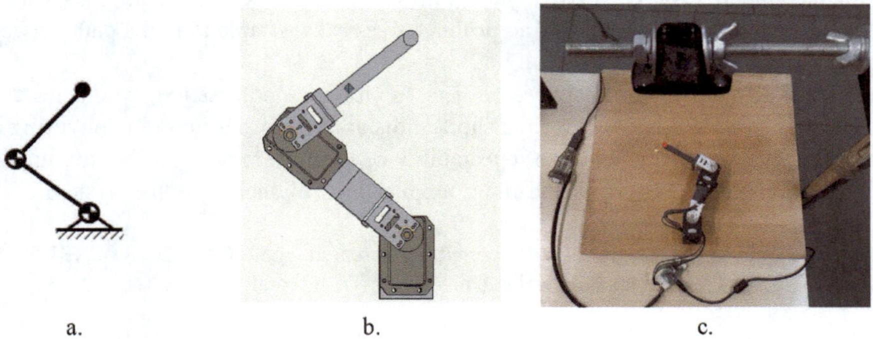

a. b. c.

**Fig. 8.** System design; a. – Kinematic diagram, b. – CAD model, c. – Demonstrator system general view

The data processing, image processing and system control is done using Python and OpenCV.

To determine the positions of important points on the arm, we are using markers of different colors. The workspace sampling was conducted in an automated process where it was imposed that the arm sweeps the entire workspace. A marker was attached to the end-effector and its center of mass coordinates were recorded in Table 1. The marker detection process is presented in Fig. 9, where Fig. 9a shows the camera image, Fig. 9b shows the isolated marker position as a mask image and Fig. 9c shows the Hue Saturation Value (HSV) settings to isolate the marker's color from the rest of the image. For the end-effector was used a purple color marker.

a. b. c.

**Fig. 9.** System design; a. – Kinematic diagram, b. – CAD model, c. – Demonstrator system general view

## 5 Conclusion

In this paper we presented our method for obtaining a path that allows the end-effector of a 2DOF arm to position itself from one location to other, avoiding obstacles. Our proposed method does not require any kinematic model of the arm and relies on a sampling step where it is made a correspondence between end-effector position and joint coordinates, to do this, the system uses an overlooking video-camera. We use a marker based tracking method to track the end-effector's position, in an attempt to mimic the human's ability to visually track his members. In Sect. 3 we have shown that by using search and path planning algorithms, the system is able to find a path between the start and target position.

The work done in this paper is subject for further optimization by the research group, mainly in deriving an absolute shortest distance from start to target, focusing on obtaining a smooth path. Also the repeatability of the system needs to be investigated in order to conclude on the usable future applications of such a control system.

**Acknowledgment.** This work was supported by research grant GNaC2018 - ARUT, no. 1364/01.02.2019, financed by Politehnica University of Timisoara.

## References


1. Mocan, B., Fulea, M., Brad, E., Brad, S.: State-of-the-art and proposals on reducing energy consumption in the case of industrial robotic systems. In: Proceedings of the 2014 International Conference on Production Research – Regional Conference Africa, Europe and the Middle East; 3rd International Conference on Quality and Innovation in Engineering and Management, Cluj-Napoca, Romania, 1–5 July, pp. 328-334 (2014). ISBN 978-973-662-978-5
2. Mocan, B., Fulea, M., Olaru, M., Buchmüller, M.: From intuitive programming of robotic systems to business sustainability of manufacturing SMEs. Amfiteatru Econ. **18**(41), 215–231 (2016)
3. Moldovan, C., Ciupe, V., Crastiu, I., Dolga, V.: Model free control of a 2DOF robotic arm using video feedback. In: 6-th International Symposium on Electric and Electronics Engineering, Galati (2019)
4. Moldovan, C., Ciupe, V., Crastiu, I., Dolga, V., Lovasz, E.-C., Maniu, I.: Design and novel control solution for a modular mechatronic demonstrator with video feedback used in research and education. In: ICMV 2019 International Conference on Machine Vision, Amsterdam (2019)
5. Murray, S., Floyd-Jones, W., Qi, Y., Sorin, D., Konidaris, G.: Robot Motion Planning on a Chip, Robotics: Science and Systems (2016)
6. Konsoulas, I.: Adaptive Neuro-Fuzzy Inference Systems (ANFIS) Library for Simulink (2020). https://www.mathworks.com/matlabcentral/fileexchange/36098-adaptive-neuro-fuzzy-inference-systems-anfis-library-for-simulink. MATLAB Central File Exchange. Accessed 21 Apr 2020
7. Jang, R., Sun, C.-T., Mizutani, E.: Neuro-Fuzzy and Soft Computing: A Computational Approach to Learning and Machine Intelligence. Prentice Hall, Upper Saddle River (1997)
8. Leven, P., Hutchinson, S.: A framework for real-time path planning in changing environments. Int. J. Robot. Res. **21**(12), 999–1030 (2002)
9. Kormushev, Y., Demiris, D., Caldwell, G.: Kinematic-free position control of a 2-DOF planar robot arm. In: 2015 IEEE/RSJ International Conference on Intelligent Robots and Systems (IROS), Hamburg (2015)
10. Chaumette, F., Hutchinson, S.: Visual servo control, Part I: Basic approaches. IEEE Robot. Autom. Mag. **13**(4), 82–90 (2006). [6]F
11. Chaumette, F., Hutchinson, S.: Visual servo control, Part II: advanced approaches. IEEE Robot. Autom. Mag. **14**(1), 109–118 (2007)
12. https://docs.opencv.org/master/d1/dc5/tutorial_background_subtraction.html. Accessed 21 Apr 2020

# Mechanical Transmissions and Machine Parts

# Design of an IR Objective (5 μm - 10 μm)

Corina M. Gruescu[1(✉)] and Renata D. Bodea[2]

[1] University Politehnica of Timisoara, Timişoara, Romania
corina.gruescu@upt.ro
[2] Universitatea din Oradea, Oradea, Romania

**Abstract.** Optics in IR imagining require special objectives, made of materials which are transparent within the IR range. The difficulty in design resides in the limited number of such materials and the mathematical compatibility of their characteristics in the development of design algorithms. The paper proposes the design of an objective as an air-spaced doublet, corrected for spherochromatism and coma. The mathematical algorithm was introduced into original software application and the results were analyzed with the program OSLO. The numerical application provided a solution for a diffraction-limited objective with f′ = 40 mm and f′/# = 5.

**Keywords:** IR imaging · IR objective · Optical algorithm and software · Optical analysis

## 1 Introduction

At present, there are a lot of applications using images based on sensing infrared radiation in all fields of activity. The design of the optics associated to such applications depends on the specific purpose, which may be thermal imaging of scenes of objects, infrared photography, detection of specific targets and many others. For each type of application is useful a sub-domain of the infrared range, which is very wide (700 nm–1 mm in wavelength).

The optical design of IR objectives is difficult because of the limited number of materials available with transparency within this spectral region. Furthermore, the broad band of working wavelengths is a problem because of the large variation in refractive and dispersive characteristics [1]. The temperature of the targets to be visualized is also a problem, because the heat is transferred to the optical and mechanical systems and produces variations in optical and mechanical properties, which lead to defocusing and image quality loss [2–4].

The design of IR objectives does not base on specific algorithms, created for the IR region. The design uses algorithms developed for the visible range (which is very narrow, e.g. 360 nm… 760 nm), which may result in mathematical incompatibility of materials. The objectives are commonly achromatic doublets [5], Petzval doublets [6] or complex systems [7] often with aspherized surfaces. Some applications, such as the astronomic ones, require special conditions regarding variable power and field of view [8, 9].


E.-C. Lovasz et al. (Eds.): MTM&Robotics 2020, MMS 88, pp. 275–283, 2021.
https://doi.org/10.1007/978-3-030-60076-1_24

The present paper proposes an IR objective for thermography cameras, computed for the range (5 μm…10 μm), as an air-spaced doublet, corrected to spherochromatism and coma. The choice went to an air-spaced doublet because it ensures four degrees of freedom and thus correction of three aberrations.

## 2 Algorithm and Software Application

The design of the air-spaced doublet (Fig. 1), made of a positive and a negative lens, based on thin-lens theory, develops with the following steps [10]:

- Input data: f′ – focal length, D – aperture, $2\sigma_p$ – field of view, material of the lenses ($n_{\lambda 1}$, $n_{\lambda 2}$, $n_{\lambda 3}$, ν), pixel size of the sensor
- Calculus of lenses 'curvatures' $c_a$ and $c_b$

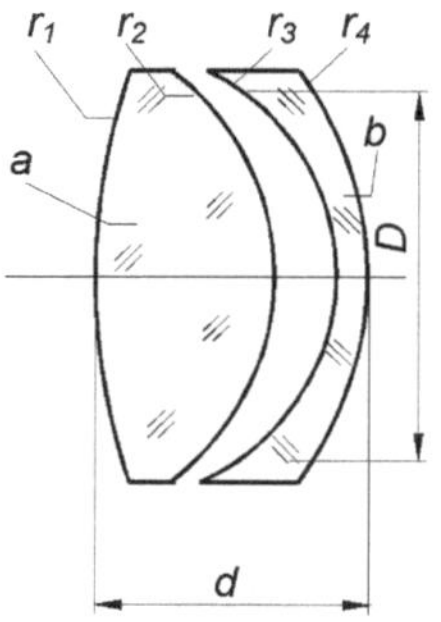


**Fig. 1.** The air-spaced doublet made of the positive lens a and negative lens b

Considering lenses a and b separated by a thin layer of air, the equation of optical power of the system and the achromatisation condition provide the focal lengths and curvatures of the lenses:

$$f'_a = \frac{f'(\nu_a - \nu_b)}{\nu_a}, \tag{1}$$

$$f'_b = \frac{f'(\nu_b - \nu_a)}{\nu_b} = -f'_a \frac{\nu_a}{\nu_b}, \tag{2}$$

$$c_a = \frac{1}{f'_a(n_a - 1)}, \tag{3}$$

$$c_b = \frac{1}{f'_b(n_b - 1)}. \tag{4}$$

- Paraxial tracing of the principal and marginal ray
- Calculus of G-sums [10]:

$$G_1 = \frac{1}{2} n^2 (n-1), G_2 = \frac{1}{2}(2n+1)(n-1), \tag{5}$$

$$G_3 = \frac{1}{2}((3n+1)(n-1), G_4 = \frac{1}{2n}(n+2)(n-1), \tag{6}$$

$$G_5 = 2\frac{(n^2-1)}{n}, G_6 = \frac{1}{2n}(3n+2)(n-1), \tag{7}$$

$$G_7 = \frac{G_2}{n}, \quad G_8 = \frac{G_1}{n}. \tag{8}$$

- Calculus of curvatures $c_1$ and $c_3$ from the conditions of correction on spherical aberration and coma:

$$k_3 c_1^2 + k_4 c_1 + k_5 = 0, \tag{9}$$

$$c_3 = k_1 c_1 + k_2 \tag{10}$$

where

$$k_1 = -\frac{G_{5a} c_a}{G_{5b} c_b}, \tag{11}$$

$$k_2 = \frac{1}{0.25 G_{5b} c_b}\left(G_{7a} c_a p_a + G_{8a} c_a^2 + G_{7b} c_b p_b + G_{8b} c_b^2\right). \tag{12}$$

$$k_3 = G_{4a} c_a + G_{4b} c_b k_1^2, \tag{13}$$

$$k_4 = -G_{5a} c_a p_a - G_{2b} c_b^2 k_1 + 2 G_{4b} c_b k_1 k_2 - G_{5b} c_b k_1 p_b - G_{2a} c_a^2, \tag{14}$$

$$\begin{aligned} k_5 = {} & G_{1a} c_a^3 + G_{3a} c_a^2 p_a + G_{6a} c_a p_a^2 + G_{3b} c_b^2 p_b + G_{6b} c_b p_b^2 \\ & - G_{2b} c_b^2 k_2 + G_{4b} c_b k_2^2 - G_{5b} c_b p_b k_2 + G_{1b} c_b^3. \end{aligned} \tag{15}$$

Equation (9) provides two, one or no real solutions, depending on the materials 'characteristics.

- Calculus of curvatures $c_2$ and $c_4$:

$$c_2 = c_1 - c_a, \tag{16}$$

$$c_4 = c_3 - c_b. \tag{17}$$

- Establishing of lenses 'geometry considering the center thickness
- Analysis of the resulted system.

The algorithm usually provides systems with a good correction of spherical aberration, very good correction of chromatism and coma.

The algorithm shortly presented above was used in developing a VB application, whose graphical interface is shown in Fig. 2.

The software application allows the user to choose materials available for IR systems in the catalogue II-VI [11], respectively ZnSe, ZnS, ZnSMs, Ge and GaAs. The program includes a database with refractive indexes for the wavelengths 5 μm, 7.5 μm and 10 μm. The reference wavelength is 7.5 μm and the dispersion number ν is computed accordingly.

The input data required by the program are the focal length, f', the aperture, D and the object abscissa (working distance), s.

The calculus provides the two solutions, which result starting from Eq. (9) or displays the warning "There are no real solutions for the chosen pair of materials".

Beside the full geometry of the system, the program performs a paraxial tracing and calculates the focal length for each solution.

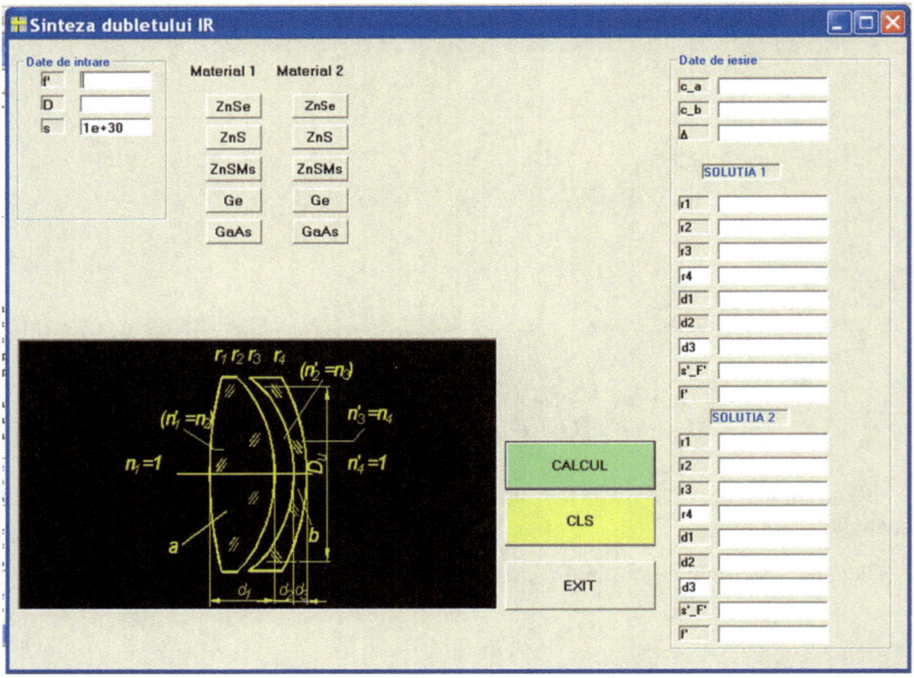


**Fig. 2.** Graphical interface of the VB application

The resulted systems must be analyzed in order to evaluate the image quality.

## 3 Numerical Application

The VB application was used to design an objective with f′ = 40 mm, FOV = 8° and aperture D = 8 mm.

The geometry and optical characteristics of the resulted system are presented in Table 1.

Data in the lower line of the table were obtained by means of analysis using the application OSLO LT [12]. Figure 3 shows the drawing of the doublet and a ray fan, in the reference wavelength (denoted W1).

**Table 1.** Geometry of the system (diffraction limited)

| | | | | | | | |
|---|---|---|---|---|---|---|---|
| Radius [mm] | 32.43 | | 47.06 | | 19.24 | | 17.62 |
| Thickness [mm] | | 2.25 | | 2.00 | | 2.25 | |
| Material | | GaAs | | Air | | ZnSe | |
| System characteristics | f'= 40.01 mm; $s'_{F'}$=32.63; D = 8 mm (f'/#=5)<br>P-V OPD = 0.14; RMS OPD = 0.04; Strehl ratio = 0.93<br>PSF = 0.96 | | | | | | |

**Fig. 3.** Drawing of the system and a tangential ray fan in W1

Image quality parameters given in Table 1 result from wavefront analysis (Fig. 4) and point spread function (Fig. 5).

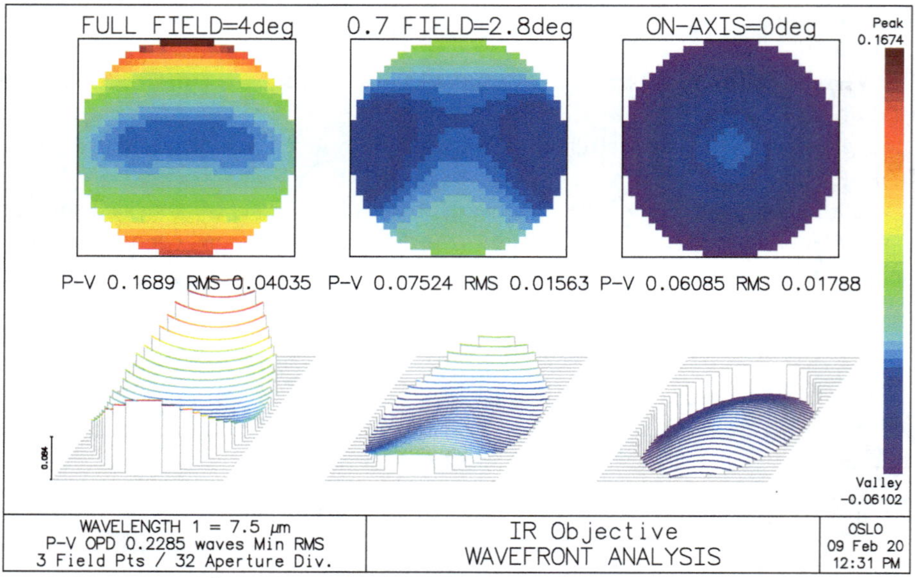


**Fig. 4.** Wavefront analysis (P-V OPD = 0.06λ and RMS OPD = 0.01λ) of the system

Beside the image quality, a very important parameter is the resolution, which was evaluated through MTF (Fig. 5) and spot diagram (Fig. 6). Considering the pixel size equal to 25 μm, the Nyquist frequency of the detector is 20 cycles/mm. The curves in Fig. 5 show that this criterion is accomplished.

The spot diagram roughly shows the shape and size of the spot onto the image plane on-axis, at 0.7 FOV and full FOV, in comparison to the Airy circle. Figure 6 indicates that the system is diffraction-limited over the full field.

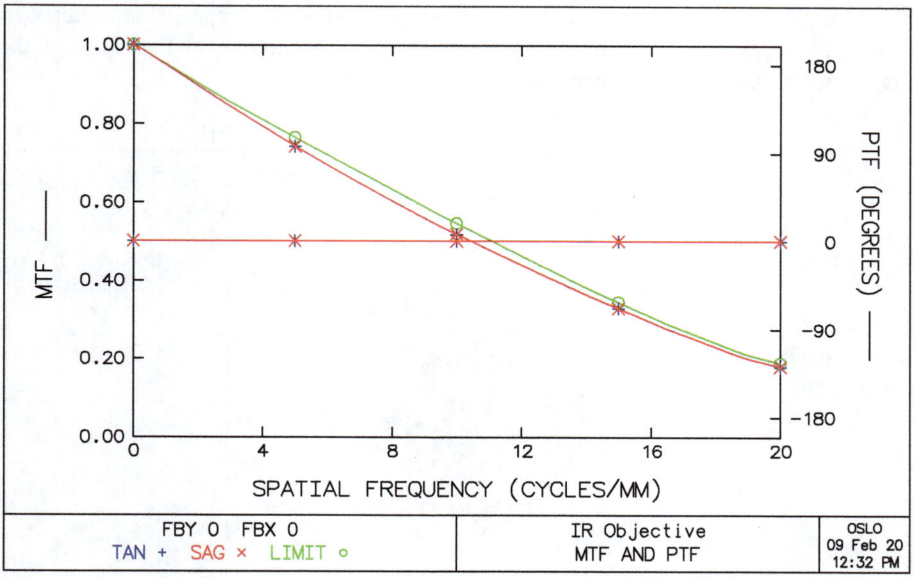


**Fig. 5.** MTF of the system drawn up to the spatial frequency of 20 cycles/mm (the Nyquist frequency of the system is ~40 cycles/mm)

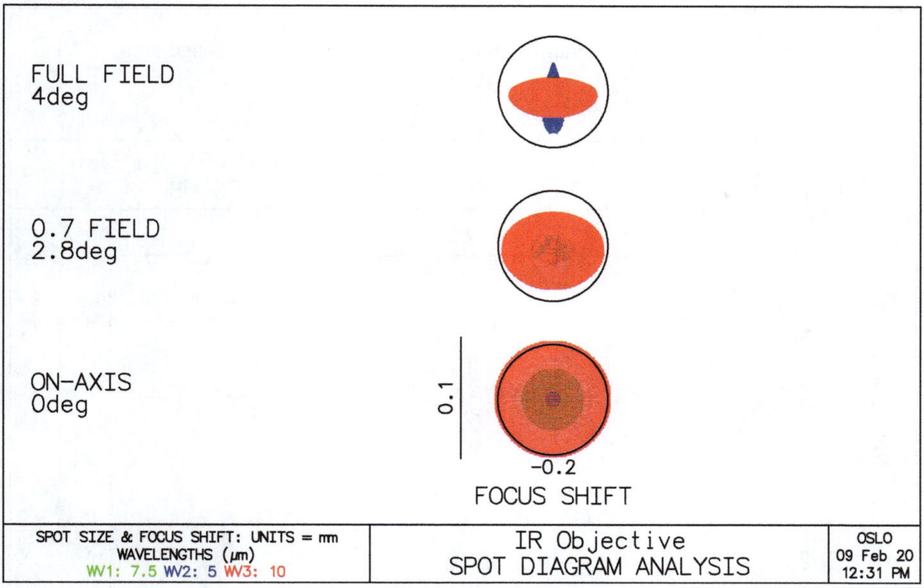


**Fig. 6.** Spot diagram of the system with emphasized Airy circle

More precise, the effective diameter of the spot onto the image plane is approximately 28 μm, as results from the single recipolar spot diagram in Fig. 7. Thus the image spot approximately covers the surface of one pixel.

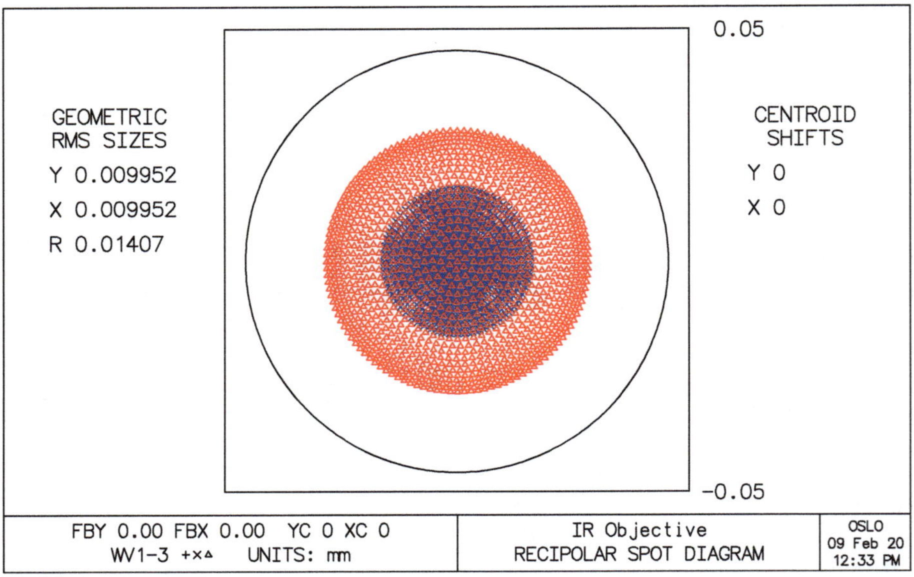


**Fig. 7.** Single spot diagram showing the spot size onto the image plane

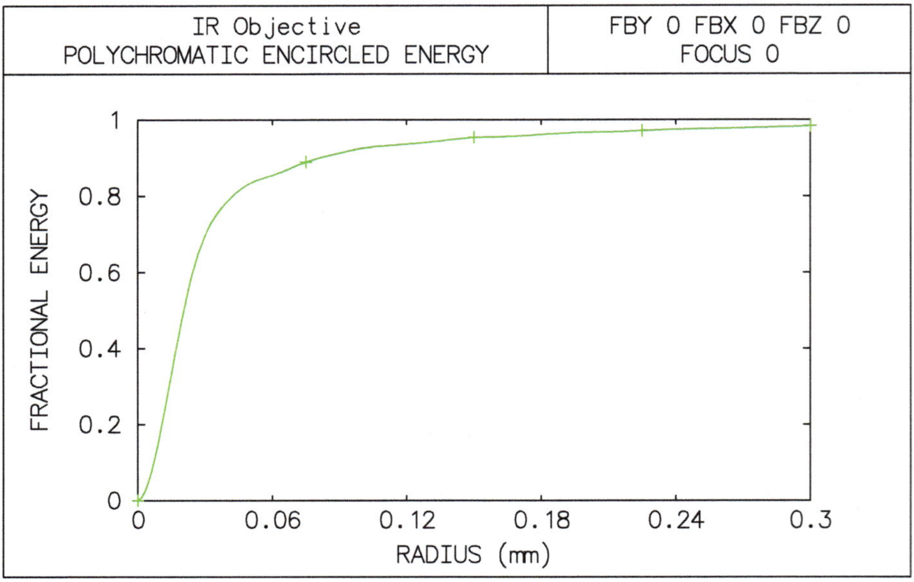


**Fig. 8.** Distribution of encircled energy

The distribution of energy within the image spot leads to similar conclusions. Figure 8 displays the encircled distribution of energy, showing that ~75% of the energy is spread over the surface of a pixel.

## 4 Conclusions

The optical design of the IR objective presented above is novel regarding:

- The choice for the structure of the IR objective (air-spaced doublet, consisting of two lenses, a positive and a negative one, in this order)
- The application of an algorithm developed for the visible range to the infrared one
- Software development for performing the algorithm
- The numerical application, which provided an objective with f = 40 mm, f/# = 5 and very good quality of the image. The system is diffraction-limited over the entire aperture, as results from wavefront analysis and spot diagram. The energy distribution is appropriate for the pixel size, so that the resolution of the sensor is ensured by the optical system.

## References


1. Walker, B.H.: Lens Design for the near IR. Correction of primary chromatic aberration. Appl. Opt. **34**, 8072–8073 (1995)
2. Spencer, H.M.: Thermal considerations in the design of a long focal length, low f-number infrared imager. In: International Optical Design Conference and Optical Fabrication and Testing, OSA Technical Digest (CD), paper ITuD7 (2010)
3. Tyagur, V.M., Kucherenko, O.K., Muravev A.V.: Passive optical athermalization of an IR three-lens achromat. J. Opt. Technol. **81**, 199–203 (2014)
4. Edmund Optics. https://www.edmundoptics.com/knowledge-center/application-notes/optics/an-introduction-to-passive-athermalization/. Accessed 10 Jan 2020
5. Vasilieva, L.V., Lebedev, O.A., Nuzhin, V.S., Solk, S.V.: Design and fabrication of lens objectives for operation in the IR region. J. Opt. Technol. **70**, 280–283 (2003)
6. Riedl, M.J.: The design of an IR Petzval objective using an aspheric and diffractive element: an exercise with MOE for beginners. In: Proceedings of SPIE 5865, Tribute to Warren Smith: A Legacy in Lens Design and Optical Engineering, vol. 586501 (2005)
7. Aboobida, Y.A., Haj noor, F.M.B., Elshafia, Ahmed, M.M.: IR optical system design of uncooled thermal imaging camera in long band (8–12 μm). IOSR J. Appl. Phys. **6**(5), 32–40 (2014)
8. Akram, M.N.: Design of a step-zoom dual field-of-view IR telescope. In: International Optical Design Conference, 2002 OSA Technical Digest Series, paper ITuD4 (2002)
9. Zhang, X., Zhang, B., Luan, Y., Sun, T., Yang, H., Li, Y., Chang, W.: Dual-waveband optical design for IR adaptive system. In: Proceedings of SPIE 9300, International Symposium on Optoelectronic Technology and Application 2014: Infrared Technology and Applications, vol. 93001D (2014)
10. Smith, W.G.: Modern Optical Engineering. Wiley, New York (1995)
11. II-VI. https://www.ii-vi.com/ir-optics-materials-bu/. Accessed 10 Jan 2020
12. LAMBDA Research Corporation. https://www.lambdares.com/oslo. Accessed 10 Jan 2020

# Meshing Limit Line of Offset Normal Arc-Toothed Cylindrical Worm Drive

Qingxiang Meng and Yaping Zhao(✉)

School of Mechanical Engineering and Automation, Northeastern University, Shenyang 110819, China
zhyp_neu@163.com

**Abstract.** In this paper, the theory of solving the meshing limit line is built for the offset normal arc-toothed cylindrical worm drive, and some fundamental and important results are obtained, for example the tooth surface equations of the worm drive, the meshing function, and the meshing limit function. The computation of the meshing limit line of the worm drive is boiled down to solve a nonlinear equation set. After solving the nonlinear equation set iteratively by means of the elimination method and the geometric construction, the numerical study concerning the meshing limit line of the worm drive is conducted. The numerical results demonstrate that there is always a meshing limit line on the worm helicoid and this will cause the length of the worm cannot be fully used. The conjugate line of the meshing limit line is roughly located in the middle area of the worm gear tooth surface.

**Keywords:** Offset normal arc-toothed cylindrical worm drive · Meshing limit line · Meshing function · Nonlinear equation set · Tooth surface

## 1 Introduction

The offset normal arc-toothed cylindrical worm drive is a type of concave-convex worm-gear drive [1–3]. The helicoid of an offset normal arc-toothed cylindrical worm is turned by a turning tool with a convex arc blade, whose cutting edge is located in the offset plane of the normal section of the worm. Therefore, the helicoid of the worm is a track surface. The worm gear is generated by a cylindrical hob, whose generating surface is identical to the helicoid of the mating worm.

For the investigation of the arc-toothed cylindrical worm with a track surface, the existing literature focused primarily on the axial arc-toothed cylindrical worm [4–6], i.e. the cutting edge of the turning tool is located in the axial section of the worm. However, the investigation of the meshing limit line for an offset normal arc-toothed cylindrical worm drive has never been implemented as far as the author has known.

Like other types of the cylindrical worms [7, 8], the meshing limit line of the offset normal arc-toothed cylindrical worm drive may also come into the worm helicoid and can divide the worm helicoid into two parts: working area and non-working area. Therefore, determining the meshing limit line of the worm drive is of great significance for investigating its meshing performance in the following works.


E.-C. Lovasz et al. (Eds.): MTM&Robotics 2020, MMS 88, pp. 284–294, 2021.
https://doi.org/10.1007/978-3-030-60076-1_25

The determination of the meshing limit line of the worm drive can be come down to solve a nonlinear equation set by means of the elimination method and the geometric construction [9–11]. Based on this, we establish the theory of the meshing limit line for the offset normal arc-toothed cylindrical worm drive in this paper. The numerical case investigation is also conducted.

# 2 Mathematic Model of Offset Normal Arc-Toothed Cylindrical Worm Drive

## 2.1 Equation of Helicoid of Offset Normal Arc-Toothed Cylindrical Worm

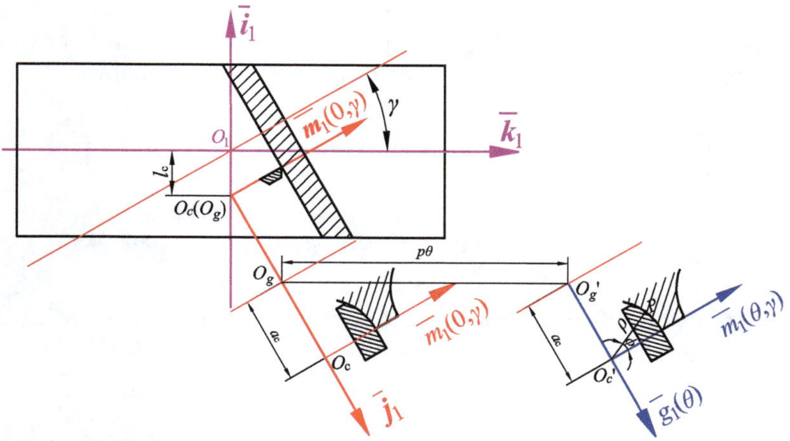


**Fig. 1.** Coordinate systems during generating offset normal arc-toothed cylindrical worm helicoid.

During the generation of an offset normal arc-toothed cylindrical worm helicoid by using a turning tool with an arc-shaped cutting edge whose radius is marked as $\rho$, all the coordinate systems related to the worm and the turning tool are drawn in Fig. 1. Connected with the worm is the coordinate system $\sigma_1\left\{O_1;\vec{i}_1,\vec{j}_1,\vec{k}_1\right\}$ whose unit vector $\vec{k}_1$ lies in the direction of the worm axial line. The point $O_1$ lies at the middle point of the thread length of the worm. At the initial position, the center point $O_c$ of the arc-shaped cutting edge of the turning tool is on the vector $\vec{j}_1$. The point $O_g$ is on the vector $\vec{i}_1$, and the distance between $O_g$ and $O_c$ is the technological distance $a_c$ during machining the worm. The distance between $O_1$ and $O_c$ is the offset distance $l_c$ of the cutting edge. The angle between the offset plane of the cutting edge and the plane $O_1-\vec{j}_1\vec{k}_1$ is the lead angle $\gamma$ of the worm. The current position of the turning tool as shown in Fig. 1 is that the turning tool has rotated the angle $\theta$ around the axial line of the worm, and the moving distance of the turning tool is $p\theta$. Here, the symbol $p$ is the helix parameter of the worm, which should take a positive value for a right-handed worm. For an arbitrary point $P$ on the worm helicoid, its position in $\sigma_1$ can be presented by the vector $\vec{r}_1 = \overrightarrow{O_1P} = \overrightarrow{O_1O'_g} + \overrightarrow{O'_gP}$. According to the spherical vector function [12], the vector equation of the offset normal arc-toothed cylindrical worm helicoid $\Sigma_1$ can be attained in $\sigma_1$ as

$$\begin{aligned}(\vec{r}_1)_1 = \left(\overrightarrow{O_1O'_g}\right)_1 + \left(\overrightarrow{O'_gP}\right)_1 &= -l_c\vec{e}_1(\theta) - (\rho\sin\phi - a_c)\vec{g}_1(\theta) + p\theta\vec{k}_1 \\ &+ \rho\cos\phi\vec{m}_1(\theta,\gamma) = x_1\vec{i}_1 + y_1\vec{j}_1 + z_1\vec{k}_1\end{aligned}, \quad (1)$$

where $x_1 = (\rho\sin\gamma\cos\phi - l_c)\cos\theta + (\rho\sin\phi - a_c)\sin\theta$, $z_1 = p\theta + \rho\cos\gamma\cos\phi$, and $y_1 = (\rho\sin\gamma\cos\phi - l_c)\sin\theta - (\rho\sin\phi - a_c)\cos\theta$.

Equation (1) contains two variables, $\theta$ and $\phi$, which decide the coordinates of an arbitrary point $P$ on the worm helicoid. Moreover, the first order partial derivatives of $(\vec{r}_1)_1$ with respect to $\theta$ and $\phi$ can be respectively worked out as follows

$$\frac{\partial(\vec{r}_1)_1}{\partial\theta} = (\rho\sin\phi - a_c)\vec{e}_1(\theta) + (\rho\sin\gamma\cos\phi - l_c)\vec{g}_1(\theta) + p\vec{k}_1, \quad (2)$$

$$\frac{\partial(\vec{r}_1)_1}{\partial\phi} = -\rho\sin\gamma\sin\phi\vec{e}_1(\theta) - \rho\cos\phi\vec{g}_1(\theta) - \rho\cos\gamma\sin\phi\vec{k}_1. \quad (3)$$

On the basis of differential geometry [13], the unit normal vector of $\Sigma_1$ can be attained in $\sigma_1$ as

$$(\vec{n}_1)_1 = \frac{\frac{\partial(\vec{r}_1)_1}{\partial\theta} \times \frac{\partial(\vec{r}_1)_1}{\partial\phi}}{\left|\frac{\partial(\vec{r}_1)_1}{\partial\theta} \times \frac{\partial(\vec{r}_1)_1}{\partial\phi}\right|} = \frac{n_e}{D_0}\vec{e}_1(\theta) + \frac{n_g}{D_0}\vec{g}_1(\theta) + \frac{n_k}{D_0}\vec{k}_1, \quad (4)$$

where $n_e = p\cos\phi - \cos\lambda sin\phi(\rho\sin\gamma cos\phi - l_c)$, $n_g = \sin\phi(\rho\cos\gamma sin\phi - C_2)$, $n_k = -\cos\phi(\rho\cos^2\gamma\sin\phi - a_c) - l_c\sin\gamma\sin\phi$,
$D_0 = \sqrt{C_3 sin^2\phi - 2C_2\rho\cos\gamma\sin\phi - 2l_cC_1\sin\phi cos\phi + a_c^2 + p^2}$,
$C_1 = a_c\sin\gamma - p\cos\gamma$, $C_2 = a_c\cos\gamma + p\sin\gamma$, and $C_3 = \rho^2\cos^2\gamma - C_1^2 - l_c^2$.

## 2.2 Equation of Tooth Surface of Worm Gear

During the meshing of an offset normal arc-toothed cylindrical worm drive, as shown in Fig. 2, two fixed coordinate systems $\sigma_{o1}\left\{O_1;\vec{i}_{o1},\vec{j}_{o1},\vec{k}_{o1}\right\}$ and $\sigma_{o2}\left\{O_2;\vec{i}_{o2},\vec{j}_{o2},\vec{k}_{o2}\right\}$ are respectively used to designate the initial positions of the worm and the worm gear. Two vectors $\vec{k}_{o1}$ and $\vec{k}_{o2}$ are along the axes of the worm and the worm gear, respectively. The two vectors $\vec{k}_{o1}$ and $\vec{k}_{o2}$ are vertical, and their common vertical line $O_1O_2$ is along the collinear vectors $\vec{i}_{o1}$ and $\vec{i}_{o2}$. The length of $\overrightarrow{O_1O_2}$ is equal to the center distance $a$ of the worm drive. Moreover, attached with the worm gear is a rotating coordinate system $\sigma_2\left\{O_2;\vec{i}_2,\vec{j}_2,\vec{k}_2\right\}$ whose rotational angle around the worm gear axis is $\varphi_1/i_{12}$. Here, the angle $\varphi_1$ is the rotational angle of the worm relative to its initial position during the meshing, and the symbol $i_{12}$ is the transmission ratio of the worm drive.

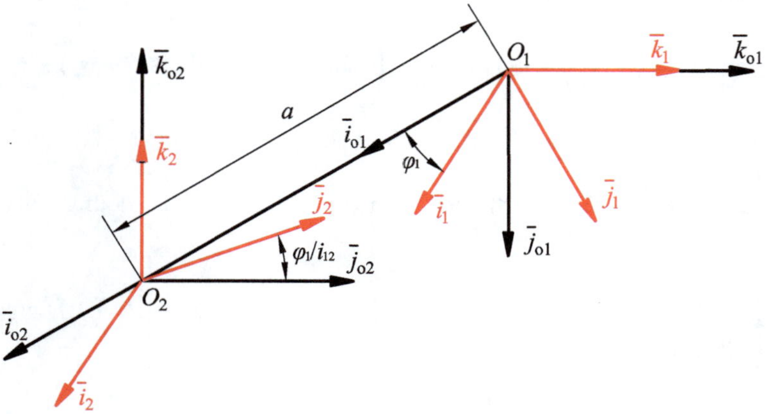


**Fig. 2.** Coordinate systems during meshing of offset normal arc-toothed cylindrical worm drive.

Adding the angle $\varphi_1$ into Eq. (1) leads up to the formation of the family of worm helicoids, which can be expressed in $\sigma_{o1}$ as

$$(\vec{r}_1)_{o1}= R\left[\vec{k}_{o1},\varphi_1\right](\vec{r}_1)_1= x_{o1}\vec{i}_{o1}+y_{o1}\vec{j}_{o1}+z_1\vec{k}_{o1}, \tag{5}$$

where $x_{o1}=(\rho\sin\gamma\cos\phi-l_c)\cos(\theta+\varphi_1)+(\rho\sin\phi-a_c)\sin(\theta+\varphi_1)$ and $y_{o1}=(\rho\sin\gamma\cos\phi-l_c)\sin(\theta+\varphi_1)-(\rho\sin\phi-a_c)\cos(\theta+\varphi_1)$.

Likewise, the unit normal vector of the family of worm helicoid can be obtained in $\sigma_{o1}$ as follows

$$(\vec{n}_1)_{o1}=R\left[\vec{k}_{o1},\varphi_1\right](\vec{n}_1)_1=\frac{n_x}{D_0}\vec{i}_{o1}+\frac{n_y}{D_0}\vec{j}_{o1}+\frac{n_k}{D_0}\vec{k}_{o1}, \tag{6}$$

where $n_x=n_e\cos(\theta+\varphi_1)-n_g\sin(\theta+\varphi_1)$ and $n_y=n_e\sin(\theta+\varphi_1)+n_g cos(\theta+\varphi_1)$.

Without loss of generality, it is assumed that the angular velocity of the worm rotating around the axis $\vec{k}_{o1}$ is $|\vec{\omega}_1|=1\,\mathrm{rad/s}$. Considering the rotational angle of the worm gear is equal to $\varphi_1/i_{12}$, the angular velocity of the worm gear can be acquired as $|\vec{\omega}_2|=1/i_{12}\,\mathrm{rad/s}$. According to the meshing theory for gearing [12], the relative angular velocity vector and the relative velocity vector of the worm drive can be respectively acquired in $\sigma_{o1}$ as

$$(\vec{\omega}_{12})_{o1} = \frac{1}{i_{12}}\vec{j}_{o1} + \vec{k}_{o1},\ \left(\vec{V}_{12}\right)_{o1} = V_{12}^{(x)}\vec{i}_{o1} + V_{12}^{(y)}\vec{j}_{o1} + V_{12}^{(z)}\vec{k}_{o1} \tag{7}$$

where $V_{12}^{(x)} = \frac{z_1}{i_{12}}\cos\varphi_1 - y_{o1}$, $V_{12}^{(y)} = x_{o1} - \frac{z_1}{i_{12}}\sin\varphi_1$, and $V_{12}^{(z)} = \frac{a - x_{o1}}{i_{12}}$.

By definition [12], the meshing function of the worm drive can be attained on the basis of Eqs. (6) and (7), which can be expressed as

$$\Phi(\phi, \theta, \varphi_1) = \left(\vec{V}_{12}\right)_{o1} \bullet (\vec{n}_1)_{o1} = \frac{1}{i_{12}D_0}[A\sin(\theta + \varphi_1) + B\cos(\theta + \varphi_1) + C], \tag{8}$$

where $A = -n_g z_1 - n_k(\rho\sin\phi - a_c)$, $B = n_e z_1 - n_k(\rho\sin\gamma\cos\phi - l_c)$, and $C = n_k(a - i_{12}p)$.

By means of Eqs. (5) and (8), the equation of the worm gear tooth surface $\Sigma_2$ can be expressed in $\sigma_2$ as follows

$$\begin{cases} (\vec{r}_2)_2 = R\left[\vec{k}_2, -\frac{\varphi_1}{i_{12}}\right]\left\{R\left[\vec{i}_{o2}, -90^\circ\right]\left[(\vec{r}_1)_{o1} + \left(\overrightarrow{O_1O_2}\right)_{o1}\right]\right\} = x_2\vec{i}_2 + y_2\vec{j}_2 - y_{o1}\vec{k}_2 \\ \Phi(\phi, \theta, \varphi_1) = 0 \end{cases}, \tag{9}$$

where $x_2 = (x_{o1} - a)\cos\frac{\varphi_1}{i_{12}} + z_1\sin\frac{\varphi_1}{i_{12}}$, $y_2 = -(x_{o1} - a)\sin\frac{\varphi_1}{i_{12}} + z_1\cos\frac{\varphi_1}{i_{12}}$.

Moreover, the meshing limit function of the worm drive can be acquired according to its definition [12] as below

$$\Phi_{\varphi 1}(\phi, \theta, \varphi_1) = \frac{1}{i_{12}D_0}[A\cos(\theta + \varphi_1) - B\sin(\theta + \varphi_1)]. \tag{10}$$

## 3 Computation of Meshing Limit Line of Worm Drive

Generally, the meshing limit line of a cylindrical worm is throughout the whole working tooth depth. Therefore, we should firstly obtain the points of intersection between the meshing limit line and the boundaries of the worm and the worm gear, respectively. The preceding two points are respectively located on the worm addendum and the worm gear addendum. The nonlinear equation sets to determine them are respectively built as below

$$\begin{cases} f_{wa}(\phi) = \sqrt{(\rho\sin\gamma\cos\phi - d_0)^2 + (\rho\sin\phi - a_c)^2} - r_{a1} = 0 \\ \Phi(\phi, \theta, \varphi_1) = 0 \\ \Phi_{\varphi 1}(\phi, \theta, \varphi_1) = 0 \end{cases}, \tag{11}$$

$$\begin{cases} f_{ga}(\phi,\theta,\varphi_1) = \sqrt{(x_{o1}-a)^2+z_1^2}+\sqrt{r_{g2}^2-y_{o1}^2}-a=0 \\ \Phi(\phi,\theta,\varphi_1)=0 \\ \Phi_{\varphi 1}(\phi,\theta,\varphi_1)=0 \end{cases}, \tag{12}$$

where $f_{wa}(\phi)=0$ and $f_{ga}(\phi,\theta,\varphi)=0$ are the equation of the worm addendum and the equation of the worm gear addendum, respectively. The symbols $r_{a1}$ and $r_{g2}$ are the tip radius of the worm and the throat circle radius of the worm gear, respectively.

Since Systems (11) and (12) include three unknowns, they can be solved by means of the elimination method and the geometric construction in order to ensure the solving process is accurate and reasonable. Combining the two equations $\Phi(\phi,\theta,\varphi_1)=0$ and $\Phi_{\varphi 1}(\phi,\theta,\varphi_1)=0$ of Systems (11) and (12) can eliminate the variable $\varphi_1$ and the obtained outcome is

$$f(\phi,\theta)=A^2+B^2-C^2=0. \tag{13}$$

On the other hand, the value of $\phi$ on the worm addendum can be directly calculated from the first expression of System (11), and is marked as $\phi_{wa}$. Using $\phi_{wa}$ to replace $\phi$ of Eq. (13) leads up to a nonlinear equation with one unknown $\theta$ which is equivalent to System (11) as

$$\begin{aligned} f_w(\theta;\phi_{wa}) = {} & z_1^2\left(n_e^2+n_g^2\right)+n_k^2\left[(\rho\sin\phi-a_c)^2+(\rho\sin\gamma\cos\phi-l_c)^2-(a-i_{12}p)^2\right] \\ & +2n_k z_1\left[n_g(\rho\sin\phi-a_c)-n_e(\rho\sin\gamma\cos\phi-l_c)\right]=0 \end{aligned}. \tag{14}$$

Similarly, combining Eq. (13) and the first expression of System (12) can result in a binary nonlinear equation set with unknowns $\phi$ and $\theta$ which is equivalent to System (12) as

$$\begin{cases} f_{g1}(\phi,\theta)=\sqrt{(\hat{x}_{o1}-aC)^2+C^2z_1^2}+\sqrt{C^2r_{g2}^2-\hat{y}_{o1}^2}-aC=0 \\ f_{g2}(\phi,\theta)=A^2+B^2-C^2=0 \end{cases}, \tag{15}$$

where $\hat{x}_{o1}=-B(\rho\sin\gamma\cos\phi-l_c)-A(\rho\sin\phi-a_c)$ and $\hat{y}_{o1}=-A(\rho\sin\gamma\cos\phi-l_c)+B(\rho\sin\phi-a_c)$.

After determining the initial values by means of the technique of geometric construction, Eqs. (14) and (15) can all well be solved iteratively. Moreover, the value range of $\theta$ can also be obtained for the whole meshing limit line located on the working tooth surface of the worm drive, and then we can assign some values for $\theta$ in its value range to determine other points on the meshing limit line via solving Eq. (13) iteratively. Finally, the meshing limit line can be portrayed by means of the method of interpolation.

## 4 Numerical Case

The main design parameters of the offset normal arc-toothed cylindrical worm drive are: the center distance $a = 180$ mm, the transmission ratio $i_{12} = 5.8$, the number of worm threads $Z_1 = 5$, the modulus $m = 9.5$ mm, the reference radius of worm $r_1 = 35.5$ mm, the modification coefficient of worm gear $X_2 = 0.7105$, the tip radius of worm $r_{a1} = 45$ mm, and the throat circle radius of worm gear $r_{g2} = 26$ mm. The main processing parameter during machining the offset normal arc-toothed cylindrical worm are: the radius of the cutting edge of turning tool $\rho = 76$ mm, the offset distance of cutting edge $l_c = -10$ mm, and the technological distance during machining the worm $a_c = 63$ mm.

According to the preceding parameters of the offset normal arc-toothed cylindrical worm drive, the obtained meshing limit line and its conjugate line are respectively drawn in the axial sections of the worm and the worm gear by employing the projection method as shown in Figs. 3 and 4.

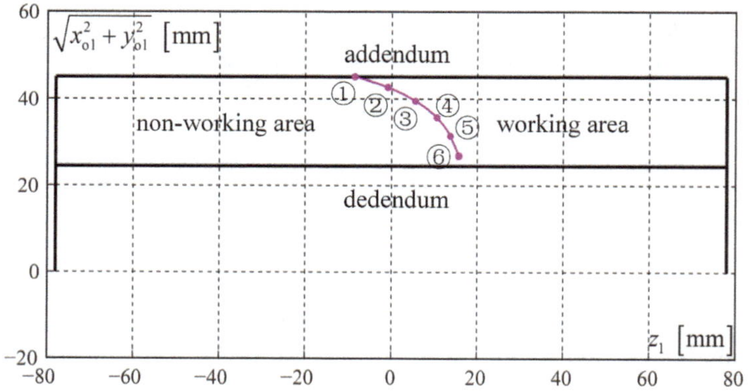


**Fig. 3.** Meshing limit line in axial section of worm.

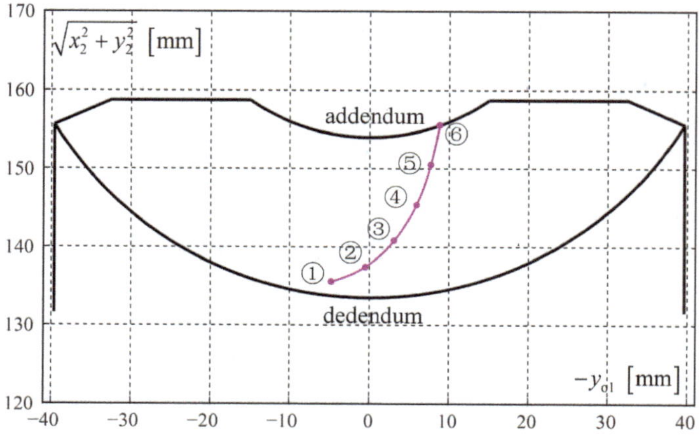


**Fig. 4.** Conjugate line of meshing limit line in axial section of worm gear.

The point ① is the point of intersection between the meshing limit line and the worm addendum. In order to determine the point ①, the curve of the function $f_w$ is drawn in Fig. 5 according to Eq. (14). The image shows that there is a point of intersection between the curve of $f_w$ and the abscissa axis in the vicinity of the point $\theta = -2.3$ rad and this point can be used as the initial value to solve the nonlinear equation $f_w(\theta) = 0$. Then the point ① is determined and the numerical results concerning this point are listed in Table 1.

The point ⑥ is the point of intersection between the meshing limit line and the worm gear addendum. For this point, based on System (15), the curves of $f_{g1} = 0$ and $f_{g2} = 0$ curve are drawn in Fig. 6 and this figure displays that these two curves have a point of intersection within the given solving domain. The coordinates of the point $(\phi, \theta) = (1.2\text{ rad}, -0.4\text{ rad})$ which is close to the preceding point of intersection can be utilized as the initial value to solve the nonlinear equation set (15), and then the point ⑥ can be determined. The numerical results of the point ⑥ are also listed in Table 1.

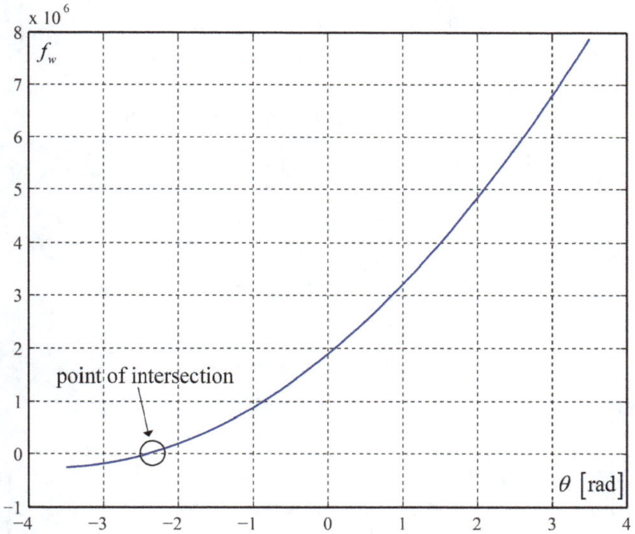


**Fig. 5.** Curve of function $f_w$.

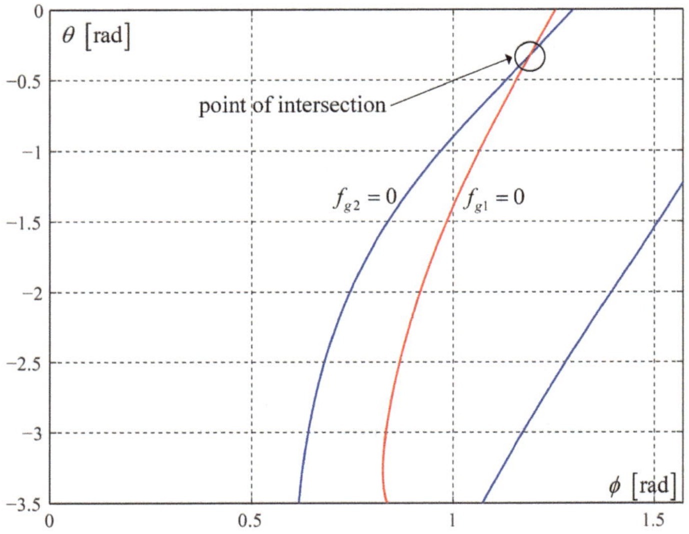


**Fig. 6.** Curves of $f_{g1}=0$ and $f_{g2}=0$.

**Table 1.** Numerical results of meshing limit line.

| Meshing point | $\phi$ [°] | $\theta$ [°] | $\varphi_1$ [°] |
|---|---|---|---|
| ① | 39.5767 | −138.0105 | 125.2155 |
| ② | 42.7231 | −114.1207 | 99.2303 |
| ③ | 47.1131 | −90.2310 | 74.9772 |
| ④ | 52.9836 | −66.3412 | 53.0079 |
| ⑤ | 60.2557 | −42.4515 | 33.5832 |
| ⑥ | 68.2707 | −18.5617 | 15.8130 |

Likewise, the points ②–⑤ are determined from Eq. (13) after the values of $\theta$ are assigned according to its value range as shown in Table 1.

After observing Fig. 3, it can be discovered that the meshing limit line is roughly located in the middle of the worm helicoid, and therefore the length of the worm cannot be fully used. The meshing limit line actually is a helical line on the worm helicoid as shown in Fig. 7. The points in the working area shown in Fig. 7 will mesh with the worm gear tooth surface. The working area and the non-working area can be judged by the plus or minus of the value of $A^2+B^2-C^2$ at the meshing point. The value of $A^2+B^2-C^2$ is greater than 0 in the working area while the value of $A^2+B^2-C^2$ is less than 0 in the non-working area.

On the other hand, Fig. 4 indicates that the conjugate line of the meshing limit line is also located in the middle of the tooth surface of the worm gear, and the meshing performance of the worm drive in the neighborhood of the conjugate line of the meshing limit line may be poor.

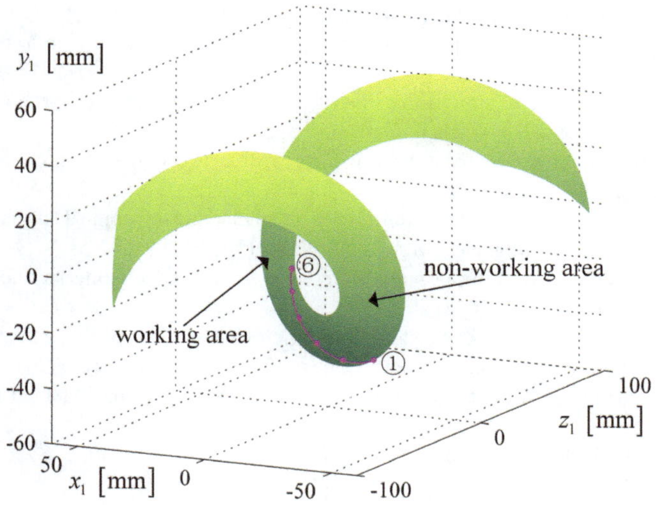


**Fig. 7.** Meshing limit line on worm helicoid.

## 5 Conclusions

The mathematic model of the offset normal arc-toothed cylindrical worm drive is established, including the equations of the tooth surfaces, the meshing function, and the meshing limit function. The computation of the meshing limit line of the worm drive is boiled down to solve a nonlinear equation set, and the computing method is introduced systematically.

The numerical study of the meshing limit line of the offset normal arc-toothed cylindrical worm drive is implemented. The numerical results show that a meshing limit line always exists on the worm helicoid and therefore only part of length of the worm can be used. The conjugate line of the meshing limit line is roughly located in the middle area of the worm gear tooth surface and will divide the meshing zone into two sub-meshing zones on the worm gear tooth surface.

**Acknowledgments.** This study was funded by the Open Fund of the Key Laboratory for Metallurgical Equipment and Control of Ministry of Education in Wuhan University of Science and Technology (2018B05) and the National Natural Science Foundation of China (51475083).

## References

1. Crosher, W.: Design and Application of the Worm Gear. ASME Press, New York (2002)
2. Radzevich, S.P.: Dudley'S Handbook of Practical Gear Design and Manufacture. CRC Press, Taylor & Francis Group, Boca Raton (2012)
3. Yang, L.: Circular Arc Tooth Cylindrical Worm Drive. Shanxi People's Publishing House, Taituan (1984)
4. Wu, H., et al.: Design of Worm Drive, vol. 1. Mechanical Industry Press, Beijing (1986)

5. Wang, S.: Circular Arc Cylindrical Worm Drive. Tianjin University Press, Tianjin (1991)
6. Litvin, F.L., et al.: Gear Geometry and Applied Theory, 2nd edn. Cambridge University Press, Cambridge (2004)
7. Zhao, Y., Sun, X.: On meshing limit line of ZC1 worm pair. In: European Conference on Mechanism Science 2018, Aachen, Germany, pp. 292–298 (2018)
8. Zhao, Y., Mu, S., et al.: Meshing limit line of involute worm drive. Mech. Mach. Sci. **73**, 1129–1138 (2019)
9. Zhao, Y., Meng, Q.: Curvature interference characteristic of conical surface enveloping conical worm. Forsch. Ingenieurwes. **81**(4), 409–419 (2017)
10. Zhao, Y., Kong, X.: Meshing principle of conical surface enveloping spiroid drive. Mech. Mach. Theory **123**, 1–26 (2018)
11. Meng, Q., Zhao, Y., Yang, Z: Meshing limit line of the conical surface enveloping conical worm pair. Proc. Inst. Mech. Eng. Part C J. Mech. Eng. Sci. **234**(2), 693–703 (2020)
12. Dong, X.: Foundation of Meshing Theory for Gear Drives. China Machine Press, Beijing (1989)
13. Kristopher, T.: Differential Geometry of Curves and Surfaces. Springer, Switzerland (2016)

# Optimized FE Model for System-Level Solder Joint Reliability Analysis of a Flip-Chip Ball Grid Array Package

Iulia-Eliza Ţinca[1,2(✉)]

[1] Department of Mechatronics, Politehnica University from Timişoara, Bd. Mihai Viteazu, nr. 1, 300222 Timişoara, Romania
iulia.eliza.tinca@gmail.com

[2] Continental Automotive Romania SRL, str. Siemens nr. 1, 300704 Timişoara, Romania

**Abstract.** When developing solder constitutive models or investigating effects such as aging, stress triaxiality due to underfilling, thermal and power cycling combined or just optimizing integrated circuits geometry, researchers use board-level finite element (FE) models. At board-level, the FE model comprises the printed circuit board (PCB) and the integrated circuit in symmetrical models (octant, quarter geometry). At the system-level, we evaluate integrated circuits' reliability in the context of the electronic device they constitute. We investigate the effects of stress triaxiality due to the thermal expansion mismatch of the housings and PCB constraints. One of the biggest challenges in system-level solder joint reliability assessment is creating an efficient FE model. There are hundreds to thousands of components on a PCB and analyzed integrated circuits such as flip-chip ball grid array (FCBGA) have hundreds to thousands of solder balls. Creating a full product model by following literature guidelines results in complex, but inefficient FE models, solving in days. The present paper focuses on the FEM component of the solder joint reliability assessment through the theoretical approach combining FEA and empirical life prediction models. The scope of the paper is to deliver FE modeling guidelines for lifetime prediction of FCBGAs, to the end goal of efficiently predicting cycles to failure in product configuration. Optimized FE models solve in half time, while lifetime prediction varies with 1%. However, while the complex FE model accurately predicts the crack location, the optimized model fails to do so.

**Keywords:** Solder joint · Reliability · System-level · Finite element analysis

## 1 Background

### 1.1 State of the Art

In solder joint reliability (SJR) assessment, solder material characterization has been a topic of interest for more than half a century. At the same time, the last two decades focused on characterizing lead-free solders due to the lead ban imposed by the EU back in 2006 [1]. Clech et al. note in recently published literature review a lack of solder constitutive models for modern solders, including constituents such as Bi, Sn, and In.


E.-C. Lovasz et al. (Eds.): MTM&Robotics 2020, MMS 88, pp. 295–302, 2021.
https://doi.org/10.1007/978-3-030-60076-1_26

Solder model development also needs to approach the phenomena taking place in the material at the microstructural level [2] as in [3].

Research on microelectronics reliability focuses on effects such as aging [4], stress triaxiality due to underfilling [5], thermal and power cycling combined [6], or integrated circuits geometry optimization [7]. In board-level reliability testing, we subject stand-alone PCBs to thermal cycling. Failures of solder joints consist of cracks occurred at the first or second level interconnect due to CTE mismatch of the materials. The FE model comprises the printed circuit board (PCB) and the integrated circuit in symmetrical models (octant, quarter geometry).

At product/system-level, thermal expansion mismatch of the housings and PCB constraints represent additional mechanical loading in the solder joints, which translates into a change in the lifetime of the solder joint. Schempp et al. argue that with the additional mechanical loads causing both bending and stretching of the PCB, the Coffin-Manson acceleration model, which correlates the change in lifetime between test temperature profiles and field profile, does not hold. Thus, they are advancing state of the art by developing a system metric based on PCB strain, PCB curvature, and component warpage for fast and accurate reliability prediction [8, 9].

In own investigations [10], well-known fatigue models such as [11] or [12] prove to be far too much on the safe side. We observe a good correlation with accelerated thermal cycling test results when the Coffin-Manson acceleration model is applied to the strain energy density ratio between a baseline and system configuration under different thermal loadings [10].

## 1.2 Scope

The scope of the paper is to deliver FE modeling guidelines for lifetime prediction of FCBGAs under the strain energy density ratio between a baseline and system configuration approach. We consider an FCBGA with 421 leads for which the first failure in board-level configuration under thermal loading TC1 is known. We consider the board-level configuration the baseline of the lifetime prediction of the same FCBGA-421 in a product configuration, under thermal loading TC2. We employ the Coffin-Manson fatigue model for lifetime prediction, and we correlate TC1 with TC2 by using the Coffin-Manson acceleration factor. Through the FE method, we determine the failure criteria, dissipated creep strain energy density.

In the first part of the work, we create the FE models for both baseline and product configurations following guidelines in [12]. These models are complex FE models.

The dissipated creep strain energy density is a system metric in the approach employed. If the FE model of the FCBGA and PCB are identical in baseline and product, we can optimize the FE models for efficient calculation.

This paper presents the steps taken for optimizing the component package and PCB FE models involved in lifetime prediction under the strain energy density ratio between a baseline and product configuration (SED-R) approach.

### 1.3 Solder Constitutive Model and Strain Energy Density Ratio Approach

Solder model implemented in the current work is the hyperbolic sine model, also known as the Garofalo law, for SAC alloys, as developed by Schubert et al. in [11]. In Eq. (1), $\alpha$ prescribes the stress level at which the strain rate dependency breaks down, $Q_a$ is the apparent activation energy, and $C_{III,\ IV}$ is constant [11].

$$\tfrac{d\varepsilon_s}{dt} = C_{III,IV}[sinh(\alpha\sigma)]^n exp\left(\tfrac{-Q_a}{kT}\right) \quad (1)$$

The study presented here uses a Coffin – Manson type fatigue model for life prediction based on the dissipated creep strain energy density during one thermal cycle, $W_{cr}^{acc}$. In Eq. (2), cycles to failure, $N_f$, are predicted.

$$N_f = \Theta_2\left(W_{cr}^{acc}\right)^{-c2} \quad (2)$$

Considering the board-level reliability test, under condition TC1 as the baseline, we define the time to failure as $N_f = N_{FailBaseCyc,\ TC1}$. We use the FE method, to compute $W_{cr}^{acc} = W_{cr,FailBaseCycle,TC1}^{acc}$. By substituting the terms in Eq. (2), $\Theta_2$ is as in Eq. (3) [10].

$$\Theta_2 = \frac{N_{FailBaseCyc,TC1}}{\left(W_{cr,FailBaseCycle,TC1}^{acc}\right)^{-c2}} \quad (3)$$

A Coffin – Manson criterion is adequate to account for the acceleration factor between different temperature profiles since the coefficient of thermal expansion (CTE) does not vary significantly in the operating range of electronic devices [13]. The acceleration factor under Coffin – Manson criteria between two thermal profiles, TC1 and TC2, AF, is defined as in Eq. (4), where $\Delta T$ is the temperature range of the thermal profile, and n is a material coefficient, ranging between 2 and 3 for lead-free alloys [10].

$$AF = \left(\frac{\Delta T_{TC2}}{\Delta T_{TC1}}\right)^{-n} \quad (4)$$

By substituting $\Theta_2$ with Eq. (3) in Eq. (2), we predict the lifetime in the new configuration. A new thermal loading, TC2, is accounted for by applying the acceleration factor, AF. Rearranging the terms, lifetime prediction using SED-R, under TC2, is defined in Eq. (5) [10].

$$N_{f,TC2} = N_{FailBaseCyc,TC1} \cdot \left(\frac{W_{cr,TC2}^{acc}}{W_{cr,FailBaseCycle,TC1}^{acc}}\right)^{-c2} \cdot AF \quad (5)$$

Schubert et al. fitted the Coffin – Manson inverse power law exponent to $c2 = 1.02$ in [11] under the dissipated $W_{cr}^{acc}$ criteria for SAC alloys [10].

## 2 Investigated Assemblies

We consider a flip-chip ball grid array integrated circuit with 421 leads. The baseline of the calculations is the FCBGA-421 in board-level configuration under thermal load condition TC1 (see Fig. 1A and Fig. 2A). The product assembly model (see Fig. 1B) is loaded with the bolt loads in the first step. Subsequent loading follows the thermal profile TC2 (see Fig. 2B). Equation (4) yields an acceleration factor between TC1 and TC2 of 2.76.

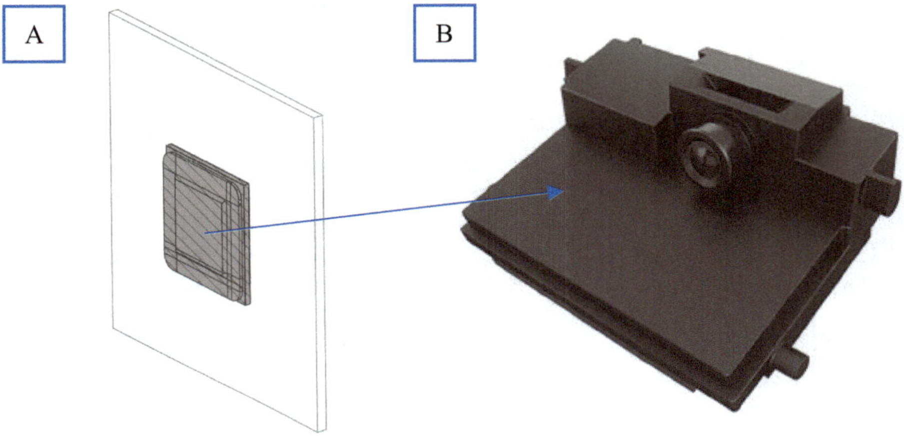


**Fig. 1.** Baseline assembly, BL (A) and product assembly, SL (B).

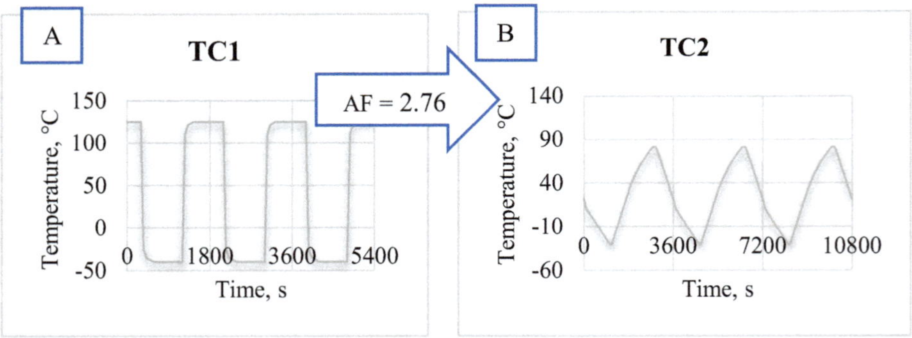


**Fig. 2.** Thermal loading for baseline (A) and system (B).

## 3 Finite Element Models

### 3.1 Complex Models

Syed removes the modeling assumptions from the prediction model, publishing FE modeling guidelines in [12] for octant or quarter of the package. FE modeling of BL and SL assemblies following Syed's guidelines leads to FE models further referred to as complex. Except for the FCBGA-421 lid, all package parts and the PCB have four hexahedral quadratic elements through-thickness (see Fig. 3A).

All solder balls have identical mesh with hexahedral elements. Critical solders have 20-nodes quadratic elements, while non-critical solders have 10-nodes linear elements. The dissipated creep strain energy is averaged on a 25-micron layer with two elements through-thickness at the package and PCB interfaces (see Fig. 3B).

In the baseline model, the open connection in the test is the critical solder ball. In the product model, the corner solders and die shadow solders are rendered critical.

Although in [12] Syed strongly recommends avoiding using MPC constraints between solder balls and package and solder balls and PCB, this guideline cannot be respected here due to the complexity of both full package and full product models. Surface-to-surface ties constrain the solder – package, and solder – PCB interfaces.

## 3.2 Optimized Models

The principle used for optimization relies on the fact that the lifetime prediction parameter is a system metric in the SED-R approach. If the FE models of the FCBGA and PCB are identical in baseline and product, we can optimize the FE models for efficient calculation. In the optimized model, the types of elements and geometrical order are identical as in the complex model, except for the solder balls, which are now all meshed with linear elements, see Table 1.

**Table 1.** Element types in complex and optimized FE models

| Part | Element type | Geometric order (complex models) | Geometric order (optimized models) |
|---|---|---|---|
| IC package – lid | Tetrahedral | Quadratic | Quadratic |
| IC package | Hexahedral | Quadratic | Quadratic |
| Critical solders | Hexahedral | Quadratic | Linear |
| Non-critical solders | Hexahedral | Linear | Linear |
| PCB | Hexahedral | Quadratic | Quadratic |

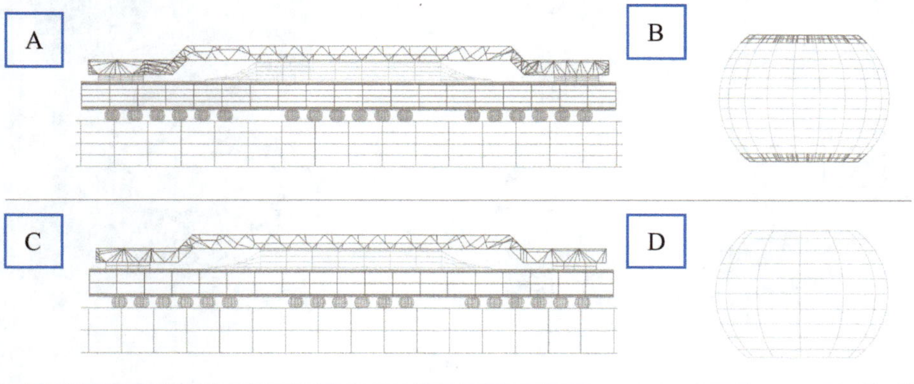


**Fig. 3.** Cross-section through package and board for complex FE model (A) and optimized FE model (B). Complex (C) and optimized (D) model of solder balls.

A simple design of experiments of a board under bending load solved with different meshes shows that two hexahedral quadratic elements through-thickness yields optimum solving time without cutting back on the quality of field outputs. The package parts and PCB are thus meshed with two quadratic elements through-thickness (see Fig. 3C). We want a uniform mesh throughout all the solders to quickly identify solders most susceptible to cracks. Hence, we choose linear elements for all solder balls, five elements from the neutral axis to each interface (see Fig. 3D). Lifetime prediction under the SED-R approach does not require absolute value for the field outputs, only outputs relative to the baseline, so the dissipated creep strain energy density can be averaged on the full solder volume.

## 4 Results and Conclusions

From FE analysis and lifetime prediction point of view, by optimizing the finite element models, we reduce the BL model by 65%, from 498952 to 173014 nodes, and the SL model by 39%, from 900095 to 549202 nodes. Considering the optimized model, the solution time decreases by half for the system model, from 10 to 5 h using a professional workstation. Comparing the outputs of the two models, we observe the following:

- The strain distribution in the complex FE models is finely captured at the end of the third loading cycle (see Fig. 4A and Fig. 4B).
- The optimized models render rough results following the pattern in complex models (see Fig. 4C and Fig. 4D).
- Equation (5) predicts 4779 cycles to first crack when the complex FE model is employed.
- Equation (5) predicts 4822 cycles predicted when the optimized model is employed.

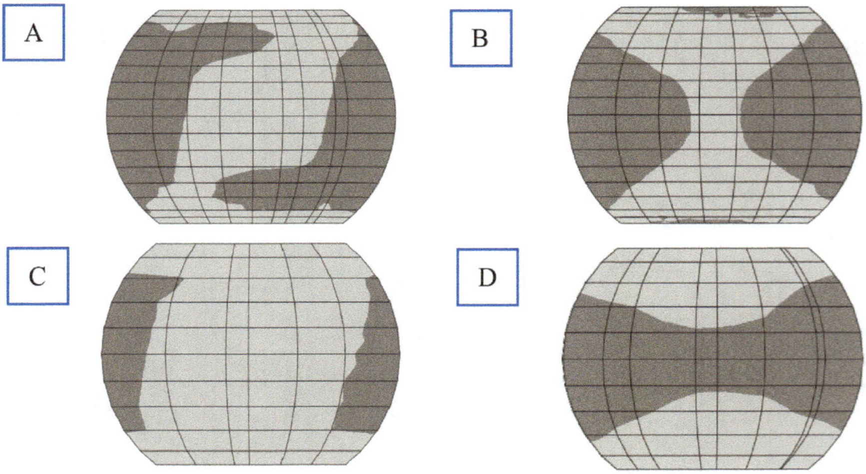


**Fig. 4.** Creep strain distribution in solder joints of the complex BL model (A), complex SL model (B), optimized BL model (C), optimized SL model (D).

We further analyze the results by identifying the failure location using FEA. The FCBGA-421 package in SL configuration under TC1 loading after 3300 cycles in the physical test presents a crack of 40% solder length in solder ball AB22 at package interface (see Fig. 5A). The complex FE model accurately predicts the crack location (see Fig. 5B), unlike the optimized model (see Fig. 5C).

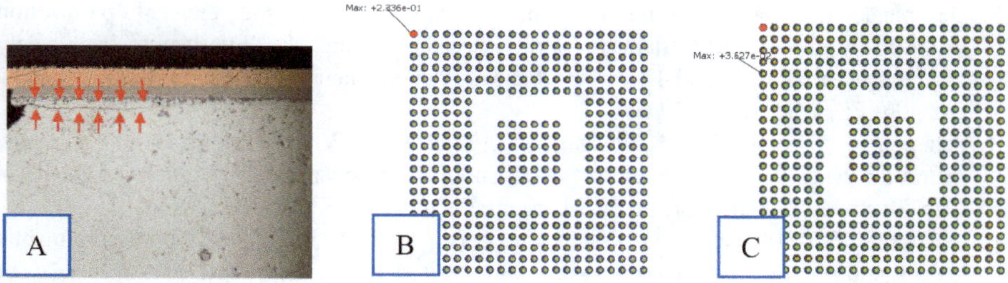


**Fig. 5.** A crack of 40% solder length in solder ball AB22 at package interface (A). Maximum creep strain location after the third loading cycle in complex FE model (B) and optimized FE model (C). Red marks the AB22 solder ball in B and C.

For lifetime prediction under the SED-R approach, when the reliability of package under system effects is in scope, guidelines proposed in Sect. 3.2 provide fast and accurate results. The present work proves that under the SED-R approach, lifetime prediction is independent of the FE model. The predicted cycles to failure vary with 1% between the optimized and complex model. This result shows that the optimized models comprise accurately the strain ratio between baseline and product-level configuration.

However, the optimized models do not capture the failure location. For the presented case study, the failure location was known, but generally we want this to be an output of the analysis.

The use of tetrahedral mesh would allow common-node mesh between solder and the connected parts, hence better strain distribution. Future research efforts focus on analyzing the possible use of common-node tetrahedral mesh to the end scope of correlating system interactions and solder joint lifetime.

## References


1. Dudek, R.: Characterization and modelling of solder joint reliability. In: Mechanics of Microelectronics, pp. 377–418. Springer, Dordrecht (2006)
2. Clech, J.-P.M., Coyle, R.J., Arfaei, B.: Pb-free solder joint thermo-mechanical modeling: state of the art and challenges. Miner. Metals Mater. Soc. **71**(1), 143–157 (2018)
3. Thambi, J.L.: Reliability assessment of lead- free solder joint, based on high cycle fatigue & creep studies on bulk specimen, Ph.D. thesis, Berlin (2018)
4. Basit, M., Motalab, M., Suhling, J.C., Hai, Z., Evans, J., Bozack, M.J., Lall, P.: Thermal cycling reliability of aged PBGA assemblies - comparison of weibull failure data and finite element model predictions. In: 65th Electronic Components and Technology Conference, San Diego (2015)

5. Serebreni, M., McCluskey, P., Hillman, D., Blattau, N., Hillman, C.: Experimental and numerical investigation of underfill materials on thermal cycle fatigue of second level solder interconnects under mean temperature conditions. In: ASME, San Francisco (2018)
6. Denria, J., Rajmane, P., Agonafer, D.: Board level solder joint reliability assessment study of megtron 6 Vs FR-4 under power cycling and thermal cycling. In: 17th IEEE Intersociety Conference on Thermal and Thermomechanical Phenomena in Electronic Systems (ITherm), San Diego (2018)
7. Deshpande, A., Khan, H., Mirza, F., Agonafer, D.: Global-local finite element optimization study to minimize BGA damage under thermal cycling. In: Fourteenth Intersociety Conference on Thermal and Thermomechanical Phenomena in Electronic Systems (ITherm), Orlando, FL, USA (2014)
8. Schempp, F., Dressler, M., Kraetschmer, D., Loerke, F., Wilde, J.: Introduction of a new metric for the solder joint reliability assessment of BGA packages on system level. In: IEEE 68th Electronic Components and Technology Conference, San Diego (2018)
9. Schempp, F., Dressler, M., Kraetschmer, D., Loerke, F., Wilde, J.: Experimental investigation of the correlation between a load-based metric and solder joint reliability of BGA assemblies on system level. In: 69th Electronic Components and Technology Conference, Las Vegas (2019)
10. Ţinca, I.E., Faller, D., Rai, P.: System-level flip-chip ball grid array solder joint reliability assessment under different methodologies and correlation with accelerated thermal cycling experimental data. In: Electronics System-Integration Technology Conferences, Vestfold (in press)
11. Schubert, A., Dudek, R., Auerswald, E., Gollbardt, A., Michel, B., Reichl, H.: Fatigue life models for SnAgCu and SnPb solder joints evaluated by experiments and simulation. In: Proceedings of the 53rd Electronic Components and Technology Conference, 2003, New Orleans (2003)
12. Syed, A.: Updated life prediction models for solder joints with removal of modeling assumptions and effect of constitutive equations. In: 7th International Conference on Thermal, Mechanical and Multiphysics Simulation and Experiments in Micro-Electronics and Micro-Systems, EuroSimE, Como (2006)
13. Kuper, F., Fan, X.: "Reliability Practice," in Mechanics of Microelectronics, pp. 35–64. Springer, Dordrecht (2006)

# Relationship Between Rotational Angle and Time-Synchronous-Averaged Meshing Vibration of Plastic Gears

Yusuke Tsutsui, D. Iba(✉), B. H. Kien, A. Kajihata, N. Miura, T. Iizuka, A. Masuda, A. Sone, and I. Moriwaki

Kyoto Institute of Technology, Mechanical and System Engineering, Kyoto 606-8585, Japan
iba@kit.ac.jp

**Abstract.** The purpose of this study is to clarify the relationship between rotational angle and meshing vibration of a plastic gear mating with a metal gear. A newly developed data acquisition system attached to a gear operation test rig enables us to measure the meshing vibration and to capture the side-view images of the gear mating part at synchronized timing. The gear meshing vibration was acquired by an acceleration sensor installed on the bearing housing of the rotating shaft of the plastic gear. A Gaussian filter eliminated the short-wavelength components in the meshing vibration. The filtered data was cut out at every pitch rotation by a zero-crossing method. After that, a time-synchronous averaging method was performed for each rotation of the plastic gear. As a result of data analysis and checking the side-view images of the captured gear mating part, it was found that a healthy plastic gear had significant amplitude modulation in the averaged meshing vibration. The amplitude increased when the teeth near the two pinholes for transmitting torque meshed, and decreased when the 90-degree phase advanced. In addition, the large magnitude of the modulation decreased with the growth of the tooth root cracks.

**Keywords:** Plastic gears · Meshing vibration · Time-synchronous averaging · Tooth root crack · Amplitude modulation

## 1 Introduction

Gears are one of the essential mechanical elements for power transmission of transport machinery such as cars and airplanes. Therefore, failure prevention of power transmission systems is desirable for accident avoidance of machines. Thus, research on the health monitoring of gears has been carried out, and many methods have been proposed for metal gears. For example, a method for detecting abnormalities at an early stage by using an AE sensor to capture elastic energy caused by gear deformation and damage [1], and the Mahalanobis-Taguchi method for classifying the gear meshing vibration signal into healthy or un-healthy [2]. In particular, the accelerometer for measuring the meshing vibration has low installation costs and few restrictions on mounting. Therefore, the health monitoring method based on meshing vibration measurement is prevalent for metal gears.


E.-C. Lovasz et al. (Eds.): MTM&Robotics 2020, MMS 88, pp. 303–312, 2021.
https://doi.org/10.1007/978-3-030-60076-1_27

In recent years, due to the development of plastic materials and processing technology, attention to plastic productions has an increasing tendency [3]. For example, running tests of the internal gears made of POM (Polyoxymethylene) showed results with very high load capacity [4]. In the future, it is more expected that the demand for lightweight and inexpensive plastic gears will increase, as technology development related to the durability and strength of the plastic progresses. However, there are few examples of health monitoring research on plastic gears compared to research on material development. Therefore, the long-time objective of our research is to develop a damage detection system for plastic-gears based on meshing vibration analysis of gear pairs using an accelerometer.

From now, our laboratory has carried out many running tests of POM spur gears using the developed gear running test rig, and measured a large amount of meshing vibration data for healthy and unhealthy. However, the data acquisition system developed by our previous research [5] could not measure the gear meshing vibration and the gear side-view images in synchronization. Therefore, the damage features that appeared in the vibration data have not been clarified.

In this paper, a new data acquisition system that can acquire the vibration and image data in synchronization is introduced, and running tests are carried out. Then, the time-synchronous averaging method reduces the noise components from the meshing vibration. After averaging, the features of the damage appearing in the time series vibration data are cleared. Finally, by checking the image and vibration data acquired synchronously, the features of crack in meshing vibration are extracted.

## 2 Experimental Device, Test Gear and Measured Data

In this section, the test rig and the test gear are explained. In addition, the measured data is introduced.

### 2.1 Gear Operating Test Rig

Firstly, the power-absorption-type gear running test rig is shown in Fig. 1. In this figure, power is supplied to the driving shaft via the v-belt from (1) (the driving motor). (2) is a gear pair that transmits power from the driving shaft to the driven shaft. After that, power is finally absorbed by (5) (the braking motor). The meshing vibration data of driven gear is measured by (3) (the accelerometer) mounted on the bearing housing that supports the driven shaft, and the load torque is measured by (4) (the torque meter) implemented on the driven shaft. In addition, the driven gear side-view images are captured by (6) and (7) (the high-speed-camera and the strobe). The vibration data and the driven gear side-view images can be acquired at the same timing by giving the camera a trigger signal synchronized with the accelerometer.

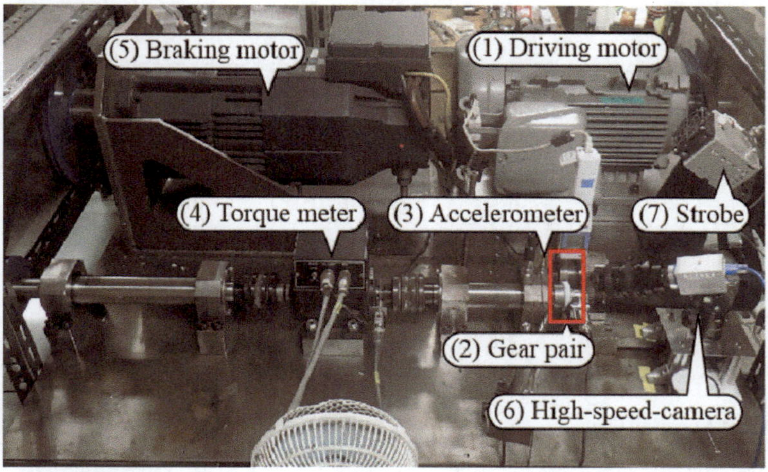


**Fig. 1.** Gear operating test rig.

## 2.2 Test Gear

In this subsection, the plastic gears used in this experiment are explained. Figure 2 shows the drawing of the plastic test gear, and Table 1 shows the specification of the driven test gear and driving master gear (the plastic gear and the metal gear). The test gear is POM (Polyoxymethylene) spur gear with 1 mm module and 48 number of teeth. The material of driven master gear is 15CrMo4 with 67 the number of teeth.

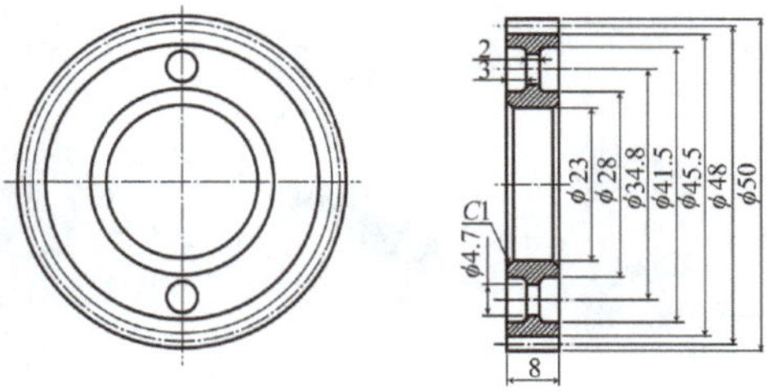


**Fig. 2.** Test gear.

**Table 1.** Specification of master and test gear.

| | Master gear | Test gear |
|---|---|---|
| Module (mm) | 1.0 | |
| Pres. angle (deg) | 20.0 | |
| Number of teeth | 67 | 48 |
| Helix angle (deg) | 0 | 0 |
| Prof. shift coef. | 0 | 0 |
| Facewidth (mm) | 10.0 | 8.0 |
| Tip diameter (mm) | 69.0 | 50.0 |
| Root diameter (mm) | 64.5 | 45.5 |
| Material | 15CrMo4 | POM |

### 2.3 Data Acquisition

The running tests are carried out under test conditions (the load torque is 7 Nm, and the rotational speed is 1,000 $min^{-1}$). The sampling frequency of meshing vibration data measurement by the accelerometer is 100 kHz. The frame rate of the high-speed-camera is 2,500 fps, and the side-view images for one rotation of the test gear can be acquired. An analog USB I/O terminal (CONTEC: AIO-160802GY-USB) is used to measure synchronized data, and a NI Multifunction I/O device (National Instruments: NI USB-6341) is used to output the camera trigger signal. Figure 3 shows a schematic diagram of the constructed data collection system.

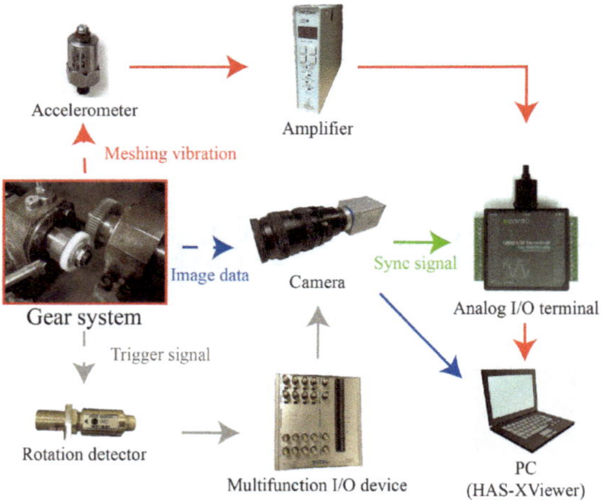


**Fig. 3.** Schematic view of automatic gear side-view images and meshing vibration data acquisition system.

Figure 4 shows the examples of the meshing vibration data measured by the accelerometer and the corresponding gear side-view images captured by the high-speed camera.

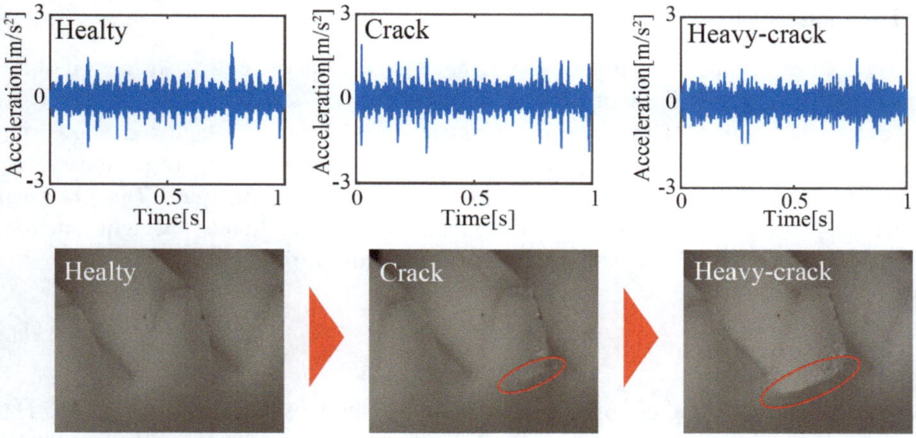


**Fig. 4.** Measured meshing vibration by accelerometer and captured side-view images by high-speed camera.

## 3 Time-Synchronous Averaging of Meshing Vibration

The measured vibration data includes various frequency components from not only the gear meshing but also the surrounding ambient. It means that it is difficult to extract the features of gear damage mode from the time-series vibration data acquired directly from sensors. A time-synchronous averaging is a powerful filtering technique that can extract periodic signals from signals containing large amounts of noise. Figure 5 shows the flowchart of the time-synchronous averaging used in this research. The detail procedure of the averaging is explained in this section.

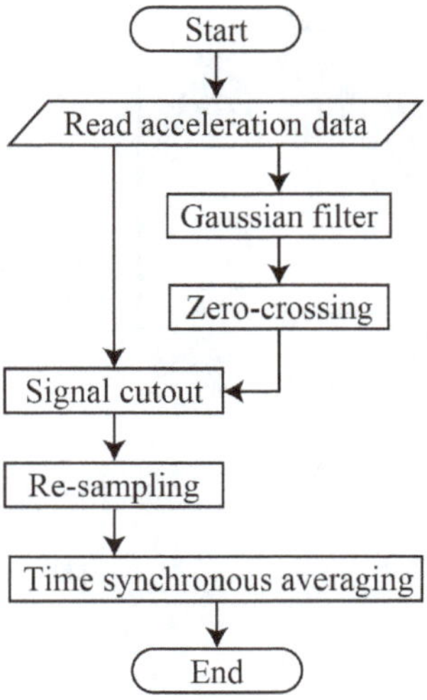


**Fig. 5.** The flowchart of time-synchronous averaging.

### 3.1 Gaussian Filter

In order to use a time-synchronous averaging, it is necessary to extract the period of the meshing frequency or the shaft rotation. In our research, a zero-crossing method is used to extract the period of the meshing frequency. However, as mentioned above, the measured data includes various frequency components. Therefore, the short-wavelength components are firstly removed using a low-pass filter. The Gaussian filter is a phase compensation filter having the normal distribution as a weight function. Equation (1) shows the definition of the Gaussian filter [6].

$$s(t) = \frac{f_c}{\alpha} e^{-\pi(\frac{f_c}{\alpha}t)^2} \tag{1}$$

In this study, $f_c$ (the cut-off frequency) is set to the meshing frequency of 800 Hz, and the coefficient $\alpha$ is defined by the following Eq. (2) and has a 50% amplitude transmissibility.

$$\alpha = (\frac{\ln 2}{\pi})^{0.5} \tag{2}$$

From Eq. (1), the measured vibration data is filtered by performing the convolution operation. Figure 6 shows the meshing vibration data for one rotation of the test gear (solid blue line) and the waveform data after filtering (solid orange line).

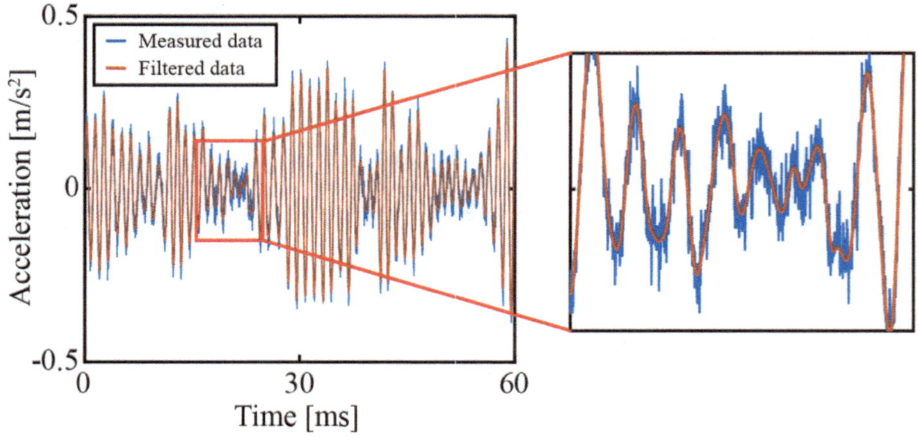


**Fig. 6.** Measured (blue line) and filtered (orange line) meshing vibration data.

### 3.2 Signal Cutout and Re-sampling

The extracted periods of the meshing frequency have fluctuation due to various factors, even if the rotation speed of the driven gear shaft is set to the constant value of 1000 $\text{min}^{-1}$. Therefore, the data length for each pitch rotation angle should be re-sampled appropriately.

The filtered meshing vibration is cutout by zero-crossing. The data length during one pitch rotation $l$ is calculated by Eq. (3). Here, $n$ is the rotational speed, $z$ is the number of teeth, and $f_s$ is the sampling frequency.

$$l = \frac{60}{n \times z} f_s \tag{3}$$

In this signal processing, the zero-crossing points were extracted based on the k-nearest neighbor method, and $l$ was 125. Therefore, the data length should become 125 by using a least-squares FIR filter that compensates the linear phase.

### 3.3 Time-Synchronous Average

The resampled vibration data is time-synchronous averaged every rotation of the gear. The time-synchronous averaging method is proposed as shown in the following Eq. (4) [7]. Here, $x_{\mathrm{TSA},k}$ is the vibration data obtained by the time-synchronous averaging, $x_k(t)$ is the original vibration data cutout for each rotation of the gear, $k$ is the number of rotations, $t$ is the time.

$$x_{\mathrm{TSA},k} = \frac{1}{N} \sum_{k=1}^{N} x_k(t) \tag{4}$$

## 4 Features of Tooth Root Crack in Averaged Acceleration Data

In this section, the averaged vibration data is investigated, and the features are discussed.

### 4.1 Influence of Tooth Root Crack on Amplitude Modulation

First, the influence of the tooth root crack on the averaged vibration waveform is confirmed. Figure 7 shows the examples of the time-synchronous-averaged vibration waveform and the corresponding gear side-view images. In this paper, 160 times averaging was carried out. The horizontal axis shows the time for 48 meshings. Figure 7 has three types of tooth conditions: healthy, middle crack, and heavy crack.

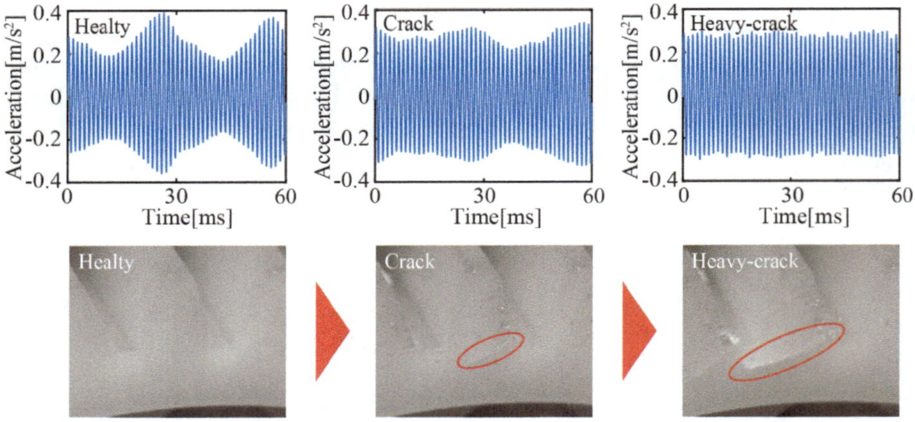


**Fig. 7.** Averaged meshing vibration and captured side-view images.

From Fig. 7, it is confirmed that the vibration of the healthy gear has amplitude modulation that has two large parts under one rotation of the gear. However, the fluctuation of the envelope disappears with the progression of the tooth crack at the root of the plastic test gear. The plastic gears used in our research are manufactured based on specifications of JIS B 1759. As shown in Fig. 2, the test gear has the two pinholes that are spaced 180° apart. The power is transmitted through these two pinholes. Therefore, it is considered that the large amplitude modulation of the healthy gear is confirmed for every 180° rotation. However, it is not clear whether the amplitude increases or decreases at the position of the holes.

### 4.2 Features of Damage in Gear Meshing Vibration and Rotation Angle Estimated from Side-View Images

In this subsection, we add a new experiment and analyze the results to clarify the cause of the amplitude modulation. Additionally, the reason why the amplitude modulation disappears as the cracks propagate is considered. In this experiment, numbers were assigned to the teeth of the test gear to identify the cause of the amplitude modulation occurring during which teeth are engaged with the metal gear. Figure 8 shows the time-synchronized averaged vibration data and the corresponding three phases of the gears in the healthy and unhealthy.

From this figure, it was confirmed that the amplitude increased when the gear teeth near the holes meshed with the metal gear in the healthy condition. On the other hand, the amplitude decreased when the teeth having the root crack meshed with the metal gear. It is considered that the amplitude was reduced because the response of the vibration isolation region as the vibration system was reduced due to the decreased stiffness of the cracked teeth.

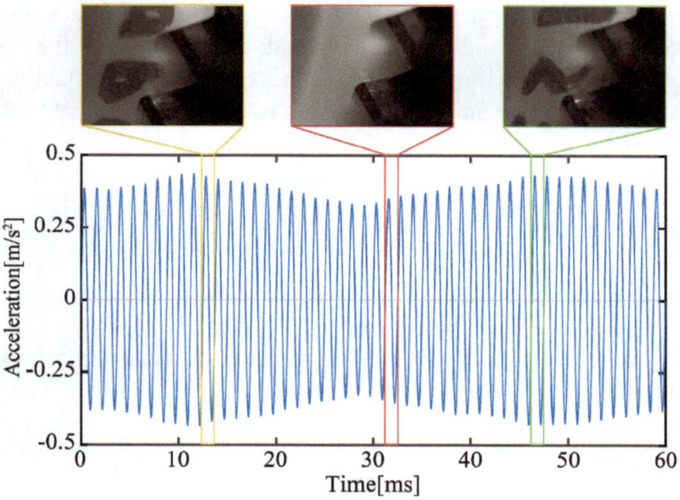


(a) Healthy

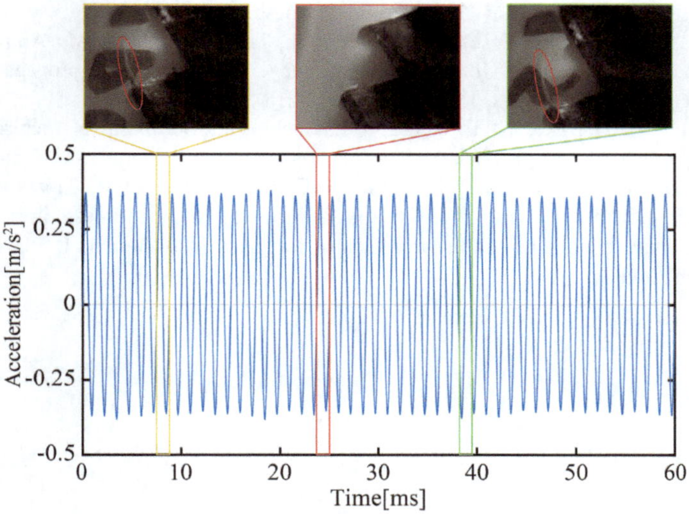


(b) Toot root crack

**Fig. 8.** Relationship between amplitude modulation of vibration data and rotation angle estimated from gear meshing part side-view images.

# 5 Conclusion

In this paper, an I/O data terminal was introduced into the automatic data acquisition system developed by our previous research. The meshing vibration data and the side-view images of the plastic gears during running tests were acquired in synchronization. The meshing vibration data was time-synchronous averaged, and frequency components that were not synchronized with rotation have been removed. The remaining vibration components had large amplitude modulation when the plastic test gear was a healthy condition. It was found that the amplitude increased when the teeth near the two pinholes that transmitted to the gear meshed, and decreased when the 90-degree phase advanced. In addition, the large amplitude of the modulation decreased with the growth of the root cracks.

## References


1. Satoshi, O., Kouitsu, M., Takao, K., Kenichi, A.: Acoustic emission in bending fatigue process of spur gear teeth. J. JSME **58**, 2219–2225 (1992). (in Japanese)
2. Wataru, S., Takao, K., Atsutaka, T., Tatsuya, S., Kazuki, N.: Abnormality detection of gear system using vibration measurement and Mahalanobis Taguchi system. In: Proceeding on JSME, Japanese, pp. 1–2 (2015)
3. Erhard, G.: Konstruieren mit Kunststoffen, 4th edn. Carl Hanser Verlag GmbH & Co. KG, Munich (2008)
4. Ichiro, M., Tomoya, M., Kohta, T., Daisuke, I., Akio, U.: Load capacity of polyacetal (POM) internal gears-issues of JIS B 1759 appeared in running tests for internal gears. J. Precis. Eng. Soc. **85**, 591–596 (2019). (in Japanese)
5. Yunosuke, I., Daisuke, I., Yusuke, T., Nanako, M., Takashi, I., Arata, M., Akira, S., Ichiro, M.: Acquisition of meshing vibration data and image pictures of crack propagation at tooth root of plastic gears. J. Trans. JSME **85**, 1–19 (2018). (in Japanese)
6. Myungsoo, M., Daisuke, I., Tomohiro, T., Junichi, H., Morimasa, N., Ichiro, M.: New measuring method for single pitch deviation of hobbed gears (Using measured helix form data conditioned with Gaussian filter). J. Trans. JSME. **82**, 1–10 (2016). (in Japanese)
7. Junichi, H., Daisuke, I., Ichiro, M., Takao, K.: Proposal of encoder-less time synchronous averaging method utilizing nonlinear oscillator for gears in operation. J. JSME. **12**, 1–11 (2016). (in English)

# Aspects Regarding the Evaluation of the Gearing Tooth Stiffness

Daniela Ghelase(✉) and Luiza Daschievici

"Dunarea de Jos" University of Galati, Galați, Romania
daniela.ghelase@ugal.ro

**Abstract.** It is known that in any kinematic chain there are gear drives with the manufacturing and assembling errors, besides, the gears bodies are not rigid. If during the meshing, the teeth flanks deform above the admissible limit, then the low of the meshing and gear ratio don't respect any longer. Therefore, speed variations of the driven shaft, vibrations, shocks, noise, power loss, low durability of gears appear in the meshing. All these negative consequences due to gearing tooth deformation influence the accuracy of the kinematic chain and the working of the machines. This paper is a study regarding the actual stage of the researches done in the field of the gearing tooth stiffness, including the approach of calculating the stiffness of the gearing tooth. An algorithm developed by the authors of the paper, for the evaluation of the worm gearing tooth stiffness, is also presented in this paper. Based on modeling and simulation, many studies can be performed on improving the performance of the gears.

**Keywords:** Stiffness · Gearing · Meshing · Precision · Machines

## 1 Introduction

The problem of the gearing tooth stiffness is very complex and depends on many factors, such as: geometry of gearing tooth, flank profile, line of contact or surface of contact (for three-dimensional gearing), material properties, load distribution along the line of contact to gearing, with or free errors, etc.

The study of the stiffness for the worm-gearing tooth is much more complicated than in the case of the plane-gearing tooth because the worm-gearing is a three-dimensional (spatial) gearing. Theoretically, line contact exists between the worm thread and gear teeth. The radii of curvature are continuously changed depending on the meshing position along the contact line. Therefore, the specialized literature presents numerous studies regarding the approach of calculating the stiffness of the gearing tooth, especially for the spur gearing tooth, less for the worm gearing tooth. It is obviously that the approach of calculating the stiffness of the worm gearing tooth may be applied for any types of cylindrical worm gearing and for spur gearing and bevel gearing.

It should be mentioned that in many studies, the modeling and the simulation of the meshing have helped to find calculation methodologies for the stiffness of the gearing tooth, which is very different from the theoretical one.


E.-C. Lovasz et al. (Eds.): MTM&Robotics 2020, MMS 88, pp. 313–323, 2021.
https://doi.org/10.1007/978-3-030-60076-1_28

## 2 Contact of the Deformable Gearing with Errors

Considering that a spatial gearing consists of more plane-gearings (pinion-rack drives), that in fact are cross sections perpendicular to worm wheel axis, this problem is studied on a worm gearing.

If the worm gearing teeth are rigid theoretical and assembling and processing errors are null, then the contact between worm and wheel is a line (Fig. 1). The lines of contact can be determined based on the enveloping conditions and enveloped surface equation:

$$\begin{aligned} &\bar{n} \cdot \bar{v}_{12} = 0; \\ &f = 0, \end{aligned} \tag{1}$$

where: $\bar{n}$ is the normal vector at the contact point of the two surfaces; $\bar{v}_{12}$ - is the relative velocity vector of the two surfaces in the contact.

The investigation of the contact conditions, the errors effect on them and the gears optimizing have difficulties because the contact lines move continuously and change shape during meshing. At the same time, the kinematic relations influenced by many design parameters and processing are also different during meshing. Another problem of the meshing research is real gearing, which is different from the theoretical model, such as a real worm gearing can't be treated like a perfect conjugated pair of wheels without errors and with rigid bodies. Solving these problems can only be done by developing mathematical and physical model and applying modern numerical methods and computer techniques. The mathematical model must include error components, consider deformations that occur under the load. This path was followed by the authors of [1].

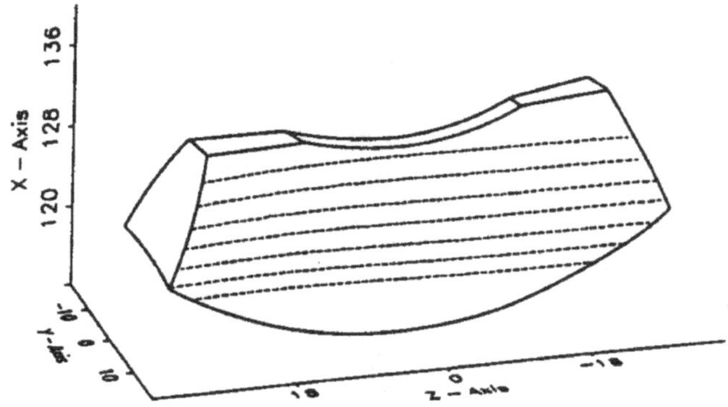


**Fig. 1.** Theoretical worm gearing contact [2]

Starting from the tool surface $r_{S1,2}$ equations, considering errors tool parameter and machine tools, considering the generation mechanism in accordance with the processing technology applied, these equations were obtained for teeth surfaces [1]:

$$r_1 = r_{10} + r_{1\Delta} = M_{1S} \cdot r_{S1}(u, \vartheta) = (M_{1S0} + M_{1S\Delta})[r_{1S0}(u, \vartheta) + r_{1S\Delta}(u, \vartheta, \Delta)] \quad (2)$$

$$r_2 = r_{20} + r_{2\Delta} = M_{2S} \cdot r_{S2}(v, \psi) = (M_{2S0} + M_{2S\Delta})[r_{2S0}(v, \psi) + r_{2S\Delta}(v, \psi, \Delta)] \quad (3)$$

and enveloped surfaces equations:

$$f_1(u, \vartheta, \Delta, \varphi_1) = n_1 \cdot v_1^{(S1)} = n_S \cdot v_S^{(S1)} = (n_{10} + n_{1\Delta})(v_{10}^{(S1)} + v_{1\Delta}^{(S1)}) = 0; \quad (4)$$

$$f_2(v, \psi, \Delta, \varphi_2) = n_2 \cdot v_2^{(S2)} = n_S \cdot v_S^{(S2)} = (n_{20} + n_{2\Delta})(v_{20}^{(S2)} + v_{2\Delta}^{(S2)}) = 0 \quad (5)$$

If in the Eqs. (2) and (3)

$$r_{1,2\Delta} = 0,$$

then the effect of the tool errors and the adjustment errors is cancelled, resulting theoretical teeth surfaces, $r_{1,20}$.

Obviously, that can make an analysis of the effect of the errors tool and the machine tools adjustment on the precision obtained teeth.

Also due to these errors, authors of [1] found it necessary to exclude the possibility of intersection of the contact surfaces. In the case of determining the contact points, the surfaces do not intersect if the sign of the reduced $\chi^{(p)}$ normal curvature remains the same throughout all normal intersections.

Therefore, at the Eqs. (2), (3), (4) and (5) adds the Eqs. (6) and (7) [1]:

$$f_3(u, \vartheta, v, \psi, \Delta_1, \Delta_2, \Delta_{1,2}, \varphi_1, \varphi_2 = i_{21}\varphi_1) = 0 \quad (6)$$

$$sign|\chi^{(p)}| = sign|\chi^{(1)} - \chi^{(2)}| = const. \quad (7)$$

To determine the contact between worm and worm wheel, a finite element computer system and a special algorithm were used. This algorithm allowed the assessment of contact pressure distribution and the effect of the error on the contact pattern. It is a complex approach that is a cause-effect analysis in the case of each error or a combination of errors, both worm and wheel.

## 3 Load Distribution Along the Line of Contact

The methodology for determining the load distribution along the line of contact between the worm tooth and wheel tooth is based on the following relation:

$$W(Z_D) = \int_{L_{it}} K_d(Z_D, Z_F) \cdot p(Z_F) \cdot dz + K_c(Z_D) \cdot p(Z_D) = \Delta y_n - e_n(Z_D) + s(Z_D) \quad (8)$$

where:
$L_{it}$ is the length of the line of contact between the worm tooth and wheel tooth;
$K_d(z_D, z_F)$ - influence factor of the load action on the tooth in the point F of the worm gearing tooth surface to the deformation of the worm tooth and wheel tooth there are in contact point D. Factor $K_d$ includes the bending and shear deformations of the worm tooth and wheel tooth, the bending and torsion deformations of the wheel, the axial deformation of the worm and support shaft deflections;
$K_c(z_D)$ - influence factor of the contact between the worm tooth and wheel tooth, including the contact deformation of the both teeth in the point D under the load action in the same point;
$p(z_F)$, $p(z_D)$ - loads on the tooth in the point F, respectively point D;
$w(z_D)$ - total deformation in the point D;
$\Delta y_n$ - projection of the wheel rotational delay to the worm on the normal to gearing teeth surface;
$e_n(z_D)$ - error in the point D, which is the sum of the processing and assembling errors;
$s(z_D)$ - relative geometric separation between worm tooth and wheel tooth in the point D.

The calculus relations of $K_d$, $K_c$ influences factors have been determined based on the analysis of the worm tooth and wheel tooth stresses, using finite element analysis.

The variables w, $e_n$ and s are measured along the normal to the conjugated surfaces. In the case of theoretical worm gearing, the contact is linear, so that the relative separation of the geometric surfaces of the worm tooth and wheel tooth is zero. But the unconjugated worm gearing, due to the introduction of changes to the surfaces of teeth or because processing cutter wheel with a diameter greater, its contact is punctual [3]. The investigations have shown that this contact transforms in a linear contact under load, but not along the whole width of the tooth wheel.

To be able to achieve a load distribution of the worm gearing with the geometric errors must know separations of the surfaces in the contact along these potential lines.

In [4] it is presented the method of determining the line with minimum separations on the contact surfaces, so this line is entirely or partially contact line. The method is based on minimizing function which determines the separation of the conjugated surfaces. The separation is defined as the distance between the points on the corresponding surfaces, obtained as points of the intersection of the common normal to the conjugated surfaces. Mathematically means minimizing function

$$s = \sqrt{(x_g^{(g)} - x_w^{(w)})^2 + (y_g^{(g)} - y_w^{(w)})^2 + (z_g^{(g)} - z_w^{(w)})^2}. \tag{9}$$

The torsion moment is given by the Eq. (10):

$$T = \sum_{i_t=1}^{i_t=N_t} \int_{L_{it}} r_F \cdot [\bar{p}(z_F) \cdot \bar{t}_{0F}] dz, \tag{10}$$

where:
$r_F$ is the distance from the load application point F on the wheel axis;
$\bar{t}_{0F}$ - unit vector tangent to the circle of radius $r_F$, passing through the point F in the transverse plane of the wheel;
$N_t$ - the number of teeth of the wheel are instantly in contact.

The distribution of the load on each line of contact is determined by solving nonlinear system of Eqs. (8), (9) and (10).

Some results of the program running developed in [3] are shown in the Fig. 2 which presents the distribution of the load on the meshing teeth at a time, in the case of the grinding worm.

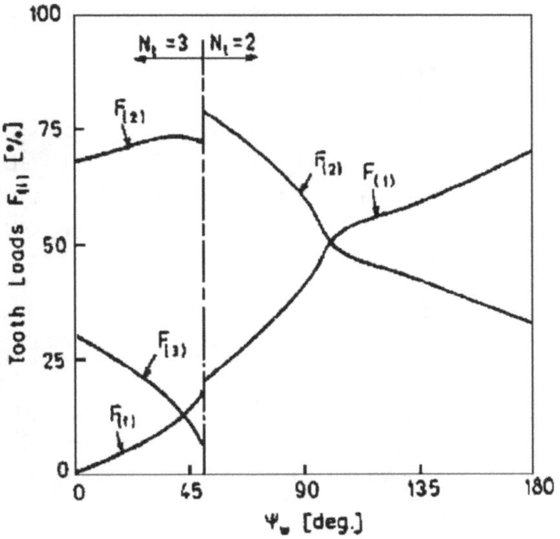


**Fig. 2.** The distribution of the load on the meshing teeth at a time [3]

As may be seen, during meshing, on the teeth are in contact, the load is distributed unequally and with discontinuities by changing the number of pairs of teeth are in the meshing. This will lead to a variable deformation of the teeth, the appearance of transmission error, which means the change of the transmission ratio.

Already in the field of the plan gearing gear were made important studies regarding reduction of these transmission errors. In this regard, there have been analyzes on the load distribution depending on the size of the load, the type of teeth (spur or bevel teeth), the width of the teeth, the position of a contact point (initial or final) [5]. In order to obtain a certain distribution of the load on the contact area, to ensure error reduction was realized a methodology for changing the profile teeth [5]. It was to develop a profile modification calculation maintaining constant stiffness of the teeth during meshing [6].

# 4 Gearing Tooth Stiffness Calculation

An algorithm for the evaluation of the worm gearing tooth stiffness has been developed [7] and it consists of:

1. Geometrical model of the worm;
2. Determination of the worm gear profile and geometry. The profile is obtained numerically by the discreting of helical surface with constant pitch.

The worm gear tooth surface is generated by rolling.

The enveloping condition is interpreted in "discrete way" by the "minimum distance method", applied in the case of "discrete representation" of the enveloping profiles. The "minimum distance method" was devised within the framework of the Machine-Building Technology Department, "Dunarea de Jos" University of Galati [8];

3. Determination of the enveloping profiles;
4. Determination of the contact ratio;
5. Calculus of one pair of elementary teeth stiffness by means of the relation for springs serially connected, considering the conjugated teeth are two springs (see bases of design). Thus, if $k_1$ is the stiffness of elementary worm tooth and $k_2$ is the stiffness of elementary wheel tooth, then the stiffness in the contact point will be given by the following relation:

$$\frac{1}{k} = \frac{1}{k_1} + \frac{1}{k_2} \tag{11}$$

6. Calculus of elementary gearing tooth stiffness (pinion-rack drive) by means of the relation for springs simultaneously (parallel) connected, considering those springs are materialized by the teeth which there are in the meshing at the same time:

$$k_H = \sum_{i=1}^{n} k_i \tag{12}$$

7. Calculus of worm-gearing tooth stiffness summing up the stiffness of the all elementary gear-tooth systems.

Based on presented algorithm, an original and special computing program for determining of the stiffness for any kind of the worm-gearing tooth system or for spur gearing and bevel gearing was developed. In addition, based on modeling and simulation, many studies can be performed for improving the performance of the gears.

## 4.1 Bases of Design

The mathematical model is based on the following bases of design [7]:

- It is assumed that worm-gearing is errors free and the gears are rigid except the teeth;

- It is taken into consideration only the bending produced by the meshing normal force;
- It is considered that the worm-gearing consists of more plane-gear drives (pinion-rack drives), named "elementary gear drives", that in fact are cross sections perpendicular to axis of rotation of the worm-gear (Fig. 3);
- The tooth of the elementary gear drive is considered to be a beam fixed at one end in the body of gear;
- The assembly of the plane-gear drives into the worm-gear drive was made provided that the teeth of the elementary gear drives to deform together and not separately under the same load.

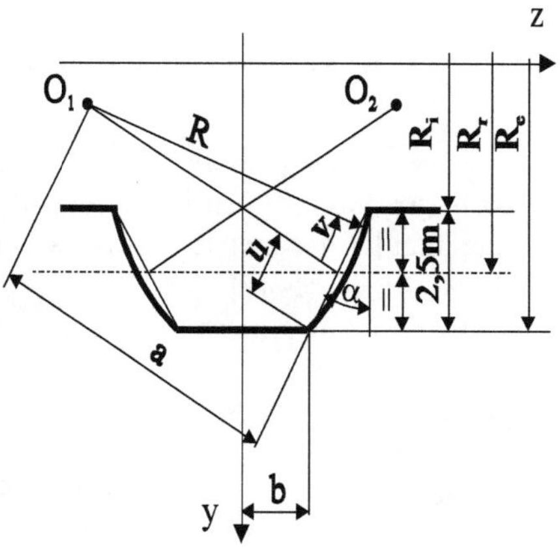


**Fig. 3.** Arch profile of worm in the axial section (X = 0) [7]

**Numerical Results and Simulations.** In this paper it is presented the numerical results performed for a cylindrical worm-gear drive having arch profile with the following parameters (Fig. 4):

- number of worm threads: $z_1 = 1$;
- number of wheel teeth: $z_2 = 53$;
- axial module: $m_x = 10$ mm;
- diametral quotient: $q = 10$;
- constructive parameter: $a = 70$ mm;
- angular increment: $\Delta\varphi = \pi/3240$;
- profile angle: $\alpha = 20°$.

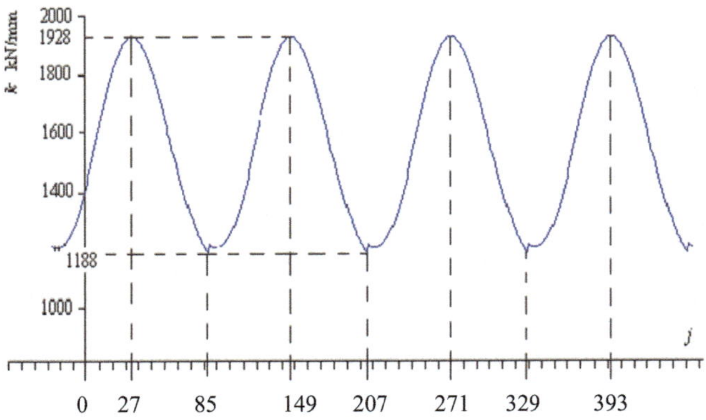


**Fig. 4.** Elastic characteristic of the worm-gearing tooth [7]

*Elasticity Characteristic of Worm-Gearing Tooth.* The elasticity characteristic represents the variation of the worm-gearing tooth stiffness depending on the rolling angle ($j \cdot \Delta\varphi$), where "j" is the rolling angular parameter. It is a cvasisinusoidal curve with the high jumps when a tooth binds or recesses [7].

The maxim stiffness is 1928.25 KN/mm at j = 27, j = 149, j = 271 and so on (j = 122k, k = 0, 1, 2… [1]). The minim stiffness is 1188.626 KN/mm at j = 85, j = 207 and so on. The amplitude of the stiffness variation is 739.62 KN/mm, the jump frequency is determined by the j = 122k.

In order to perform a complete research on elastic system, for example any kind of the gear drive, the elasticity characteristic must be studied together the static characteristic and frequency characteristic.

It is known that, the static characteristic (Fig. 5) represents the applied force in function of the deformation.

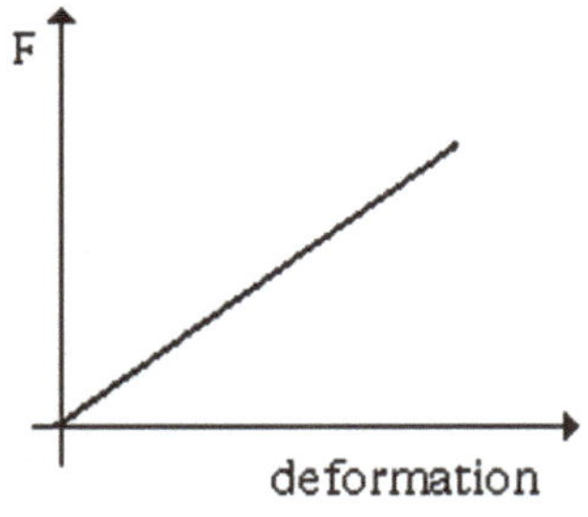


**Fig. 5.** Static characteristic

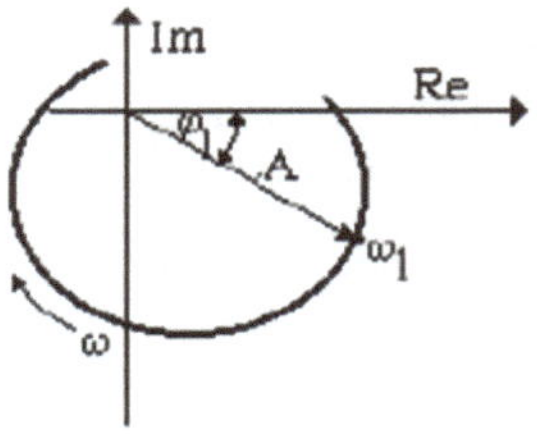


$$A(\omega) = \sqrt{\mathrm{Re}^2(\omega) + \mathrm{Im}^2(\omega)};$$
$$\varphi(\omega) = arctg(\mathrm{Im}/\mathrm{Re}).$$

**Fig. 6.** Frequency characteristic

The information about the dynamic parameters of the elastic system and about the stability ratio is given by the frequency characteristic (Fig. 6).

## 5 Evaluating Gearing Stiffness Using Contact Analysis

In [9], to examine gear mesh stiffness, it is assumed that the gears are elastic bodies and then the contact between them is modeled. The paper's author performed a stationary parametric analysis to determine the mesh stiffness of the gears for different positions in a mesh cycle. A mesh cycle is defined as the amount of gear rotation after which the next tooth takes the position of the first one.

Using the penalty contact approach, the paper's author modeled the contact between the teeth of the two gears. The boundaries of the two gears in contact with each other are shown below (Fig. 7).

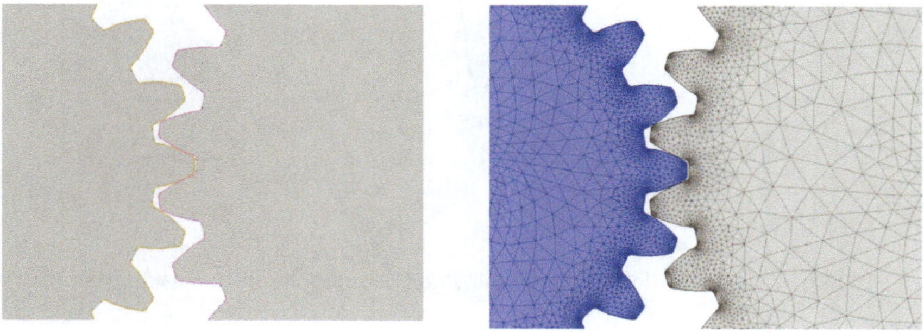

**Fig. 7.** The contact pair boundaries and the finite element mesh in the gear pair [9]

The wheel is given a twist, $\theta_t$, and the required twisting moment, $T$, is evaluated on the hinge joint. Hence, the torsional stiffness of the gear pair is prescribed as:

$$k_t = T/\theta_t \tag{13}$$

Once we know the torsional stiffness, we can define the stiffness along the line of action as:

$$k_g = 4k_t/\left(d_{pw}cos\alpha\right)^2 \tag{14}$$

where $d_{pw}$ is the pitch diameter of the wheel and $\alpha$ is the pressure angle.

Figure 8 shows the variation of computed gear mesh stiffness with the rotation of the pinion for two mesh cycles. We can see that the gear mesh stiffness is periodic in each mesh cycle as well as across multiple mesh cycles, increasing in the beginning and then later decreasing. This happens due to the changing contact ratio. In the beginning of a mesh cycle, the contact ratio increases from 1 to 2, but then drops back down to 1.

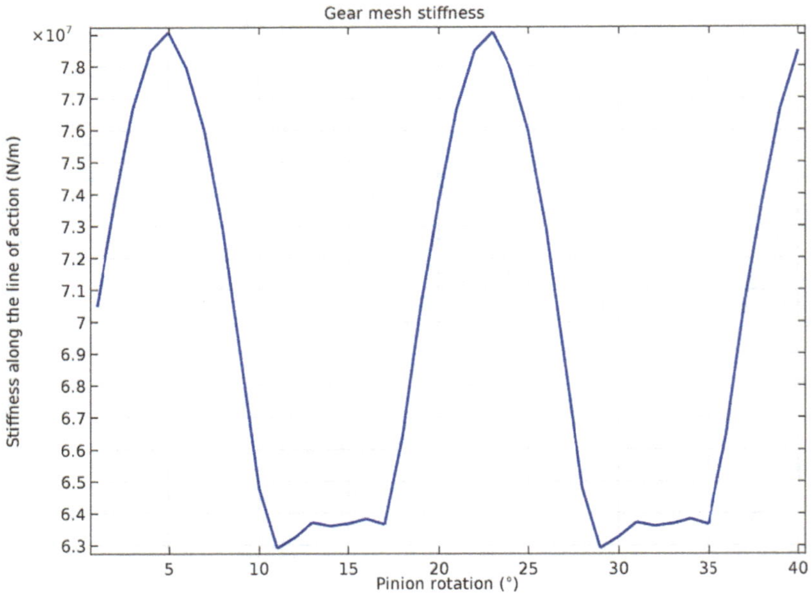


**Fig. 8.** The variation of gear mesh stiffness with the pinion rotation [9]

In the paper [9] is also developed a study about the dependence of gearing stiffness on several parameters (such as: tooth parameters, geometrical and material parameters) to improve the performance of the gears.

## 6 Conclusions

As the result of investigation above, we can make the following conclusions:

1) Two methods to evaluate the stiffness of gearing tooth were presented;
2) The first proposed approach may be applied for any types of cylindrical worm gearing, for spur gearing and bevel gearing;
3) Both methods developed enable to obtain numerical solutions and simulations;
4) The proposed methods, carrying out simulations of the meshing, allow the geometry optimization and study of the meshing for various geometrical parameters of the gear drives;
5) But, more important is the fact that the parameters which influence the improvement of stiffness gearing tooth are determined;
6) The introduction of "elasticity characteristic" concept contributes to the completion of study for the used gearing;
7) his study leads to increase of the accuracy of the mechanisms, machine-tools or robots.

## References

1. Bercsey, T.: Error analysis of worm gear pairs. In: Proceeding of 4th Word Congress on Gearing and Power Transmission, Paris, pp. 419–424 (1999)
2. Seol, I.H.: Computerized design, generation and simulation of meshing and contact of worm-gear drives with improved geometry. Comput. Methods Appl. Mech. Eng. **138**, 73–103 (1996)
3. Simon, V.: Characteristics of a new type of cylindrical gear drive. In: Proceeding of Power Transmission and Gearing Conference ASME. DE, San Diego, California, vol. 88, pp. 133–140 (1996)
4. Simon, V.: Tooth contact analysis of mismatched worm gears. In: Proceeding of 9th Word Congress on Theory of Machines and Mechanisms, Milano, Italia, pp. 530–534 (1995)
5. Regalado, I: Profile modifications for minimum static transmission error in cylindrical gears. In: Design Engineering Technical Conference ASME, DETC/PTG-5781, Atlanta, 13–16 September 1998
6. Chun, L.: Automatic design method of minimizing gear transmission error with torque insensitivity. M.S. thesis, Graduate School The Ohio State University (1995)
7. Ghelase, D.: Modelling and assisted design of the worm-gearing. In: proceeding of System Science and Simulation in Engineering, Japan, pp. 161–165, 4–6 October 2010
8. Oancea, N.: Metode numerice pentru profilarea sculelor, vol. II. "Dunarea de Jos" University of galati, Romania (1992)
9. Soami, P.: How to Evaluate Gear Mesh Stiffness in a Multy Dynamics Model. https://www.comsol.com/blogs/how-to-evaluate-gear-mesh-stiffness-in-a-multibody-dynamics-model/. Accessed 9 Dec 2016

# Tolerance Analysis of an Over-Constrained Assembly with Press-Fit Solderless Electric Contact Pins

Mihai Drienovsky(✉), Arjana Davidescu(✉), and Dmytro Rozputniak

Politehnica University of Timisoara, Timişoara, Romania
mihai.drienovsky2@student.upt.ro,
arjana.davidescu@upt.ro

**Abstract.** Bearing in mind current lead-free regulations, solder-less press-fit technology has proven itself to be of use and interest owing to its low-cost assembly process, as far as automotive products are concerned.

Usually, cost effective solutions stand in need for over-constrained type assemblies. Whilst there is abundant research in the field of over-constrained assemblies, worst case calculation remains the preferred method for evaluating design decisions, due to its easy-to-apply nature. The objective of this paper is to highlight the benefits of avoiding overengineering when using an easy-to-program method that implements elastic deformation of the components in comparison to a worst-case calculation. Applying the two methods on a multiple pin solderless press-fit assembly, shows that worst case calculation would overestimate the variation by 150% in comparison to a more detailed analysis.

**Keywords:** Tolerance analysis · Over-constrained · Interference · Flexible assembly

## 1 Introduction

In an industrial context, where assemblies are manufactured on a large scale, design architecture should accommodate for component manufacturing and assembling achievable tolerances. Simulating mechanical system behavior as a result of interaction between parts with defects is done through tolerance analysis.

When selecting design architecture, a go to method, for determining design robustness, for given process tolerances, is the Worst-Case calculation. Due to its easy to learn and apply workflow, this method would be replaced by more accurate calculation methods only when there are no architecture alternatives that fulfill design functionality under this evaluation. This type of engineering workflow may have left some cost-effective design alternatives not investigated only due to lack of time and know-how of alternative analysis methods. One potential group of assembly's which may prove to produce this type of cost-effective architectures are the over-constrained assemblies.

A large number of studies have been conducted to approach the subject of the over-constrained assembly summarized by Cao et al. [1] and Homri et al. [2]. Different models were developed that can study also flexible assemblies, for example:


E.-C. Lovasz et al. (Eds.): MTM&Robotics 2020, MMS 88, pp. 324–337, 2021.
https://doi.org/10.1007/978-3-030-60076-1_29

Tolerance-maps [3], Skin model [4, 5], Vector loop models [6–9], Polytope based models [10], worth mentioning Teissander and Delos developed a software tool PolitoCAT and Politopix that allows tolerance analysis based on polytopes [11–13].

At the same time, there are studies that help analysis of multiple pin-hole assembly, from both modeling and measuring point of view. Load distribution has been studied experimentally and analytically on multi-bolt assemblies [14–18]. Li et al. [19] proposes a global sensitivity model that helps rank sources of uncertainty in a multi-bolt composite joint. Mounaud et al. presents a model to analyze assembly sequence impact on final assembly results [20]. Li et al. proposes a model to analyze deviation propagation using linear combination of part defects [21]. Schlather et al. presents a model for analyzing assembly that combines FEA and tolerance analysis using Monte Carlo Simulation [22]. Yumbla et al. creates a database of different cable connectors, maximum movement between the female and male connectors, by accurately measuring the tolerance [23]. Gouyou et al. proposed a method based on polytope analysis of multiple pin-hole assembly, where the interferences are identified, and deformations are simulated in a FEA tool [24].

It can be observed that all these studies deal with different aspects concerning multiple pin-hole over constrained assembly, which tolerance analysis methods to use, how to integrate FEA results into the tolerance analysis and how to best represent the data. For a particular case of the pin-hole assembly group, that is the press-fit technology in electrical connections, the Worst-Case calculation method used to choose between design architectures remains unchallenged.

Starting with 2006, RoHS (Restriction of Hazardous Substances Directive) has announced the installment of an advanced electronic equipment that will be lead-free [25]. This means that solder alloys would have to be completely free of lead and solder melting point will rise, having a great impact on the mounting process. A possible solution for solder-less mounting of power electronics is press-fit technology.

Press-fit technology for electrical connection [26, 27] is not a new idea. Integrating it into automotive industry requires it to be almost always assembled in an over-constrained type of assembly, that includes elastic and plastic deformation of the pins.

The high growing usage of this type of electrical connections in new designs arises the need for engineers to evaluate such assemblies with easy to apply and accurate methods.

Tolerance analysis of an over-constrained assembly is not an easy task, this is mainly because of the complexity of mathematical tools and methods that need to be deployed.

This paper presents the benefits in comparison with the challenges that an accurate tolerance analysis of an over-constrained assembly would bring. This is done by comparing the results from a Worst-Case type of calculation and a vector loop model that accounts for elastic and plastic deformation to analyze an assembly composed of compliant parts. For the presented case study, it is proposed to analyze the impact of press-fit type assembly on position and orientation variation of an IC-sensor (integrated circuit sensor) relative to the BLDC-motor (brushless direct current motor).

# 2 The Definition of the Tolerance Chain

## 2.1 Assembly Description

Due to its principle of functioning BLDC-motor (brushless direct current motor) requires constant information on rotor angular position so that the proper coils are energized. This information can be obtained through several methods, some of these methods require a sensor that detects the angular position (examples may include optical and Hall effect) [28], this means that the final design of a BLDC should incorporate such a sensor, usually the sensor is a standard IC-sensor (integrated circuit sensor). Performing position and orientation tolerance analysis simulation of such sensor relative to the BLDC motor assembly would be a prerequisite for evaluating design robustness.

The combination of RoHS regulation with the constant drive to lower cost, part count and manufacturing optimization have led to a design proposal "Fig. 1 shows the design to be analyzed from tolerance point of view". To easily convey the idea, rotor is not represented in the following illustrations, and is considered to be centered relative to the stator.

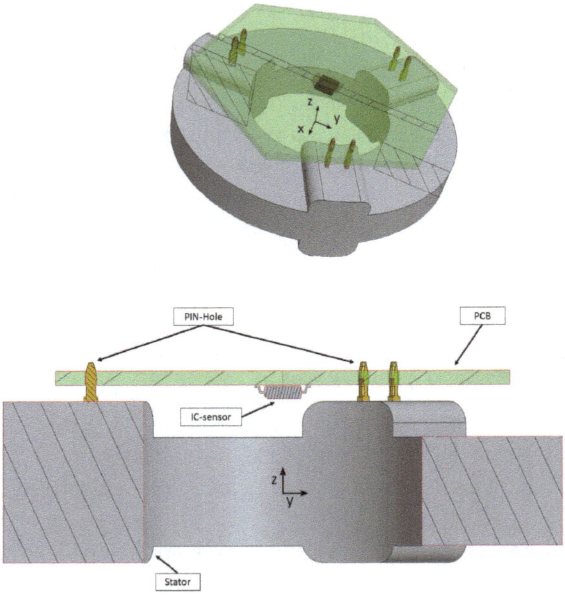


**Fig. 1.** Assembly: IC-sensor; PCB; press-fit Pin-Hole; Stator.

## 2.2 Tolerance Chain Description

The IC-sensor is a component mounted on the PCB, positioned so that it may read the rotor's orientation. The PCB is mechanically fixed to the stator component through six press-fit pins, that also provide the electrical connection from the supply of electrical energy to the coils. The pins are welded to the coil winding and an over-molding process fixes the pins into position. The pins are acting also as a mechanical structure, supporting and centering the PCB relative to the stator.

The IC-PCB-Stator tolerance chain would show how the individual defects from the components, contribute to the IC-sensor and BLDC-rotor relative position misalignment.

In order to solve the tolerance chain model, a simplifying hypothesis was assumed for both methods, which involves reducing the assembly to X-Y 2D plane, thus neglecting the depth of the assembly.

## 2.3 Relevant Positions and Tolerance Definition

The studied assembly has the positions and tolerances of the binding features presented in "Table 1 showing pin position tolerances", "Table 2 showing hole position tolerances", "Table 3 showing IC position tolerances". All input tolerances are considered to belong to a uniform distribution.

**Table 1.** Pin position relative to the stator

| Pin no. | X [mm] Position | Y [mm] Position | ±X [mm] Tolerance | ±Y [mm] Tolerance |
|---|---|---|---|---|
| 1 | 2.50 | 27.39 | ±0.10 | ±0.10 |
| 2 | 24.97 | −11.53 | ±0.10 | ±0.10 |
| 3 | 22.47 | −15.86 | ±0.10 | ±0.10 |
| 4 | −22.47 | −15.86 | ±0.10 | ±0.10 |
| 5 | −24.97 | −11.53 | ±0.10 | ±0.10 |
| 6 | −2.50 | 27.39 | ±0.10 | ±0.10 |

**Table 2.** Hole position relative to the PCB

| Hole no. | X [mm] Position | Y [mm] Position | ±X [mm] Tolerance | ±Y [mm] Tolerance |
|---|---|---|---|---|
| 1 | 2.50 | 27.39 | ±0.10 | ±0.10 |
| 2 | 24.97 | −11.53 | ±0.10 | ±0.10 |
| 3 | 22.47 | −15.86 | ±0.10 | ±0.10 |
| 4 | −22.47 | −15.86 | ±0.10 | ±0.10 |
| 5 | −24.97 | −11.53 | ±0.10 | ±0.10 |
| 6 | −2.50 | 27.39 | ±0.10 | ±0.10 |

**Table 3.** IC sensor position and angle relative to the PCB

| Sensor | X [mm] Position | Y [mm] Position | ±X [mm] Tolerance | ±Y [mm] Tolerance | Nominal angle | Angular [°] Tolerance |
|---|---|---|---|---|---|---|
| | 0.00 | 0.00 | ±0.15 | ±0.15 | 0.00 | ±1.00° |

## 3 Worst-Case Calculation (WC)

Due to the symmetrical nature of the problem, all pins and holes are equally distant in nominal position, from the center of the stator and all the pins have the same tolerance range as do the holes "Table 1 and 2".

If all the pins have the same tolerance range in one direction, for the presented case of ±0.1 mm, the maximum deviation of the stator would be 0.1 mm in one direction and 0.1 mm in another direction, if all pins would move to their most extreme points in the tolerance range. The assumption that the pins and holes move as two separate groups transforms the over constrained problem in to a simple 2D deviation stack-up. "Figure 2 shows the Worst-Case tolerance chain, where the groups of pins is represented by the position vector $P_{Stator}$ and holes by $P_{PCB}$".

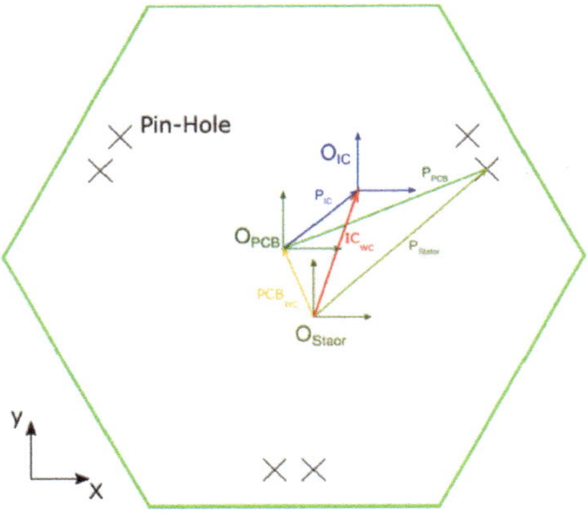


**Fig. 2.** Worst-case method reduces the model to a simple deviation stack up.

Cumulating the deviations "Eq. (1) will result in ±0.35 mm worst case position deviation of the IC relative to the stator", "Eq. (2) will result in ±0.2 mm worst case position deviation of the PCB relative to the stator".

$$IC_{WC} = P_{IC} + P_{PCB} + P_{Stator} \tag{1}$$

$$PCB_{WC} = P_{PCB} + P_{Stator} \tag{2}$$

Worst case angle deviation of the stator could also be obtained by moving all pins to their most extreme positions. Using "Eq. (3) an angular deviation ±0.21° is obtained for the Stator and the PCB". Cumulating the angular deviations "Eqs. (1) and (2)" will result in a deviation of ±1.42° for the IC and ±0.42° for the PCB relative to the Stator (Table 4).

$$Angle = arcsin\left(\frac{Deviation}{Radius}\right) \tag{3}$$

**Table 4.** Worst-case calculation results for the PCB and IC-sensor relative to the Stator

| | X [mm] | Y [mm] | Angle [°] |
|---|---|---|---|
| $\pm WC_{PCB}$ | ±0.200 | ±0.200 | ±0.416 |
| $2 * \lvert \pm WC_{PCB} \rvert$ | 0.400 | 0.400 | 0.832 |
| $\pm WC_{IC}$ | ±0.350 | ±0.350 | ±1.416 |
| $2 * \lvert \pm WC_{IC} \rvert$ | 0.700 | 0.700 | 2.832 |

# 4 Deviation Analysis with Monte Carlo Simulation

## 4.1 Over-Constrained Virtual Spring (VS) Model Definition

It can be observed that the over-constrained behavior of the assembly comes from the six pins that are fixed to the stator and are always press-fitted into the PCB holes. This type of assembly forces the pins to pull one another, until the PCB reaches an equilibrium point relative to the stator, given by the "Eq. (1) sum of all forces equal 0" and "Eq. (2) sum of all torques equal 0".

$$\sum \vec{F} = \vec{0} \tag{4}$$

$$\sum \vec{\tau} = \vec{0} \tag{5}$$

Because the pins are press-fitted into the PCB, there is no play between the parts. For simplification, it has been considered that the pin symmetry planes will align to the center of the hole due to symmetrical forces that are coming out of the press fit. The entire assembly is considered to be rigid, while the pins were allowed to deform.

All above mentioned conditions will reduce the pin-hole connection to a virtual spring model in the 2D plane connecting the center of the hole with the center of the pin, "Fig. 3 illustrates a random position of the PCB, highlighting the displacement vectors $\overrightarrow{\Delta x}$ for all pin-hole couples, which generate the PCB positioning forces".

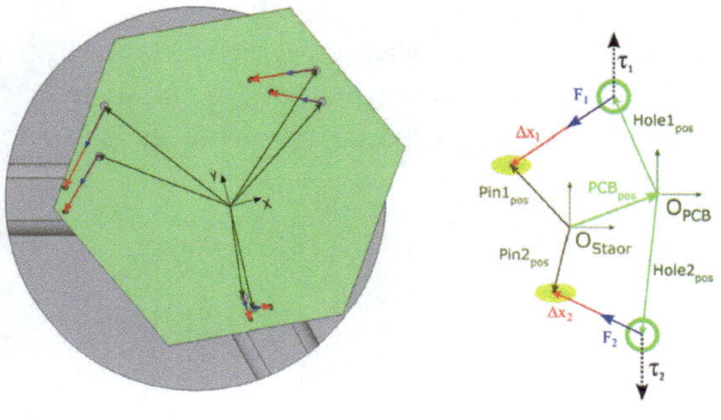


**Fig. 3.** Shows $\overrightarrow{\Delta x}$ as a distance between pin and hole position

In this model, only the elastic force is considered for each pin-hole couple "Eq. (6)" and "Eq. (7)", frictions and other contact phenomenon are not modeled.

$$\vec{F} = f\left(\overrightarrow{\Delta x}\right) \tag{6}$$

$$\overrightarrow{\Delta x} = \overrightarrow{Pin_{pos}} - \overrightarrow{Hole_{pos}} \tag{7}$$

The elastic force function $f(\Delta x)$ is defined as a transfer function coming from FEA analysis of the pin reaction force when displacement is applied as shown in "Eq. (8)" linear interpolation of "Table 5 that shows pin reaction force for a certain displacement".

$$f(Disp) = Force(Disp) \tag{8}$$

Torque generated by each pin-hole couple, acting on the PCB is calculated as follows:

$$\vec{\tau} = \overrightarrow{Hole_{pos}} \times \vec{F} \tag{9}$$

A program was developed to determine the PCB position in the 2D plane, relative to the stator position. The program finds the equilibrium point with a precision of $10^{-4}$ for position and $10^{-5}$ for angles. The simulation program was written in Processing sketchbook [29], which is open source.

## 4.2 Definition of Numerical Model

**Finite Element Analysis (FEA) Boundary Conditions.** During the assembly process, due to the misalignment tolerances between the pins and the hole, the pins will be subjected to a bending phenomenon.

For the bending process of the pin, first assumption was that a node displacement would be most representative model for the behavior where all pins would be pulled by the PCB holes. Second assumption was made that the pin will be fixed in the stator assembly, therefore a fix support was applied at the bottom of the pin "Fig. 4 shows simulation constraints and graphical representation of the results".

The material used in the simulation was considered as bilinear, in order to adjust the model according to the plastic deforming property of the material "Fig. 4 shows the outcome of using bilinear behavior of the material with the curve that has a changing slope angle".

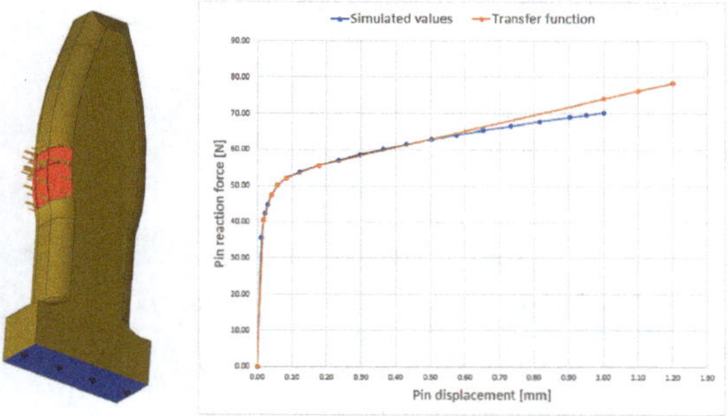


**Fig. 4.** Simulation conditions and simulation results.

**FEA Results.** "Table 5 presents values to create the transfer function, based on the FEA simulation results that show force values required to produce different amount of pin displacement".

**Table 5.** Transfer function used data.

| Displacement [mm] | Force [N] |
|---|---|
| 0.000000 | 0.000 |
| 0.016313 | 40.438 |
| 0.040359 | 47.422 |
| 0.057445 | 50.062 |
| 0.083074 | 52.117 |
| 0.177530 | 55.561 |
| 1.000000 | 74.000 |
| 1.100000 | 76.242 |

## 5 Results from Monte Carlo Simulation

Monte Carlo simulation is a frequently used method in the scientific domain for statistical analysis of data, mainly due to its ease of use. The method suggests creating a large sample data, by generating random input values from predefined statistical distributions (in this case the individual part tolerances). Pass the input data through a mathematical model that best represents the problem at hand (in this case the PCB position is determined by finding the equilibrium point resulted from the virtual spring forces). The final step is to analyze the output data with statistical tools (in this case PCB positions and orientation of all the individual samples are being plotted into histograms and evaluated)

After evaluating more than 10 000 virtual samples, the following results were obtained.

### 5.1 Position Tolerances of PCB Relative to Stator

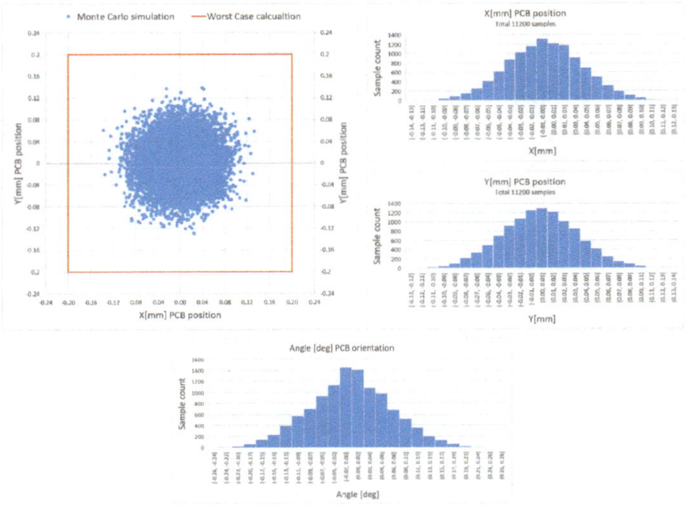


**Fig. 5.** PCB position and orientation statistical distribution results.

Data from "Fig. 5 where statistical results of the PCB position relative to the stator are shown compared with the worst-case calculation" suggests that the PCB probabilistic distribution of positions and orientations tends towards normal distribution rather than uniform, in spite of the input data (pin and hole positions) being of uniform distribution, this can be explained through the Central Limit Theorem. "Table 6 shows a statistical analysis of the MC simulation".

### 5.2 Position Tolerances of IC Relative to Stator

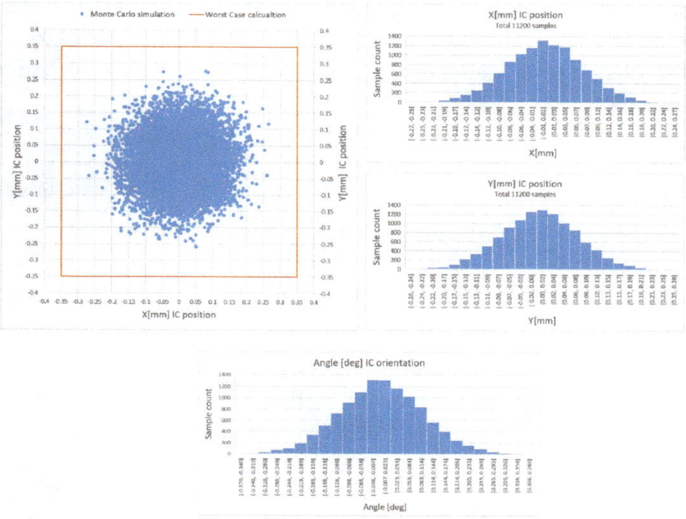


**Fig. 6.** IC position and orientation statistical distribution results.

Data from "Fig. 5 where statistical results of the IC position relative to the stator are shown compared with the worst-case calculation" suggests that the IC probabilistic distribution of positions and orientations tends towards normal distribution rather than uniform, in spite of the input data (pin and hole positions) being of uniform distribution, this can be explained through the Central Limit Theorem. "Table 6 shows a statistical analysis of the MC simulation".

### 5.3 Comparing Results

Comparing tolerance evaluation resulted from Virtual Spring Model to Worst-Case Model, suggests that the real assembly would behave in an improving way from statistical point of view.

To give an approximate value by how many times the Worst-Case Model would overestimate the real behavior, "Eq. (10) and (11)" are proposed as measuring method and the real behavior is considered to be best represented by the Virtual Spring Model.

It is proposed to study the two model's performance through an overestimated degree ε shown in "Eq. (10) and (11)":

$$\varepsilon = \frac{WorstCase}{VirtualSpring} \tag{10}$$

$$\varepsilon = \frac{2 * |\pm WC|}{Maximum\langle (Max - Min); (6 * \sigma) \rangle} \tag{11}$$

"Equation (11)" is made so that it shows the smaller overestimation degree to prevent display of false benefits from the assembly.

**PCB Position and Orientation Distribution.** From the data, "Fig. 5" and "Table 6", it is obtained an $\varepsilon = 1.49$ for positions and an $\varepsilon = 1.54$ for orientation, showing that the worst-case calculation will overestimate by almost 1.5 times the error cumulation that will occur in reality.

**Table 6.** PCB position statistical evaluation and WC (worst case results)

| Statistic/WC info | X [mm] | Y [mm] | Angle [°] |
|---|---|---|---|
| Max | 0.133 | 0.138 | 0.279 |
| Min | −0.136 | −0.130 | −0.261 |
| μ | 0.000 | 0.001 | −0.003 |
| σ | 0.037 | 0.037 | 0.074 |
| ±3σ | 0.111 | 0.112 | 0.222 |
| 6σ | 0.223 | 0.224 | 0.445 |
| ±WC | ±0.200 | ±0.200 | ±0.416 |
| 2 * \|±WC\| | 0.400 | 0.400 | 0.832 |

**IC Position and Orientation Distribution.** From the data, "Fig. 6" and "Table 7", it is obtained a $\varepsilon = 1.3$ for positions and $\varepsilon = 3.74$ for orientation, showing that the worst-case calculation will overestimate by almost 1.3 times the error cumulation that will occur in reality.

**Table 7.** IC position statistical evaluation and WC (worst case results)

| Statistic/WC info | X [mm] | Y [mm] | Angle [°] |
|---|---|---|---|
| Max | 0.266 | 0.276 | 0.386 |
| Min | −0.273 | −0.260 | −0.370 |
| μ | 0.001 | 0.003 | 0.000 |
| σ | 0.074 | 0.075 | 0.105 |
| ±3σ | 0.223 | 0.224 | 0.316 |
| 6σ | 0.446 | 0.449 | 0.632 |
| ±WC | ±0.350 | ±0.350 | ±1.416 |
| 2 * \|±WC\| | 0.700 | 0.700 | 2.832 |

The overestimated degree changes from $\varepsilon = 1.49$ to $\varepsilon = 1.3$ when the tolerance chain took in consideration also the contribution of the IC-sensor position variation, suggesting that the overestimated degree will be much closer to 1 for bigger stack-up chains, and above 1, when including over-constrained assemblies in the tolerance chain. This shows that the beneficial impact of an over-constrained assembly is reduced with the increase of the tolerance chain.

Please keep in mind that the above-mentioned results are obtained despite efforts of getting the overestimation degree as close as possible to 1, by making "Eq. 11" with the biggest possible denominator and using uniformly distributed input data instead of normal distribution which a standard approach in tolerance calculation.

## 6 Conclusions

The cost benefits of such an assembly reside in the idea that the function of mechanical fixing, structural rigidity, electrical contact and conductivity, can be fulfilled by the same individual components. Although there are cost benefits to such an assembly, the mechanical behavior at first glance looks uncertain if it is improved or worsened in comparison to other architectures, especially regarding the tolerance chain.

Based on the Worst-Case calculation, chances are that the presented design architecture would have been rejected for something more expensive that allowed the sensor to have better position tolerances relative to the BLDC-motor. Applying a little more complex calculation, may have saved designers some costs that come with overengineering.

In all the tolerance studies a high level of simplification occurs, allowing for different software's in assisting the designer to perform the tolerance analysis. For more complex architectures, where over-constrained behavior occurs, lack of a general solution, forces the designers to overcome the challenge by creating and applying methods and tools only to a particular problem. Without the analysis of the system as a whole and in large sample sizes, where it can be observed how the elements influence one another, through emerging probabilistic patterns, the final results may be ambiguous.

The worst-case method calculation, seen in chapter 3, was presented as a reference, because when it comes to fast and cost-effective evaluations it is the most widely used method for evaluating tolerances in engineering. Nonetheless, applying it to this problem was easy because of the many symmetrical properties that the problem had. Application of the method in more asymmetric circumstances would prove difficult to apply and may provide unreliable information.

This paper's contribution is that of clearly highlighting the benefits of investing more energy and time in building or using a tool for a more detailed over-constrained tolerance calculation when faced with a design change decision. Although anyone can argue that the virtual spring model is far from a perfect representation of the real behavior, the $\varepsilon = 1.5$ times overestimation of the worst-case model suggests that higher costs and overengineering would burden the design to achieve smaller tolerances, if those decisions were based only on a worst-case calculation. It should be kept in mind, that the high overestimation value was obtained, despite all efforts of getting it as close as possible to 1.

From the results of the present study, it can be stated that more research of the over-constrained assembly's may be rewarded with robust cost-effective design, as their system like behavior may tend to reduce the overall defect cumulation. The next steps for tolerance study of press-fit assemblies would be to analyze what happens if the assembly becomes more complex with more parts included, parts that would have different elastic behaviors. An interesting topic would be to find if there is a transfer function that allows applying of Worst-Case calculation and correcting the result with an adjustment coefficient.

## References


1. Cao, Y., Liu, T., Yang, J.: A comprehensive review of tolerance analysis models. Int. J. Adv. Manuf. Technol. **97**(5–8), 1–31 (2018)
2. Homri, L., Dantan, J.-Y., Levasseur, G.: Comparison of optimization techniques in a tolerance analysis approach considering form defects (2016)
3. Jaishankar, L.N., Davidson, J.K., Shah, J.J.: Tolerance analysis of parallel assemblies using tolerance-maps® and a functional map derived from induced deformations. In: International Design Engineering Technical Conferences and Computers and Information in Engineering Conference, vol. 55898. American Society of Mechanical Engineers (2013)
4. Schleich, B., Wartzack, S.: Approaches for the assembly simulation of skin model shapes. Comput. Aided Des. **65**, 18–33 (2015)
5. Yan, X., Ballu, A.: Tolerance analysis using skin model shapes and linear complementarity conditions. J. Manuf. Syst. **48**, 140–156 (2018)
6. Chase, K.W.: Tolerance analysis of 2-d and 3-d assemblies. ADCATS Report 94.4 (1999)
7. Jiang, Y., et al.: Kinematic error modeling and identification of the over-constrained parallel kinematic machine. Robot. Comput. Integr. Manuf. **49**, 105–119 (2018)
8. Cammarata, A.: A novel method to determine position and orientation errors in clearance-affected overconstrained mechanisms. Mech. Mach. Theory **118**, 247–264 (2017)
9. Bedoustani Babazadeh, Y., Taghirad, H.D., Aref, M.M.: Dynamics analysis of a redundant parallel manipulator driven by elastic cables. In: 2008 10th International Conference on Control, Automation, Robotics and Vision. IEEE (2008)
10. Gouyou, D., et al.: Statistical tolerance analysis applied on overconstrained mechanisms with form deviations. J. Comput. Design Eng. **7**, 308–322 (2020)
11. Teissandier, D., Couetard, Y., Delos, V.: Operations on polytopes: application to tolerance analysis. In: Global Consistency Of Tolerances, pp. 425–434. Springer, Dordrecht (1999)
12. Delos, V., Teissandier, D.: Minkowski sum of polytopes defined by their vertices. arXiv preprint arXiv:1412.2564 (2014)
13. Ledoux, Y., Teissandier, D., Delos, V.: Fast analysis of compliant assembly. Procedia CIRP **70**(1), 138–143 (2018)
14. Lecomte, J., et al.: An analytical model for the prediction of load distribution in multi-bolt composite joints including hole-location errors. Compos. Struct. **117**, 354–361 (2014)
15. McCarthy, M.A., Lawlor, V.P., Stanley, W.F.: An experimental study of bolt-hole clearance effects in single-lap, multibolt composite joints. J. Compos. Mater. **39**(9), 799–825 (2005)
16. Xiong, Y.: An analytical method for failure prediction of multi-fastener composite joints. Int. J. Solids Struct. **33**(29), 4395–4409 (1996)
17. Fan, W.-X., Qiu, C.-T.: Load distribution of multi-fastener laminated composite joints. Int. J. Solids Struct. **30**(21), 3013–3023 (1993)
18. McCarthy, C.T., McCarthy, M.A., Lawlor, V.P.: Progressive damage analysis of multi-bolt composite joints with variable bolt–hole clearances. Compos. B Eng. **36**(4), 290–305 (2005)
19. Li, H.-S., Gu, R.-J., Zhao, X.: Global sensitivity analysis of load distribution and displacement in multi-bolt composite joints. Compos. B Eng. **116**, 200–210 (2017)
20. Mounaud, M., et al.: Assembly sequence influence on geometric deviations propagation of compliant parts. Int. J. Prod. Res. **49**(4), 1021–1043 (2011)

21. Li, Y., Zhao, Y., Yu, H., Lai, X.: Modeling deviation propagation of compliant assembly considering form defects based on basic deviation fields. Assembly Autom. **39**(1), 226–242 (2019)
22. Schlather, F., et al.: Tolerance analysis of compliant, feature-based sheet metal structures for fixtureless assembly. J. Manuf. Syst. **49**, 25–35 (2018)
23. Yumbla, F., et al.: Tolerance dataset: mating process of plug-in cable connectors for wire harness assembly tasks. Intell. Serv. Robot. **13**(1), 159–168 (2020)
24. Gouyou, D., et al.: Tolerance analysis of overconstrained and flexible assemblies by polytopes and finite element computations: application to a flange. Res. Eng. Design **29**(1), 55–66 (2018)
25. Directive, R.: Directive 2002/95/EC of the European Parliament and of the Council of 27 January 2003 on the restriction of the use of certain hazardous substances in electrical and electronic equipment. Official J. Eur. Union **13**, L37 (2003)
26. Thoben, M., et al.: Press-fit technology, a solderless method for mounting power modules. In: Proceedings of PCIM, Nuremberg (2005)
27. Stolze, T., et al.: Reliability of PressFIT connections. Infineon Technologies AG (2008). www.bodospower.com
28. Yedamale, P.: Brushless DC (BLDC) motor fundamentals. Microchip Technology Inc., vol. 20, pp. 3–15 (2003)
29. Processing. https://www.processing.org/. Accessed 25 Mar 2020

# Contact Stress of the Contact Region of a Cam Mechanism with a Flat-Faced and Spherical-Faced Follower

Jiří Ondrášek(✉)

VÚTS, a.s., Svárovská 619, Liberec, Czech Republic
jiri.ondrasek@vuts.cz

**Abstract.** The aim of this paper is to provide information about the contact stress and pressure distribution in the contact region of a general kinematic pair of a cam mechanism with a flat-faced follower and eventually with a spherical-faced follower. The contact region is composed of the contact of the working surfaces of a cam and a follower. The knowledge of the distribution of the contact stress and pressure is a determinant for the service life estimation of the contact surfaces.

**Keywords:** Cam mechanism · Flat-faced follower · Spherical-faced follower · Contact stress · Contact pressure

## 1 A Cam Mechanism with a Flat-Faced and Spherical-Faced Follower

In many technical applications, a cam mechanism is advantageously used, of which a general kinematic pair is formed by the contact of a cam and a flat-faced follower and eventually a spherical-faced follower. Both cases are schematically represented in Fig. 1. Their application can be found, for example in metal forming, textile, printing and sewing machines, etc. In the case of the flat-faced and spherical-faced follower, the cam working surface slides over the follower working surface and the frictional forces occur in the course of the relative motion between the cam and the follower. The contact of the cam with the follower takes place in the contact area, in which the normal reaction $N$ and tangential forces $T$ are transferred in a general kinematic pair of a cam mechanism. The relationship between the frictional forces $T$ and the normal reaction $N$ is given by Coulomb's law of sliding friction complied with the equation

$$T = \mu \cdot N \tag{1}$$

where $\mu$ is the coefficient of sliding friction with no distinction between static and kinetic friction. The normal reaction $N$ is variable. It is the result of the weight and the load of the follower and the inertia, resistance and elastic forces. If the displacement law $v(\psi)$ of the cam mechanism is known, the time course of the normal reaction $N$ can easily be found based on the quasi-dynamic analysis. The displacement law $v(\psi)$ expresses a functional dependency of the driven body motion $v$ on the driving member


E.-C. Lovasz et al. (Eds.): MTM&Robotics 2020, MMS 88, pp. 338–345, 2021.
https://doi.org/10.1007/978-3-030-60076-1_30

motion $\psi$ of the given cam mechanism. In Fig. 1, the variable $\phi$ denotes the pressure angle. It is defined by the angle between the normal to the pitch profile of the cam and the direction of the follower motion. This variable expresses the cam force effect on the driven member of the given cam mechanism. The variables $m_v$ a $I_\psi$ represent the mass and inertia properties of the follower and the cam, respectively. Detailed information about the problematics of cam mechanisms is given in [1–3].

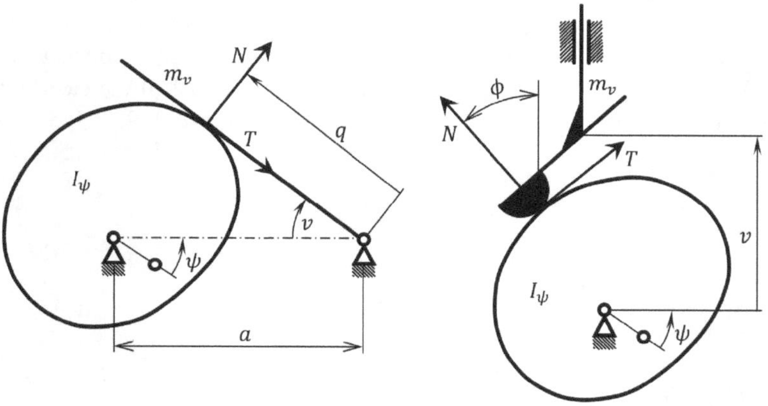


**Fig. 1.** A cam mechanism with a flat-faced oscillating and spherical-faced translating follower.

The disadvantage of all cam mechanisms in which sliding occurs between the cam and the flat-faced and spherical-faced followers is the occurrence of increased wear of the cam and follower contact surfaces. Wear often occurs due to insufficient lubrication of the point of contact. This causes the possibility of creating clearances and increasing deviations of the movement of the working member, relative to the desired displacement law of the cam mechanism of a mentioned type. This results in the formation of impacts and an increase in the noise of an entire machine, which includes a cam mechanism whose the follower slides over the cam during the working cycle. For these reasons, these cam mechanisms are only advantageous for medium-speed machinery. These afore mentioned effects are caused by the fact that loads are often supported on a small surface area of the components in the contact. Contact pressures and stresses therefore have a tendency to be high. Excessive contact stress or deformation can lead to the fatigue damage of the component contact area. According to this, the stress analysis of the parts in contact can usefully predict the service life of the components.

The state of deformation and stress existing between two quite simple component geometries in contact under load can be established based on the Hertzian contact stress theory [4]. The contact stress and pressure analysis is frequently difficult and therefore analysis by numerical methods is required for complex shapes of components and for loads. In connection with the rapid development of numerical mathematics, informatics and computer technology the finite element method – FEM can be applied with advantage, but only under certain assumptions for its effective use.

## 2 Contact Modelling Using FEM

Friction is inherent between contacting parts in relative motion with the relative velocity $\dot{x}$, thus the frictional forces $T$ have to be considered in the computational analysis in the case of contact of a cam with a flat-faced or spherical-faced follower.

When two nonconforming elastic bodies are brought into contact they touch initially along a line; in the case of the contact of a cam with a flat-faced follower or at a single point in the case of the contact of a cam with a spherical-faced follower. They will deform under the action of the slightest loads $N$ and $T$ and contact is made over a finite area, which is up to three times smaller than in the case of the dimensions of both bodies. The remote parts $P_1$ and $P_2$ of the bodies are approached the each other by a distance of $\delta$. The described case is schematically shown in Fig. 2.

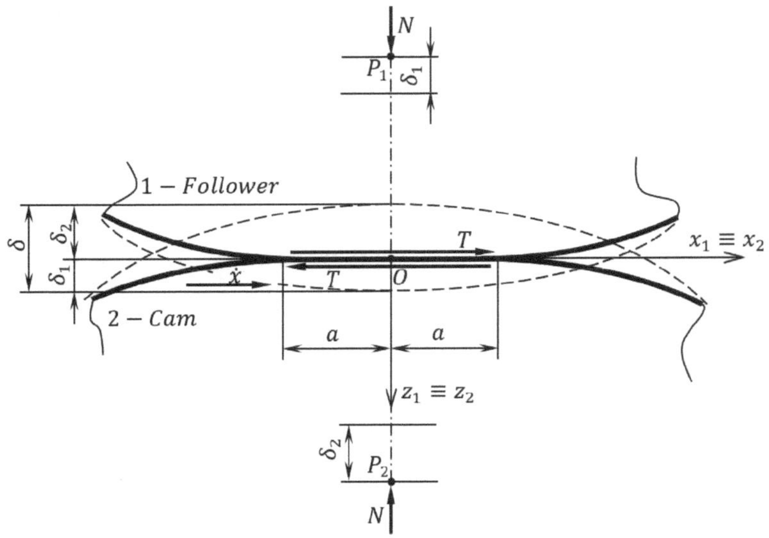


**Fig. 2.** Tangential loading and sliding contact of two non-conforming bodies after elastic deformation.

The transmitted loads tend to be relatively high and the contact area increases in the size as the loads increase. The contact stress is highly concentrated in the vicinity of the contact area and decreases rapidly with an increasing distance from it, therefore the stress area is close to the contact of the bodies. In view of the fact that the contact regions are dimensionally much smaller than the characteristic dimensions of the bodies in contact, the stresses around the contact area are dependent neither on the bodies in the contact, nor on the way of bearing the bodies. This hypothesis is well-known as Hertzian theory of contact which is described in detail in publication [4] or basic the information is given in [5, 6]. It simplifies the definition of boundary conditions and allows the application of the theory of elasticity of large bodies using the finite element method to analyze the contact stress distribution.

In order to estimate the stress in the contact areas of various bodies, the finite element method can be used with some advantages. However, it is well-known that in the case of contact tasks, this method is time-consuming as far as calculations are concerned. To achieve the relevant results, a dense finite element mesh in the contact regions is required. This requirement leads to a large number of solved linear algebraic equations. The computational body contact algorithm is based on a numerical iteration to finding the elements in contact, so that the numerical solution of the equations of the assignment takes place in several steps [5]. The intent of generating a computational model of a general kinematic pair for FEM purposes is to minimize the size of the solution task of two bodies which are in mutual contact. In this case, the general kinematic pair is formed by the contact of a cam with a flat-faced follower or a spherical-faced follower. This means that the necessary needed number of equations based on the model is established, the solution of which will produce satisfactory results.

In the case of cam mechanisms with a flat-faced follower, the contact is approximated by a the contact between a cylinder and a flat plane. In the case of cam mechanisms with a spherical-faced follower, the contact is approximated by the contact between a cylinder and a three-dimensional body, which is described locally with an orthogonal radii of curvature $\rho_{x1}$, $\rho_{y1}$ as shown in Fig. 3. The cam that is approximated by the cylinder with the width $l_e$ is locally described with radius of curvature $\rho_{x2}$. The cam and follower come into contact at the identical tangential $Oxy$ plane of the $Oxyz$ coordinate system with the origin in the centre of the contact areas.

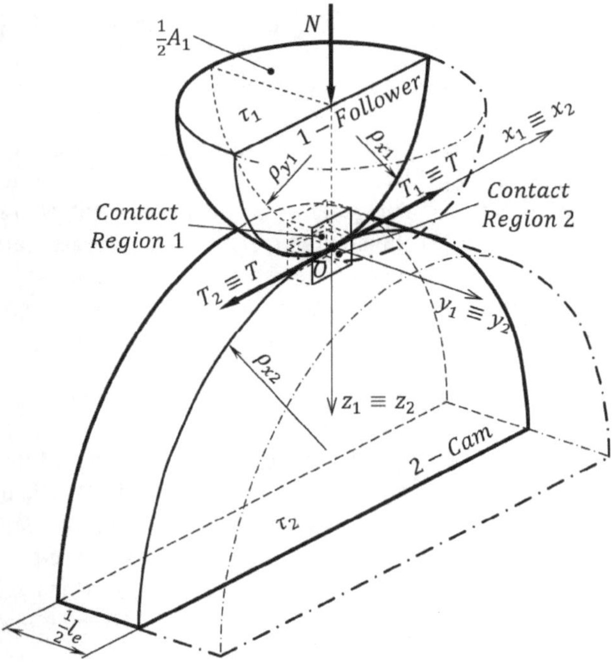


**Fig. 3.** Contact model of a normal and tangential loaded general kinematic pair.

According to Fig. 3, the contact task of a cam and a follower can be considered to be symmetrical with the *Oxz* symmetry plane with respect to the geometry of the bodies, loads of bodies and gearing of bodies. Given the validity of the assumptions and conclusions of the Hertzian contact theory and the symmetry of the given contact task, a quarter of each of the bodies in contact can be used for the purposes of the computational analyses. In the *Oxz* symmetry plane, shown in Fig. 3, the symmetry boundary conditions of the solved task are defined accordingly. The upper $\tau_1$ plane of the body that has a follower meaning is allowed the displacement in the direction of the applied force *N*, i.e. in the direction of the $z$ – axis of the *Oxyz* coordinate system, by the boundary conditions. In $\tau_1$ plane, the acting force *N* is also evenly distributed as follows

$$p_N = \frac{N}{A_1} \; [\mathrm{Nm}^{-2}] \tag{2}$$

where $A_1$ denotes the area of the follower segment. The lower $\tau_2$ plane of the body that has a cam meaning is fixed. Frictional forces $|T| = |T_1| = |T_2|$ are defined in the same *Oxy* plane, frictional force $|T_1|$ acts on *body 1* and frictional force $|T_2|$ acts on *body 2*. Both forces are transmitted through the appropriate contact surface in which they are distributed uniformly. Based on the Hertzian theory of contact [4–6], the shape and size of the contact region of two three-dimensional bodies in contact is predicted to the definition of an even and fine mesh of elements of an appropriate size in the vicinity of contact. With increasing distance from this area, the size of the elements can be gradually increased. Furthermore, the shape and size of the each contact region is determined depending on the load size. In Fig. 3, the appropriate contact regions are denoted with *Contact Region 1* and *Contact Region 2*.

In order to estimate the operating lifetime of the contact areas, it is necessary to determine the magnitude of the reduced stress $\sigma_{red}$ for example by using FEM. This value is compared with the limit value of the $\sigma_h$ stress. In the case of general kinematic pairs, the contact load on the contact surface of the bodies and under it has a periodic course. At the contact areas of the general kinematic pair and below these points, contact stress becomes a periodical magnitude related to the angular cam displacement $\psi$ [5]. Strength conditions under a variable load are given by the stress limit $\sigma_h$ which is in agreement with the disturbance caused by the transitory stress. The reduced stresses $\sigma_{red}$ $(\psi, z)$ are limited by the actual strength condition, written as [1]

$$\max(\sigma_{red}(\psi, z)) < \sigma_h, \quad \psi \in \langle 0, 2\pi), \quad z \geq 0 \tag{3}$$

where variable $z$ is the depth under the contact surface of the cam or follower. Based on the tests of steels with ultimate tensile strength $R_m \in \langle 500, 1500 \rangle$ MPa the fatigue limit of the material loaded by the tension is supposed as $\sigma_C \approx 0.35 \cdot R_m$ and the yield strength as $R_e \approx (0.55 \div 0.8) \cdot R_m$, see [1]. Since the limit, value of the transient stress can be approximately determined according to $\sigma_h \approx \sigma_C \approx 0.7 \cdot R_m$ then it is possible to replace the condition Eq. (3) by the inequality [1]

$$\max(\sigma_{red}(\psi, z)) < R_e, \quad \psi \in \langle 0, 2\pi), \quad z \geq 0 \tag{4}$$

Conditional inequality Eq. (4) describes the fact that during the operation of the cam mechanisms, no destructive action of elastic deformation occurs in the general kinematic pair [5].

The application of the mentioned procedure of the contact modelling using FEM is demonstrated in the example of analyzing the contact stress distribution in the cam and the flat-faced follower, which are in sliding contact with the sliding friction coefficient $\mu = 0.2$ under action normal force $N = 1000$ N. According to Eq. (1), the tangential force takes the value $T = 200$ N. The cylindrical segment that has the cam meaning has nominal dimensions: $\rho_{x2} = 10$ mm; $l_e/2 = 1.5$ mm. The thickness of the segment that has the flat-faced follower meaning is 5 mm for computational simulations. The characteristic material parameters of the cam and the flat-faced follower are stated in Table 1.

**Table 1.** Material characteristics of steels.

| | | Follower: 90MnCrV8 | Cam: 16MnCr5 |
|---|---|---|---|
| Mass density | $\rho$ [kg · m$^{-3}$] | 7850 | 7850 |
| Young's modulus of elasticity | $E$ [GPa] | 200 | 206 |
| Shear modulus | $G$ [GPa] | 77.821 | 79 |
| Poisson's ratio | $\nu$ [–] | 0.285 | 0.3038 |

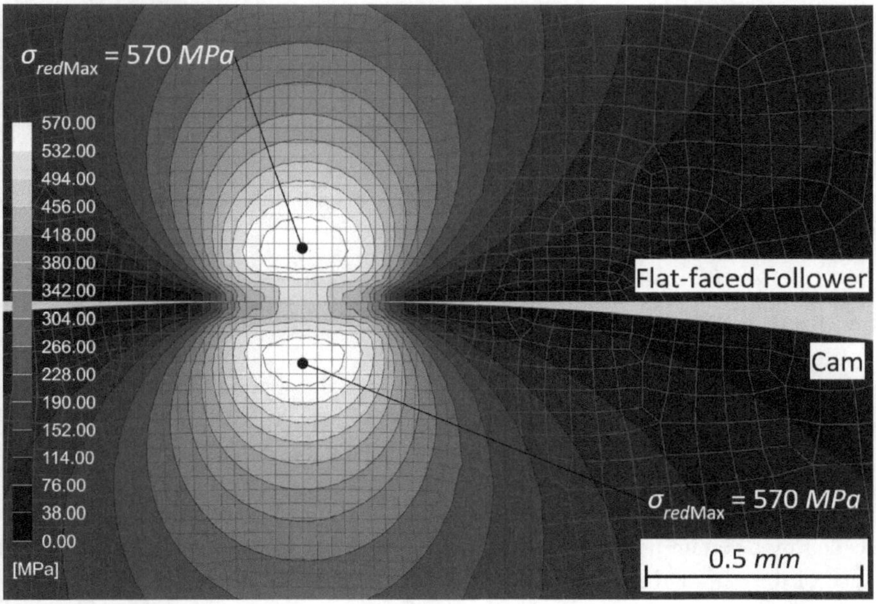


**Fig. 4.** Von Mises stress field of the normal loaded contact regions.

Two types of contact tasks are considered here: analysis of the contact stress field of the normal loaded contact regions (see Fig. 4) and analysis of the contact stress field of the normal and tangential loaded contact regions (see Fig. 5).

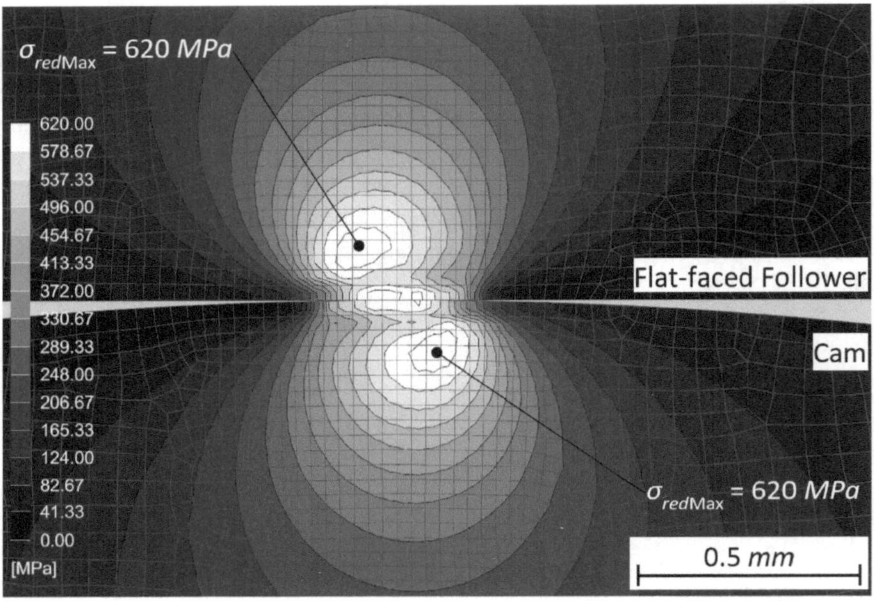


**Fig. 5.** Von Mises stress field of the normal and tangential loaded contact regions.

Based on the performed analyses it is evident, that the frictional load has a significant effect on the stress field, which is distorted by this tangential loading in the course of sliding and the maximum reduced stress $\sigma_{red}$ is nearer to the surface and increases in value. The start of plastic yielding of the bodies in contact will be determined by the maximum value of this stress, which has to be compared with the value of the yield strength $R_e$ of the relevant material.

## 3 Conclusions

The contact of the cam and the flat-faced follower or the spherical-faced follower itself takes place in the contact area, which puts a strain on the cyclic loading during the working cycle and the bodies in contact slide over the contact surfaces. Thus, the transmission of normal and tangential forces is realized. These phenomena will cause the deformation of the two bodies in contact and cause contact stresses, which, if exceeded by a certain limit of this stress, may result in fatigue damage of the working surface of the cam and the follower in the cam mechanism operation.

An effective way of analysis of the dynamic contact strain of the general kinematic pair of the cam mechanism consists in the interaction of the knowledge and conclusions of the Herzian contact theory between two bodies and the advantages of using the finite element method. Using the Hertzian contact theory, the shape and size of the contact region of the bodies in contact can be predicted. In this way, the region to create an even and fine mesh of elements of an appropriate size in the vicinity of contact is predicted. Taking into account the assumptions of the Herzian theory, the definition of the FEM-model of a general kinematic pair is considerably simplified and this reduces the number of algebraic equations needed to solve this problem thereby the computational time is reduced. This method leads to the achievement of relevant results compared to the real state. The method is presented in the task of determining the stress field of the contact region of the general kinematic pair, which is formed by the cam and the flat-faced follower.

**Acknowledgments.** This paper was created within the work on the project "National Centre of Competence ENGINEERING": TN01000015/16 – Project supported by the Technology Agency of the Czech Republic.

## References

1. Koloc, Z., Václavík, M.: Cam Mechanisms. Elsevier, Amsterdam, (1993). ISBN 0-444-98664-2
2. Norton, R.L.: Cam Design and Manufacturing Handbook. Industrial Press Inc., New York (2009). ISBN 978-0-8311-3367-2
3. Rothbart, A.H.: Cam Design Handbook, 1st edn. McGraw-Hill Professional, New York (2003). ISBN-10: 0-07-137757-3
4. Johnson, K.L.: Contact Mechanics. Cambridge University Press, Cambridge (1985). ISBN 0 521 34796 3
5. Ondrášek, J.: The General Kinematic Pair of a Cam Mechanism, Kinematics - Analysis and Applications, Joseph Mizrahi, IntechOpen, 1 July 2019. https://doi.org/10.5772/intechopen.86682. https://www.intechopen.com/books/kinematics-analysis-and-applications/the-generalkinematic-pair-of-a-cam-mechanism. ISBN 978-1-78984-491-7; Print ISBN 978-1-78984-490-0; eBook (PDF) ISBN 978-1-78985-700-9; Copyright year: 2019
6. Ondrášek, J.: The stress distribution in the contact region of a cam mechanism general kinematic pair. In: Mechanism and Machine Science 52: Mechanisms, Transmissions and Applications, Proceedings of the 4th. MeTrApp Conference 2017, pp. 99–108. Springer (2018). ISBN 978-3-319-60701-6
7. Ondrášek, J.: The elastic compression in the contact region of a cam mechanism general kinematic pair. In: Advances in Mechanisms Design II: Proceedings of the XII International Conference on the Theory of Machines and Mechanisms, pp. 57–63, Springer, Cham (2016). ISBN 978-3-319-44086-6, ISSN 2211-0984

# SIG Machining Center Renovation

Pavel Fišer(✉), Pavel Pátek, and Pavel Dostrašil

VÚTS, a.s., Svárovská 619, Liberec XI, 460 01 Liberec, Czech Republic
{pavel.fiser, pavel.patek, pavel.dostrasil}@vuts.cz

**Abstract.** The functional prototype of the CNC machine will be created by the reconstruction of the existing SIG CF2 machining center. Including the development of a new single purpose control system based on HW and SW components from OMRON, which results in faster, more efficient and easier to operate and service than the old control method. This also involves the replacement of the individual drives of the interpolation and positioning axes, the necessary external encoders and the machining spindles, together with following structural and mechanical changes. This makes it possible to achieve higher accuracy and speed during machine working movements. The new system also required unavoidable intervention in the electrical installation, its reorganization and replacement of the old nonfunctional components. All modifications are summarized in detailed documentation. The paper describes the current state of renovation of the machine, which will then be able to be involved in the ongoing serial production of radial and axial cams, either as a backup or as a replacement for one of the existing SIG machines at the research institute.

**Keywords:** Axis · Drive · Servomotor · Replacement · Control system · Documentation · Wiring

## 1 Introduction

The machining center SIG CF 2 has been in the machine equipment of VÚTS for several decades, during which time it underwent several minor renovations and modifications both in the field of wiring and in the form of design modifications on the machine itself. After years of using the machine, a system error was occurred, that greatly reduced the functionality of the machine itself. Outdated unknown control system and unavailability of the programing software condemned the machine as unrepairable. Due to the increasing demand for cam types that can only be produced on the same or dimensionally similar machine together with the mechanical unreliability and with existing machinery, that cannot be fixed, after careful consideration of all options and risks has led us to self-reconstruction of the machine, because in the field cam machining VÚTS has many years of experience, including specialized software for creating custom displacements [1, 2] and [5].

A thorough analysis helped us to choose control platform offered by OMRON AUTOMATION based on which the control system will be programmed. Unique programs for the production of radial and axial cams will be created in the offered programing environment called Sysmac Studio. The SW change goes hand in hand with the hardware one, which includes the actual replacement of the motion controller


E.-C. Lovasz et al. (Eds.): MTM&Robotics 2020, MMS 88, pp. 346–355, 2021.
https://doi.org/10.1007/978-3-030-60076-1_31

and the frequency inverters powering the spindle and the individual servomotors. So intervention in the electrical wiring was inevitable, in addition to that it was also necessary to modify the connection points on the machine so that the new motors could be mounted. All changes were carefully described and documented (Fig. 1).

**Fig. 1.** Machine SIG CF 02

## 2 Drive Replacement and Design Changes

At first, the protective and functional covers were professionally dismantled to allow access to existing motors and their mechanical and electrical connections. These had to be removed for incompatibility with the OMRON platform. Only spindle drive has been retained since it is a classic asynchronous motor that can be easily controlled by any Inverter. After labeling (as same as axes; sliding V, Z, U and rotary C) and removing the motors, electromechanical parameters were deducted from their identification plates. We figured out that all axes have servomotors with the same parameters from one model series with a rated torque of 18.5 Nm and a speed of 3000 rpm. With that acquisition of information, the drives with the closest higher values of the monitored parameters were selected from the supplier's catalogs [6] of the new system. When comparing the proportions of newly selected drives with the connecting points of the positioning mechanisms and with the available space for mounting on the machine, we found, that it would not be possible to mount the motors directly due to differences between them or collision with other parts of the machine. That's why we decided for following individual adjustments.

## 2.1 Axis C

In the case of a single rotary axis C, the value of the drive torque was determined from given already existing transmission 1:180 (worm-gear combination transmission). After that, it was found that the original motor was oversized. The maximum nominal servomotor speed of drive was chosen with consideration of bearable speed over the surface of the minimum diameter workpiece which can be machined with current clamping chuck. The 3 kW OMRON R88M-K3K020-S2 engine has been selected with a rated torque of 14.3 Nm and a speed of 2000 rpm [3]. It will be more than enough to accomplish future tasks on the machine.

**Fig. 2.** Rotational C axis before and after replacement

The connection between transmission and old motor was performed by cogwheel fully attached to its shaft, which had to be removed, modified and complemented with clutch that helped us to mount it to new motor.

Another problem with replacement of current motor was position of mounting holes for fixing. This with the existing gear on the shaft and with location of cables and other distributions (see Fig. 2), caused impossibility to select any dimensions of flange. The flange had to be precisely manufactured so that the gears on the motor shaft and on the input gear reducer could fit precisely with minimal backlash. It resulted in slight rotation of mount position of servomotor (15°) from the original one.

## 2.2 Axis V

This axis carries the support mechanisms, the clamping chucks, the worm-gear reducer, and the C-axis drive itself. Servomotor is mounted on the linear guidance carrier moving by rotation of the ball screw with a pitch of 10 mm per revolution.

For this purpose, a 4 kW R88M-K4K020F-S2 engine with a rated torque of 19.1 Nm and a rated speed of 2000 rpm was selected [3]. The shaft revolutions is now reduced with angled planetary gearbox by the ratio of 1:4, because the flange size of the new engine prevented direct assembly. The chosen gear ratio reduced the moment of inertia of the load. It was allow us to adjust the control loop parameters to perform a minimum position error value can be achieved when the axis moves. The output from the gearbox is connected to an existing lead ball screw via a fixed coupling (Fig. 3).

**Fig. 3.** Axis V before and after replacement

## 2.3 Axis Z

Like the axis V was the Z-axis is geared by an angled planetary gearbox by the same gear ratio and the ball screw has the same pitch per revolution. However, new servomotor has lower torque 14.3 Nm with type designation R88M-K3K020-S2 [3], because it is a positioning axis. It doesn't reach such dynamic load as interpolation axes. After replacing, was found that the newly mounted gearbox interferes with the active area of the original axis travel, therefore the positions of the hardware limit switches on the linear guidance had to be adjusted and thus the maximum axis travel was shortened. The output of the gearbox is also mounted with a custom fixed coupling (Fig. 4).

**Fig. 4.** Z axis before and after replacement

## 2.4 Axis U

The last remaining interpolation axis is the axis U with the new R88M-K4K020F-BS2 servomotor with rated torque value of 19.1 Nm and rated speed 2000 rpm [3]. The mounting flange and the fixed coupling connecting the motor shaft to the end of the ball screw had to be extended, because the cylinder of the hydraulic piston used for hydraulic balancing of the vertical axis was in the way. Therefore, the following design was adopted, with an elongated coupling and flanged parts (see Fig. 5).

**Fig. 5.** Vertical axis U before and after replacement with extended coupling

The U-axis is the only one equipped with an electromagnetic brake to lock the position. The brake is active in the off state.

# 3 Electrical Adjustments

Parallel to the replacement of the servomotors, the inseparable modification of the electrical box, including the description in the electro documentation, took place.

## 3.1 Wiring and Replacements

The first step of this phase was to remove the original servopacks and their accessories (chokes, EMC filters and resistors). Furthermore, the removal of the old control unit, unused terminals and unnecessary electronics for support drives that no longer exist. After the removal of excess components the space for the new OMRON system was created.

Followed by sorting of all listed and carefully marked input, output and other signal wires connected to the removed components. Redundant ones were discarded without replacement, others replaced or added. For example, it has been found that the signals from the hardware limit switches can be newly wired directly to the servopacks and some of the monitored outputs can be replaced by sending the same information between the devices via the EtherCAT® communication bus.

OMRON NJ501-5300 with sufficient memory capacity and possibility of direct connection of additional I/O cards was chosen as the motion controller.

For chosen servomotors and unchanged asynchronous motor for spindle were picked suitable servopacks and frequency invertor with supplement EMC filters. Transformer based electrical DC (24 V) sources were replaced by electronic ones with internal protection.

The components were placed in enclosure according to a new mechanical scheme, connected with new wires with newly dimensioned circuit breakers, motor starters and protection components. At that time it was possible to recognize the new and the original part of the electrical wiring. The interconnection of both parts and their smooth cooperation was realized together with the revision of functionality of the unchanged control, signaling and protection elements for supporting technologies during grinding and milling like lubrication, cooling, filtration and chip removal etc.

**Safety.** The motion controller is directly connected via the internal communication bus to the safety controller supplemented with safety cards of safety inputs and outputs, this replace original safety relay that was disconnected. The safety inputs were also supplied with signals from the inverters securing drive shutdown even when the main PLC failure happens. The E-STOP inputs and mechanical barriers, error signals from supporting technologies were brought to safety inputs.

**Position Determination.** The new motors are mounted straight to the lead ball screws by the couplings. So the position obtained from the servomotor encoder does not match with the exact position of the mechanism end points. The error is caused by mechanical and assembly inaccuracies. For that reason, each individual axis are supplemented by external linear encoder, which are wired directly into the servopacks. The rotary axis C with worm-gear transmission is mounted with a measuring ring. Upon consideration, the current external encoders were kept on the machine. The only problem was the shape and type of the output signal transmitting the position information. It is an analog current signal with an amplitude of 11 μA, but the servo inverters need to be supplied by a digital voltage signal TTL [3] for its correct function. The solution was the purchase of A/D converters and the necessary modification of the wiring and connectors (Fig. 6).

**Fig. 6.** External encoders

**Secondary Switchboard.** Besides the main electric box, the machine is equipped with smaller one with an operator panel and basic controls along with a pendant. The original obsolete industrial computer was removed and replaced by a combination of a compact HMI and an external waterproof touch screen (see Fig. 7). The pendant is now plugged directly into dedicated digital I/O cards, which are communicating with the motion controller in the main enclosure via EtherCAT® bus.

**Fig. 7.** New electrical components

## 3.2 Electrical Design Documentation

The modifications are accurately and clearly described in the new updated electrical documentation created for this purpose in the QElectroTech program environment (Figs. 8 and 9).

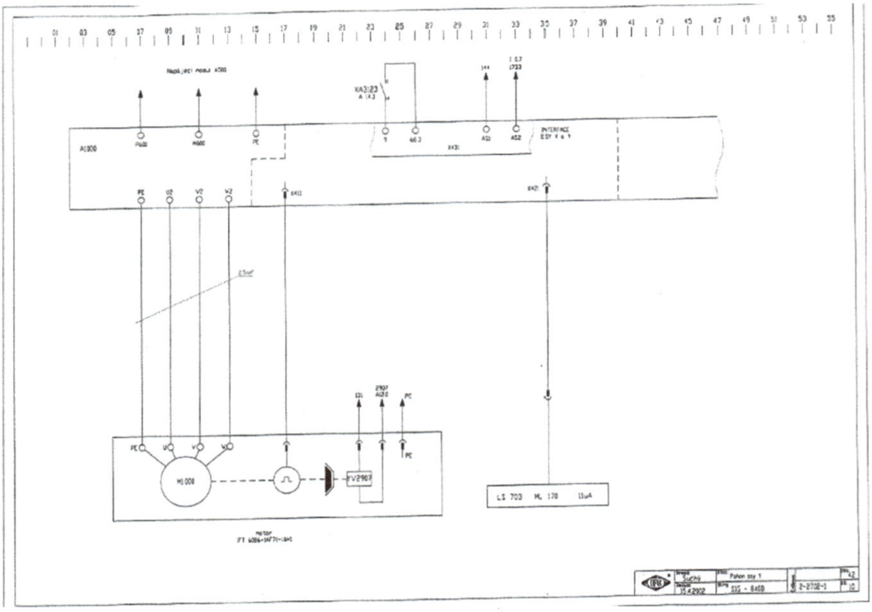

**Fig. 8.** Original electrical design documentation

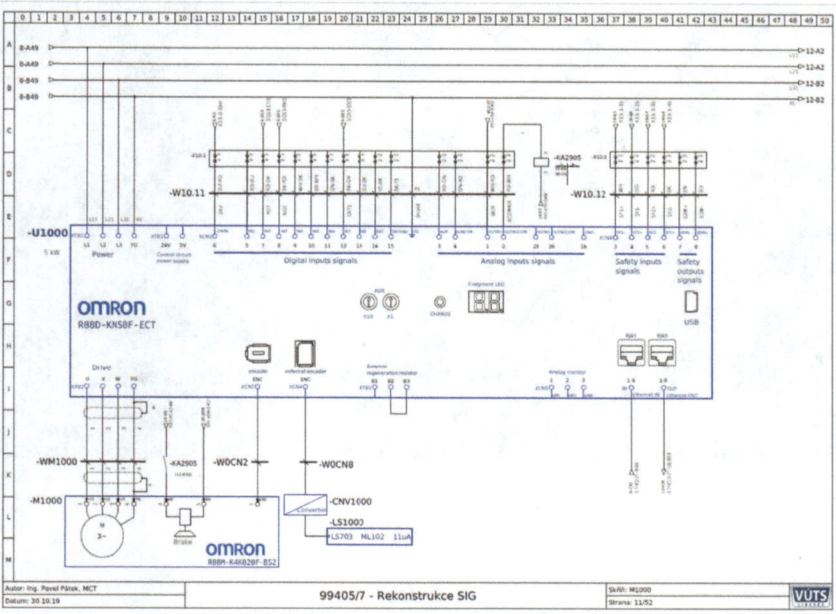


**Fig. 9.** New electrical design documentation

The original components were re-marked. Their labels now correspond to the new ones. Furthermore, unique schematic symbols of converters, power supplies and other electrical elements of the distribution system had to be created.

## 4 Test Control System

A special system for positioning of individual axes was created in the program environment supporting the selected platform to check the correct function and wiring of new and original components.

It also includes the demand of start up the necessary technologies, such as balancing or lubricating of the linear guidance.

### 4.1 Manual Mode

In manual mode, only one selected axis can be controlled and movement can be terminated at any time to prevent possible collision. The supported functions are linear axis movement - *JOG*, travel and preselected increment - *STEP*, move to absolute position - Move Abs and the associated with home position setting and returning to it (*HOME* and *Set Pos*). Speed and acceleration can be set for each function, the maximum limits are selected with respect to the mechanical structure of the machine and the parameters of the selected drives (Fig. 10).

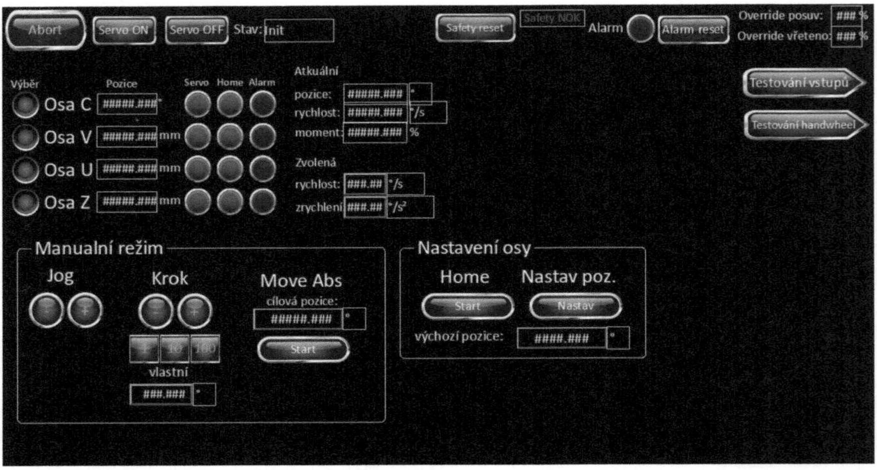


**Fig. 10.** Test manual mode

## 4.2 Technologies Control Mode

The reconstructed machine uses most of the original technologies. It's required to test their correct function or control sequence of starting and stopping in temporary states as well as set their parameters. Most of them have their own control interface, however it is necessary to set communication between them and with the master system. This all is provided by this control test mode.

## 4.3 Test CAM Mode

The CAM mode provides possibility to achieve a simulated movement of grinding of workpiece by interpolation of the axes. We are able to calculate the path of the tool center movement from the input CAM data by using the theoretical knowledge [4]. Final movement is then divided into individual ones of the selected axes and with respect to the constant movement of the tool on the workpiece surface we calculate their velocities.

Subsequent testing will help us to monitor the machine's behavior during working movements and eliminate some errors.

This mode will serve as the foundation for the future automatic machining mode where final procedures will be prepared based on input production data and selected technological processes.

## 5 Conclusion

The machine was turned on after completing testing of the functionality of newly connected devices, setting their parameters and communication. By using the test system programmed in motion controller and HMI we can verify of machine basic functions (movement of axes, control of inputs and outputs, pendant, spindle control etc.).

Thereafter, the individual technologies will started up be one by one to that level, which is necessary for the machine to run in automatic grinding mode. This will help us identify any temporary, collision, error and operating states of the machine that need to be properly handled. Then it will be possible to create the first procedures for the production of the radial and axial cam. The full functionality of the machine will be tested first without the workpiece, followed by any correction of the prepared input theoretical data. The positive result will then conclude in the creation of sophisticated fully editable databases of procedures required for CAM production.

All of this is planned in next months of the project solution, which will end with a fully functional reconstructed SIG machining center.

## References


1. Jirásko, P., et al.: Mechatronika pohonů pracovních členů mechanismů. VÚTS, Liberec (2015). ISBN 978-80-87184-63-9
2. Fišer, P.: FDL-K 001 wooden strip milling machine. In: Special Mechanisms and Their Drives IV, pp. 220–221. VÚTS, Liberec (2018). ISBN 978-80-87184-87-5
3. G5-series WITH BUILT-IN EtherCAT® COMMUNICATIONS: User's Manual. In: Omron: Industrial Automation. OMRON EUROPE B.V, Hoofddorp, The Netherlands (2020). https://assets.omron.eu/downloads/manual/en/v1/i576_accurax_g5_servomotors_drives_with_ethercat_control_users_manual_en.pdf. Accessed (2011)
4. Koloc, Z., Miroslav, V.: Cam Mechanisms: Studies in Mechanical Engineering. Elsevier Science, Amsterdam (1993). ISBN 0-444-98664-2
5. Dostrašil, P.: Effective design and implementation of specific displacement diagrams to control kinetic sculptures. In: MATEC Web Conference Palma de Mallorca (2018). https://doi.org/10.1051/matecconf/201821004004. ISSN 2261-236X
6. ACCURAX G5 SERVO SYSTEM. In: OMRON: Industrial Automation. OMRON EUROPE B.V., Hoofddorp, The Netherlands. https://assets.omron.eu/downloads/brochure/en/v2/accurax-g5_brochure_en.pdf

# The Stress and Life Tests of Coatings of Linear Dovetail Guide

Martin Bušek(✉)

VÚTS a.s. Liberec, Svárovská 619, Liberec, Czech Republic
martin.busek@vuts.cz

**Abstract.** The subject of the paper is the evaluation of stress and life tests performed on the final parts used in the monitored transformation mechanism - linear dovetail guide. The supporting material of the parts are primarily steels of ČSN 14220.3 and ČSN 19573 class with applied coatings - selected thin layers based on TiAlN (trade name Balinit Hadlube), amorphous carbon ta-C and classical nitriding.

**Keywords:** Mechanisms · Linear guide · Coating · Life test

## 1 Introduction

The stress and life tests of the individual parts of the linear dovetail guide were preceded by basic testing on laboratory samples in the laboratory of tribology (Department of Materials, TUL). Subsequently, a suitable coating material applicable to exposed linear guide pieces was selected. Thin layers based on TiAlN (trade name Balinit Hadlube), amorphous carbon ta-C, and classic nitriding were selected for surface pieces. Furthermore, the surface roughness analysis, contact angle measurement, coating adhesion evaluation, coating hardness evaluation and tribological and abrasion and surface wear resistance [1, 2] also performed.

The long-term tests carried out on the final parts of the mechanism and the complete linear dovetail guides with selected thin layers are performed at VUTS a.s. Liberec. Above all, the wear of the exposed parts of surfaces of the mechanism, including pitting, were monitored after several millions of cycles elapsed under a defined mechanism load. Further testing focuses on identifying and changes of positioning accuracy, energy consumption, vibrations and inaccuracy in the mechanisms throughout the testing period. Three some thin layers (TiAlN/WC/C (trade name: Balinit Hadlube), ta-C coatings and classic nitriding), which have a low coefficient of friction and are suitable for applications with insufficient lubrication, have been prepared for testing [3]. Each of these thin layers was applied to pieces of one representative linear guide (see Fig. 1, 2). The production, assembly and adjustment of the linear guides held in early 2019. Furthermore, the guides were electrically put into operation and submitted to the load tests, which are extend continuously.


E.-C. Lovasz et al. (Eds.): MTM&Robotics 2020, MMS 88, pp. 356–363, 2021.
https://doi.org/10.1007/978-3-030-60076-1_32

**Fig. 1.** Linear dovetail guides with thin layers applied to the surfaces of the exposed part

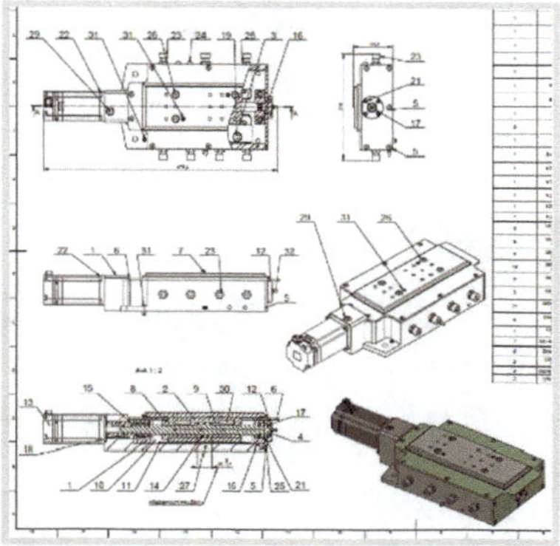

**Fig. 2.** Linear dovetail guide – model

## 2 The Stress and Durability-Life Tests

Since 06/2019, long-term tests have been carried out on the assembled and adjusted linear dovetail guides (optionally with three different coatings - nitriding surface treatment unit selected for the paper) in the Mechatronics laboratory at VÚTS a.s. The units are loaded with an unbalanced weight of 31.4 kg and perform a periodic movement with an 18 mm stroke in a cycle of 96 strokes/min. The tests preceded the documentation of the basic state of the mechanisms key parts before their final assembly and adjustment. This was a standard output control of the production process, along with photo documentation provided by a DNT's DigiMicroProfi microscope. Which will further serve to compare wear rates after several millions of operating cycles (Fig. 3).

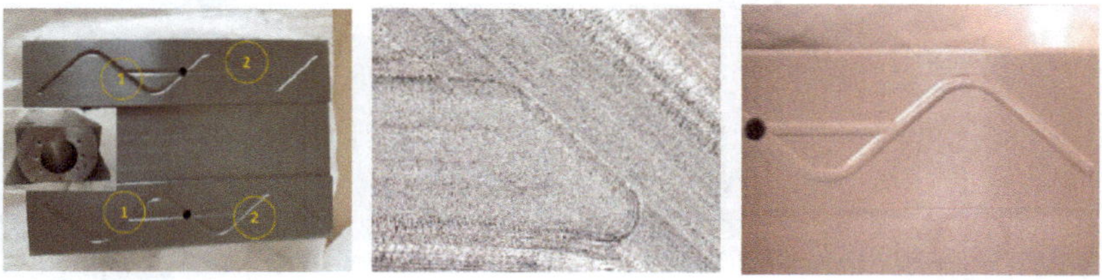

**Fig. 3.** Photo documentation of the selected parts of the mechanisms - dovetail guidance

## 2.1 Analysis of Pitting Formation

Pitting is a form of concentrated corrosion on the surface of a work piece that leads to the formation of small holes on the surface. The occurrence of pitting is manifested in several ways, especially by increasing the vibrations of the device, i.e. increased noise. The determination of the pitting presence is preceded by a thorough examination of the damaged surface under a stereomicroscope or other suitable microscope with low magnification. The degree and the nature of the damage can be ascertained [4]. The DigiMicro Profi microscope from the DNT is used to assess the pitting at VUTS. This microscope camera allows a personal computer to be connected, thus significantly comforts the basic analysis and facilitating data transmission. Furthermore, it is advisable to expose the surface to high magnification optical and electron microscopy analyzes (Figs. 4, 5 and 6).

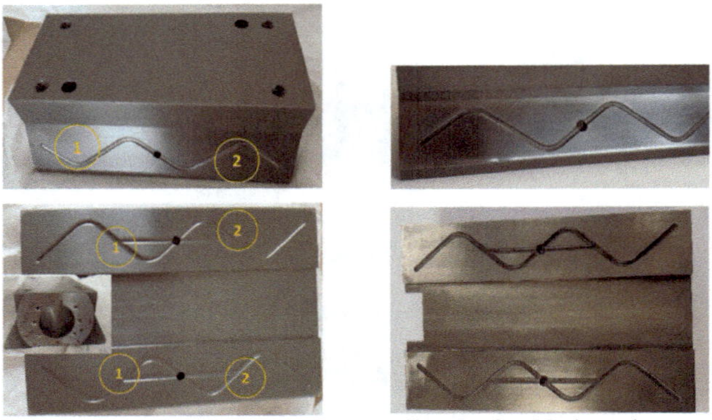

a) new part b) part after 35mio cycles

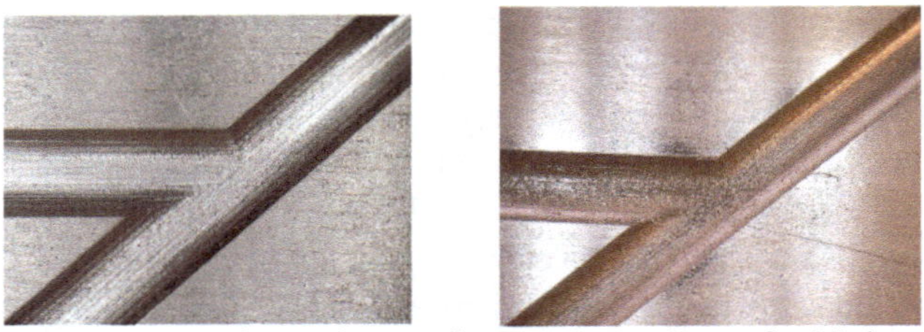

c) new part no. bottom 2max d) part no.bottom 2max after 35mio. cycles

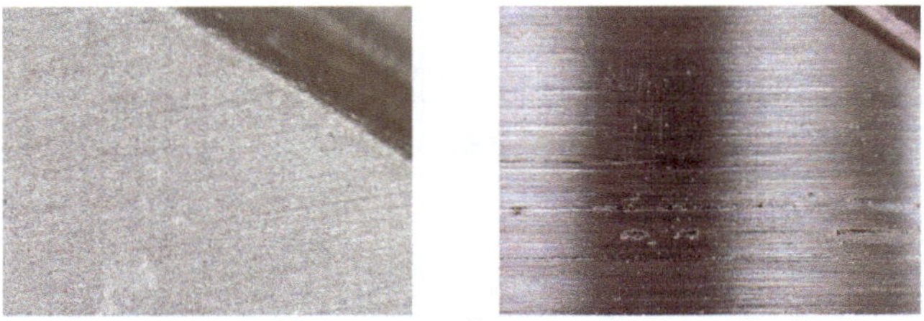

e) new part no. side 2max f) part no. side 2max after 35mio. cycles

**Fig. 4.** The main part of the linear guide comparison with the DigiMicro microscope

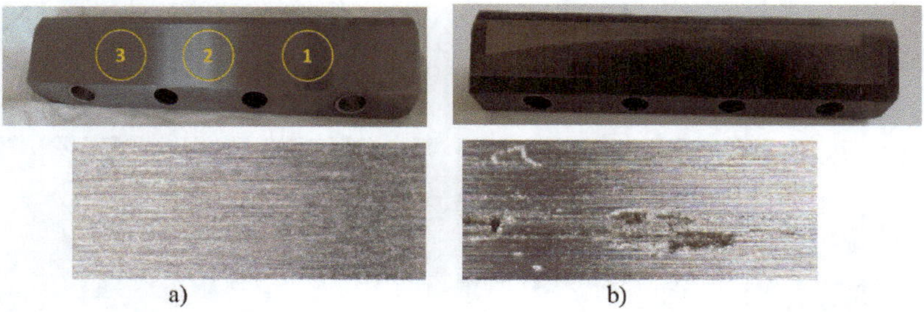


**Fig. 5.** The inclined part of the linear guide comparisons with the DigiMicro microscope a) new part no. 1_2max, b) part no. 1_2max after 35mio. cycles

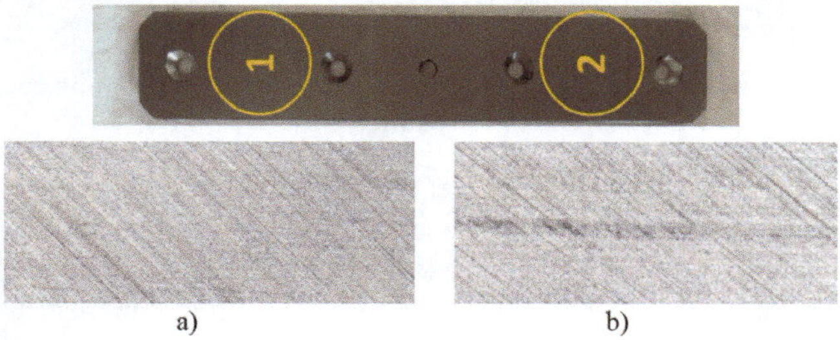


**Fig. 6.** The flat part of the linear guide comparison with the DigiMicro microscope a) new part no. 2_1max, b) part no. 2_1max after 35 mio. cycles

The obtained information can be used to estimate the lifetime of the mechanism in the application, or it is possible to propose changes of the base material, applied coating – thin layers, etc., which can be eliminate or delayed the occurrence of pitting.

## 2.2 The Positional Accuracy and Mechanism Clearance Analysis

The source of vibrations and clearances in the mechanisms are closely related. Their analysis is performed using standard laboratory measurement techniques at VUTS a.s. The target is to detect to minimal damage of the transformation mechanism parts after the manufacturing and assembling by the usual technology and in operation at a defined load. It is necessary to focus on the exposed parts such as friction surfaces of the dovetail guide, trapezoidal screw and others. Furthermore, due to operating wear in the mechanisms, there may be the occurrence of unacceptable inaccuracies in the positioning of the transformation mechanism. The position accuracy analysis (adherence to the prescribed position/stroke/step) aims to monitor to the range and evolution of the position error as a function of the mechanism operating cycles during long-term life tests.

The analysis used the following measuring instruments:

- Laser position sensor Micro-Epsilon ILD2220-100 (2–3 points)
- System Renishaw (head, ring RESM and Signum/Quantic Interface)

- Strain gauge force sensor HBM U9B/2kN, amplifier KISTLER
- Measurement system Dewetron DEWE 5000

**Fig. 7.** The measurement of clearance and positional accuracy by optical sensors

Linear dovetail guides with unbalanced load were measured for clearance detection after 35 mio cycles. The measurement places were at both extreme (L1 and L2) stroke positions. An external force was applied to the left (L1L, L2L) and right side (L1P, L2P) of the load (see Fig. 7). The values of the external force applied, the changes in longitudinal position Y-axis (the stroke axis of the unit) and the changes in the transverse positions axes X1 and X2 were recorded.

The results are shown in Fig. 8 and Fig. 9 and they are recorded in Table 1 and Table 2.

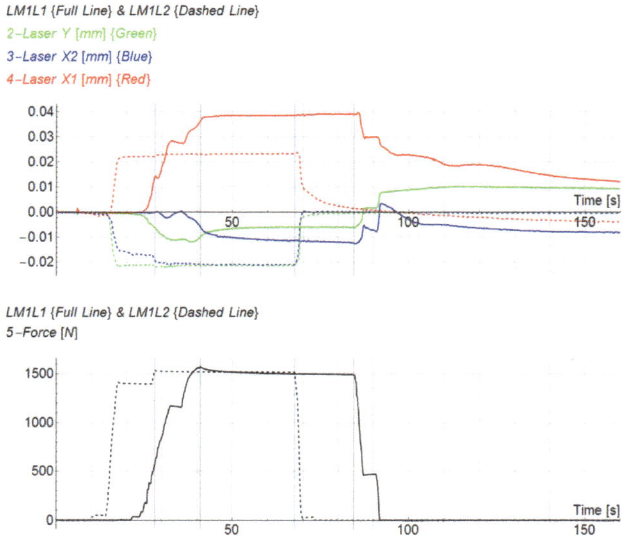


**Fig. 8.** The clearance measurement of the linear dovetail guide with coating - nitriding, measurement points L1L and L2L

**Table 1.** The clearance measurement of the linear dovetail guide with coating - nitriding, measurement points L1L and L2L

| | Time[s] | L1: Y[mm] | L2: Y[mm] | L1: X2[mm] | L2: X2[mm] | L1: X1[mm] | L2: X1[mm] | L1: Force[N] | L2: Force[N] |
|---|---|---|---|---|---|---|---|---|---|
| 1 | 0. | 0.00031799 | 0.000074 | -0.0000133 | 0.0000888 | $3.78 \times 10^{-6}$ | 0.00059044 | 0.00903514 | 0.14509232 |
| 2 | 10. | $5.9 \times 10^{-6}$ | -0.00001215 | -0.0000697 | -0.0000686 | -0.00008101 | -0.00058443 | -0.04734309 | 11.85467589 |
| 3 | 20. | -0.00036578 | -0.0212348 | -0.000012 | -0.015755 | -0.00009592 | 0.02208529 | 2.3737354 | 1400.089606 |
| 4 | 30. | -0.00898571 | -0.02153793 | -0.0009568 | -0.0201692 | 0.01842654 | 0.02300578 | 850.0401447 | 1525.068075 |
| 5 | 40. | -0.01085159 | -0.02130603 | -0.0042885 | -0.0207962 | 0.03462657 | 0.02309814 | 1557.670771 | 1520.201311 |
| 6 | 50. | -0.00643052 | -0.0212633 | -0.0099651 | -0.0209347 | 0.0385254 | 0.02324754 | 1517.826354 | 1518.531624 |
| 7 | 60. | -0.00585688 | -0.02131313 | -0.0109386 | -0.0208173 | 0.03862747 | 0.02329249 | 1505.230191 | 1517.394648 |
| 8 | 70. | -0.0057794 | -0.00178936 | -0.0113882 | 0.0005342 | 0.03915605 | 0.00978058 | 1498.615586 | 24.13476318 |
| 9 | 80. | -0.00577179 | -0.00023078 | -0.0118251 | 0.0004221 | 0.03917565 | 0.0024875 | 1493.934069 | 0.34845043 |
| 10 | 90. | 0.00187635 | -0.00013626 | -0.0071203 | 0.0002528 | 0.0300182 | 0.0010684 | 470.9926104 | 0.25377242 |
| 11 | 100. | 0.00897308 | -0.0001325 | -0.002948 | 0.0001314 | 0.02295563 | -0.0002538 | 0.12258526 | 0.11420204 |
| 12 | 110. | 0.0101107 | -0.00004356 | -0.0049457 | -0.0001085 | 0.01890796 | -0.00126555 | 0.18487639 | 0.07571965 |
| 13 | 120. | 0.0103113 | 0.00007138 | -0.006226 | -0.0002469 | 0.01910898 | -0.00199614 | 0.19061534 | 0.17561878 |
| 14 | 130. | 0.01005716 | 0.0000392 | -0.0071192 | -0.0001826 | 0.01714545 | -0.00257669 | 0.29490471 | 0.18366857 |
| 15 | 140. | 0.00983599 | $4.3 \times 10^{-6}$ | -0.0075604 | -0.0002657 | 0.0147124 | -0.0031817 | 0.28226426 | 0.12816795 |
| 16 | 150. | 0.00962108 | $-1.53 \times 10^{-6}$ | -0.0077843 | -0.0002135 | 0.01316822 | -0.00358318 | 0.18970709 | 0.13629842 |
| 17 | 160. | 0.00944881 | -0.00004053 | -0.0079176 | -0.0002323 | 0.0121642 | -0.0038532 | 0.1592814 | 0.16441985 |

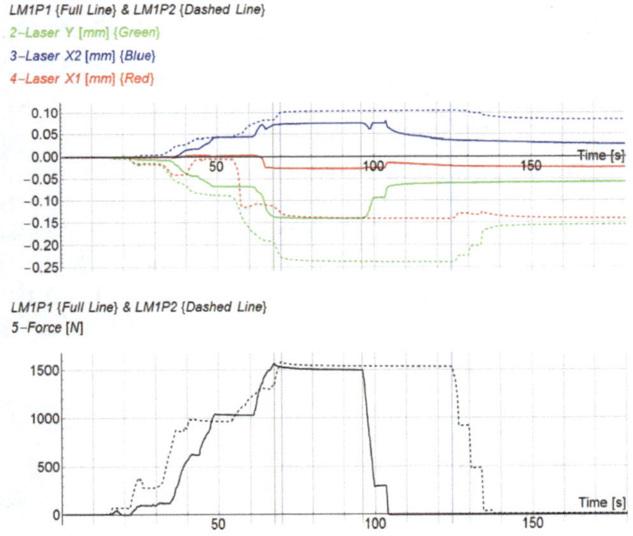


**Fig. 9.** The clearance measurement of the linear dovetail guide with coating - nitriding, measurement points L1P and L2P

**Table 2.** The clearance measurement of the linear dovetail guide with coating - nitriding, measurement points L1P and L2P

| | Time[s] | P1: Y[mm] | P2: Y[mm] | P1: X2[mm] | P2: X2[mm] | P1: X1[mm] | P2: X1[mm] | P1: Force[N] | P2: Force[N] |
|---|---|---|---|---|---|---|---|---|---|
| 1 | 0. | 0.00009996 | 0.0002669 | 0.0002012 | -0.0000123 | 0.00008927 | 0.00014878 | 0.060252 | -0.03295031 |
| 2 | 10. | -0.00002846 | -0.0001023 | -0.0000502 | 0.0000419 | 0.0001778 | 0.00004243 | -0.00518886 | -0.02488059 |
| 3 | 20. | -0.00066243 | -0.00286495 | -0.0005336 | 0.0000704 | -0.00045347 | -0.00091837 | -3.67987502 | 67.94894187 |
| 4 | 30. | -0.00630688 | -0.01659261 | -0.0002249 | 0.0038764 | -0.00074098 | -0.01508521 | 98.76771754 | 281.967464 |
| 5 | 40. | -0.03657055 | -0.06324329 | 0.0150088 | 0.0277747 | 0.00246053 | -0.03966162 | 569.7639747 | 911.9918629 |
| 6 | 50. | -0.06659399 | -0.08628497 | 0.0430061 | 0.0436651 | 0.00192458 | -0.00479149 | 1036.446519 | 964.9348097 |
| 7 | 60. | -0.0669975 | -0.16323224 | 0.0444851 | 0.0695141 | 0.00339673 | -0.11315556 | 1029.878708 | 1208.060462 |
| 8 | 70. | -0.13913314 | -0.22687216 | 0.0731082 | 0.1003182 | -0.02573772 | -0.12683316 | 1530.113284 | 1584.078275 |
| 9 | 80. | -0.1406598 | -0.23589218 | 0.0749345 | 0.1036931 | -0.02640449 | -0.13878687 | 1508.822931 | 1547.008429 |
| 10 | 90. | -0.14126204 | -0.23748788 | 0.0755881 | 0.1043595 | -0.02659884 | -0.14061847 | 1501.716881 | 1540.317847 |
| 11 | 100. | -0.09339521 | -0.23840259 | 0.0758763 | 0.1044988 | -0.02505995 | -0.14156095 | 292.0576572 | 1536.584268 |
| 12 | 110. | -0.06223259 | -0.23901076 | 0.0484736 | 0.1047416 | -0.01528751 | -0.14223954 | -8.54954204 | 1533.939625 |
| 13 | 120. | -0.0597369 | -0.23949948 | 0.0396924 | 0.1050492 | -0.0190874 | -0.14271188 | -8.62476924 | 1531.954295 |
| 14 | 130. | -0.05868103 | -0.22171445 | 0.0359806 | 0.1041845 | -0.02171435 | -0.12952362 | -8.65541717 | 794.7901634 |
| 15 | 140. | -0.05809354 | -0.16002262 | 0.034247 | 0.0892226 | -0.02229427 | -0.13547335 | -8.62539468 | 1.23368354 |
| 16 | 150. | -0.05754476 | -0.1573516 | 0.0322902 | 0.0860517 | -0.02306078 | -0.13941514 | -8.60452244 | 0.20361177 |
| 17 | 160. | -0.05724817 | -0.15656229 | 0.0305733 | 0.0849083 | -0.0230963 | -0.14073456 | -8.5549289 | 0.21226849 |
| 18 | 170. | -0.05696188 | -0.15608792 | 0.0292949 | 0.0843204 | -0.02360905 | -0.14159695 | -8.56038431 | 0.28921039 |
| 19 | 180. | -0.05678736 | -0.15578147 | 0.0281786 | 0.0841745 | -0.02368063 | -0.14165605 | -8.61425381 | 0.22711663 |

The measurement results are continuously compared with other linear units (with other coatings). The final comparison of the units after 60 mio cycles will be made to select the most suitable thin layer for the application.

### 2.3 Energy Performance Analysis

The life tests monitoring of transformation mechanisms also includes an assessment of the energy intensity. The mechanisms are monitored in each phase of the operation – start running/standard run/standard run after 10 mio cycles. In particular, the torque on the shaft of the individual mechanisms drive is monitored, because the changes in torque can predict the moving parts degradation. The torques on the drives are recorded and stored to intervals and evaluated in the maximum, minimum and rms values. The changes in these quantities predict changes in the mechanism and may indicate damage to the surfaces and respectively the coatings. Any damage can be detected earlier than during normal visual and listening controls. It is also possible to monitor the electric power input of the servo drives of the individual mechanisms.

The analysis used the following measuring instruments:

- PLC system– trace function, position/torque tracking
- Voltage measurement - direct measurement on servomotor using DAQP-HV 0 .. ± 1400 V high voltage module, analysis by the Dewetron measuring system
- Measurement of electric current - direct measurement on servo drive with Chauvin Arnoux E3N clamp current probe connected to DAQP-STG 0.1 .. 1000 mV/V module, analysis by the Dewetron measuring system (Figs. 10 and 11)

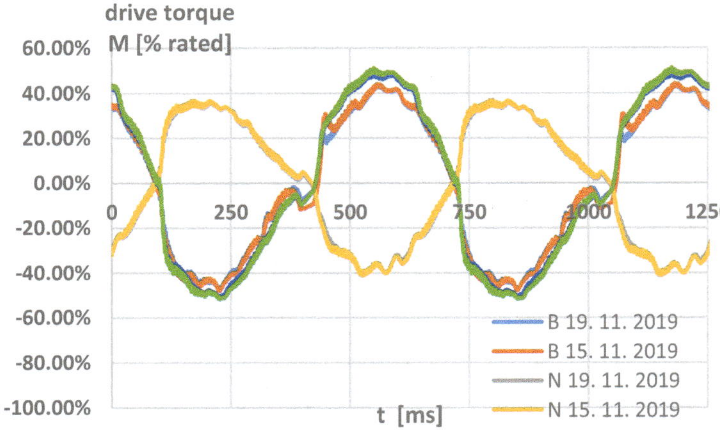


**Fig. 10.** Linear dovetail guide – actual servomotor torque (B: coating TiAlN/WC/C, N: nitriding, TUL: coating ta-C)

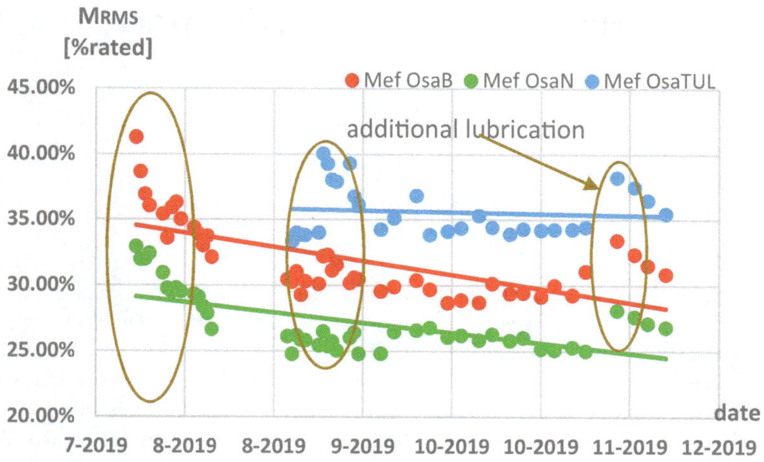


**Fig. 11.** Linear dovetail guide – effective servomotor torque (B: coating TiAlN/WC/C, N: nitriding, TUL: coating ta-C)

## 3 Conclusion

The experiments performed after 35 million cycles show minimal changes in the mechanical properties of the mechanisms applied with a thin layers (TiAlN/WC/C, nitriding, coating ta-C). There are not significant clearances that prevent linear guides from running. There is no significant deformation of the sliding surfaces of the linear dovetail guide. Furthermore, there is a decreasing trend in the torque of the linear guide drive, with no signs of seizing. The experimental activities of the stress and life tests of the presented mechanisms will continue to a significant point of 70 mio cycles, respectively 100 mio cycles.

## References

1. Bakalova, T.: Problematika mazání kluzných stykových ploch v kombinaci s povrchovou úpravou, roční výzkumná zpráva FV 20547 2018/2019
2. Kavánová, A.: Hodnocení užitných vlastností tenkých vrstev a jejich aplikace v průmyslu, Diplomová práce. Liberec, Technická univerzita v Liberci, Fakulta strojní, Ing. Totka Bakalova, Ph.D, Vedoucí práce (2019)
3. Bušek, M., Bakalová, T.: Ověřená technologie výroby kluzného lineárního rybinového (trojbokého) vedení, projekt TRIO II FV 20547, Liberec (2018)
4. Hejnová, M.: Assessment of the rolling contact fatigue. In: Mechanisms, Transmissions and Applications. Proceedings of the Fourth MeTrApp Conference 2017, pp. 89–98. Springer (2018). https://doi.org/10.1007/978-3-319-60702-3. ISBN 978-3-319-60701-6

# Sliding at the Cam Mechanisms

Monika Gromadova(✉)

VÚTS, a.s., Svarovska 619, Liberec, Czech Republic
monika.gromadova@vuts.cz

**Abstract.** Sliding can occur on the various places at the cam mechanism (e.g. roller placing on the pin as a plain bearing). However, this paper deals with sliding between cam and cam follower. There are two types of followers at the cam mechanisms. The roller follower has more complicated construction as other types of followers. On the other hand, it has an advantage that in contact surfaces does not arise a high friction force. Roller follower has the advantage of lower friction (rolling) than the sliding contact of the other types of followers, but it can be more expensive. A ratio describing value of the sliding is called slide/roll ratio. It is dependence of mutual velocities of the sliding/rolling surfaces. A lot of parameters influences on sliding between cam and roller follower. Very important is friction between contact surfaces. A value of the SRR has influence on the service life of the contact surfaces. Sliding can lead to surface damage. The most common is abrasive and adhesive wear. The paper deals short with this two types of damage and protection against them.

**Keywords:** Sliding · Cam mechanism · Cam follower · Surface damage

## 1 Introduction

Sliding can occur on the various places at the cam mechanism (e.g. roller placing on the pin as plain bearing). However, this paper deals only with sliding between cam and cam follower.

Usually, we use Hertzian contact theory [1] at the estimation of the cam mechanism service life. This theory is quite simple for determining of the stresses occurring in loaded parts. A disadvantage of this method is that it can be used only for cases, where is no sliding between rolling parts (cam and roller follower). At cases with sliding should be this fact taken into account or used other suitable method for service life estimation (FEM analysis, Johnson-Kendall-Roberts model etc.).

This paper deals with basic information introducing problematics of the sliding in the cam mechanisms. In the following research will be this information refined and supplemented.


E.-C. Lovasz et al. (Eds.): MTM&Robotics 2020, MMS 88, pp. 364–374, 2021.
https://doi.org/10.1007/978-3-030-60076-1_33

## 2 Slide/Roll Ratio SRR

A ratio describing value of the sliding is called slide/roll ratio. It is dependence of mutual velocities of the sliding/rolling surfaces. We can calculate it according to Eq. (1), where $v_1$ and $v_2$ are velocities of the surfaces. Description of these parameters is shown in Fig. 1.

$$SRR = \frac{|v_1 - v_2|}{\frac{1}{2}(v_1 + v_2)} \tag{1}$$

*SRR* can be write in form of number from 0 to 2 or, more common in practice, is value in percent. Then is Eq. (1) multiplied by 100. There can occur two boundary states at *SRR* calculation:

1. if $v_1 = v_2$, then $SRR = 0\%$ (pure rolling),
2. if $v_1 = 0$, then $SRR = 200\%$ (pure sliding).

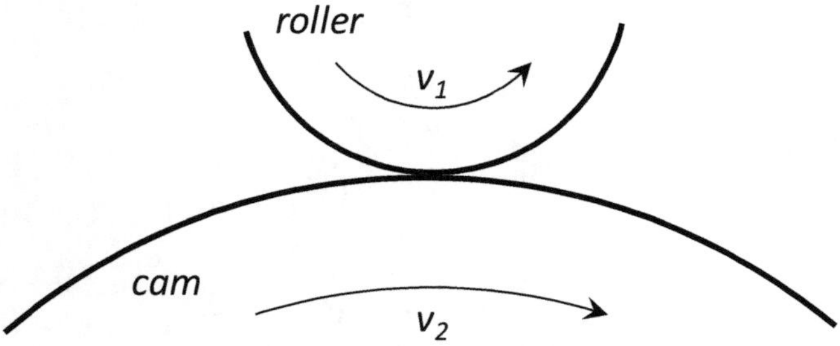


**Fig. 1.** Parameters of the slide/roll ratio.

Both states will be described below. In some arrangement of the cam mechanisms is *SRR* constant equal to 200%. In other cases can take a value from 0 to 200% depending on the follower position on the cam circuit. In practice exists cases when sliding is negative. Then it is necessary to use another suitable parameter for description.

## 3 Cam and Roller Follower

There are two types of follower at cam mechanisms. The roller follower has more complicated construction as other types of followers. On the other hand, it has an advantage that in contact surfaces does not arise a high friction force. The cam with oscillating roller follower is shown in Fig. 2. There are marked characteristic parameters. The cam follower has the shape of the lever.

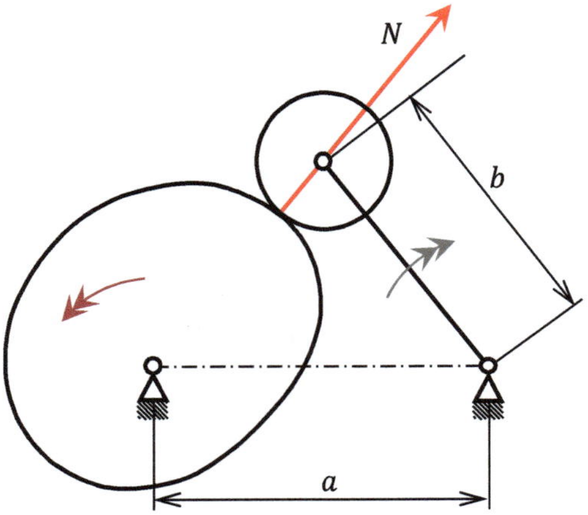


**Fig. 2.** Radial cam with oscillating roller follower

There is shown translating follower in Fig. 3. It derives a reciprocating translational movement from cam.

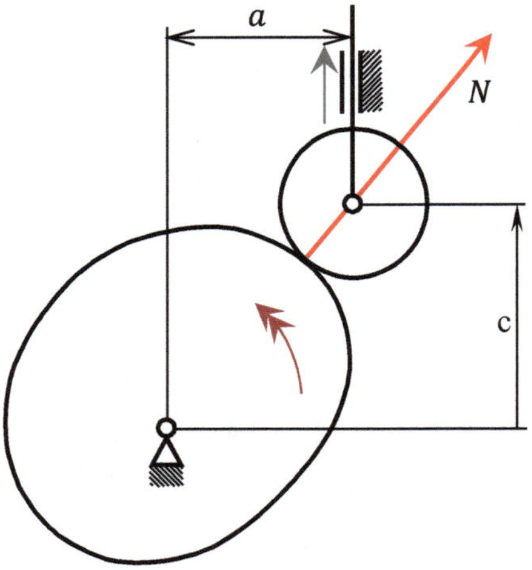


**Fig. 3.** Radial cam with translating roller follower

Concrete application of oscillating follower is at e.g. indexing gear. There can be used 2 × 3 or 2 × 4 roller followers with conjugate cams (see Fig. 4). This type of indexing gear is produced in our company and we have large experiences with its production and service.

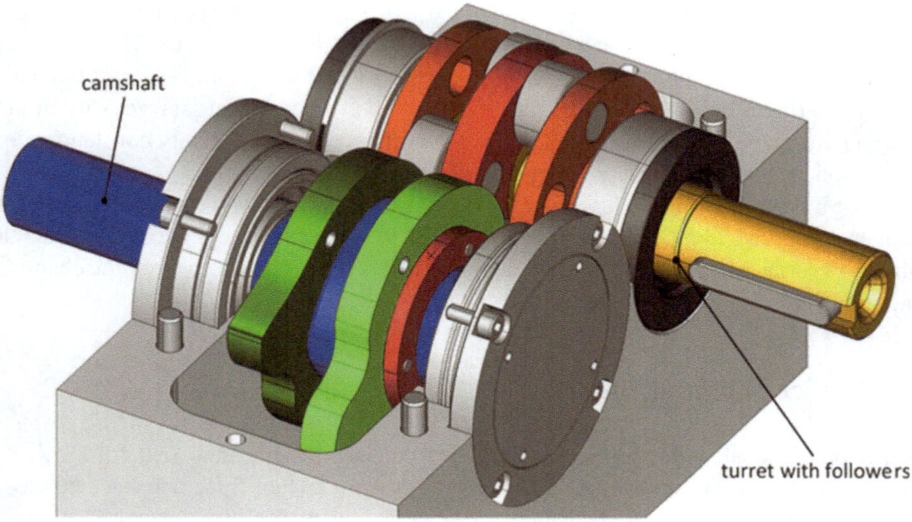


**Fig. 4.** Parallel indexing cam gear, generating stepwise output motions with planar cams – model

## 4 Principle of the Sliding Formation

Theoretically, we calculate *SRR* from mutual velocities of contact surfaces. In fact, it is much more difficult to determine a real value of the sliding. Sliding does not have the constant value at whole circuit of the cam. It is related to shape of the cam, respectively with displacement diagram of the cam. There is no sliding in places, where the radius of the cam is constant, $SRR = 0\%$. Sliding occurs in places where the cam radius changes (non- zero acceleration in the acceleration diagram). This implies that sliding is specific for every shape of the cam and we need to know exact parameters of the cam (and follower).

Another circumstance influencing on sliding are properties of the contact surfaces, especially surface roughness and coefficient of friction. Important are parameters of lubrication, too.

It is good to realize that friction coefficient between the cam and the roller everywhere along the cam surface is less than the limiting friction coefficient. Therefore, there is only a small velocity difference between the roller and the cam surface. The sliding friction created by this velocity difference together with the rolling friction is sufficient to drive the roller rolling on the cam surface [2].

The roller angular speed is a function of the working frictional forces at the cam–roller and roller–pin contact and inertia torque caused by angular acceleration of the roller itself [3]. The roller rotational speed is obtained by balancing the tractive torque (acting at the outer surface of roller) with the combined torques due to roller–pin friction and roller inertia force, by iteratively adjusting the roller rotational speed [4].

An important role plays the camshaft rotational frequency, the *SRR* and the roller inertia moment. Literature [2] says that during the part of cam angle, where the acceleration of the cam surface is the highest, a larger friction coefficient than the limiting value is required to keep the roller rolling on the cam surface. The paper [2] describes negative slide between the cam and the roller on the cam falling flank. This can be interpreted from the inertial moment principle. When the surface velocity of the cam reduces rapidly, the roller tries to keep the roller rotational velocity constant under the action of inertia, therefore a negative slide results [2].

Last but not least has value of the load between roller follower and cam significant influence on the sliding. The mutual velocity of the contact surfaces is connected to all above described parameters and it should be evaluate for every cam mechanism individually.

## 5 Service Life and SRR

Morrison [5] dealt with the dependence between service life and *SRR* for concrete cases. Curves for various values of the *SRR* are shown in Fig. 5 for one selected case.

Curves shown in Fig. 5 are obtained from testing of the heat treated 4150 steel running against the same material but phosphate-coated (obtained from test made by Morrison [5]). These curves describe dependence of load-stress factor K on number of stress cycles for which the factor K is required.

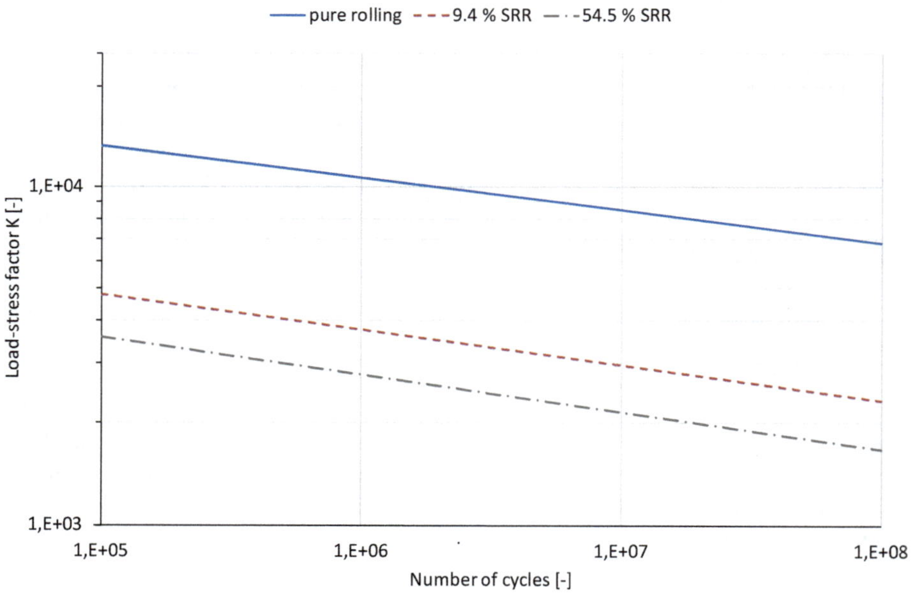


**Fig. 5.** Load/life curves for various SRR values [5]

The Eq. (2) for the factor K as a function of service life is derived from the lines in Fig. 5. Constants A and B are parameters of the concrete material, they are set from service life tests. The list of these parameters is given in the literature [5].

$$\log_{10} K = \frac{B - \log_{10} N}{A} \tag{2}$$

From Eq. (2), useful reference curve of K versus any desired range of stress cycles can easily be plotted.

Curves in Fig. 5 show that relatively small sliding can occur significant reduction of the service life compared to pure rolling. For this reason we have to determine a value of the *SRR* at the estimation of the cam mechanism service life.

## 6 Cam and Other Types of Follower

Roller follower has the advantage of lower friction (rolling) than the sliding contact of the other types of followers, but it can be more expensive [6]. The other types of followers is shown in Table 1. All these types can be used with oscillating or translating movement (see Fig. 6 and Fig. 7).

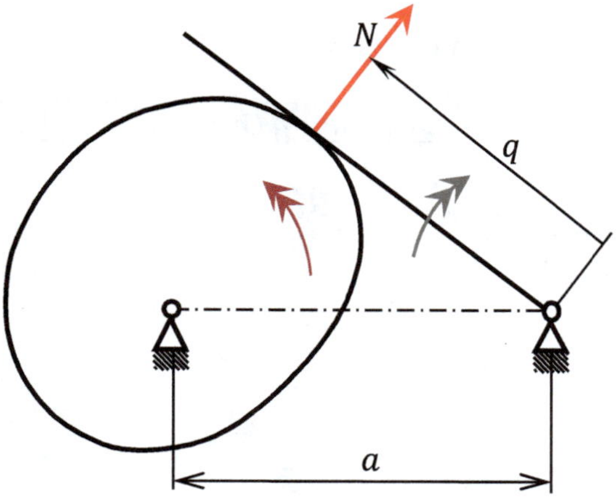


**Fig. 6.** Radial cam with oscillating follower

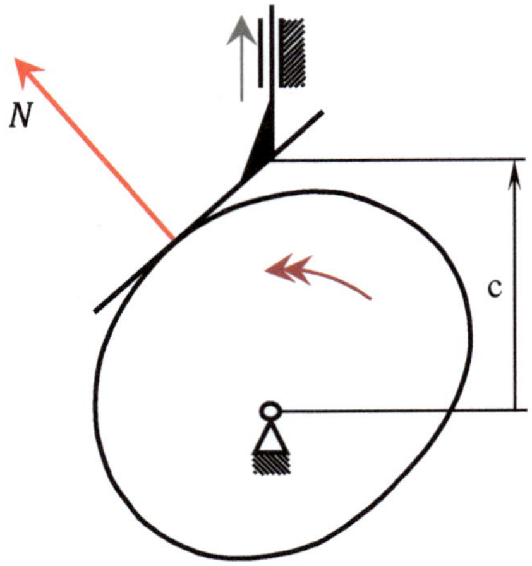


**Fig. 7.** Radial cam with translating follower

For this arrangement is characteristic value of *SRR* = 200%. It means, that one body (cam) is rotating and second body (follower) is not moving with the respect to the first body.

**Table 1.** Most common types of followers [1].

| ***oscillating follower*** | ***translating follower*** | ***name*** |
|---|---|---|
|  |  | knife edge follower |
|  |  | mushroom follower |
|  |  | flat faced follower |

Typical using of this type follower is in the engine valve train, concrete with flat faced follower.

## 7 Surface Damage

Pitting (as a result of the rolling contact fatigue) typically arises at the pure rolling contact (without sliding). It is manifested by the presence of pits on the material surface in the contact. In the case of sliding or combination of sliding and rolling can occur other types of surface damage- galling, scuffing etc. Of course, an influence on their formation can have fatigue wear, too. More information is in literature e.g. [7].

## 8 Wear Mechanisms

**Abrasive Wear**

Abrasive wear can occur when a hard surface or hard particles pass over a soft surface, causing loss of material. However, the abrasive grooves can be found on the wear tracks of the sliding friction between similar metals. It means that abrasive particles may be formed during the wear process due to work hardening, phase transformations and third body formation at the interface [8].

A significant influence has the lubricant layer. When it is not thick enough, solids come into contact and abrasive wear can occur.

Abrasive wear can be divided into two groups: two-body and three-body abrasive wear. In the three-body abrasion, the particles formed during the operation or coming from outside the system cause the abrasive wear, whereas in two body, one of the materials plays the role of the abrasive [8].

The shape of the abrasive particle, the hardness, the load and the shear strength at the contact interface determine the state of the abrasive wear. The abrasive wear coefficients of metals are in the range of depending on the conditions, which is larger compared to adhesive wear. Thus, abrasive mode is mainly a severe type of wear [8].

**Adhesive Wear**

Adhesive wear usually occurs as a next phase of abrasion of two bodies. If the lubricant film disappears between two lubricated surfaces that move each other, the abrasion of both surfaces begins. This abrasion leads to friction, which generates heat. If the friction and heat are sufficient, adhesion of both surfaces may begin (friction welding) [9].

Adhesive wear occurs when two solids are in a sliding contact and the atomic bonding forces occurring between the materials on the interface are stronger than the strength of the surrounding area in either of the materials. In this case, shear starts in the weakest material and the fragment of a surface is detached from one body and transferred to another. Once the fragment is transferred to another surface due to high bonding strength on the interface it may remain at the surface and form a ‘transfer film’, or a loose wear particle may form [10].

**Scuffing**

Scuffing is type of the adhesive wear. It is usually defined as excessive damage characterized by the formation of local welds between sliding surfaces. For metallic surfaces to weld together the intervening films on at least one of them must become disrupted and subsequently metal–metal contact must take place through the disrupted film. [11] The main reason for this type of wear is surface adhesion and material fatigue. The process can lead to seizure of surfaces.

**Galling**

After scuffing is galling the second level of the adhesive wear [7].

In metal-to-metal wear tests, high stresses can result in catastrophic galling and eventual seizure even after a single cycle. High loads, poor lubrication, high temperatures, high sliding speeds may results in galling. In contrast to many other forms of wear, galling occurs spontaneously. Galling results in a catastrophic failure and accompanied by materials transfer and appears as a groove, or score mark, terminating in a mound of metal [12].

Galling is an adhesion phenomenon which tends to be exaggerated with materials that exhibit a high degree of plastic behavior. It tends to occur in situations involving high loads, large apparent areas of contact, and sliding. The use of hard and dissimilar materials and a lubricant tends to reduce adhesion under these conditions [12].

## 9 Protection Against Surface Damage

The suitable method protecting against fatigue wear is to reduce a coefficient of friction between two contact surfaces. This value should be such that it is not sufficient to delamination at sliding. Another very important aspect at the contact fatigue control is the material purity. For rolling and sliding should be chosen materials with a few inclusions and imperfections as possible. In some cases can increasing of hardness also help, but this method is limited by the embrittlement of the hardened material. Materials for mutual sliding should be chosen very thoroughly. We should avoid the sliding between two identical materials, as their identity can lead to adhesive wear.

Between cam and follower occur sliding. Moreover, sliding occurs at the changes of mutual velocities between cam and roller follower. This sliding can lead to surface damage. Therefore it is appropriate to modify the contact surfaces by such technology to prevent wear.

There are some possibilities how to ensure resistance to wear:

1. **Materials with increased abrasion resistance:** This possibility has the disadvantage that mostly it is about special expensive materials which do not have to meet other requirements for the production of cam mechanisms.
2. **Abrasion resistant surface layers:** This surface treatment consists in modification of the chemical composition (or structure) of the surface layer. For example, boriding, nitriding, carbonitriding, etc.
3. **Abrasion resistant coatings:** The area of coating is quite extensive, there are several technological processes on the market for coating the exposed parts. Each coating has typical parameters and limitations. Basically, coating is very thin layer

which has small influence on base material, but it has significant influence on behavior with surroundings.

Very important is the choice of the suitable lubricant. This problem is extensive and in this paper is not paid attention to it. Information you can find in the relevant literature.

## 10 Conclusions

Problematics of the sliding at the cam mechanisms is very large and includes great amount of the factors which have the influence on the result value of the sliding.

This paper submitted only rough overview of the most important information which we have to take into account at the cam mechanism design. Problematics of sliding is divided into two areas-for roller follower and for other types of followers. This second case is simpler because *SRR* is constant during the whole working cycle time ($SRR = 200\%$). In the case of the roller follower more parameters influences on the result value of the *SRR*.

In the end of the paper is chapter dedicated to surface damage which can occur in the presence of the sliding. There are introduced possibilities how to prevent the damage or to delay it.

The aim of the following research is to achieve exact mathematical model of sliding in the specific cam mechanisms.

## References


1. Koloc, Z., Vaclavik, M.: Cam Mechanisms. Elsevier, Amsterdam (1993). ISBN 0-444-98664-2
2. Ji, F., Taylor, C.M.: A tribological study of roller follower valve trains. Part 1: a theoretical study with a numerical lubrication model considering possible sliding. In: Dowson, D., et al. (eds.) Tribology for Energy Conservation, pp. 489–497. Elsevier, Amsterdam (1998)
3. Alakhramsing, S., Rooij, M., Schipper, D., Drogen, M.: A full numerical solution to the coupled cam–roller and roller–pin contact in heavily loaded cam–roller follower mechanisms. Proc. Inst. Mech. Eng. Part J J. Eng. Tribol. **232**, 1273–1284 (2017). https://www.researchgate.net/publication/321714616_A_full_numerical_solution_to_the_coupled_cam-roller_and_roller-pin_contact_in_heavily_loaded_cam-roller_follower_mechanisms. Accessed 22 Mar 2020
4. Alakhramsing, S., Rooij, M., Schipper, D. Drogen, M.: Lubrication and frictional analysis of cam–roller follower mechanisms. Proc. Inst. Mech. Eng. Part J J. Eng. Tribol. **232**, 347–363 (2017). https://www.researchgate.net/publication/318022632_Lubrication_and_frictional_analysis_of_cam-roller_follower_mechanisms. Accessed 22 Mar 2020
5. Morrison, R.A.: Load/life curves for gears and cam materials. Mach. Des. **40**, 102–108 (1968)
6. Norton, R.L.: Cam Design and Manufacturing Handbook. Industrial Press, New York (2009). ISBN 978-0-8311-3367-2
7. Hejnova, M.: Materialova problematika vackovych mechanismu. VÚTS, Liberec (2018). ISBN 978-80-87184-79-0

8. Akchurin, A.: Abrasive wear. https://www.tribonet.org/wiki/abrasive-wear/. Accessed 22 Mar 2020
9. Šest druhů opotřebení hydraulických součástí. https://e-konstrukter.cz/novinka/sest-druhu-opotrebeni-hydraulickych-soucasti. Accessed 24 Mar 2020
10. Adhesive wear. https://www.tribonet.org/wiki/adhesive-wear/. Accessed 22 Mar 2020
11. Stolarski, T.A.: Lubrication and efficiency of involute gears. In: Tribology in Machine Design. Elsevier (1999). ISBN 978-0-08-051967-8. https://www.sciencedirect.com/science/article/pii/B9780080519678500116. Accessed 21 Mar 2020
12. Galling. https://www.tribonet.org/wiki/galling/. Accessed 21 Mar 2020

# Bending Critical Speed Determination of a Shaft Carrying Multiple Rotors by Vibroacustical Signal Analysis

Ion Crastiu, Dorin Simoiu, and Liviu Bereteu(✉)

Mechanics and Materials Strength Department, Politehnica University of Timişoara, Bd. Mihai Viteazul, nr. 1, 300222 Timişoara, Romania
ion_crastiu@yahoo.com, dorin.simoiu@yahoo.com, liviu.bereteu@upt.ro

**Abstract.** The main purpose of this paper is to estimate the first natural frequency of a mechanical system consisting of a shaft that carries multiple rotors by analyzing a vibroacoustic signal obtained from a mechanical impulse applied to the system. From the first frequency it is easy to determine the critical bending speed. Experimentally, the vibroacoustic signal is taken over by a condenser microphone that acts as an acoustic sensor. The Finite Element Method (FEM) is used to numerically determine the natural frequencies and natural modes of a multi-rotor shaft specimen. The mechanical impulse is applied in the boundary conditions used for numerical analysis. Two cases of border conditions are analyzed. In the first case the simple supported shaft (SS) is considered, and in the second case the shaft is considered with free ends. From an experimental point of view for the SS case, this is done by supporting the shaft on two sharp prismatic edges. If the ends are free, the shaft is suspended by two elastic threads by a support bracket. The vibroacoustic response is analyzed using the algorithm given by Fast Fourier Transform (FFT). An analytical method based on the Rayleigh-Ritz method is also used to compare the results.

**Keywords:** Multiple rotors · Critical speed · Impulse excitation method · FE method · FFT algorithm

## 1 Introduction

The problem of determining the natural frequencies of a bar or a shaft that carries multiple rotors has been put in several works. The introduction of the Dirac function to specify the position of the concentrated masses is used in the works [1] and [2]. The analysis of the dynamic behavior of a slender bar carrying a concentrated mass in different positions is given in the paper [3]. The free vibrations of a Timoshenko type bar, having concentrated masses elastic fixing of the bar are analyzed in the paper [4]. This paper analyzes the dynamic behavior of a shaft carrying many rotors. The analytical solution is given by the Rayleigh method, considering the shaft elastic and the disks rigid. The numerical analysis is done by MEF. In this case the entire system shaft-multiple-rotors were considered to be made of the same steel, and are therefore


E.-C. Lovasz et al. (Eds.): MTM&Robotics 2020, MMS 88, pp. 375–383, 2021.
https://doi.org/10.1007/978-3-030-60076-1_34

deformable. From an experimental point of view, the natural frequencies contained in the spectrum are extracted from the acquired signal.

## 2 Analytical Development

The analytical study considers the free vibrations of a flexible shaft that carries multiple concentrated masses or rotors, considered rigid disks. The differential equation that governs the bending vibrations of the shaft is the well-known Euler-Bernoulli equation

$$EI\frac{\partial^4 y(x,t)}{\partial x^4} + \rho A\frac{\partial^2 y(x,t)}{\partial t^2} + \sum\nolimits_{i=1}^{n} m_i \delta(x - x_i)\frac{\partial^2 y(x_i,t)}{\partial t^2} = 0. \tag{1}$$

The first two terms of the equation characterize the vibrations of the shaft considered as a continuous body, while the third term characterizes the inertia of concentrated masses. They act as forces concentrated at the points of the abscissa $x_i$, which is why the Dirac function $\delta(x - x_i)$ must be introduced. Here, y(x, t) is the shaft deformation, x is the axial coordinate, t is the time, EI is the flexural rigidity of the uniform shaft, $\rho$ is the density, A is the cross-sectional area, and $m_i$ are the masses of the rotors considered as disks.

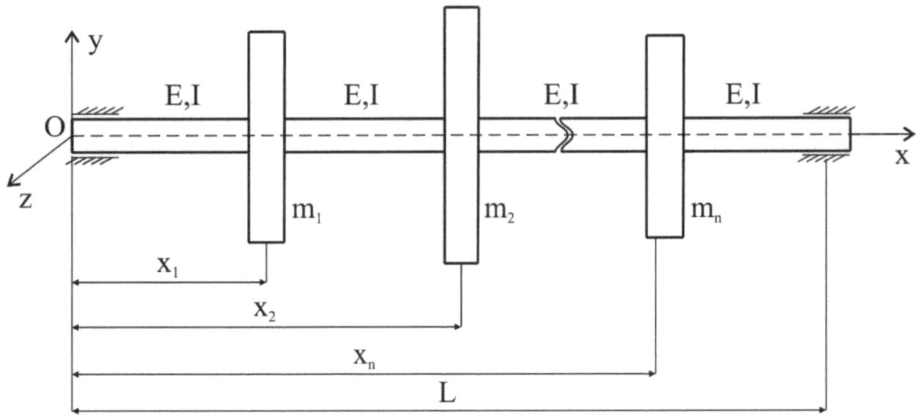


**Fig. 1.** The shaft carrying multiple rotors

The boundary conditions for the shaft shown in Fig. 1 are as follows:

$$y(0,t) = y(L,t) = 0, \text{ and } \frac{\partial^2 y(0,t)}{\partial x^2} = \frac{\partial^2 y(L,t)}{\partial x^2} = 0, \tag{2}$$

To generate a model with n degrees of freedom for a continuous system, the classic solution given by the Rayleigh method is used, where the deformation of the shaft will take the form:

$$y(x,t) = \sum_{i=1}^{n} \phi_i(x) q_i(t) = \{\phi\}^T\{q\} \tag{3}$$

where $\phi_i(x)$ are the mode shape functions assumed here to satisfy all geometric boundary conditions of the shaft. The time functions $q_i(t)$ correspond to the generalized coordinates.

Concentrated masses, located on the abscissa $x_i$, will have velocities

$$v_i = \frac{\partial y(x_i,t)}{\partial t} = \{\phi\}^T\{\dot{q}\}. \tag{4}$$

With these notations it was possible to calculate the kinetic energy of the system, given by:

$$\begin{aligned} E_c &= \frac{1}{2}\int_0^L \rho A \left(\frac{\partial y(x,t)}{\partial t}\right)^2 dx + \frac{1}{2}\sum_{i=1}^{n} m_i \left(\frac{\partial y(x_i,t)}{\partial t}\right)^2 \\ &= \frac{1}{2}\{\dot{q}\}^T[M_1]\{\dot{q}\} + \frac{1}{2}\{\dot{q}\}^T[M_2]\{\dot{q}\} \end{aligned} \tag{5}$$

where

$$[M_1] = \rho A \int_0^L \{\phi\}\{\phi\}^T dx,\ [M_2] = \sum_{i=1}^{n} m_i \left[\{\phi(x_i)\}\{\phi(x_i)\}^T\right] \tag{6}$$

and

$$[M] = [M_1] + [M_2] \tag{7}$$

is the mass matrix.

Similarly, the potential deformation energy of the shaft can be calculated

$$E_p = \frac{1}{2}\int_0^L EI\left(\frac{\partial^2 V(x,t)}{\partial x^2}\right)^2 dx = \frac{1}{2}\{q\}^T[K]\{q\} \tag{8}$$

where

$$[K] = EI\int_0^L \frac{\partial^2}{\partial x^2}\{\phi\}\frac{\partial^2}{\partial x^2}\{\phi\}^T dx \tag{9}$$

is the stiffness matrix.

From the mechanical energy conservation theorem can be obtained the differential equation of the free bending vibrations of the multi-rotor shaft system.

$$[M]\{\ddot{q}\} + [K]\{q\} = \{0\} \tag{10}$$

It can be seen that the mechanical structure consisting of shaft and rotors has two planes of symmetry therefore a bending displacement can take place both in the OXY plane and in the OXZ plane. Consequently, the two functions of the same shape, which determine the bending displacement, can be written as follows:

$$y(x,t) = \sum_{i=1}^{n} Y_i(x)q(t)_i \text{ and } z(x,t) = \sum_{i=1}^{n} Z_i(x)q(t)_i. \tag{11}$$

The number of acceptable functions chosen in Eq. (3) is determined by the number of degrees of freedom considered sufficient to describe the bending vibrations of the shaft. This paper aims to determine the first natural vibration frequency, from which the first critical speed is obtained. Therefore a single function is sufficient. Consequently, the displacements in the two planes OXY and OXZ will be given by the function:

$$y(x,t) = z(x,t) = \sin\frac{x\pi}{L} \cdot \cos(pt-\varphi) \tag{12}$$

where p is the first natural circular frequency.

For the approximate analytical method, the Rayleigh ratio is obtained applying the theorem of conservation of mechanical energy in OXY plane:

$$p^2 = \frac{EI\left(\frac{\pi}{L}\right)^2 \frac{L}{2}}{\frac{\rho AL}{2} + m_1 + m_2 + m_3} \tag{13}$$

and the critical speed will have the expression:

$$n = \frac{30}{L}\sqrt{\frac{EIL}{\rho AL + 2m_1 + 2m_2 + 2m_3}}. \tag{14}$$

Therefore, taking into account that from machining there may not be a perfect symmetry of the multi-rotor shaft, it is necessary to measure the fundamental frequency in the two planes, OXY and OXZ.

## 3 Modal Analysis

In order to correlate the experimentally determined frequencies with the forms of the vibration modes and the corresponding natural frequencies, it is indicated to make a modal analysis by FEM. A mechanical system with n degrees of freedom is governed by the matrix differential equation:

$$[M]\{\ddot{q}\} + [K]\{q\} = \{Q(t)\}. \tag{15}$$

Here, by $\{q\}$ was denoted the vector of displacements, by $\{Q(t)\}$ the external applied forces, and by [M] and [K] the matrices of inertia and rigidity. In the case of free vibrations of the multi-rotor shaft system, the characteristic equation is:

$$\left|-p^2[M] + [K]\right| = 0 \quad (16)$$

This equation leads to a problem of eigenvalues and eigenvectors. In extracting a large number of eigenvectors and corresponding eigenvalues the most recommended method is Block Lanczos. It is also efficient in the case of composite structures, giving good results in a frequency band.

## 4 Experimental Measurement

For the experimental measurement, a stand was made in which the structure subjected to a mechanical impulse consists of a shaft and three rotors. The shaft has a length L = 294 mm, and the three rotors have a bore diameter equal to the diameter of the shaft, i.e. D = 20 mm.

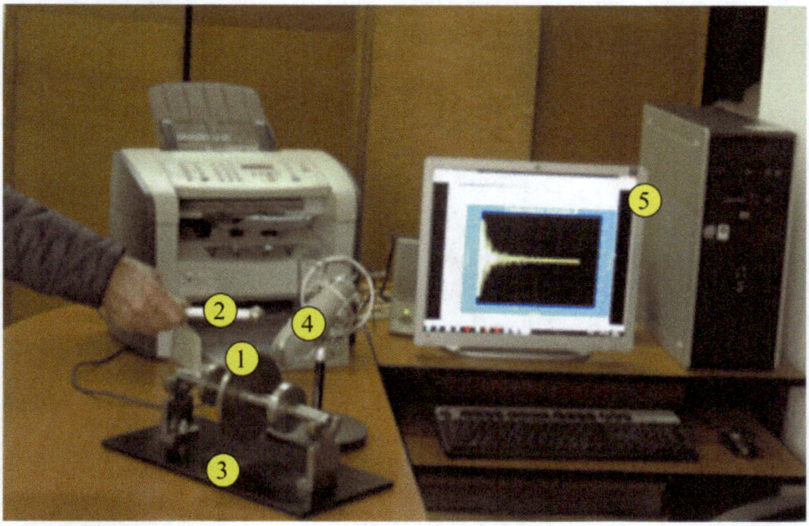


**Fig. 2.** Experimental stand.

For the experimental measurements of free vibrations, a stand was made taking into account the methodology given in [5]. This stand allows the non-contact measurement of the free vibrations of the samples and is shown in Fig. 2. It is composed of the shaft with the three rotors 1, a mini impulsive hammer 2; shaft support 3; the acoustic sensor consisting of a condenser type microphone 4 and the computer which also has the data acquisition board 5. The measuring principle and the equipment, was also used for the works [6–8].

## 5 Results and Discussions

In the modal analysis a three-dimensional element of type Solid 187 was chosen. The number of nodes was 4231, and the number of elements 1821, and the number of modes 28. The mass of the shaft and discs was measured by a digital scale with an accuracy of one hundredth gram. The following values were obtained: mass of the shaft $m_s = \rho AL = 706.24\,\text{g}$, mass of the first disk $m_1 = 214.31\,\text{g}$; the mass of the second disc $m_2 = 807.57\,\text{g}$ and the mass of the third disc $m_3 = 334.27\,\text{g}$. The material from which the entire shaft-rotor system was made was an ordinary hotel, having the density $\rho = 7850\ \text{kg/ m}^3$ and the modulus of elasticity E = 210 GPa. Two cases of boundary conditions were analyzed, namely: free-free ends (FF) and simply supported ends (SS).

The fundamental frequencies of the sample obtained by FEM are given in Table 1 and in Figs. 3 and 4 for F-F ends, and in Figs. 5 and 6 for S-S ends.

**Table 1.** The fundamental frequencies of the shaft in two planes by FEM analysis

| No. | Type of boundary | Fundamental frequency in OXY plane [Hz] | Critical speed in OXY plane [rpm] | Fundamental frequency in OXZ plane [Hz] | Critical speed in OXZ plane [rpm] |
|---|---|---|---|---|---|
| 1 | F-F | 722.57 | 43,354.20 | 722.45 | 43,347 |
| 2 | S-S | 485.61 | 29,136.6 | 485.45 | 29,127 |

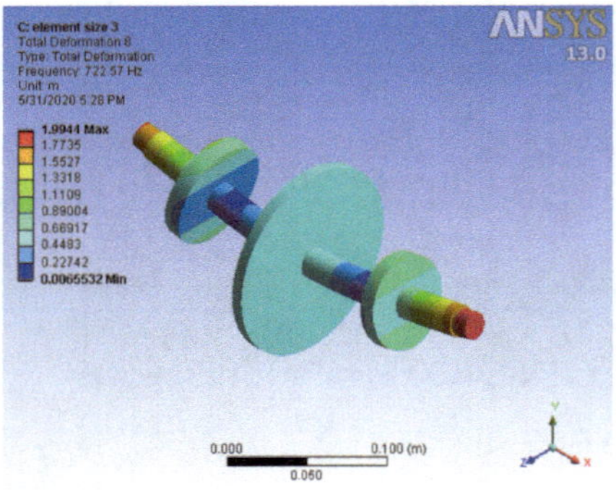


**Fig. 3.** First mode in OXY plane for free-free disk

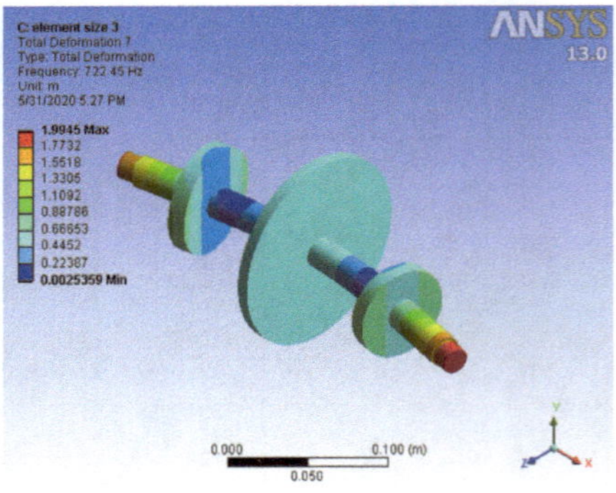


**Fig. 4.** First mode in OXZ plane for free-free disk

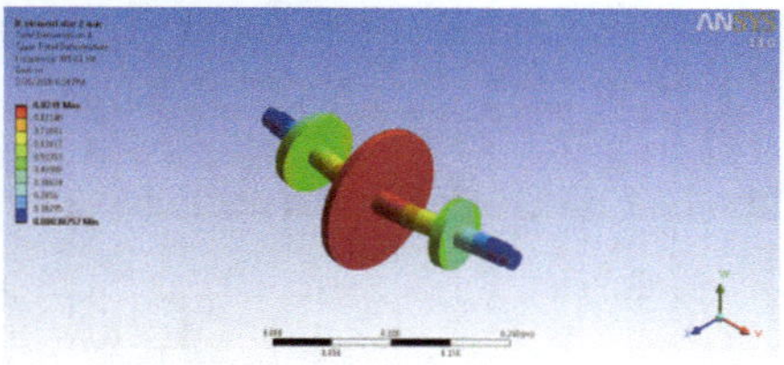

**Fig. 5.** First mode in OXY plane for simple supported disk

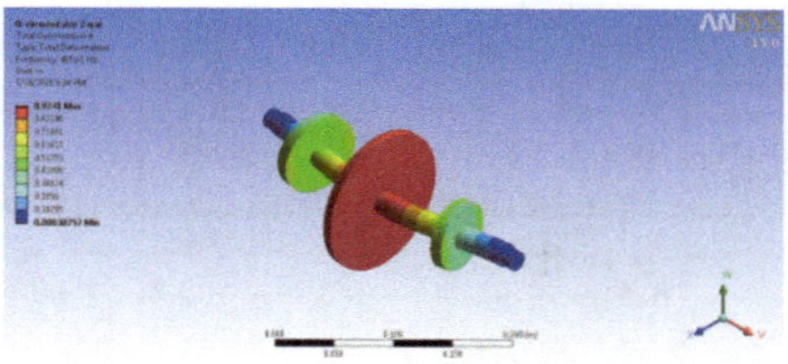

**Fig. 6.** First mode in OXZ plane for simple supported disk

To determine the critical speeds of the three-rotor shaft in the two planes, four signals were acquired: two for F-F ends and two for S-S ends. The analysis of the signals was done with FFT.

The acquired signals have the shape of the one that appears on the computer display, and the fundamental frequency from spectrum of the same signal is given in Fig. 7.

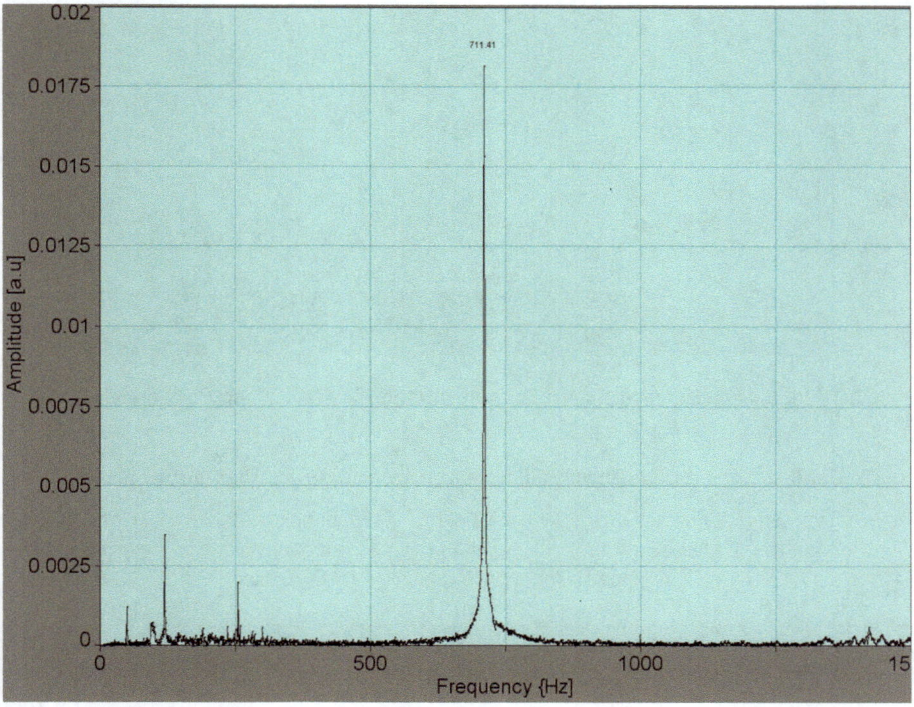


**Fig. 7.** Fundamental frequency from spectrum in OXY plane for F-F ends

If the shaft has supported ends (SS), a vibroacoustic signal is also acquired. After filtering, allowing only components below 800 Hz to pass, the Fourier spectrum is shown in Fig. 8.

The results obtained from the all signals after their processing, as an average value, are given in the Table 2.

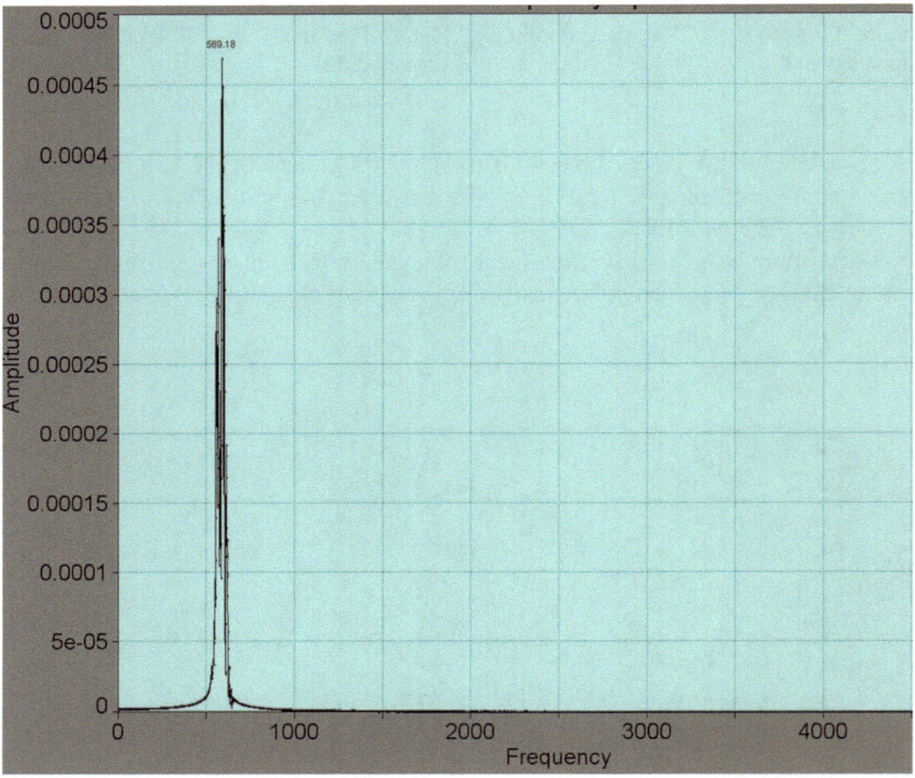


**Fig. 8.** Fundamental frequency from spectrum in OXY plane for S-S ends

**Table 2.** The experimental fundamental frequencies of the shaft in two planes

| No. | Type of boundary | Fundamental frequency in OXY plane [Hz] | Critical speed in OXY plane [rpm] | Fundamental frequency in OXZ plane [Hz] | Critical speed in OXZ plane [rpm] |
|---|---|---|---|---|---|
| 1 | F-F | 711.41 | 42,684.6 | 711.73 | 42,703.8 |
| 2 | S-S | 589.18 | 35,350.8 | 589.81 | 35,388.6 |

## 6 Conclusions

The results obtained by using the FEM modal analysis in two planes are very close to those obtained by the experimental measurement of the fundamental frequencies, which validates the proposed method. There are also larger differences in the case of measurements made with the SS ends, which means that the bearing introduces measurement errors due to the frictions in it. The case of the shaft with free ends gives very good results precisely due to the fact that the support has been replaced with elastic threads. So there were no frictions. Peaks appear in the spectrum that are caused by the network's influences on the microphone. One last conclusion is related to the way the tree is considered. If we take into account the rotational inertia and the effect of shear forces, maybe the results would be much closer. This is the purpose of future work.

## References

1. Chen, Y.: On the vibration of beams or rods carrying a concentrated mass. J. Appl. Mech. **30** (2), 310–311 (1963)
2. Kang, M.G.: Effect of rotary inertia of concentrated masses on the natural vibration of conveying pipes. J. Korean Nucl. Soc. **31**(2), 202–2011 (1999)
3. De Rosa, M.A., Franciosi, C., Maurizi, M.J.: On the dynamic behaviour of slender beams with elastic ends carrying a concentrated mass. Comput. Struct. **58**(6), 1145–1159 (1996)
4. Yesilce, Y., Demirdag, O., Catal, S.: Free vibrations of a multi-span Timoshenko beam carrying multiple spring-mass systems. Sadhana **33**(4), 385–401 (2008)
5. ASTME 1876-15: Standard test method for dynamic Young's modulus, shear modulus and Poisson's ratio by impulse excitation of vibration (2015)
6. Suciu, O., Bereteu, L., Draganescu, G., Ioanovici, T.: Determination of the elastic modulus of hydroxiapatite doped with magnesium through nondestructive testing. Adv. Mater. Res. **814**, 115–122 (2013)
7. Suciu, O., Ioanovici, T., Bereteu, L.: Mechanical properties of hydroxiapatite doped with magnesium, used in bone implants. Acoust. Vibr. Mech. Struct. **430**, 222–229 (2013)
8. Suciu, O., Bereteu, L., Draganescu, G.: Determination of mechanical properties of hydroxiapatite doped with magnesium. In: AIP Conference Proceedings, vol. 1564, pp 132–137 (2013)

# Yarn Tension Control During Weaving Process

Aleš Bílkovský(✉)

VÚTS a.s., Liberec, Czech Republic
ales.bilkovsky@vuts.cz

**Abstract.** The aim of this paper is to provide information about a yarn tension control of the air-jet weaving machine. The stability and robustness of the controlling method of tension is important for the continuous weaving process. The measuring and controlling the tension of warp's thread is not simple during weaving, where the shed is moving and the batten mechanism generates impulses to yarn. However, the quality of the fabric is dependent on the weaving tension. Finally, the goal is to suppress the shed moving tension impact and the quality of yarn spinning and does not react to the batten mechanism. The results are from integration of control to PLC and measured on the real machine.

**Keywords:** Tension control · Weaving · PID regulator · Air-jet weaving machine · PLC

## 1 Introduction

The weaving machines are mechatronics machines that produce different kinds of fabrics. The production of kinds of fabrics is very wide. The weaving process defines the quality of a fabric, which is dependent on the technical solutions of the weaving machines. The basic mechanisms should be controlled, then the tension of the yarn is controlled and moving of the take-up mechanism for the density of the fabric is controlled. The basic mechanisms are the batten mechanism, the shed mechanism, the take-up, and the warp mechanism. The quality of the fabric is dependent on the pick of the weft during the weaving. The crossing of the yarn and weft threads is made fabric.

The weft thread way is moving along the reed, it is narrow but straight. Otherwise, the yarn way influences the weft way by the tension of yarn because of low tension yarn in collision with wefts. When the collision happens then the machine is stopped.

The tension control is also applied in printing, packaging, textiles and paper production.

## 2 Weaving Machine

The described weaving machine is Camel Adaptive by VUTS [1]. It is the air-jet machine to produce leno fabric. It is a mechatronics machine with electronic cam to control mechanisms. Due to electronic cam control, it is possible to change the displacement law of batten mechanism and shed mechanism by the machine speed in rpm.


E.-C. Lovasz et al. (Eds.): MTM&Robotics 2020, MMS 88, pp. 384–390, 2021.
https://doi.org/10.1007/978-3-030-60076-1_35

The schema of the air-jet weaving machine is shown in Fig. 1 and consists of several mechanisms [3]. The basic of the machine is consists of batten mechanism with the reed (a) with accumulation energy in springs, shed mechanism with the harness (b) accumulated the energy in springs and mechanism with needle rod (c). The density of the fabric is defined by a cylinder (d) that take-up the fabric out of the machine. The warp (e) is source of yarns (f). The tension of the yarn is measured by full-width load cell measurement (g).

The yarn is going to reed from warp over full-width load cell measurement, then it is divided into needle way and the shed way. The shed way is going over metal spring compensator (h), which solves the geometric changes due to shed moving. Shed moving and batten moving to load the yarn.

In the case of the Camel Adaptive weaving machine, the density of leno fabric is low and batten mechanism does not generate a peak in yarns. The weft is inserted in reed and is tied in fabric without significant influence of tension.

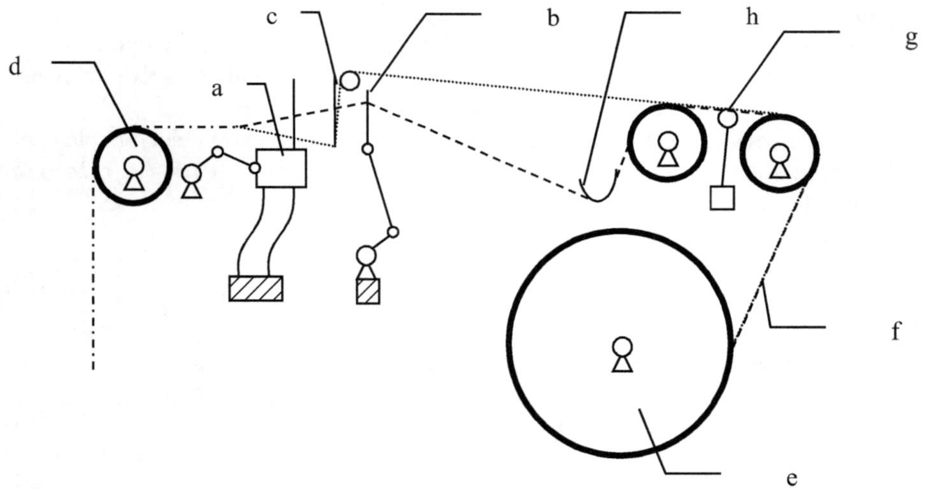


**Fig. 1.** Schema of air-jet weaving machine Camel Adaptive

## 3 Yarn Tension Control

The regulated system is consists of the two-controlled axis. First is the take-up cylinder of fabric, which moves by the machine speed and density of the fabric. The take-up is synchronized with the master axis of the machine and moving constantly. The second axis is warp which moves by the circumferential velocity of take-up.

The basic part of regulated system is full-width load cell measuring of yarn tension. The whole yarn slides over the full-width measuring of tension. If there is some lower tension part of warp that it can be compensated by increasing the tension in other parts in full-width measuring. The tension is define by 1 m of width of warp and the actual tension is average value of the warp tension per 1 m.

If the tension in warp is low, there may be a collision with the weft and the machine will often stop. However, at a higher tension, various fabric or warp thread defects occur. One such defect is warp hairiness when passing through needles or reed. There may also be more breaks that are frequent and thus stopping the machine.

For higher accuracy of tension, it is better measured the diameter of warp. If the position is not sense, the cylinder diameter is calculated by unwinding the spiral from the wound warp of defined length.

### 3.1 Warp Controller to Control Tension

The controlled value is fabric velocity, which is the circumferential velocity of cylinders. Each cylinder has a different diameter and different angular velocity, but the fabric velocity is the same. Finally, the fabric velocity of the controlled warp cylinder is

$$v_{WD} = v_{TU} - \Delta v, \tag{1}$$

where $v_{WD}$ is the circumferential velocity of the warp cylinder, $v_{TU}$ is the circumferential velocity of the take-up cylinder and $\Delta v$ is the control output of the regulator. It represents that the fabric velocity of the warp cylinder is solved as fabric velocity of take-up cylinder and regulation by tension.

As can be seen in the block diagram below (Fig. 2), this function block implements a PI transfer function with the following controller parameters: *Gain* ($K_p$), *IntegrationTime* ($T_i$). The transfer function G(s) is

$$G(s) = K_p\left(1 + \frac{1}{T_i s}\right) \tag{2}$$

The function block output is calculated as follows:

$$e(t) = Set\ Tension - Act\ Tension \tag{3}$$

$$PP = K_p e(t) \tag{4}$$

$$IP = \frac{K_p}{T_i}\int e(t)dt, \tag{5}$$

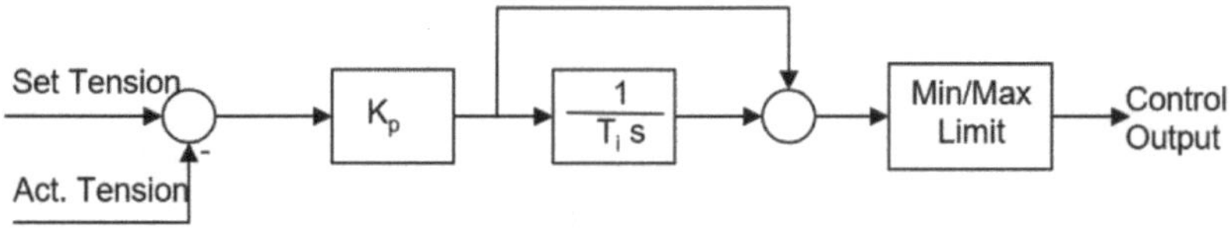


**Fig. 2.** Block diagram of PI regulator

where e(t) is a deviation, $K_p$ is a proportional gain and $T_i$ is an integration time, PP is a proportional part and IP is an integration part. The Control Output is limited by minimal and maximal value.

### 3.2 Axis Mode

The fabric velocity of the warp cylinder is the input value for the angular velocity of the axis. The axis is defined as an infinite axis with a period of 1 turn that is defined with 360000 points. The factor is set to 1000 and the final setting is 360.000 degrees per revolution.

The control mode of the axis is set as the cyclic velocity control [2], where the value is defined every cycle time and changing of diameter is gradual. The angular velocity of warp cylinder is

$$\omega = \frac{v_{WD}}{D_{act}}, \tag{6}$$

where $D_{act}$ is the actual diameter of the warp cylinder. The yarns are unwinding out of warp and the diameter is slowly going to the basic limit. The diameter is not measured and it is calculated.

### 3.3 Diameter Calculation

For the diameter calculation was select the model of a spiral. The yarns are winded on the cylinder. The thickness of the yarn is very low but the number of yarn's layers is significant. The basic values of the warp diameter are the length of the threads L, the initial diameter of full warp $D_1$ and diameter of warp tube $D_0$. The actual diameter is

$$D_{act} = \left(D_1 - \frac{D_1^2 - D_0^2}{4L}\varphi_{WD}\right) \tag{7}$$

$$\varphi_{WD} = 2\Pi\left(N_{WD} + \frac{ActPos}{360}\right), \tag{8}$$

where $N_{WD}$ is the number of revolutions from the beginning of warp and ActPos is actual position.

## 4 Results of Control Tension

The value of actual tension is measured for four separate moving conditions. The first is the synchronization of the machine, second is positioning, third is slow forward moving and last is running.

### 4.1 Synchronization

The machine is state "zero", where the axis power is off and the machine is safe. After set Enable to on the axis power is changed to power on and waiting for synchronization. The pressing synchronization the machine is set to control mode and then it is ready for weaving. The tension in threads of the warp is controlled during synchronization. The result of tension control is on Fig. 3.

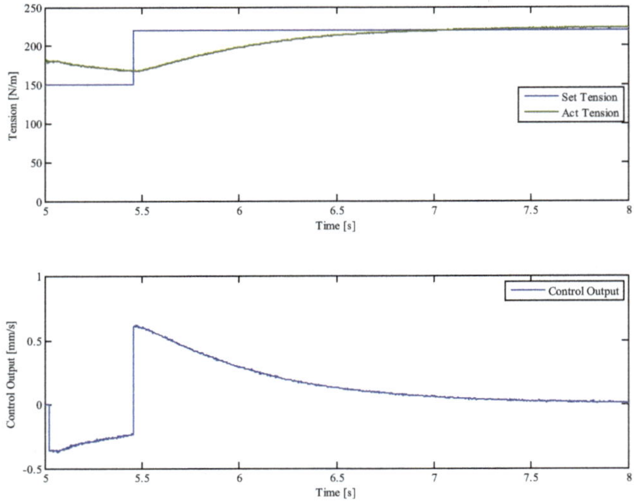


**Fig. 3.** Tension during synchronization

At the first time, the control output is negative, because the initial value of set tension value is lower than the actual value. During the synchronization's steps, the set tension value is changed and the regulation is working.

### 4.2 Positioning

One of the moving of the machine is positioning by Go to position. From 0° position move to 180° position and then back (Fig. 4). Due to the geometry of the machine, the first part of the shed moving causes the lower tension in yarns and the control output winds up. Then the tension is rising and control output unwinds.

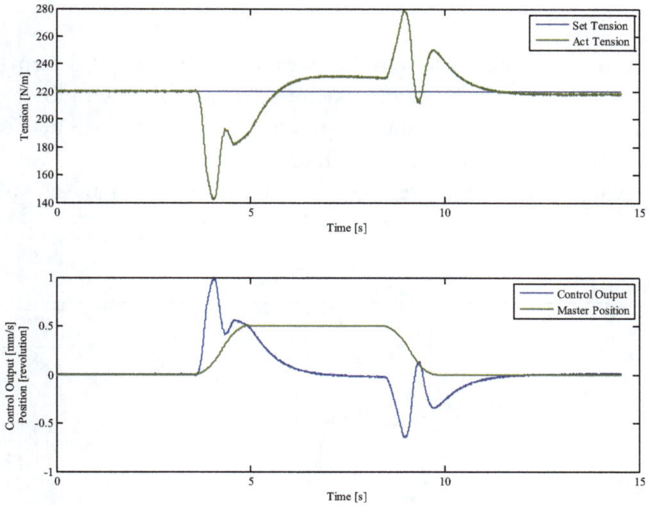


**Fig. 4.** Tension during positioning to 180° and back to 0°

## 4.3 Slow Forward Moving

Another move of the machine is slow forward moving. The tension is significantly influenced by shed (Fig. 5) and the tension value repeat in each turn.

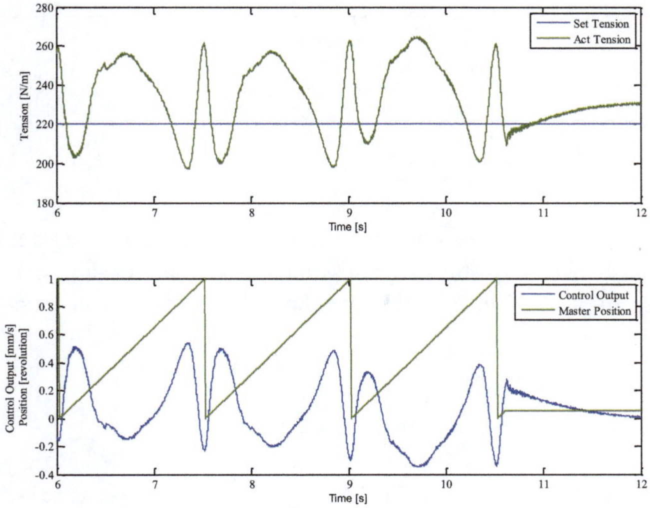


**Fig. 5.** Tension during slow forward moving 40 rpm

### 4.4 Run

The results for running are measured for weaving speed of 600 rpm (Fig. 6). The system reacts to moving all parts of the machine, mainly the shed moving. Take-up of fabric is continuously moving and the warp is moving with correction. Control output is negative and the warp is moving faster then take-up.

The mean value during the weaving is a little higher than set value and it is 239 N/m.

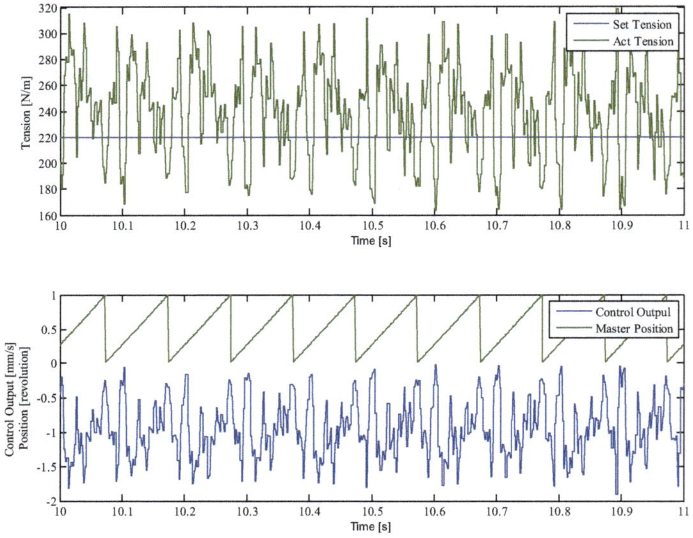


**Fig. 6.** Tension during running 600 rpm

## 5 Conclusion

Separate moving conditions were tested. The tension control of the yarn is influenced by machine mechanisms, quality of yarn winding, winding tension and by the calculated diameter of warp. The diameter of the warp is significant parameter of the regulation.

## References

1. Prototyp - Tryskový tkací stroj CAM EL ADAPTIVE. VÚTS, Liberec. https://www.vuts.cz/prototyp-tryskovy-tkaci-stroj-cam-el-adaptive.html. Accessed 08 Apr 2020
2. B&R Industrial Automation GmbH. Automation Studio: B&R Help Explorer, Eggelsberg, March 2017
3. Dvořák, J., Kastner, V., Karel, P.: Weaving machine. EP1675988 Patent. Date of publication and mention of the grant of the patent, 14 May 2014

# IC Engine: Increasing Efficiency by Using Epicyclic Non-circular Gear Train

Dijana Čavić, Maja Čavić, Marko Penčić(✉), Jovan Dorić, and Milan Rackov

Faculty of Technical Sciences, University of Novi Sad, Trg Dositeja Obradovića 6, 21000 Novi Sad, Serbia
{dijana.cavic, scomaja, mpencic, jovan_d, racmil}@uns.ac.rs

**Abstract.** Increasing the efficiency of IC engines is a frequently discussed topic, and one of the approaches taken is to modify the thermodynamic work cycle of the engine so that it resembles the ideal cycle more closely. This paper attempts to achieve that by altering the kinematics of the piston-crank mechanism by introducing an epicyclic non-circular gear train between the crank and the flywheel. The addition of the non-circular gears alters the work cycle by introducing a state of dwell during key points of the cycle. Possible gear shape solutions were calculated using a known, periodic function to express the gear transmission ratio. The final result was an epicyclic non-circular gear train that has a noticeable effect on the kinematics of the piston-crank mechanism resulting in an increase of the engines work cycle efficiency.

**Keywords:** IC engine · Increasing efficiency · Epicyclic non-circular gear train

## 1 Introduction

The chain of thermodynamic reactions occurring in the piston chamber of an IC engine is known as the work cycle. Due to the complexity of these reactions, the work cycle is analyzed through different mathematical models. In theory, the most efficient thermodynamic cycle for IC engines is the Otto cycle, which consists of isentropic compression and expansion processes, and isochoric heat addition and expulsion processes [1].

The traditional design of IC engines with a simple slider-crank mechanism does not allow enough time for isochoric combustion which significantly reduces the cycle efficiency. The ideal scenario is to initiate and complete the combustion event while the piston remains at the TDC position. This provides the maximum thermal potential and eliminates the negative work due to early ignition which is well into the compression stroke with conventional engine strategies. One practical method of achieving this optimization is to significantly reduce the piston velocity at the TDC position to provide extra time for completing the combustion [2]. Modifications to the motion of the

The modified piston-crank mechanism described in this paper is patent pending [P-2020/0854].


E.-C. Lovasz et al. (Eds.): MTM&Robotics 2020, MMS 88, pp. 391–402, 2021.
https://doi.org/10.1007/978-3-030-60076-1_36

piston made in order to improve the efficiency of the IC engine have been studied by many researchers [3–5].

This paper explores the possibility of increasing the efficiency of IC motors by modifying the piston-crank mechanism with the addition of non-circular gears, thus changing the law of motion of the piston and forcing a period of pseudo-dwell which in turn would bring the thermodynamic work cycle of the engine closer to the ideal work cycle. The goal of this paper is to offer a viable solution for the problem of increasing IC engine efficiency in the form of non-circular gears and analyzing their effect on the whole mechanism and their overall viability.

## 2 State of the Art

Non-circular gears can be used in a wide spectrum of machinery whenever an output link of the mechanism has to move according to a specific law of motion [6]. The use of non-circular gears in the design of a mechanical press which is driven according to a kinematically optimized law of motion is shown in [7]. Another novel press mechanism is presented in [8]. The application of non-circular gears for the generation of functions is done in [9]. Novel design solutions for fishing reel mechanisms based on non-circular gears are proposed in [10]. The synthesis of transmissions with a dwell or a constant transmission ratio based on a combination of planar linkage mechanisms and non-circular gears is presented in [11]. The synthesis of mechanisms with intermittent motion which incorporate non-circular gears is shown in [12, 13]. Despite the fact that the kinematics of non-circular gears are well researched [14–16], they are not yet widely used in industrial applications, due to manufacturing issues. Different ways of tooth profile generation and preproduction processes are discussed by [17–20].

## 3 Modified Piston-Crank Mechanism

Figure 1 shows the conventional piston-crank mechanism used in IC engines, and consists of crankshaft, connecting rod and piston. While the piston is technically the driving element in this case, the mechanism is powered by the crankshaft for three out of the four strokes, by way of the kinetic energy accumulated in the flywheel. Due to this, the crankshaft will be regarded as the driving (input) link further in this paper.

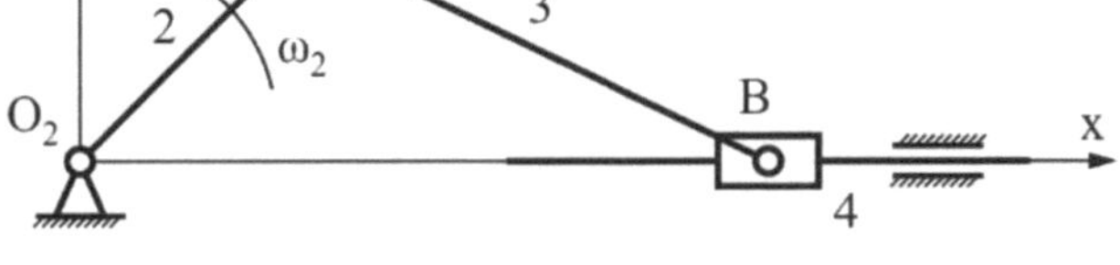


**Fig. 1.** Kinematic scheme of the conventional piston-crank mechanism: 2 – crankshaft, 3 – connecting rod and 4 – piston.

Piston laws of motion obtained by kinematic analysis are shown in Fig. 2. Following parameters were adopted as such (the kinematic parameter of most IC engines being $\lambda = r_2/r_3 = 0.25 \div 0.3$ [1]): $\omega_2 = 100\pi\ s^{-1}$, $r_2 = 0.06$ m, $r_3 = 0.2$ m. The parameters $\omega_2$, $r_2$ and $r_3$ being the velocity of the crankshaft, the radius of the crankshaft and the length of the connecting rod, respectively.

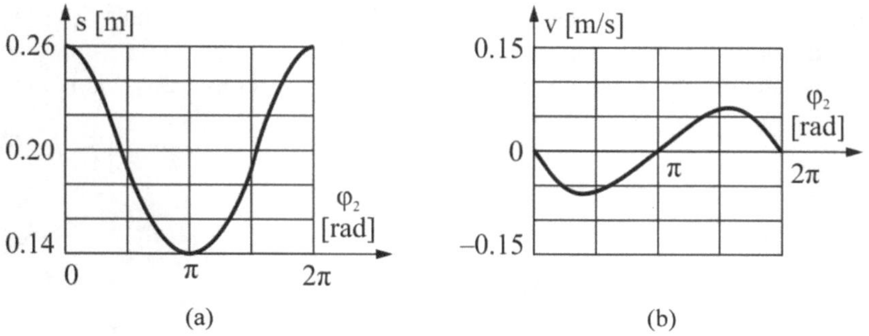


**Fig. 2.** Conventional piston-crank mechanism: (a) piston position and (b) piston velocity.

### 3.1 Modification

As is evident in Fig. 2b, the velocity of the piston is equal to zero only in the moments when it passes through the top and bottom dead centers. Figure 3 shows the difference between the law of motion of the conventional mechanism and the law of motion the piston would need to follow to achieve an ideal work cycle.

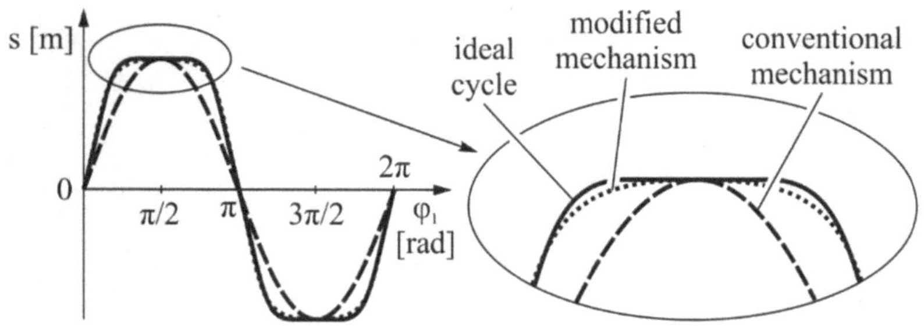


**Fig. 3.** Law of piston motion.

The conventional piston-crank mechanism operates on the assumption that the input velocity supplied by the crankshaft is approximately constant due to the flywheel. By changing this velocity from constant to the variable, it becomes possible to influence the law of motion of the piston. However, since the crank piston mechanism is only one part of the more complicated IC engine, which is again only a part of the power train of a vehicle, certain other requirements need to be taken into consideration. Therefore, the modified mechanism needs to fulfill the following requirements:

- supply the variable input speed of the crankshaft needed to achieve the prescribed law of motion of the piston,
- avoid unnecessary complications of the construction and
- avoid stopping the latter parts of the vehicle power train (the flywheel, the gearbox elements, etc.).

Figure 4 shows one possible solution that would fulfill these requirements – a pair of non-circular gears inserted between the flywheel and the crankshaft. This type of mechanism only offers the possibility of pseudo-dwell. To bring the work cycle as close to the ideal cycle as possible, a mechanism that offers true dwell is required.

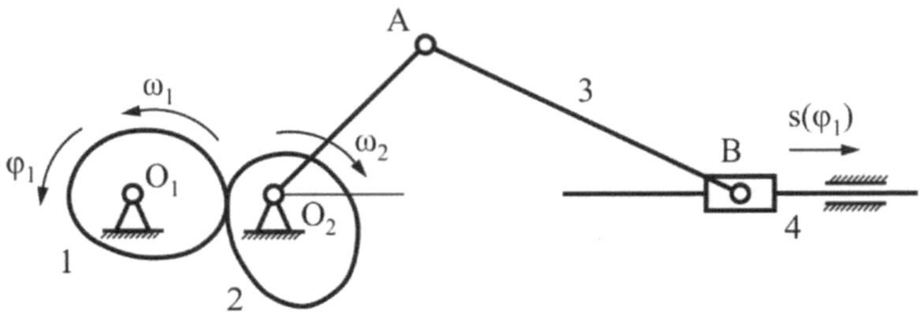


**Fig. 4.** Kinematic scheme of the modified piston-crank mechanism (non-circular gear pair): 1 – input gear (flywheel shaft), 2 – output gear (crankshaft), 3 – connecting rod and 4 – piston.

Figure 5 shows a mechanism that can achieve true dwell. The gear connected to the flywheel would be considered the input gear and would have a constant velocity. The gear connected to the crankshaft would have a variable velocity which is dictated by the shape of the gears, therefore theoretically offering the possibility of defining the motion of the crankshaft with an arbitrary function, and consequently, the motion of the piston as well.

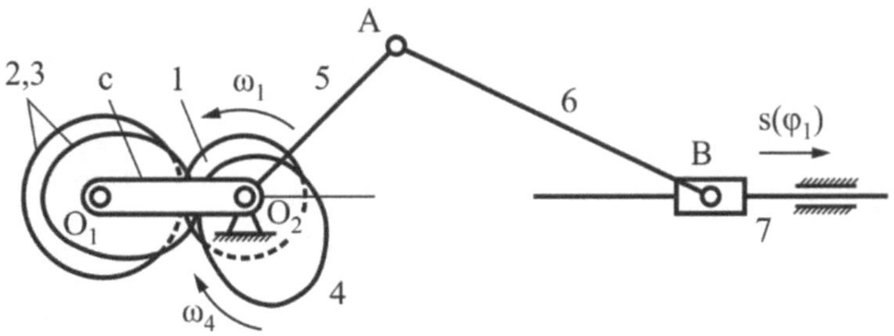


**Fig. 5.** Kinematic scheme of the modified piston-crank mechanism (epicyclic non-circular gear train): 1 – input gear (flywheel shaft), c – carrier, 2 and 3 – planetary gears, 4 – output gear (crankshaft), 5 – crank, 6 – connecting rod and 7 – piston.

### 3.2 Kinematics

Figure 6 shows an epicyclic gear train with both circular and non-circular gears. There are five elements for each one stage epicyclic gear train: carrier *c*, central gears 1 and 4, and planet gears 2 and 3 which are rigidly connected. Central gear 4 provides the output motion. Planet carrier *c* and central gears 1 and 4 are coaxial. The planet gears perform planetary motion – they revolve around their own axis $O_2O_3$ (carrier *c*) and around the common axis of rotation $O_1O_4$ of gears 1, 4 and carrier *c*. In order for this to be a mechanism with one degree of freedom, one of the central gears has to be fixed, and in this case it is gear 1, so $\omega_1 = 0$. Gears 1 and 2 are circular, while gears 3 and 4 are non-circular. Due to the variable transmission ratio of the non-circular gear pair, if the gear train has constant input $\omega_c$ = const., the output $\omega_4$ that changes according to a specific law of motion $i^1_{4c}(\varphi_c)$ can be obtained.

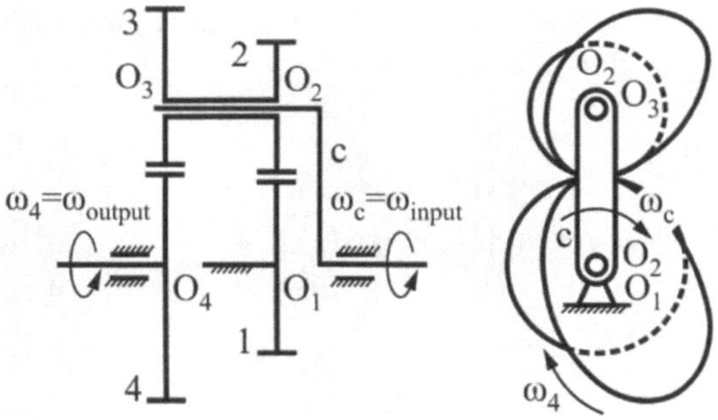


**Fig. 6.** Epicyclic gear train featuring circular and non-circular gears.

According to Willis, the following is obtained:

$$i^1_{4c} = \frac{\omega_4 - \omega_1}{\omega_c - \omega_1} = \frac{\omega_4}{\omega_c} \tag{1}$$

$$\omega_4(\varphi_c) = i^1_{4c}(\varphi_c) \cdot \omega_c(\varphi_c) \;\Rightarrow\; \varphi_4(\varphi_c) = \int_0^{\varphi_c} i^1_{4c}(\varphi_c) d\varphi_c \tag{2}$$

For the piston to achieve dwell in the top ($\varphi_c = 0$) and bottom ($\varphi_c = \pi$) dead center, the angular velocity of the crankshaft must be equal to zero in a specified range around angles $\varphi_c = 0$ and $\varphi_c = \pi$. Also, the angular velocity of the crankshaft must not change direction. In accordance with these requirements, Fig. 7 shows the function graph of the transmission ratio $i^1_{4c}$.

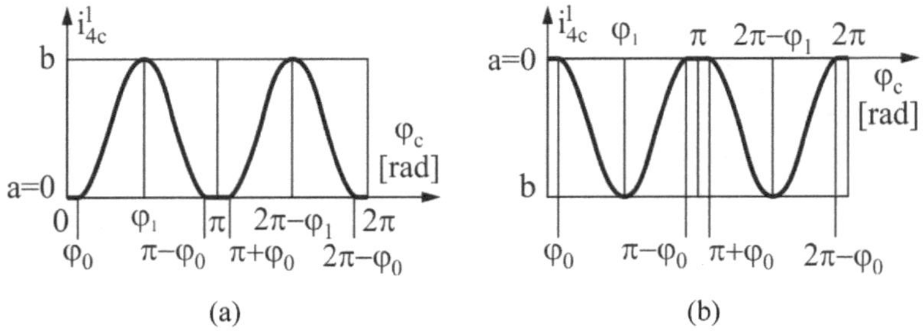


**Fig. 7.** Transmission ratio $i^1_{4c}$: (a) positive sign and (b) negative sign.

Figure 7a shows the transmission ratio $i^1_{4c}$ when it is positive – angular velocities $\omega_c$ and $\omega_4$ have the same direction, while Fig. 7b shows the transmission ratio $i^1_{4c}$ when it is negative – angular velocities $\omega_c$ and $\omega_4$ are in different directions. Accordingly, the general form of the function is as follows:

$$i^1_{4c}(\varphi_c) = \begin{cases} \pm a, & \varphi_c \in [0, \varphi_0) \\ \pm\frac{1}{2}(a+b) \mp \frac{1}{2}(b-a)\cos(\pi\frac{\varphi_c-\varphi_0}{\varphi_1-\varphi_0}), & \varphi_c \in [\varphi_0, \varphi_1) \\ \pm\frac{1}{2}(a+b) \mp \frac{1}{2}(b-a)\cos(\pi\frac{\pi-\varphi_c-\varphi_0}{\varphi_1-\varphi_0}), & \varphi_c \in [\varphi_1, \pi-\varphi_0) \\ \pm a, & \varphi_c \in [\pi-\varphi_0, \pi) \\ i^1_{4c}(\varphi_c-\pi), & \varphi_c \in [\pi, 2\pi) \end{cases} \quad (3)$$

Taking into consideration the basic functional requirements of an IC engine, for every full rotation of the crankshaft, the flywheel shaft must also complete one full rotation. Therefore:

$$\varphi_4(\varphi_c = 2\pi) = 2\pi \;\Rightarrow\; 2\pi = \int_0^{2\pi} i^1_{4c}(\varphi_c)d\varphi_c \quad (4)$$

Solving Eq. (4) results in the following equation connecting parameters $a$ and $b$:

$$b = \frac{2 - 2a\frac{\varphi_0}{\varphi_1}}{1 - \frac{\varphi_0}{\varphi_1}} - a \quad (5)$$

Since transmission ratio $i^1_{4c}$ cannot be directly expressed using the geometric parameters, by using Eq. (1) the following equation is obtained:

$$i^1_{4c} = 1 - i^c_{41} = 1 - \left(-\frac{R_3}{R_4}\right)\left(-\frac{R_1}{R_2}\right) = 1 - i^c_{43}i^c_{21} \quad (6)$$

where $i^c_{43}$ and $i^c_{21}$ are the transmission ratios for relative motion in regard to the carrier $c$.

Relevant to the previous equation is the following:

$$i_{21}^{c}(\varphi_{1r}) = -\frac{R_1}{R_2} = \frac{\omega_2 - \omega_c}{\omega_1 - \omega_c} = \frac{\omega_{2r}}{\omega_{1r}} \tag{7}$$

where:

$$\omega_{1r} = \omega_1 - \omega_c = -\omega_c \tag{8}$$

$$\omega_{2r} = -\frac{R_1}{R_2}\omega_{1r} = \frac{R_1}{R_2}\omega_c \tag{9}$$

$$\varphi_{2r} = -\frac{R_1}{R_2}\varphi_{1r} = \frac{R_1}{R_2}\varphi_c \tag{10}$$

Gears 2 and 3 are rigidly connected, meaning $\omega_{2r} = \omega_{3r}$ and $\varphi_{2r} = \varphi_{3r}$. According to this, the transmission ration between gears 3 and 4 for the motion relative to the carrier $c$ is:

$$i_{43}^{c}(\varphi_{3r}) = \left(-\frac{R_3}{R_4}\right) = \frac{\omega_4 - \omega_c}{\omega_3 - \omega_c} = \frac{\omega_{4r}}{\omega_{3r}} \tag{11}$$

$$\varphi_{4r}(\varphi_{3r}) = \int_0^{\varphi_{3r}} i_{43}^{c}(\varphi_{3r})d\varphi_{3r} \tag{12}$$

Keeping in mind that $\omega_{4r} = \omega_4 - \omega_c$:

$$\varphi_{4r}(\varphi_c) = \varphi_4(\varphi_c) - \varphi_c \tag{13}$$

If $i_{4c}^{1} \geq 0$, then $\varphi_{4r}$ changes as an oscillatory function, which is beneficial in certain situations [13]. However, it is unacceptable in this case, so the adopted condition is $i_{4c}^{1} \leq 0$. Figure 8a shows the function describing the output angle $\varphi_4(\varphi_c)$, while Fig. 8b shows the function describing the position angle of gear 4 during motion relative to the carrier $c$, $\varphi_{4r}(\varphi_c)$.

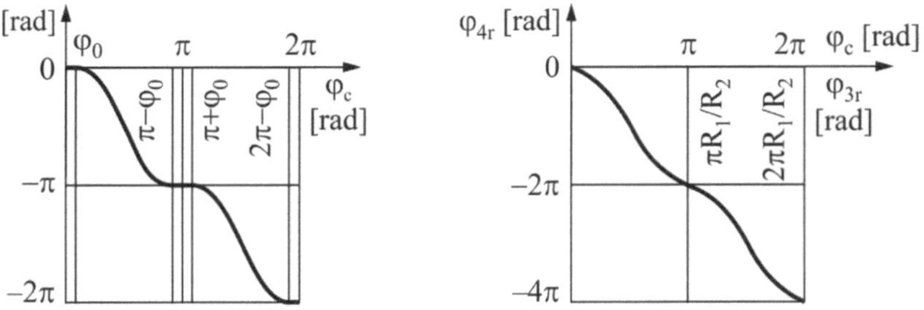


(a) (b)

**Fig. 8.** The functions describing angles: (a) output angle $\varphi_4(\varphi_c)$ i (b) position angle of gear 4 during motion relative to the carrier *c*, $\varphi_{4r}(\varphi_c)$.

The centroids of gears 3 and 4 need to be closed curves, and therefore:

$$\frac{T_{3r}}{n_{3r}} = \frac{T_{4r}}{n_{4r}} \tag{14}$$

where $T_{3r}$ and $T_{4r}$ are the time periods of the rotation of gears 3 and 4 for the motion relative to the carrier *c*, while $n_{3r}$ and $n_{4r}$ are integers.

While there are many possible solutions, since $\varphi_{4r}(\varphi_{3r})$ is a periodic function with the period being $\pi R_1/R_2$, the technologically simplest solution is with $R_1/R_2 = 2$ and $n_{3r} = n_{4r} = 1$. The following angle values are adopted as $\varphi_1 = \pi/2$ i $\varphi_0 = 10\pi/180$. Keeping in mind Eq. (12), transmission ratio $i_{43}^c$ equals:

$$i_{43}^c(\varphi_{3r}) = \frac{d\varphi_{4r}}{d\varphi_{3r}} \tag{15}$$

Figure 9 shows the function graph of transmission ratio $i_{43}^c(\varphi_{3r})$.

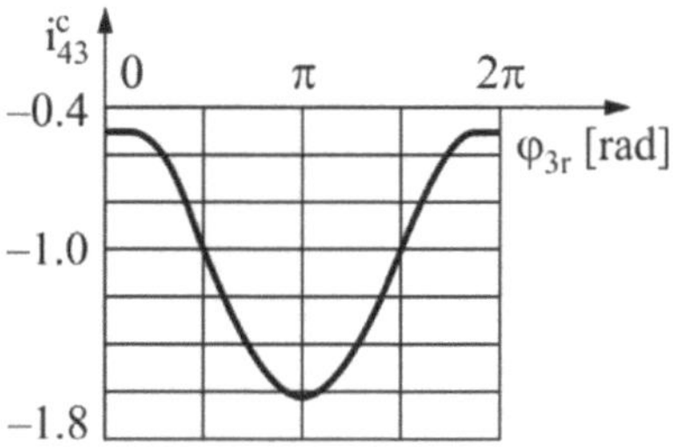


**Fig. 9.** Transmission ratio $i_{43}^c(\varphi_{3r})$.

The centroids of gears 3 and 4 are described with the following functions:

$$R_3(\varphi_{3r}) = \frac{L}{1 + \left|i_{43}^c(\varphi_{3r})\right|}\left|i_{43}^c(\varphi_{3r})\right|,\ \varphi_{3r} = 0\ldots 2\pi \tag{16}$$

$$R_4(\varphi_{3r}) = L - r_3(\varphi_{3r}),\ \varphi_{4r}(\varphi_{3r}) = \int_0^{\varphi_{3r}} i_{43}^c(\varphi_{3r})d\varphi_{3r},\ \varphi_{4r} = 0\ldots 2\pi \tag{17}$$

Figure 10, based on Eqs. (16–17) and an interaxis distance adopted as L = 100 mm, shows the obtained centroids of gears 3 and 4.

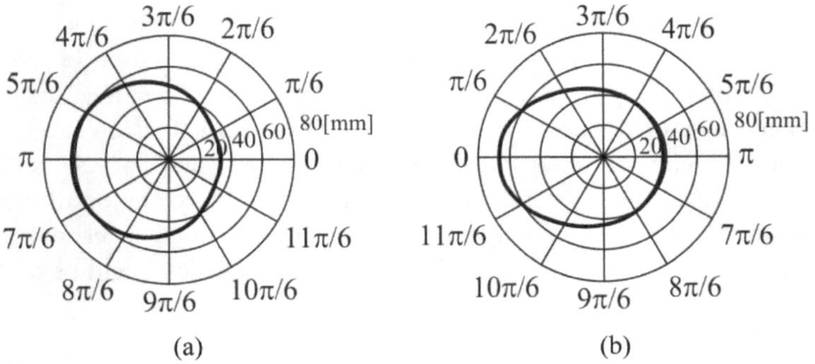


**Fig. 10.** Centroids of gears: (a) gear 3 and (b) gear 4.

## 4 Motion Simulation and Results

After conducting the kinematic analysis of this newly modified piston-crank mechanism, the law of motion of the piston is obtained. The bottom and top dead center positions and the piston stroke are adopted based on existing IC engines [1], respectively: $S_{BDC}$ = 140 mm, $S_{TDC}$ = 260 mm, h = $S_{TDC}$–$S_{BDC}$ = 120 mm.

Figure 11 shows the modified law of motion, and allows the comparison of the modified and conventional laws of motion. The piston of the modified mechanism dwells when near the top and bottom dead centers, in intervals –10÷10° of the crankshaft revolution, meaning for the 10° before and after the dead centers. It is evident that the piston velocity of the modified mechanism is zero in the stated intervals of the crankshaft revolution around the dead centres, which was the goal.

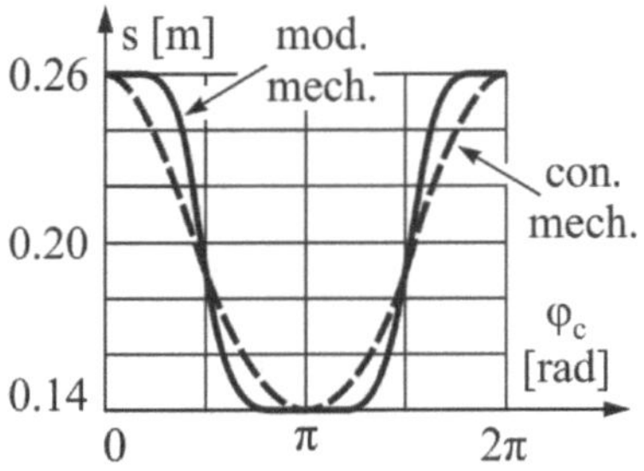


**Fig. 11.** Piston position.

Figure 12a shows the difference between the piston velocities of the conventional and modified mechanisms. Due to the fact that the input velocity is now variable instead of constant, the law of motion describing piston acceleration of the modified mechanism is different from the piston acceleration of the conventional mechanism – both piston accelerations can be seen in Fig. 12b. The modified piston acceleration is a smooth function with no sudden extreme changes in value, which is favorable. However, it is evident that, when compared to the piston acceleration of the conventional mechanism, the acceleration magnitude of the modified mechanism is much higher. This implies the existence of higher inertial loads, which could be problematic.

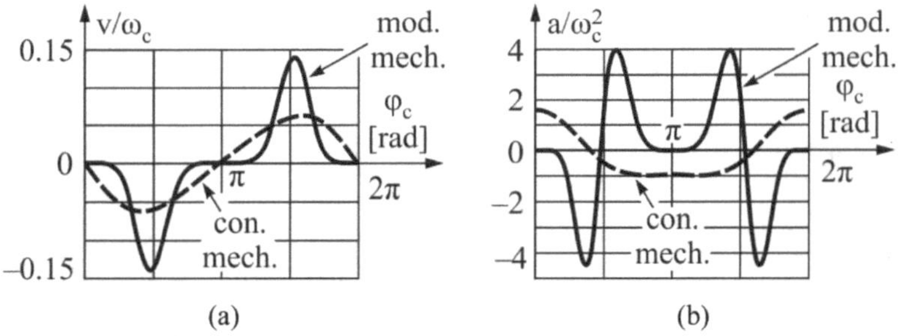


**Fig. 12.** Piston velocity (a) and piston acceleration (b).

## 5 Conclusion

This paper proposes the use of non-circular planetary gears with the goal of forcing the piston into a dwell period at the top and bottom dead centers. Changing the piston kinematics this way influences the work cycle at the engine, bringing the real work cycle closer to the ideal cycle. Piston dwell in the dead centers, mainly the top dead center, simulates the isochoric heat transfer present in the ideal cycle, which results in an increase of engine efficiency.

Earlier attempts to influence the piston kinematics with a pair of non-circular gears did not yield appropriate results, as it was impossible to create a gear shape that both achieved real dwell and was possible to make. These gear pairs made possible a period of pseudo-dwell, and a compromise had to be made between the length of the pseudo-dwell and the manufacturing difficulty of the gears. This problem has been completely circumvented with the use of non-circular planetary gears, making both a period of real dwell and machinable gear shapes possible. Also worth mentioning is that by using planetary gears, the coaxiality between the crank and the flywheel is kept intact. However, not all problems were avoided.

The kinematics analysis shows both higher values and sharp peaks of the acceleration around the dead centers, which implies high inertial loads. A dynamic analysis of the mechanism to examine the inertial loads is a topic planned for future research. The machinability of the gear teeth was not studied in detail in this paper, and deserves more attention in the future. Although the automobile industry is attempting to transition to more eco-friendly energy sources, IC engines will stay current for a long time. Due to this, any research into making them more efficient, thus making them consume less fuel and emit less pollution, stays relevant.

## References


1. Heywood, J.B.: Internal Combustion Engine Fundamentals. McGraw-Hill, New York (1988)
2. Dorić, J.Ž., Klinar, I.J.: Efficiency of a new internal combustion engine concept with variable piston motion. Therm. Sci. **18**(1), 113–127 (2014)
3. Dorić, J.Ž., Klinar, I.J.: Efficiency characteristics of a new quasi-constant volume combustion spark ignition engine. Therm. Sci. **17**(1), 119–133 (2013)
4. Pouliot, H.N., Delameter, W.R., Robinson, C.W.: A variable-displacement spark-ignition engine. SAE Tech. Paper **770114**(2), 446–464 (1978)
5. ElBahloul, M.A., Aziz, E.S., Chassapis, C.: Performance study of the hypocycloid gear mechanism for internal combustion engine applications. Int. J. Engine Res. (2019, in press). https://doi.org/10.1177/1468087419893583
6. Chironis, N.P.: Mechanisms, Linkages and Mechanical Controls. McGraw-Hill, New York (1965)
7. Mundo, D., Danieli, G.A.: Use of non-circular gears in pressing machine driving systems. IASME Trans. **1**(1), 7–11 (2004)
8. Soong, R.-C.: On the new mechanical press with a planetary gear train. Trans. Can. Soc. Mech. Eng. **40**(1), 45–58 (2016)
9. Litvin, F.L., Gonzalez-Perez, I., Fuentes, A., Hayasaka, K.: Design and investigation of gear drives with non-circular gears applied for speed variation and generation of functions. Comput. Methods Appl. Mech. Eng. **197**(45–48), 3783–3802 (2008)
10. Lovasz, E.-C., Modler, K.-H., Neumann, R., Gruescu, C.M., Perju, D., Ciupe, V., Maniu, I.: Novel design solutions for fishing reel mechanisms. Chin. J. Mech. Eng. **28**(4), 726–736 (2015)
11. Modler, K.-H., Lovasz, E.-C., Bär, G.F., Neumann, R., Perju, D., Perner, M., Mărgineanu, D.: General method for the synthesis of geared linkages with non-circular gears. Mech. Mach. Theory **44**(4), 726–738 (2009)
12. Zheng, F., Hua, L., Han, X., Li, B., Chen, D.: Synthesis of indexing mechanisms with non-circular gears. Mech. Mach. Theory **105**(1), 108–128 (2016)
13. Smelyagin, A.I., Prikhod'ko, A.A.: Structure and kinematics of a planetary converter of the rotational motion into the reciprocating rotary motion. J. Mach. Manuf. Reliab. **45**(6), 500–505 (2016)

14. Mundo, D.: Geometric design of a planetary gear train with non-circular. Mech. Mach. Theory **41**(4), 456–472 (2006)
15. Litvin, F.: Gears Geometry and Applied Theory. Prentice Hall, New Jersey (1994)
16. Tong, S.-H., Yang, D.C.H.: Generation of noncircular pitch curves. ASME J. Mech. Des. **120**(2), 337–341 (1998)
17. Chang, S.-L., Tsay, C.-B., Wu, L.-I.: Mathematical model and undercutting analysis of elliptical gears generated by rack cutters. Mech. Mach. Theory **31**(7), 879–890 (1996)
18. Chang, S.-L., Tsay, C.-B.: Computerized tooth profile generation and undercut analysis of noncircular gears manufactured with shaper cutters. ASME J. Mech. Des. **120**(1), 92–98 (1998)
19. Litvin, F.: Gear Geometry and Applied Theory. Cambridge University Press, Cambridge (2004)
20. Litvin, F.L., Gonzalez-Perez, I., Yukishima, K., Fuentes, A., Hayasaka, K.: Generation of planar and helical elliptical gears by application of rack-cutter, hob, and shaper. Comput. Methods Appl. Mech. Eng. **196**(41–44), 4321–4336 (2007)

# Interference Fits. Bearing Capacity Under Complex Loading – FEM Analysis

Corina-Mihaela Gruescu, Arjana Davidescu, Carmen Sticlaru(✉), and Erwin-Christian Lovasz(✉)

University Politehnica of Timisoara, Timişoara, Romania
carmen.sticlaru@upt.ro

**Abstract.** Interference fits are dimensioned by calculating the necessary contact pressure for transmitting axial and/or tangential loadings. The FEM simulations presented in the paper aim to learn if the vector composition law of loadings is also fit for contact pressures. An interference fit Φ30H7/r6 is analyzed. The analytical relationships for pressure and bearing capacity under simple loading are validated by FEM analysis. Under combined loading, axial and tangential, the simulations show that vector composition of pressures is not valid. For the pressure resulted from a given interference, the fit born 100% of axial (tangential) loading and 80% of tangential (axial) loading. These results recommend a FEM analysis for each case, in order to avoid unnecessary high pressure and stress of materials.

**Keywords:** Interference fit · Bearing capacity · Complex loading · FEM

## 1 Introduction

Interference fits are widely used in mechanical assemblies. Their design is based sometimes on thin-walls theory [1], but most times on thick-walls theory. Under this mathematical approach and elastic behavior hypothesis, the literature provides studies based on analytical methods regarding the pressure variation [2], the influence of a bending couple [3], the local deformation in connection with the initiation of failure [4], the contact parameters [5, 6]. The contact problem is addressed for axial loading [7], but most applications suppose complex loading, axial (force) and tangential (torque). Some authors extend the study to plastic or elastic-plastic behavior of the fit [8, 9].

Modelling and analysis with FEM may widen the area of research to aspects which are difficult or impossible to approach analytically. FEM was used to provide information on failure [10], influence of surface roughness [11], estimation of friction coefficient [12], behavior of graded hubs [13], tapered contact surfaces [14], dynamic effects [15] and others.


E.-C. Lovasz et al. (Eds.): MTM&Robotics 2020, MMS 88, pp. 403–414, 2021.
https://doi.org/10.1007/978-3-030-60076-1_37

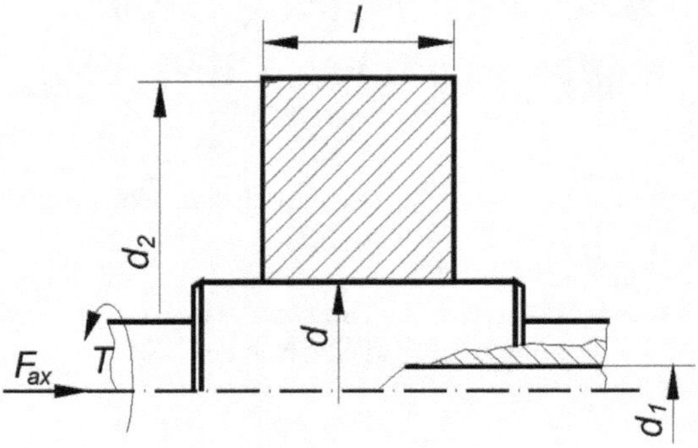


**Fig. 1.** Cylindrical interference fit - notations

Considering the case of a cylindrical interference fit (Fig. 1), the dimensioning of the fit is based on calculus of minimum pressure needed along the axial direction (Eq. 1) and minimum pressure needed on the tangential direction (Eq. 2):

$$p^*_{\min ax} = \frac{F_{ax}}{\mu \cdot \pi \cdot d \cdot l}, \quad (1)$$

where: $p^*_{\min\ ax}$ – minimum pressure needed along the axial direction
$F_{ax}$ – axial loading (force)
μ – friction coefficient
d – nominal diameter of the fit
l – length of the fit.

$$p^*_{\min t} = \frac{2T}{\mu \cdot \pi \cdot d^2 \cdot l}, \quad (2)$$

where: $p^*_{\min\ t}$ – minimum pressure needed to transmit the tangential loading
T – torque.

The literature states the Eqs. 1 and 2 and/or assumes a vector composition law for the minimum pressure needed to transmit both loadings (Eq. 3):

$$p^*_{\min} = \sqrt{p^{*2}_{\min ax} + p^{*2}_{\min t}}. \quad (3)$$

According to Lame's laws, the relationship between the pressure and the interference is [16]:

$$p^{*}_{\min} = \frac{\delta}{10^{3} d\left(\frac{C_1}{E_1} + \frac{C_2}{E_2}\right)}, \tag{4}$$

where: $\delta$ - interference
$E_1$, $E_2$ – Young modulus for the shaft and hub materials
$C_1$, $C_2$ – constants depending on the dimensions of the shaft and hub (Fig. 1):

$$C_1 = \frac{d^2 + d_1^2}{d^2 - d_1^2} - \nu_1, \tag{5}$$

$$C_2 = \frac{d_2^2 + d^2}{d_2^2 - d^2} + \nu_2. \tag{6}$$

With minimum pressure in Eq. 4 is usually calculated the bearing capacity in axial and tangential directions:

$$F_{ax\,cap} = p^{*}_{\min} \cdot \mu \cdot \pi \cdot d \cdot l, \tag{7}$$

$$T_{cap} = p^{*}_{\min} \cdot \mu \cdot \pi \cdot d^2 \cdot l/2, \tag{8}$$

where: $F_{ax\ cap}$ – bearing capacity on axial direction
$T_{cap}$ - bearing capacity on tangential direction.

The topic of this paper is to question the Eq. 3, which may lead to an unnecessary over-dimensioning of the fit.

The study is based on modelling and simulating by means of finite element method (FEM). For a chosen nominal diameter, different interference are imposed and the bearing capacity on axial, tangential and axial-tangential loading is detected and compared to analytical results foreseen by Eqs. 1–4 and 6, 7.

## 2 FEM Modelling and Simulation

### 2.1 Concept

The shaft and the hub were generated with the CAD application Creo, having nominal dimensions in Fig. 2.

The analysis was conceived in two steps and developed in the FEM application ANSYS Workbench. Considering there is an interaction between the contact pressure and stress due to loading [16], the analysis was conducted using three-dimensional elements.

In the first step, the interference is solved.

The second step evaluates the maximum transferable torque or axial force.

The following conditions were applied:

- The hub is fixed on the outer diameter.
- In the second step *Remote displacement* was chosen (0.2° around the z-axis or 0.2 mm along z-axis, respectively).
- The analysis is nonlinear due to frictional coefficient of the contact and large deflections activated. In ANSYS, two faces in frictional contact are free to separate and within the contact they may slide only when shear stress between them exceed a critical value, calculated by multiplying the normal stress by the friction coefficient [16]. ANSYS Workbench offers several contact formulations to enforce contact compatibility at the contact interface, preventing penetration of the contacting point into the target faces. Augmented Lagrange formulation was selected, considering that it is recommended for general frictional in large deformation problems [16]. Since Lagrange method takes extra computing time, the behavior of the contact was set asymmetric, to spare some computing.

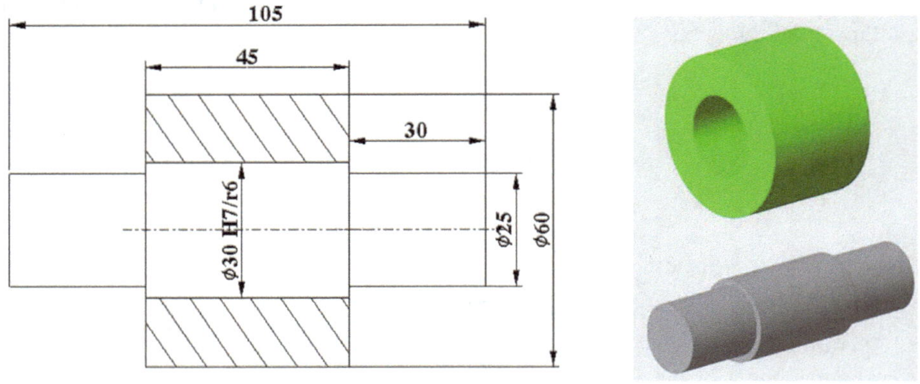


**Fig. 2.** Cylindrical interference fit – model (nominal diameter: 30, length: 45, outer diameter of the hub: 60, inner diameter of the shaft: 0). 2D image (left) and 3D image (right)

- For the interference treatment *Add Offset* and *Ramped Effects* were selected and Offset was set to 0.
- The maximum transferable torque or axial force were depicted as converged value of *Moment reaction*, respectively *Force reaction* versus time, in the second step of the analysis.

## 2.2 Step 1. Simulation of Interference and Validation of the Model

The model was generated for a fit with the nominal diameter of 30 mm and preferred medium fit H7/r6. The limit deviations for the basic hole system are $\Phi30_{0}^{+25}$ for the hub and $\Phi30_{+34}^{+50}$ for the shaft. The maximum interference (measured on the diameter) is $\delta_{min} = 50\ \mu m$ and the minimum interference is $\delta_{max} = 9\ \mu m$.

For the analysis were chosen five possible interferences within this range [9…50] (Table 1).

The material for both shaft and hub is steel with E = $2.1.10^5$ MPa and ν = 0.3.

In the first step, which simulates the interference, the validity of the model was checked by comparing the values of the contact pressure, deformations and stress got from the simulation with the values of theoretical sizes given in Eq. 4, 9–15.

**Table 1.** Interferences chosen for simulation.

| Hub diameter [mm] | Shaft diameter [mm] | Interference, δ [μm] |
|---|---|---|
| 30.000 | 30.048 | 48 |
| 30.015 | 30.045 | 30 |
| 30.010 | 30.035 | 25 |
| 30.015 | 30.038 | 23 |
| 30.020 | 30.040 | 20 |

At the outer surface of the shaft, the radial, tangential stress and radial deformation are (for the frequent case $d_1 = 0$):

$$\sigma_{r\,shaft} = -p, \tag{9}$$

$$\sigma_{t\,shaft} = -p, \tag{10}$$

$$u_{r\,shaft} = -\frac{p \cdot d}{2E}(1 - \nu). \tag{11}$$

At the inner surface of the hub, the radial, tangential stress and radial deformation are:

$$\sigma_r\, hub = -p, \tag{12}$$

$$\sigma_t\, hub = p \cdot \frac{1 + k^2}{1 - k^2}, \tag{13}$$

where: $k = d/d_2$.

$$u_{r\,hub} = \frac{pd}{2E}\left(\frac{1 + k^2}{1 - k^2} + \nu\right). \tag{14}$$

The radial interference is:

$$\frac{\delta}{2} = \left|u_{r\,shaft}\right| + \left|u_{r\,hub}\right|. \tag{15}$$

The values of radial deformations resulted from FEM analysis were depicted from images such as the ones illustrated in Fig. 3 and 4. From these figures, $u_{r\ shaft} \sim 6.2$ μm and $u_{r\ hub} \sim 6.0$ μm, which gives $\delta \sim 24.4$ μm, for the theoretical interference $\delta = 25$ μm.

Data regarding the contact pressure is presented in Table 2. Images such as the one illustrated in Fig. 5 were used to estimate the average value of the pressure at the contact surface. The errors around 5% are considered reasonable.

Thus the model is considered valid and step 2 may be performed with reliable results.

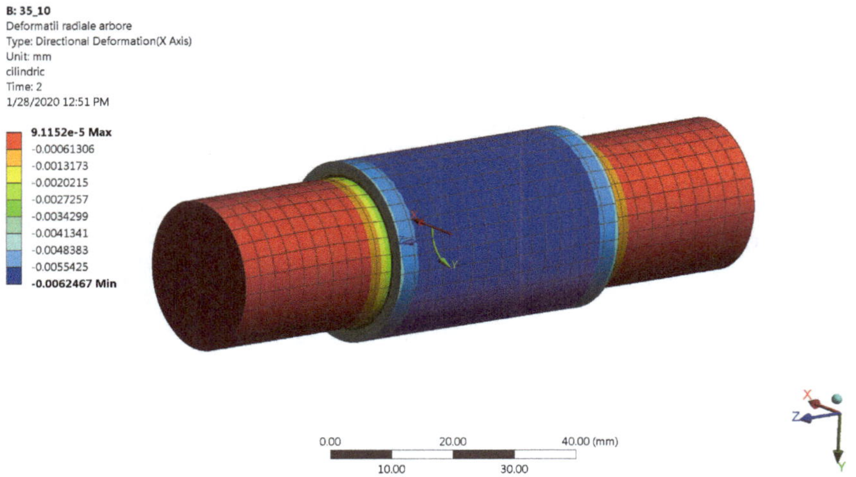


**Fig. 3.** Radial deformations of the shaft for the theoretical interference of 25 μm

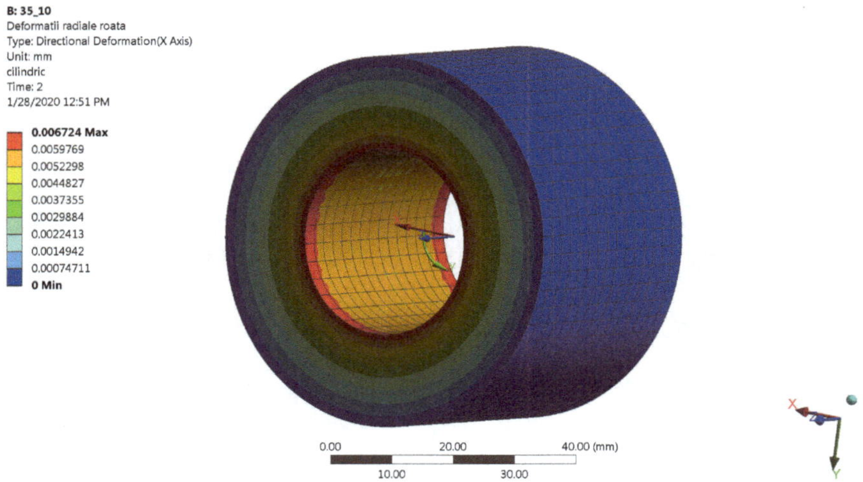


**Fig. 4.** Radial deformations of the hub for the theoretical interference of 25 μm

**Table 2.** Validation of the model (contact pressures).

| Interference, δ [μm] | Theoretical pressure [MPa] | FEM pressure [MPa] | Error [%] |
|---|---|---|---|
| 48 | 126.0 | 120.0 | −4.8 |
| 30 | 78.8 | 75.0 | −4.8 |
| 25 | 65.6 | 62.4 | −4.9 |
| 23 | 60.4 | 57.0 | −5.6 |
| 20 | 52.5 | 50.0 | −4.8 |

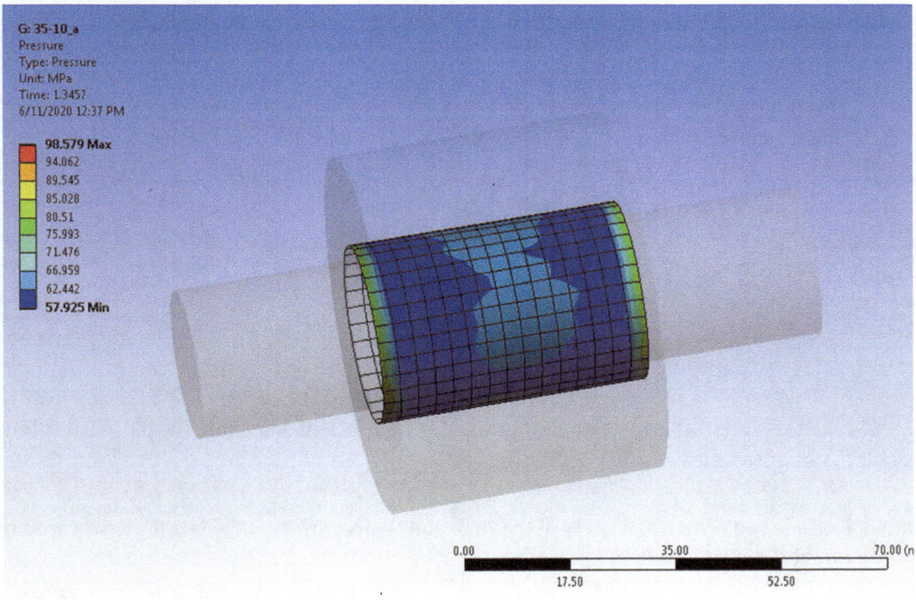


**Fig. 5.** Contact pressure distribution, for the case δ = 25 μm

## 2.3 Step 2. Simulation of Simple and Complex Loadings

### 2.3.1 Axial Bearing Capacity

Once the interference was achieved in step one, during the second step is applied the loading and is evaluated the bearing capacity of the fit.

First, an axial load was applied until the fit started to loosen (the surfaces start to mutually slide). The loading was applied gradually and the simulation was set to be finished when the surfaces have a relative displacement of 0.2 mm.

Table 3 shows a synthesis of the results got with FEM analysis in comparison with analytical data, computed on the basis of thick-walled sleeves. The radial and tangential stress according to this theory, at the contact surface and for solid shaft are:

$$\sigma_{r\,shaft} = -p, \tag{16}$$

$$\sigma_{t\,shaft} = -p, \tag{17}$$

$$\sigma_{r\,hub} = -p, \tag{18}$$

$$\sigma_{t\,hub} = p \cdot \frac{1 + \left(\frac{d}{d_2}\right)^2}{1 - \left(\frac{d}{d_2}\right)^2}. \tag{19}$$

The axial bearing capacity of the fit obtained by modelling and simulation is different with only about 1% from the theoretic numbers, which is regarded as a very good similitude. The results are necessary for the analysis under combined loading.

**Table 3.** Axial beating capacity.

| Interference $\delta$ [μm] | Analytical pressure [MPa] | Pressure FEM [MPa] | Analytical $F_{ax}$ [N] | FEM $F_{ax}$ [N] | Error $F_{ax}$ [%] |
|---|---|---|---|---|---|
| 48 | 126 | 123 | 26706 | 27569 | −3.2 |
| 30 | 79 | 77 | 16691 | 16892 | −1.2 |
| 25 | 66 | 65 | 13909 | 14049 | −1.0 |
| 23 | 60 | 58 | 12796 | 12934 | −1.1 |
| 20 | 53 | 57 | 11127 | 11243 | −1.0 |

The results regarding the radial and tangential stress of the shaft (Table 4) and hub (Table 5) confirm the validity of the model.

**Table 4.** Radial and tangential stress of the shaft at its outer surface at the bearing axial capacity (beginning of the relative sliding).

| Interference $\delta$ [μm] | Analytical pressure [MPa] | Analytical $\sigma_r$ [MPa] | FEM $\sigma_r$ [MPa] | Analytical $\sigma_t$ [MPa] | FEM $\sigma_t$ [MPa] |
|---|---|---|---|---|---|
| 48 | 126 | −126 | −120 | −126 | −120 |
| 30 | 78 | −78 | −73 | −78.75 | −72 |
| 25 | 65 | −65 | −55 | −65.62 | −61 |
| 23 | 60 | −60 | −51 | −60.37 | −56 |
| 20 | 52 | −52 | −48 | −52.50 | −49 |

**Table 5.** Radial and tangential stress of the hub at its inner surface at the bearing axial capacity (beginning of the relative sliding).

| Interference $\delta$ [μm] | Analytical pressure [MPa] | Analytical $\sigma_r$ [MPa] | FEM $\sigma_r$ [MPa] | Analytical $\sigma_t$ [MPa] | FEM $\sigma_t$ [MPa] |
|---|---|---|---|---|---|
| 48 | 126 | −126 | −119 | 210 | 208 |
| 30 | 78 | −78 | −71 | 131 | 130 |
| 25 | 65 | −65 | −64 | 109 | 108 |
| 23 | 60 | −60 | −62 | 100 | 100 |
| 20 | 52 | −52 | −51 | 87.5 | 87 |

The results in this paragraph validate Eq. 1 and Eq. 7.

### 2.3.2 Tangential Bearing Capacity

A tangential load (torque) was applied until the fit started to loosen (the surfaces begin to mutually rotate). The loading was applied gradually and the simulation was set to be finished when the surfaces have a relative rotation displacement of 0.2 deg.

Table 6 shows a synthesis of the results got with FEM analysis in comparison with analytical data. The tangential bearing capacity of the fit obtained by modelling and simulation is different with only about 1% from the theoretic numbers, which is regarded as a very good similitude. The results are necessary for the analysis under combined loading.

**Table 6.** Tangetial beating capacity.

| Interference $\delta$ [μm] | Analytical pressure [MPa] | Presure FEM [MPa] | Analytical T [Nmm] | FEM T [Nmm] | Error T [%] |
|---|---|---|---|---|---|
| 48 | 126 | 120 | 400586 | 375450 | 6.3 |
| 30 | 79 | 75 | 250366 | 234900 | 6.2 |
| 25 | 66 | 62 | 208638 | 195570 | 6.3 |
| 23 | 60 | 57 | 191947 | 179980 | 6.2 |
| 20 | 53 | 50 | 166911 | 156500 | 6.2 |

The results regarding the radial and tangential stress of the shaft (Table 7) and hub (Table 8) confirm the validity of the model.

**Table 7.** Radial and tangential stress of the shaft at its outer surface at the bearing tangential capacity (beginning of the relative rotation).

| Interference $\delta$ [μm] | Analytical pressure [MPa] | Analytical $\sigma_r$ [MPa] | FEM $\sigma_r$ [MPa] | Analytical $\sigma_t$ [MPa] | FEM $\sigma_t$ [MPa] |
|---|---|---|---|---|---|
| 48 | 126 | −126 | −124 | −126 | −112 |
| 30 | 79 | −79 | −77 | −79 | −75 |
| 25 | 66 | −66 | −65 | −66 | −57 |
| 23 | 60 | −60 | −60 | −60 | −53 |
| 20 | 53 | −53 | −54 | −53 | −50 |

**Table 8.** Radial and tangential stress of the hub at its inner surface at the bearing tangential capacity (beginning of the relative rotation).

| Interference $\delta$ [μm] | Analytical pressure [MPa] | Analytical $\sigma_r$ [MPa] | FEM $\sigma_r$ [MPa] | Analytical $\sigma_t$ [MPa] | FEM $\sigma_t$ [MPa] |
|---|---|---|---|---|---|
| 48 | 126 | −126 | −130 | 210 | 201 |
| 30 | 79 | −79 | −69 | 131 | 125 |
| 25 | 66 | −66 | −68 | 109 | 105 |
| 23 | 60 | −60 | −53 | 101 | 96 |
| 20 | 53 | −53 | −46 | 88 | 84 |

The results in this paragraph validate Eq. 2 and Eq. 8.

## 2.4 Bearing Capacity Under Axial and Tangential Loading

The purpose of the simulations in this paragraph is to learn the level of loading when the contact between surfaces is broken under both axial and tangential load.

For the simulations, the interference $\delta$ = 30 μm was chosen. From the previous paragraphs results that the contact pressure is p = 80 MPa and the fit gets loose at $F_{ax}$ = 16691 N if the loading is purely axial or at T = 250.365 Nmm if the loading is purely tangential.

The question is: what is the behavior of the fit loaded both axially and tangentially? Is it necessary a higher pressure, according to Eq. 3, in order to transmit both loadings?

In order to answer this questions two sets of simulations were performed:

- the assembly was loaded with the torques ($T_1$ = 50.000 Nmm, $T_2$ = 100.000 Nmm, $T_3$ = 150.000 Nmm, $T_4$ = 200.000 Nmm, as 20%, 40%, 60% and 80% of the tangential bearing capacity of ~250.000 Nmm) and the simulation provided the axial force at which the contact is broken and the fit yields
- the assembly was loaded with axial forces ($F_{ax\ 1}$ = 3340 N, $F_{ax\ 2}$ = 6680 N, $F_{ax\ 3}$ = 10020 N, $F_{ax\ 4}$ = 13360 N, as 20%, 40%, 60% and 80% of the axial bearing capacity of ~17.000 N) and the simulation provided the torque at which the contact is broken and the fit yields.

The results of the simulation are presented in Tables 9 and 10.

**Table 9.** Maximum axial loading born by the fit at different levels of tangential loading.

| Torque [Nmm] | $F_{ax\ cap}$ [N] | Pressure [MPa] | $\sigma_{r\ shaft}$ [MPa] | $\sigma_{t\ shaft}$ [MPa] | $\sigma_{z\ shaft}$ [MPa] | $\sigma_{r\ hub}$ [MPa] | $\sigma_{t\ hub}$ [MPa] | $\sigma_{z\ hub}$ [MPa] |
|---|---|---|---|---|---|---|---|---|
| 50000 | 15870 | 72 | −72 | −71 | −30 | −81 | 124 | −1 |
| 100000 | 15870 | 72 | −70 | −70 | −23 | −76 | 124 | 3 |
| 150000 | 15870 | 72 | −74 | −67 | −13 | −75 | 124 | −1 |
| 200000 | 15870 | 72 | −72 | −70 | −16 | −75 | 124 | 3 |

**Table 10.** Maximum tangential loading born by the fit at different levels of axial loading.

| $F_{ax}$ [N] | $T_{cap}$ [Nmm] | Pressure [MPa] | $\sigma_{r\ shaft}$ [MPa] | $\sigma_{t\ shaft}$ [MPa] | $\sigma_{z\ shaft}$ [MPa] | $\sigma_{r\ hub}$ [MPa] | $\sigma_{t\ hub}$ [MPa] | $\sigma_{z\ hub}$ [MPa] |
|---|---|---|---|---|---|---|---|---|
| 3340 | 236190 | 77 | −76 | −77 | −13 | −82 | 126 | 6 |
| 6680 | 237710 | 77 | −72 | −77 | −15 | −80 | 127 | 7 |
| 10020 | 239220 | 77 | −70 | −77 | −18 | −80 | 127 | 7 |
| 13360 | 240000 | 77 | −70 | −80 | −20 | −73 | 127 | 6 |

Considering data in Table 9, it is to notice the following observations:

- At four levels of torque, the fit born the same axial force (15870 N), whose value is ~95% of the axial bearing capacity if axial force is the only loading (16691 N)
- The contact pressure and stress, both radial and tangential do not vary within all four simulations. Only $\sigma_z$ is variable for the shaft
- The fifth simulation was not possible to run. The fit did not bear 100% of both axial and tangential loading.

Considering data in Table 10, it is to notice the following observations:

- At four levels of axial force, the fit born the same torque (~240.000 Nmm), whose value is ~95% of the tangential bearing capacity if torque is the only loading (~235.000 Nmm)
- The contact pressure and stress, both radial and tangential do not vary within all four simulations. Only $\sigma_z$ is variable for the shaft
- The fifth simulation was not possible to run. The fit did not bear 100% of both axial and tangential loading.

## 3 Conclusions

The simulations allow the statement of the following remarks:

- the interference provides a contact pressure, which is given analytically in Eq. 4 and is validated trough all simulations
- the contact pressure ensures an axial or tangential bearing capacity according to Eqs. 7 and 8, for simple loading (axial or tangential)
- the contact pressure resulted from interference ensures combined loading axial and tangential: 100% of maximum axial force and 80% of maximum torque or 100% of maximum torque and 80% of maximum axial force. These percentages are not considered to be a general rule, but they demonstrate that Eq. 3 foresees an over-interference, which is not necessary and involves high levels of stress and strain of materials
- the recommendation suggested by the authors is to perform a FEM analysis in order to determine the bearing capacity under combined loading, at a given interference (standard fit).

## References


1. Campos, U.A., Hall, D.E.: Simplified Lamé's equations to determine contact pressure and hoop stress in thin-walled press-fits. Thin-Walled Struct. **138**, 199–207 (2019). Author, F.: Article title. J. **2**(5), 99–110 (2016)
2. Wang, X., Lou, Z., Wang, X., Xu, C.: A new analytical method for press-fit curve prediction of interference fitting parts. J. Mater. Process. Tech. **250**, 16–24 (2017)
3. Radi, E., Lanzoni, L., Strozzi, A., Bertocchi, E.: Shaft-hub press fit subjected to bending couples: analytical evaluation of the shaft-hub detachment couple. Appl. Math. Model. **50**, 135–160 (2017)
4. Aleksandrova, N.: Analytical modeling in deformation analysis of interference-fit structures. Structures **6**, 30–36 (2016)
5. Xiong, L., Shang, P., Zhang, Y., Xu, Y.: Exact solution of stress and displacement of rotating elastic interference fit for a quill shaft of micro gas turbine. IOP Conf. Ser. Earth Environ. Sci. **170**, 042035 (2018)
6. Shang, P., Zhan, R., Xiong, L., Sun, Y., Zhou, J.: Exact solutions of interference fit of a high-speed coupling for micro gas turbine. IOP Conf. Ser. Earth Environ. Sci. **170**, 042030 (2018)
7. Lopes, J.P., Hills, D.A., Paynter, R.J.H.: The axisymmetric shrink fit problem subjected to axial force. Eur. J. Mech. Solids **70**, 172–180 (2018)
8. Gamer, U.: The shrink fit with elastic-plastic hub exhibiting constant yield stress followed by hardening. Int. J. Solid Struct. **23**(9), 1219–1224 (1987)
9. Guven, U.: The rotating shrink fit with elastic-plastic hub exhibiting variable thickness. Int. J. Solids Struct. **29**, 89–96 (1992)
10. Gürer, G., Gür, C.H.: Failure analysis of fretting fatigue initiation and growth on railway axle press-fits. Eng. Fail. Anal. **84**, 151–166 (2018)
11. Boutoutaou, H., Bouaziz, M., Fontaine, J.-F.: Modelling of interference fits with taking into account surface sroughness with homogenization technique. Int. J. Mech. Sci. **69**, 21–31 (2013)
12. Seifi, R., Abbasi, K.: Friction coefficient estimation in shaft/bush interference using finite element model updating. Eng. Fail. Anal. **57**, 310–322 (2015)
13. Arslan, E., Mack, W.: Shrink fit with solid inclusion and functionally graded hub. Compos. Struct. **121**, 217–224 (2015)
14. Vishwakarma, N., Renuke, A., Phalle, V.M.: Effect of tapered interference fit between impeller and shaft in turbo machines. J. Mech. Eng. **68**(3), 25–32 (2018)
15. Chen, S.Y., Kung, C., Liao, T., Chen, Y.-H.: Dynamic effects of the interference fit of motor rotor on the stiffness of a high speed rotating shaft. Trans. Can. Soc. Mech. Eng. **34**(2), 243–261 (2010)
16. Budynas, R.G., Nisbett, J.K.: Shigley's Mechanical Engineering Design, 10th edn. McGraw Hill, New York (2015)
17. Lee, H.-H.: Finite Element Simulations with ANSYS Workbench 15. Theory, Applications, Case Studies. SDC Publications, Mission (2014)

# Mechatronics

# Continuum Arm Control with Constraints on the Driving Forces via Fractional Order Models

Mircea Ivanescu[1(✉)], Mircea Nitulescu[1], and Cristian Vladu[2]

[1] Department of Mechatronics, University of Craiova, Craiova, Romania
{ivanescu,nitulescu}@robotics.ucv.ro
[2] Electrical Engineering Department, University of Craiova, Craiova, Romania
cristian.vladu@ie.ucv.ro

**Abstract.** The paper presents an extension of the Circle Criterion for the dynamic models of the continuum robots. This technique is applied to the control of the fractional order model system with constraints of the driving forces. It is demonstrated that the stability of the closed loop control system requires that the arm frequency plot does not intersect the circle of driving constraints. It is shown that nonlinear components worsen the stability of the system. The proposed methods are illustrated by numerical simulation and experimental tests.

**Keywords:** Continuum arm · Control · Stability

## 1 Introduction

Recent achievements in the interpretation of Fractional Order Operators allowed to apply these results for the processes that are better described by the Fractional Order Models rather than Integer Order Models. A class of these models is represented by the hyper-redundant robots or robots with continuum viscoelastic components in their structure. The conventional architecture of robot with continuum characteristics have been extensively studied in a large number of papers. A summary of these results can be found in [1–4]. Real time kinematic control studies are conducted in [5–8]. Dynamic control is treated in [9, 10]. Continuum robot stability considering the system non-linearities is developed in [11, 12]. The weighting techniques of distributed parameters for the control of hyper-redundant architectures is studied in [13, 14].

This paper studies a class of continuum robots for which the driving forces are subject to classes of restrictions. The viscoelastic models are used to describe the behavior of a 3-D sensor used for measuring techniques of position and orientation. A control law defined with respect to direct observable variable, with a physical meaning, is proposed. The Popov Circle Criterion is proposed and the conditions that determine the motion stability for linear models and linear models with nonlinear components, with constraints on the driving forces, are inferred. It is shown that nonlinear components worsen the stability of the system.


E.-C. Lovasz et al. (Eds.): MTM&Robotics 2020, MMS 88, pp. 417–431, 2021.
https://doi.org/10.1007/978-3-030-60076-1_38

The paper is structured as follows: Sect. 2 presents continuum arm architecture, Sect. 3 treats the arm dynamics, Sect. 4 describes the control system, Sect. 5 verifies the control techniques by numerical simulations, Sect. 6 presents the results of the experimental platform and Sect. 7 resumes the conclusions.

## 2 Continuum Arm Architecture

The mechanical architecture consists by a long flexible backbone surrounded by an elastic cylinder. A viscoelastic solution separates these components ensuring a large flexibility of the whole system (see Fig. 1). The mechanical structure is composed of several segments with individual driving systems made of antagonistic Bowden cables. This actuation system allows an independent decoupled control strategy for each arm segment. The whole arm operates in 3-D space performing a complex movement that allows to reach any position and orientation. It is very difficult to measure the 3-D position of this arm. For this reason, we propose a special technique that allows the measuring of the segment parameters by using an indirect method [15].

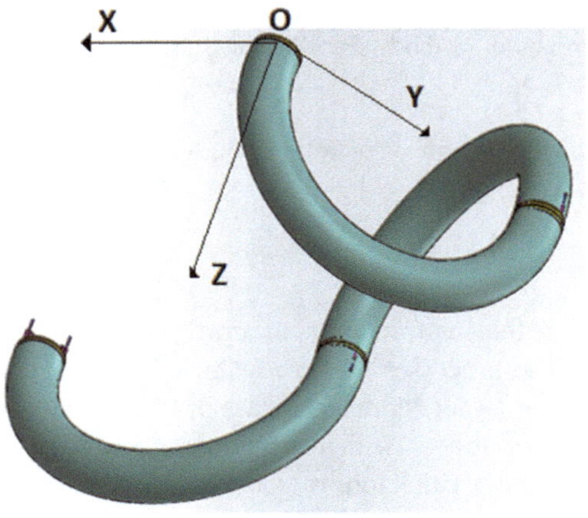


**Fig. 1.** Continuum robot arm.

The geometry of this continuum arm is characterized by the backbone curve $\Gamma$ and two angles $q_1(\mathrm{s})$ and $q_2(\mathrm{s})$ where $\mathrm{s} \in \Omega = [0, \mathrm{L}]$, and L. the arm length. The robot's orientation is given by $(q_1, q_2)$ with respect to a right-handed orthonormal basis vector $\{i_x, i_y, i_z\}$, where $\{i_x, i_y, i_z\}$ is parallel to a base coordinate frame, (see Fig. 2) [1, 2].

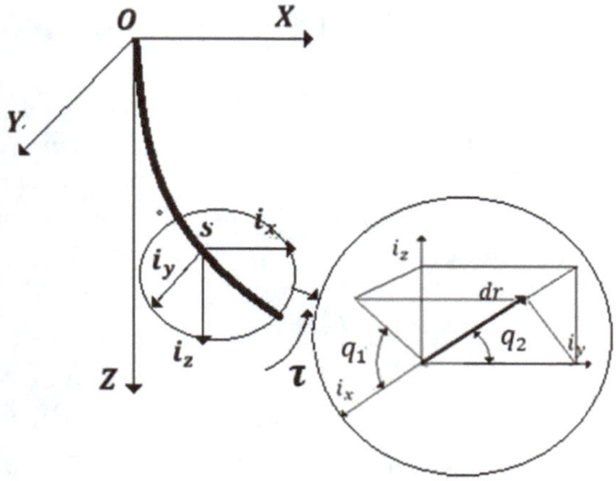


**Fig. 2.** Geometrical parameters of the arm.

A 3-D technique as described in Fig. 3 is used for orientation measuring. Three antagonist cables are implemented at the periphery of the system. In a static behavior, the orientation angles are obtained by the differential measurement of the cable lengths [10, 15].

$$q_1 = F_1(\Delta L_1, \Delta L_2, \Delta L_3);\ q_2 = F_2(\Delta L_1, \Delta L_2, \Delta L_3) \tag{1}$$

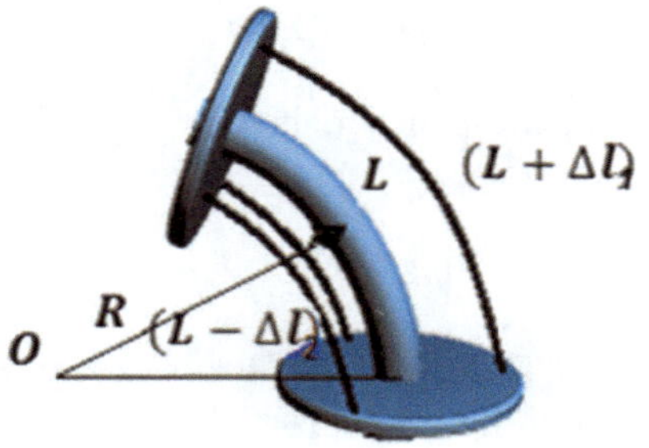


**Fig. 3.** Angular position measuring

## 3 Arm Dynamics

The dynamic behavior of the arm is defined by the fractional Kevin-Voigt model of the main arm component, in which the elastic phase with the coefficient $c_e$ is connected in parallel with a fractional viscoelastic phase that is characterized by coefficient $c_v$ and fractional exponent $\beta$ [16]. The general behavior is defined by the stress-strain relation

$$\zeta(t) = c_e \varrho(t) + c_v \left( {}_t^C D_0^\beta \varrho \right)(t) \tag{2}$$

where $\left({}_{t}^{C}D_{0}^{\beta}\varrho\right)(t)$ is the Caputo fractional derivative of order $\beta$, where $\beta$ is approximated as $\beta = 0.5$.

The whole dynamic model is obtained as in [1–3, 13] by using Lagrange equations for this class of these models described by Partial Differential Equations. We consider a mechanical architecture with a distributed mass, with inertial rotational density $I_\rho$, with equivalent elastic modulus E and moment of inertia $I$. In terms of curvature variables, these models can be rewritten as

$$\boldsymbol{I}_\rho \ddot{q} = \boldsymbol{EI} q_{ss} - \boldsymbol{C}_v D^\beta q - C_e q - \boldsymbol{h}(q) \tag{3}$$

where $\boldsymbol{q} = \boldsymbol{q}(t,s)$, $q_{ss} = \dfrac{\partial^2 q(t,s)}{\partial s^2}$, $\boldsymbol{q} = [q_1 q_2]^T, q_1, q_2 \in L_2(\Omega)$, $D^\beta \boldsymbol{q} = {}_{t}^{C}D_0^\beta \boldsymbol{q}(t,s)$, $\boldsymbol{I}_\rho = diag(I_{\rho 1}, I_{\rho 2})$, $\boldsymbol{EI} = diag(EI_1, EI_2)$, $\boldsymbol{C}_v = diag(c_{v1}, c_{v2})$ $\boldsymbol{C}_e = diag(c_{e1}, c_{e2})$. The initial conditions are

$$\boldsymbol{q}(0,s) = \boldsymbol{q}_0(s) \tag{4}$$

The boundary conditions are determined by the driving moments,

$$\boldsymbol{q}(t,0) = 0; \boldsymbol{q}_s(t,l) = \boldsymbol{\tau}(t) \tag{5}$$

where $\boldsymbol{\tau} = [\tau_1, \tau_2]^T$. Technical constraints impose the sector constraints on the driving torques,

$$\alpha \boldsymbol{q} \le \boldsymbol{\tau} \le \beta \boldsymbol{q} \tag{6}$$

The nonlinear term $\boldsymbol{h}(q)$ corresponds to the gravitational influence and satisfies the constraints [15]

$$\|\boldsymbol{q}\boldsymbol{h}(\mathrm{q})\| \le \chi \boldsymbol{q}^T \boldsymbol{q} \tag{7}$$

where $\chi$ is a positive constant. This constraint allows us to treat the nonlinear vector as a decoupled one. The output of the system is evaluated as a weighted average defined as

$$z(t) = \int_0^l \psi(s) \boldsymbol{q}(t,s) ds \tag{8}$$

where $z = [z^1, z^2]^T$ and $\psi(s)$ is an eigenfunction of the spatial operator $\frac{\partial^2}{\partial s^2}$. In relation with this output, the dynamic model (3)–(5) can be rewritten as

$$I_\rho \ddot{z}^i(t) = -C_v D^\beta z^i(t) - \left(\frac{\pi^2}{4l^2} EI + C_e\right) z^i(t) + EI\tau_i(t) - h^*(y), i = 1,2 \tag{9}$$

$$z^i(0) = z_0^i \tag{10}$$

The following fractional order variables are defined [16–19]:

$$w_1^i(t) = z^i, D^{\frac{1}{2}} w_1^i(t) = w_2^i(t), D^{\frac{1}{2}} w_2^i(t) = w_3^i(t), D^{\frac{1}{2}} w_3^i(t) = w_4^i(t), i = 1, 2. \tag{11}$$

From (9), (10), (11), the following fractional order model can be inferred,

$$D^{\frac{1}{2}} \boldsymbol{w}^i(t) = \boldsymbol{A}^i \boldsymbol{w}^i(t) + \boldsymbol{b}^i \tau^i(t) + \boldsymbol{h}^{i*}\left(\boldsymbol{w}^i\right), i = 1, 2 \tag{12}$$

$$\boldsymbol{w}^i(0) = \boldsymbol{w}_0^i \tag{13}$$

where

$$\boldsymbol{w}^i(t) = \left[w_1^i(t) w_2^i(t) w_3^i(t) w_4^i(t)\right]^T,\ i = 1, 2 \tag{14}$$

$$\boldsymbol{A}^i = \begin{bmatrix} 0 & 1 & 0 & 0 \\ 0 & 0 & 1 & 0 \\ 0 & 0 & 0 & 1 \\ -\frac{1}{I_\rho}\left(\frac{\pi^2}{4l^2} EI + C_e\right) & -\frac{1}{I_\rho} C_v & 0 & 0 \end{bmatrix} \tag{15}$$

$$\boldsymbol{b}^i = \begin{bmatrix} 0 & 0 & 0 & \frac{1}{I_\rho} \end{bmatrix}^T; \boldsymbol{h}^{i*} = \begin{bmatrix} 0 & 0 & 0 & \frac{1}{I_\rho} h^{i*} \end{bmatrix}^T \tag{16}$$

Fractional models contain virtual variables as $w_2^i(t), w_4^i(t)$ together with physical significance variables as position and velocity. The control system uses the output defined by the position variable as control variable,

$$\boldsymbol{y}^i(t) = \boldsymbol{c}^T \boldsymbol{w}^i(t), i = 1, 2 \tag{17}$$

where

$$\boldsymbol{c} = \begin{bmatrix} 1 & 0 & 0 & 0 \end{bmatrix}^T \tag{18}$$

# 4 Control System

Control system ensures the motion of the arm toward a desired position defined by the target variable $w_t$, considering the constraints imposed on the control variables (6). The general control system is shown in Fig. 4. A controller is introduced to ensure the control desired performances. Two cases will be analyzed: the control system for a light weight continuum arm and the control system with nonlinear term determined by gravitational components, that verifies the constraint (7).

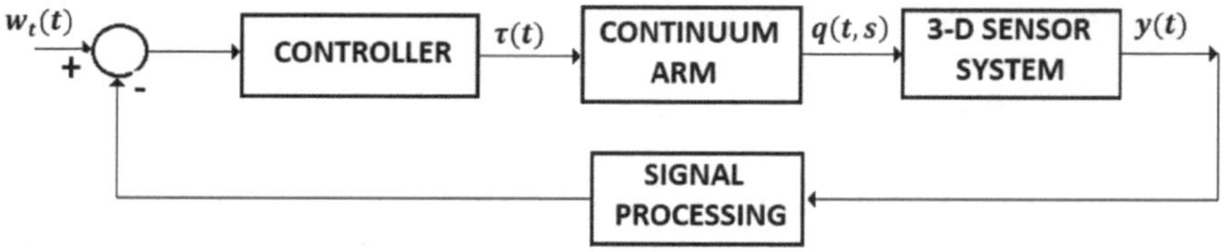


**Fig. 4.** Control system

### 4.1 Light Weight Continuum Arm

Consider the linear form of the dynamic model (12), (13) and assume a control law (for simplification of notations the index i is eliminated starting from here),

$$\tau(t) = -k\,y(t) \tag{19}$$

The driving force constraints (13) determine similar constraints on the gain $k$,

$$k \in \left[k_\alpha, k_\beta\right] \tag{20}$$

where $k_\alpha, k_\beta$ define the new sector constraints of the control law.

*Theorem 1.* The system (12), (13) with the control law defined by (19), (20) is asymptotically stable if:

a) $A^* = (A - R)$ is Hurwitz stable, where $R = R^T > 0$ (21)

b) $Re(G(j\omega)) \geq -\sqrt{\sigma}, \forall\omega \in (-\infty, +\infty)$ (22)

c) $\alpha^* = \varrho - 2\lambda_{\max(PR)} > 0$ (23)

where $G(j\omega) = c^T(j\omega I - A^*)^{-1}b$, $Q = qq^T$, $P$ is a solution of Lyapunov equation, $\varrho = \left\|(q + k\sqrt{\sigma}c)(q + k\sqrt{\sigma}c)^T\right\|$ and $\sigma$ is a constant that satisfy the condition

$$k\sigma \leq 1 \tag{24}$$

The constraints (20) determine as

$$\sigma \in \left[\sigma_\beta, \sigma_\alpha\right] \tag{25}$$

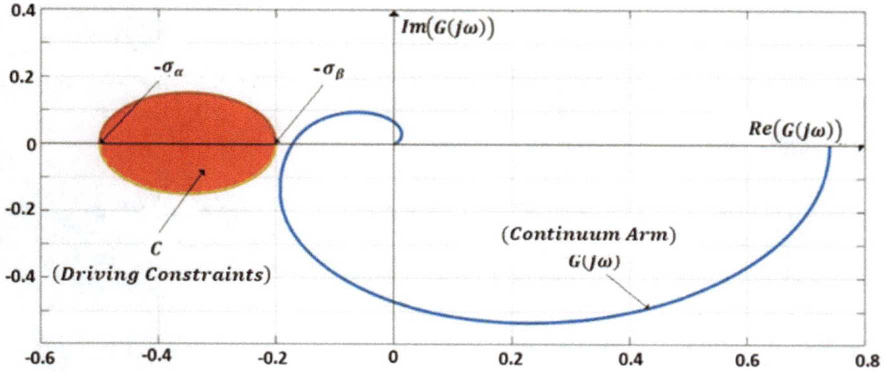


**Fig. 5.** Circle criterion for light weight continuum arm

The frequency condition (22) can be rewritten in terms of the sector constraints (20), (25) as [20],

$$Re\left[\frac{\sigma_\beta + G(j\omega)}{\sigma_\alpha + G(j\omega)}\right] > 0, \forall\omega \in (-\infty, +\infty) \tag{26}$$

If we define $C(\sigma_\alpha, \sigma_\beta)$ to be the closed circle in the complex plane whose diameter is the line segment connecting the points $-\sigma_\alpha + j0$ and $-\sigma_\beta + j0$, the condition (26) can be formulated as: all points of plot of $G(j\omega)$ must be strictly outside the circle $C(\sigma_\alpha, \sigma_\beta)$, or as the plot of $G(j\omega)$ does not intersect $C(\sigma_\alpha, \sigma_\beta)$ (Fig. 5). If we consider that $G(j\omega)$ defines the frequency behavior of the continuum arm and $C(\sigma_\alpha, \sigma_\beta)$ represents the driving constraint domain, the stability condition can be formulated as:" the both domains must not intersect".

$$G \cap C(\sigma_\alpha, \sigma_\beta) = \emptyset$$

This condition can be considered as an extension of the *Popov Circle Criterion.* For the systems described by fractional order models [20].

*Proof.* Assume the Lyapunov function

$$V(w) = w^T P w \tag{27}$$

where $P = P^T > 0$. This function satisfies conditions [16] that ensure asymptotic stability for $\alpha_1 = \lambda_{min}(P), \alpha_2 = \lambda_{max}(P)$. By using [21–24], the fractional derivative of (27) will be

$$D^\beta V(w) \le \left(D^\beta w^T\right) P w + w^T P \left(D^\beta w\right) \tag{28}$$

By evaluating (28) along the solutions of (12), yields

$$D^{\beta}V(w) \leq w^{T}\left(A^{T}P + PA\right)w + 2\boldsymbol{w}^{\boldsymbol{T}}Pbu \tag{29}$$

By using (21), (29) becomes

$$D^{\beta}V(w) \leq w^{T}\left((A-R)^{T}Pw + P(A-R)\right)w + 2\boldsymbol{w}^{\boldsymbol{T}}Pbu + 2w^{T}PRw \tag{30}$$

From (19), (20), (24), this inequality is rewritten as

$$D^{\beta}V(w) \leq w^{T}\left((A-R)^{T}P + P(A-R)\right)w + 2\boldsymbol{w}^{\boldsymbol{T}}\left(Pb - \frac{1}{2}c\right)u - \sigma u^{2} + 2w^{T}PRw \tag{31}$$

By using Yakubovici-Kalman-Popov (YKP) Lemma [20], the first two terms can be evaluated as

$$x^{T}\left((A-R)^{T}P + P(A-R)\right)x = -qq^{T} \tag{32}$$

$$Pb - \frac{1}{2}c = \sqrt{\sigma}q \tag{33}$$

Now, substituting this result into (31), considering the control law (19), after simple calculations, it follows that

$$D^{\beta}V(\boldsymbol{w}) \leq -w^{T}\left(q + k\sqrt{\sigma}c\right)\left(q + k\sqrt{\sigma}c\right)^{T}w + \rho_{PR}w^{T}w \tag{34}$$

or, by [25, 26],

$$D^{\beta}V(\boldsymbol{w}) \leq -\alpha^{*}\|w\|^{2} \qquad QED \tag{35}$$

### 4.2 Continuum Arm with Gravitational Components

Consider the model (12) with nonlinear gravitational component $\boldsymbol{h}$ that satisfies the constraints (7) and assume a control law (19), (20).

*Theorem 2*. The loop control system (12), (13), with the control law defined by (19), (20), is asymptotically stable if:

a) $A^{*} = (A-R)$ is Hurwitz stable, where $R = R^{T} > 0$ (36)

b) $Re(\bar{G}(j\omega)) \geq -\sqrt{\sigma}, \forall\omega \in (-\infty, +\infty)$ (37)

c) $\alpha^* = \varrho - 2\lambda_{\max(PR)} > 0$ (38)

where $\bar{G}(j\omega) = c^T\left(j\omega I - (A_0^* + \chi \mathrm{I})\right)^{-1}b$, I is the identity matrix, $Q = qq^T$, $P$ is a solution of Lyapunov equation and $\varrho = \left\| (q + k\sqrt{\sigma}c)(q + k\sqrt{\sigma}c)^T \right\|$.

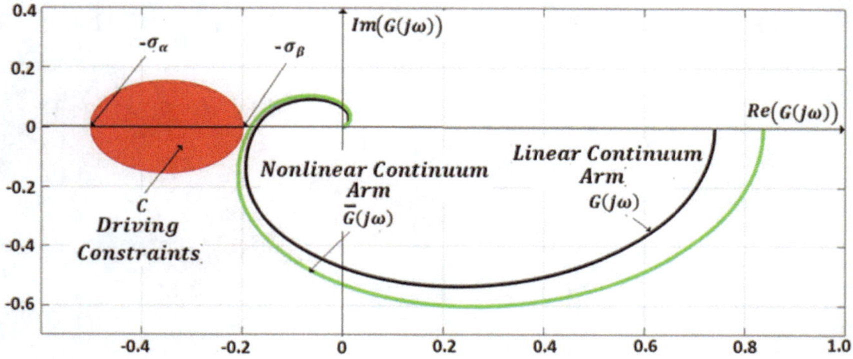


**Fig. 6.** Circle criterion for continuum arm with gravitational component

*Proof.* The proof is the same as for the previous Theorem 1.

*Remark 1.* In Fig. 6 are shown two plots of the linear and nonlinear continuum arm models, respectively. We remark that the nonlinear component moves the plot of $\bar{G}(j\omega)$ to the left, toward the circle $C$ which leads to a decrease in stability. For high value of nonlinearity coefficient $\chi$, when $\bar{G}(j\omega)$ intersects the circle $C$, the system becomes unstable.

## 5 Simulations

Consider mechanical structure of continuum arm characterized by the mechanical parameters: the length of the segment $l = 0.25$ m, the rotational inertial density $I_\rho = 0.002\, kg.m^2$ the bending stiffness $EI = 1.5\,\mathrm{N.m}^e$ equivalent Kelvin-Voigt model coefficients: $c_v = 0.22$ Nm/rad, $c_e = 2.8$ NM rad$^{-1}$.

The fractional order system (12)–(16) will be

$$A^i = \begin{bmatrix} 0 & 1 & 0 & 0 \\ 0 & 0 & 1 & 0 \\ 0 & 0 & 0 & 1 \\ -239 & -36.8 & 0 & 0 \end{bmatrix}, i = 1, 2 \tag{39}$$

$$b^i = \begin{bmatrix} 0 & 0 & 0 & 150 \end{bmatrix}^T, i = 1, 2 \quad (40)$$

$$c^{iT} = \begin{bmatrix} 1 & 0 & 0 & 0 \end{bmatrix}, i = 1, 2 \quad (41)$$

The eigenvalue distribution for the system is shown in Fig. 7. We remark that the system is fractional order stable but is not Hurwitz stable [25–27]. The matrix $R = diag(2.5\ 2.5\ 2.5\ 2.5)$ is selected to satisfy the condition (21). We simulated the behavior of the nonlinear model for $\chi = 8$.

A control law (19) is proposed with $k^i = 4, k_\alpha = 2, k_\beta = 5, \sigma_\alpha = 0.5, \sigma_\beta = 0.2$. The condition (38) is verified for $\lambda_{\max(Q)} = 12.5, \lambda_{\max(P)} = 0.625$, $\|PR\| = 2.37$, $\rho_R = 2.5$.

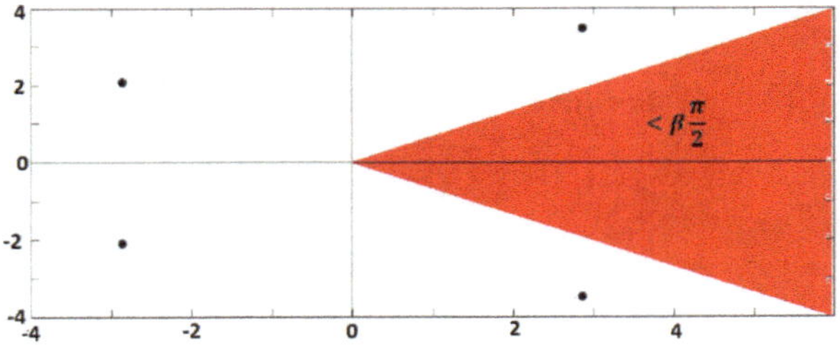


**Fig. 7.** Eigenvalue distribution

In Fig. 6 is verified the frequential condition (37). A comparative analysis of the two cases, with linear model and with gravitational component with $\chi = 8$, is presented. The two plots of $\boldsymbol{G(j\omega)}, \boldsymbol{\overline{G}(j\omega)}$ shows the verification of the stability conditions as well as their deterioration due to the nonlinear gravitational component. MATLAB/SIMULINK and techniques based on the Mittag-Leffler functions are used for the simulation [28, 29].

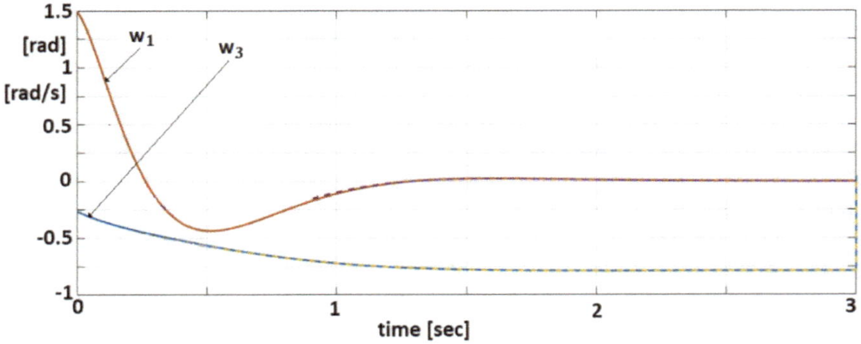


**Fig. 8.** Linear system trajectories for initial condition $w_1(0) = \frac{\pi}{6}$, $w_3(0) = -0.25$ rad/s

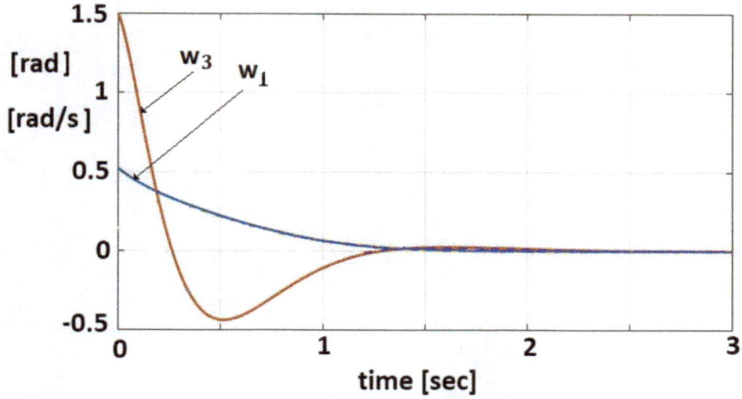


**Fig. 9.** Nonlinear system trajectories for initial condition $w_1(0) = \frac{\pi}{6}$, $w_3(0) = 0.5$ rad/s

The system trajectories for position and velocity, for the two cases: linear model and nonlinear model with gravitational components are shown in Fig. 8 and Fig. 9, respectively.

## 6 Experimental Results

A platform with a continuum arm has been used for testing (see Fig. 10). A cable driven system, with Bowden cables, is employed for the bending movement of all arm segments. The arm is made from homogeneous materials and an elastic backbone represents the support of mechanical architecture. Each arm segment is actuated by DC motors and a transmission system developed by three Bowden cables. A PC system and a Quanser platform implement the signal processing and control laws. A control law (19) with $k = 15$ and the constraints defined by $[k_\alpha, k_\beta]$=$[2, 30]$ is implemented. The angular positions are evaluated by the technique used in (1) [15].

$$q = \frac{L(\Delta l_1 + \Delta l_2)}{r\left(\frac{3}{2}L + \frac{\Delta l_1}{2} - \Delta l_2\right)} \tag{42}$$

where $\Delta l_1, \Delta l_2$ are the cable length variations.

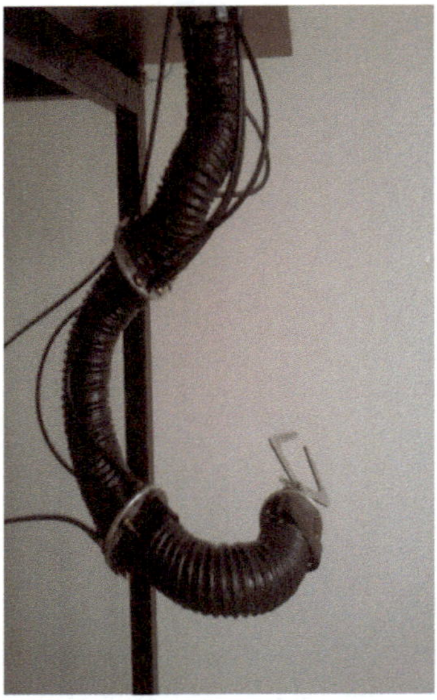

**Fig. 10.** Experimental platform

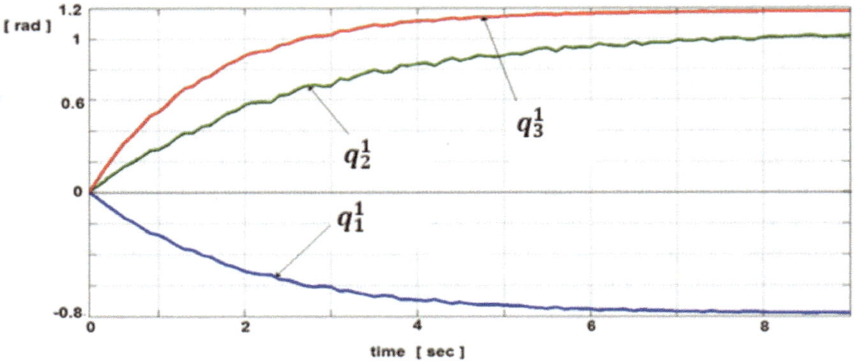


**Fig. 11.** Trajectories $q_{id}^1(t, l), i = 1, 2, 3$ for Test 1.

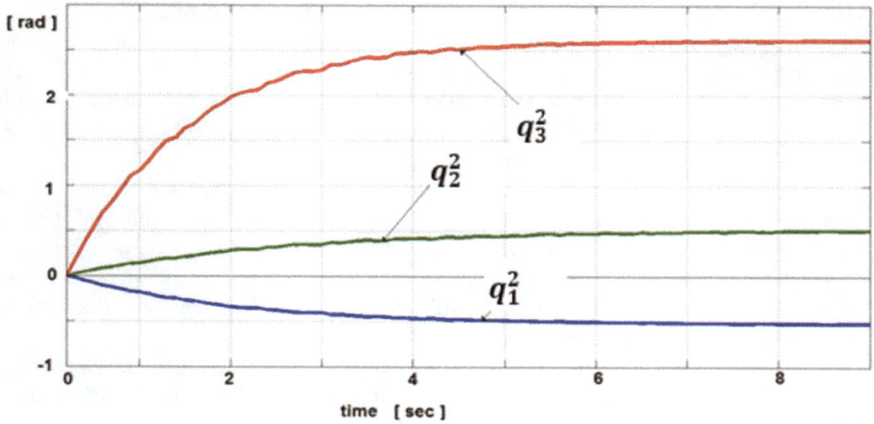


**Fig. 12.** Trajectories $q_{id}^2(t, l), i = 1, 2, 3$ for Test 2.

Two experimental tests are verified:

Test 1 for desired positions: $q_{1d}^1 = -\pi/4,\ q_{2d}^1 = \pi/3,\ q_{3d}^1 = 3\pi/8.$

Test 2 for desired positions:$q_{1d}^2 = -\pi/6,\ q_{2d}^2 = \pi/6,\ q_{3d}^2 = 5\pi/6$

The 3-D motion trajectories for specified desired positions are shown in Fig. 11 and Fig. 12, respectively. The good quality of motion can be inferred by the trajectory analyze.

## 7 Conclusions

An extension of the Circle Criterion for the dynamic models of the continuum robots is obtained. This technique is applied to the control of the fractional order model system with constraints of the driving forces. It is demonstrated that the stability of the closed loop control system requires that the arm frequency plot does not intersect the circle of driving constraints. It is shown that nonlinear components worsen the stability of the system.

## References

1. Robinson, G., Davies, G.B.C.: Continuum robots – a state of the art. In: Proceedings of the IEEE International Conference on Robotics and Automation, Detroit, May 1999, pp. 2849–2854 (1999)
2. Gravagne, I.A., Walker, I.D.: On the kinematics of remotely - actuated continuum robots. In: Proceedings of the 2000 IEEE International Conference on Robotics and Automation, San Francisco, April 2000, pp. 2544–2550 (2000)
3. Gravagne, I.A., Walker, I.D.: Kinematic transformations for remotely- actuated planar continuum robots. In: Proceedings of the 2000 IEEE International Conference on Robotics and Automation, San Francisco, April 2000, pp. 19–26 (2000)

4. Gravagne, I.A., Walker, I.D.: Uniform regulation of a multi-section continuum manipulators. In: Proceedings of the 2002 IEEE International Conference on Robotics and Automation, Washington DC, May 2002, pp. 1519–1525 (2002)
5. Gravagne, I., Walker, I.D.: Manipulability and force ellipsoids for continuum robot manipulators. In: 2001 IEEE/RSJ International Conference on Intelligent Robots and Systems, Maui, Hawaii, 29–31 October, pp. 304–310 (2001)
6. Chirikjian, G.S., Burdick, J.W.: An obstacle avoidance algorithm for hyper-redundant manipulators. In: Proceedings of the IEEE International Conference on Robotics and Automation, Cincinnati, Ohio, May 1990, pp. 625–631 (1990)
7. Mochiyama, H., Kobayashi, H.: The shape Jacobian of a manipulator with hyper degrees of freedom. In: Proceedings of the 1999 IEEE International Conference on Robotics and Automation, Detroit, May 1999, pp. 2837–2842 (1999)
8. Li, J., Xiao, J.: Determining grasping configurations for a spatial continuum manipulator. In: 2011 IEEE/RSJ International Conference on Intelligent Robots and Systems, 25–30 September 2011, San Francisco, pp. 4207–4213 (2011)
9. Walker, I., Hannan, M.: A novel elephant's trunk robot. In: AIM 1999, pp. 410–415 (1999)
10. Jones, B., Walker, I.D.: Practical kinematics for real-time implementation of continuum robots. IEEE Trans. Robotics **22**(6), 1087–1099 (2006)
11. Kapadia, A.D., Walker, I., Dawson, D.: A model – based sliding mode controller for extensible continuum robots. In: Recent Advances in Signal Processing, Robotics and Automation, ISPRA Conference, pp. 103–120 (2009)
12. Rucker, D.C., Webster III, R.J., Chirikjian, G.S., Cowan, N.J.: Equilibrium conformations of concentric-tube continuum robots. Int. J. Robot. Res. **29**(10), 1263–1280 (2010)
13. Popescu, N., Popescu, D., Ivanescu, M.: A spatial weight error control for a class of hyper-redundant robots. IEEE Trans. Robot. 29(4), 1043–1050 (2013). ISSN 1552-3098
14. Ivanescu, M., Popescu, N., Popescu, D.: The shape control of a tentacle arm. Robotica Cambridge J. **33**(03), 684–703 (2015)
15. Ivanescu, M., Popescu, N., Popescu, D.: A decoupled sliding mode control for a continuum arm. Adv. Robot. Spec. Issue Continuum Robots Manipulation **29**(13), 831–845 (2015)
16. Diethelm, K.: The Analysis of Fractional Differential Equations. Springer, London (2004)
17. Petras, I: Fractional-Order Nonlinear Systems, Modeling, Analysis and Simulation. Higher Education Press, Beijing (2011). Springer
18. Monje, C., Chen, Y.Q., Vinagre, B., Hue, D., Feliu, V.: Fractional-Order Systems and Controls. Springer, London (2010)
19. Aguila-Camacho, N., Duarte-Mermoud, M., Callegos, J.: Lyapunov functions for fractional order systems. Commun. Nonlinear Sci Numer. Simulat. **19**, 2951–2957 (2014)
20. Khalil, H.: Nonlinear Systems. Prentice Hall, New Jersey (2003)
21. Li, Y., Chen, Y.Q., Podlubny, I.: Stability of fractional order nonlinear systems: Lyapunov direct method and generalized Mittag-Leffler stability. Comput. Math. Appl. **59**, 1810–1821 (2010)
22. Agarwal, R., Hristova, S., O'Regan, D.: Lyapunov functions and strict stability of Caputo fractional differential equations. Advances in Difference Equations **2015**(1), 1–20 (2015). https://doi.org/10.1186/s13662-015-0674-5
23. Zhao, Y., Wang, Y., Liu, Z.: Lyapunov function method for linear fractional order systems. In: Proceedings of the 34th Chinese Control Conference, pp. 1457–1463 (2015)
24. Dadras, S., Malek, H., Chen, Y.: A note on the Lyapunov stability of fractional order nonlinear systems. In: Proceedings of the ASME Cleveland, Orlando, August 2017, pp. 123–129 (2017)
25. Al-Saggaf, U.M., Mehedi, I.M., Mansouri, R., Bettayeb, M.: Rotary flexible joint control by fractional order controllers. Int. J. Control Autom. Syst. **15**(6), 2561–2569 (2017)

26. Rhong, L., Peng, X., Zhang, B.: A reduced-order fault detection filter design for polytopic uncertain continuous-time Markovian jump systems with time-varying delays. Int. J. Control Autom. Syst. **16**(5), 2021–2032 (2018)
27. Khimani, D., Patil, M.: High performance super-twisting control for state delay systems. Int. J. Control Autom. Syst. **16**(5), 2063–2073 (2018)
28. Wu, B., Wang, C.-L., Hu, Y.-J.: Stability analysis for time-delay systems with nonlinear disturbances via new generalized integral inequalities. Int. J. Control Autom. Syst. **16**(6), 2772–2780 (2018)
29. Heymann, N., Podlubni, I.: Physical interpretation of initial conditions for fractional differential with Riemann-Liouville fractional derivatives. Rheologica Acta **17**(23), 45–63 (2014)

# Fusion Sensors Experiment for Active Cruise Control

Daniel Turcian(✉), Valer Dolga, Darius Turcian, and Cristian Moldovan

Politehnica University of Timisoara, Timisoara, Romania
daniel.turcian@student.upt.ro

**Abstract.** Speed control is one of the important topics when we analyze the perspective of autonomous driving. The intelligent system based on sensors is controlling the throttle or the brake to obtain the desired speed. Keeping a safe distance by the forward vehicle is one of the targets for the safety systems. To avoid lags of one type of sensor, there are used multiple sensors for the same purpose – fusion sensor. To keep the distance constant we propose a system based on camera processing and ultrasonic sensor in a simple experiment. The aim of the paper is to give an overview of the usage of a fusion sensor for the speed control system of one robot. The robot system is based on Arduino and a video camera that works together with an ultrasonic sensor. The distance is calculated using the video camera's images calibrated with the ultrasonic sensor. The process consists of the acquisition of images, detection of the vehicle ahead, and mapping of the vehicle's image size with the data obtained from the ultrasonic sensor.

**Keywords:** Active cruise control · Fusion sensor · Intelligent camera · Ultrasonic sensor · Cooperative adaptive cruise control

## 1 Introduction

In the last decades, there created the concept of automated controlled systems able to improve the safety on the roads and the comfort for the passengers. Assisting system is used growly and new mechatronics trends are involved in the automotive industry [1].

If in the early '60, there were only a few electronics used, nowadays the synergy of mechanics and electronics is largely used.

The safety and comfort system knew a boom in the last years when the idea of partial or full automated cars become popular. The transition of the decisional center from humane to the machine is focusing on studying the human skills involved in driving. The vision system is one of them. It is used in creating a link between vehicle and environment. The relevance of the vision system is seen in all levels of autonomous driving and used in very different components, including active safety of a high level of the comfort system [2].

Automated speed control is largely used in the automotive industry but it is still under development to archive fully autonomous driving. For this purpose, it needs to senses the environment and to react to different situations that involve deceleration or


E.-C. Lovasz et al. (Eds.): MTM&Robotics 2020, MMS 88, pp. 432–443, 2021.
https://doi.org/10.1007/978-3-030-60076-1_39

acceleration or emergency stop. During the time, there were found solutions and now we have different sensors used for this proposal. Fusion technology added robustness and accuracy into data provided by the sensing system and leads to increasing the trust of users in the systems [3].

The aim of this paper is to give an overview of the concept of the fusion sensor used in the vehicle's speed control system. The solution presented is the fusion data of the camera and an ultrasonic sensor to detect the distance to objects found on the route. The system is using an experimental mobile robot based on Arduino. The main subject analyzed is the possibility to get the distance using the mono-vision system through a dynamic calibration realized with the ultrasonic sensor.

The paper is structured on 4 sections, following these topics: Sect. 1 – Introduction in the topic, Sect. 2 – active cruise control based on fusion sensor, Sect. 3 – Experiment setup, Sect. 4 – Conclusion and future study approaches.

## 2 Active Cruise Control Based on Fusion Sensor

### 2.1 Speed Control Based on Intelligent Systems

One of the technologies developed in the last years is proposed to take control of the vehicle's speed. In [4] it is presented a solution based on a radar system to keep the distance constant. In [5], there is used a system that uses a Lidar system to measure the distance between two cars or a car and an object. In [6], a communication-based system is used to determine the environment around. In [7], it is proposed a system that determines the distance using a camera.

All these systems present advantages and disadvantages. The sensors used in these solutions have limitations and because it is needed to have a robust system capable to take the control, we have to overcome these limitations. For this purpose, a combination of different sensors is used under the concept of fusion sensors. The technology behind the fusion term produces data characterized by high accuracy, robustness, and real-time availability. The latency is controlled and it is minimized and fusion data provides higher independence by the environment [8].

The speed control system knew a huge development in the last years. It improves the comfort of the driver in the situation of long trips, but it also goes into the safety area. The adaptive cruise control (ACC) system is considered now a driver assistance system and it has importance from the safety point of view. The main goal of this system is to maintain a constant distance between two consecutive cars [9].

The system has been tested first time in 1980 [8] and it was created using the RADAR sensor. Nowadays, the technology used for ACC is typically based on RADAR, LIDAR, cameras or ultrasonic sensors [9]. The functionality of the ACC system is depending on some requirements. Thus, if no vehicle is in front of the car, it behaves as a cruise control system. Its focus is to provide comfort. When following another car, it should switch from the cruise control controller to the ACC. This implies the measurement of the distance to the front vehicle and depending on the distance measured, it has some options: fast or smooth decrease of the speed to maintain the safe distance. It is important to have also control over the stability and the time gap [8]. This system is not the collision avoidance system. When the distance is under a defined value, the control is given to the driver.

In Table 1 it is done an overview of the speed control system at this moment.

**Table 1.** Speed control methods bibliography

| Speed control method | Briefly description | Reference |
|---|---|---|
| Cruise control | Maintain the user's selected speed | [10–12] |
| Adaptive cruise control | Cruise control + maintain a safe distance to the following car | [13–15] |
| Communicative adaptive cruise control | Adaptive cruise control + communication with the environment | [16–18] |
| Stop&Go | Adaptive cruise control + stop function in emergency situation in urban areas | [19, 20] |
| Emergency stop | Adaptive cruise control failed system; Stop&Go in any undefined situation | [21] |

## 2.2 Distance Measurement Using Camera

Camera usage for sensing the environment has become a major topic for the automotive industry. Used under the concept of the vision system, it is used for controlling a large number of components. The potential of vision technology is under research, but at the moment is used to identify pedestrian or obstacles and for the ACC system.

To identify the objects around is not the subject of this paper, but we are going to see the methods used to measure the distance with a camera.

Vision technology has multiple usages. A field where cameras became a master technology is active safety. This is due to the possibility to measure the distance to the objects around. To measure the distance whit a camera, there are two solutions [22]:

- A single sensor called mono-vision
- Multiple cameras called stereovision

**Mono-vision Technology.** The system consists of only one camera able to record a sequence of images in real-time. To measure the distance to an object there were developed different solutions for each specific situation.

The majority of the solutions are using technics like in [23] where translation and rotation matrix is used, in [24] the distance between the camera and an object surface is detected with an arbitrary texture pattern or construction of 3D image [25]. This technique is using the fusion data, due to another sensor data for detecting the depth.

The programming technics used to find the distance to an object are not the subject of this paper.

Our proposal for this paper is to use a fixed monocular camera mounted on a mobile robot and to find out the distance to the object situated ahead of the robot.

The technique used in this experiment is based on the lens equations to calculate the distance to an object. Figure 1 is giving the principle for the Eq. (1) that defines the calculation method of distance.

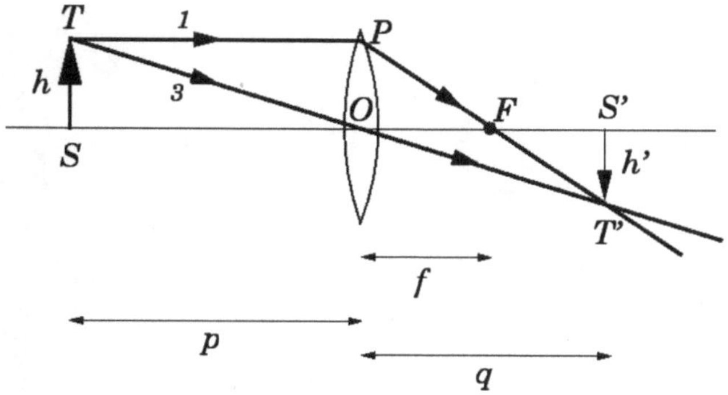


**Fig. 1.** Image formatting for a converge lens [26]

It is interesting to find the distance $p$ that defines the distance between camera and the object ahead where $h$ is the height of real object, $h'$ is the height of the virtual object and $q$ is the image distance along the optical axis to the lens.

Now, the angles SOT and S'OT' are similar and based in this we can write the Eq. (1) [23]:

$$\frac{-h'}{h} = \frac{OS'}{OS} = \frac{q}{p} \tag{1}$$

The angles OPF and S'T'F are also similar and we can express Eq. (2) for the magnification of a spherical mirror:

$$\frac{S'T'}{OP} = \frac{FS'}{OF'} \tag{2}$$

From (1) and (2) we can define Eq. (3) where M is linear magnification [22]:

$$M = \frac{h'}{h} = \frac{q}{p} = \frac{q-f}{f} \tag{3}$$

The Eq. (3) can be rearranged as in (4):

$$\frac{1}{p} + \frac{1}{q} = \frac{1}{f} \tag{4}$$

Based on these equations, a lot of solution was developed. In [24] is used the principle of the complex log mapping (CLM). This is based on mapping the first image from orthogonal coordinate system to the polar coordinate system. The principle of this theory is shown in Fig. 2.

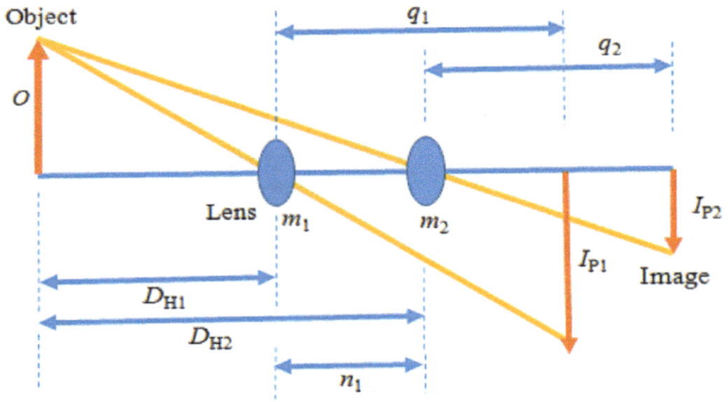


**Fig. 2.** Distance measurement between camera and object [24]

Applying Eq. (4) in this situation, we have:

$$I_{P_i} D_{H_i} = Oq_i,\ (i = 1, 2) \tag{5}$$

If there is a movement of the camera or object on the x-axis, the $I_P$ is changed. The ratio of the old and new $I_P$ is noted Q. Based on this relation, it is possible to define the distance as in (6) and (7):

$$D_{H_1} = f - \frac{Q}{1 - Q} n_1 \tag{6}$$

$$D_{H_2} = D_{H_1} - n_1 \tag{7}$$

Based on these equations, a mapping relationship for each pixel $(D_{H_i}, q_i)$.

In our experiment, we use an additional sensor for getting the first distance to the object $D_{H_1}$.

**Stereovision Technology.** In this case, two cameras are used. They are aligned at the same y coordinate, but a known distance exists between the cameras on axe x. The method used in this situation is called triangulation. This technique is explained in [27–29],

The principle of triangulation is presented in Fig. 3. It is based on calculating the gle P. Then, because the distance $C_R C_L$ is known, in the triangle $C_R OP$, we can apply the arctg function to determine the angle $\widehat{C_L P C_R}$. This gives the rapport of $C_L O/OP$ or $C_R O/OP$. But because we know the distance between the cameras, we can determine OP – the distance to the object.

This method is more accurate and it could be used in a wide range of situation. In this case, it is not needed a new sensor or data to obtain de distance. Also, this technique is used for 3D reconstruction and it provides more algorithms for object detection, distance detection, etc.

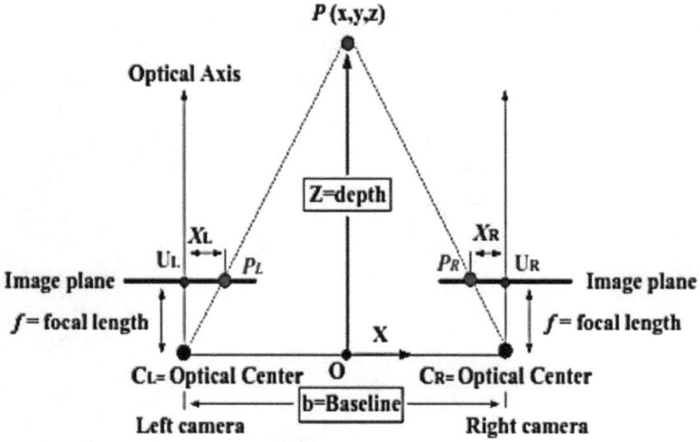


**Fig. 3.** Distance measurement using stereovision [30]

## 2.3 Distance Measurement Using Ultrasonic Sensor

The technology behind this sensor is called the time of flight. It is a system designed and used in the measurement of speed and distance. The principle that is used in this situation consists of measuring the time needed for a particle or sonic wave to "touch" an object and to return back at the sensor. The measurement of this time is depending on many parameters and it is detected using direct measurement or indirect [31].

The distance is given by this Eq. (8) [22]:

$$d = \frac{1}{2} v_{signal} t_{flight} \tag{8}$$

Where d is the distance, $v_{signal}$ is the speed of signal emitted, $t_{flight}$ is the time during emitted and reflected time.

## 2.4 Fusion Sensor

The fusion concept is based on using data from multiple sources to get a real output [32]. The complexity of the automotive industry asked for a powerful sensing system, more independent of environment or changes, with less limitation and more accurate. The rise of autonomous driving driven this technology to the maximum importance. The fusion is defined in [8] as the process to combine data from multiple sensors to predict or estimate a status. In our scenario, the data regarding the distance to the ahead object is provided by the camera and ultrasonic sensors. Our proposal for this experiment is to create a system that use the monocular camera for real-time distance measurement and an ultrasonic sensor to automatic calibration of the camera.

In Fig. 4 it is shown the schematic of our proposal for the fusion data for distance detection. The ultrasonic sensor is used only for mapping a new detected object.

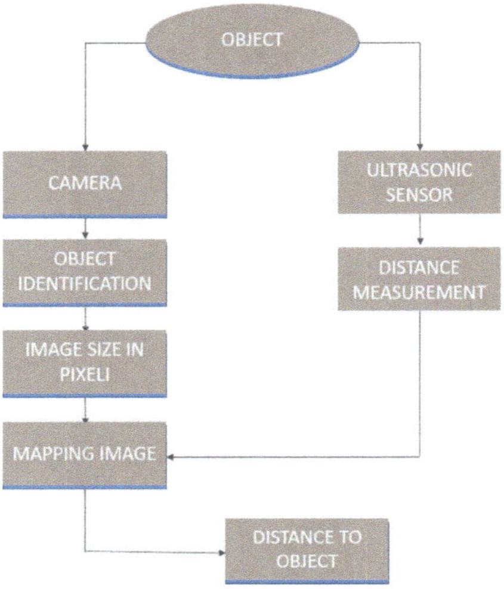


**Fig. 4.** Fusion data schematic method

The data fusion encompasses a large number of fusion process related to measurements. It uses multiple sensors to obtain information about an object state.

The data fusion is divided into these categories, considering the data usage [34]:

1. Complementary fusion: there are used different sensors that provide partial data used for obtaining the desired information;
2. Competitive fusion: the same sensors are used for accurate measurements;
3. Cooperative fusion: different sensors are used to provide independent data for the same purpose.

In this paper, we used first method. Two different sensors are used to obtain the distance to the ahead object. Camera provides information about the height and weight of the object and the ultrasonic sensor provide the distance to the object ahead. The ultrasonic sensor is used only for mapping the image of object.

On the other hand, from the algorithm point of view, the fusion is characterized on 3 levels [35]:

1. Low-level fusion – it processes the raw data from sensors to obtain the raw data set that is expected. This method is based on the principle data in – data out. The input is a raw data and the output is a more accurate data. There are multiple algorithms, named kinematic data fusion.

2. Intermediate-level, mid-level or feature level fusion – combines various features to obtain a data. The algorithms used for these levels are called pixel-, feature-, or image-level fusion.
3. High-level fusion combine decisions from intelligent systems. Algorithms used for this level are called voting, fuzzy logic or statistical methods.

In our experiment, we used a low-level fusion, based on the input data from camera and ultrasonic sensor. Data from the ultrasonic sensor is used at the moment of new object detection to measure the distance to this one. The distance is saved. In the same time a picture is taken. The picture is later processed and the dimension of the object ahead saved.

# 3 Experiment Setup

The desired output of this experiment is to validate the active cruise control system for a mobile robot using the technology of distance measurement based on a fusion sensor. The solution we propose is to use a monocular camera for distance detection and an ultrasonic sensor to calibrate the camera for each new object detected. We started with the theory explained in [21], but instead of mapping all the images, the ultrasonic sensor is giving primary information regarding the distance between camera and object ahead. Based on that information, the system is able to maintain the safe distance constant to the ahead vehicle.

## 3.1 Intelligent Camera

The intelligent camera used in the project is a machine vision camera from JeVois. The characteristics of this camera are presented in Table 2 [33]:

**Table 2.** Characteristics of camera

| Camera JeVois | Briefly description |
|---|---|
| Processor | Allwinner A33 quad core ARM Cortex A7 processor |
| Memory | Cruise control + maintain a safe distance to the following car |
| Camera sensor | 1.3 MP camera with:<br>SXGA (1280 × 1024) up to 15 fps (frames/second)<br>VGA (640 × 480) up to 30 fps<br>CIF (352 × 288) up to 60 fps<br>QVGA (320 × 240) up to 60 fps<br>QCIF (176 × 144) up to 120 fps<br>QQVGA (160 × 120) up to 60 fps<br>QQCIF (88 × 72) up to 120 fps |
| Camera field of view | 65° horizontal, F2.8, 4:3 sensor aspect ratio |
| Power | 3.5 W maximum from USB port. Requires USB 3.0 port or Y-cable to two USB 2.0 ports |

This camera is prepared for running the most of machine vision frameworks available. The identification of an object is easily implemented with Open CV or other libraries. All the operations were realized with the help of the OpenCV library.

### 3.2 Arduino Robot with Camera and Ultrasonic Sensor

The mobile robot is built upon Arduino Uno board. It consists of 4 wheels, each of them with a separate motor. On each wheel, it has a speed optic sensor. The motors are controlled with an L298P shield motor driver. More than that, it has a Bluetooth module to set the desired speed for a mobile robot.

The architecture of the robot is shown in Fig. 5.

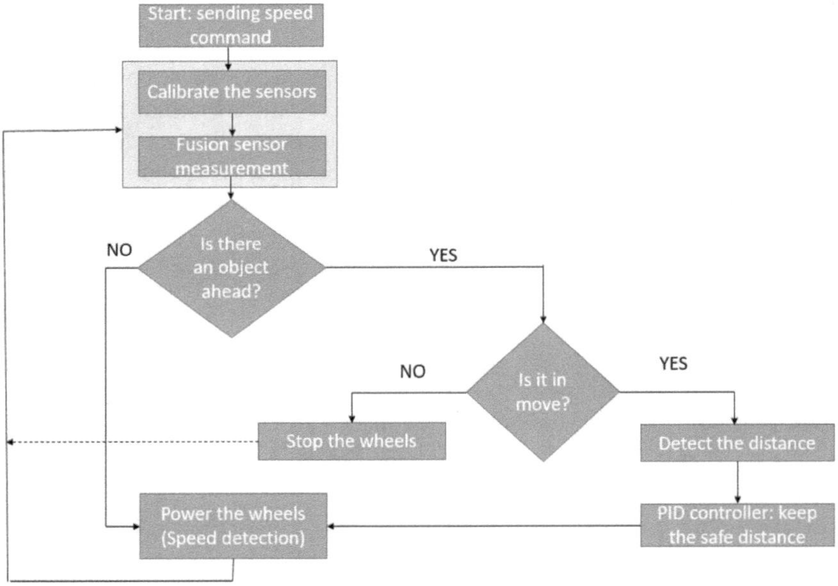


**Fig. 5.** The function diagram of the robot

The user sends the start command that consists of a PWM (pulse-width-modulation) from a user interface created in Matlab. The command is sent via the Bluetooth module. This is the desired speed. Before the command arrives at the electric motor, the fusion sensor is calibrated. This supposes that the ultrasonic sensor measures the distance to the ahead object. If this is missing, then it will measure the environment around. The camera is making a record. Then, the robot is moved within a short distance. This creates the opportunity to use Eqs. (3) and (4) as well. Using the focal length and the ultrasonic sensor to detect the initial value of distance p, we can detect the real value of object's size, h. This can be used to map the size of the object. The relative speed between the robot and another robot will create a resize of the h'.

It is followed by the Eq. (6) that gives the relative displacement between the robots or between the robot and on a static object. From Eq. (7), the new distance can be found. In this mode, the system needs only the initial distance between the robot and the ahead object and to memory the value of the real h. It is memory resource-saving and more precise than using only one sensor.

After the calibration and first measurement, the code makes a first check of the environment. It checks if there is an object ahead or no. If there is no object, the robot will speed up with the command from the user. If there is an object ahead, it will wait a while and it will measure if the distance is changed. This gives the information if the object ahead is moving or not. Then the robot will maintain a safe distance to the object ahead. If it is moving, the robot will control the speed of the motor with the PID controller to maintain the distance constant or if the object is static, the robot will stop the wheels.

## 4 Results and Conclusion

The robot was tested in three situation. First scenario, the robot was in standstill and the object fixed. It has been measured manually, with ultrasonic sensor and with camera. The test was repeated in situation when the robot was in movement and the object fixed. For the last test, the robot was in movement and the object was also in movement. The results of the experiment are shown in Table 3. The measurements of 2nd and 3rd scenario were analyzed in a video recorder.

**Table 3.** The result of the measurements

| Scenario | Real value [mm] | Camera [mm] | Ultrasonic [mm] |
|---|---|---|---|
| Static robot, static object | 120 | 119.958 | 120 |
| Robot moves, static object | 132 | 133.56 | 133 |
| Robot moves, object moves | 125 | 126.78 | 127 |

The system is working properly for a simple experiment. The idea of only one object ahead of the robot is not suitable for a real traffic situation. Therefore, it needs further research in the domain of multi-object identification. The ultrasonic sensor is sensitive to the noise and it might be used only in some situation. The camera system had a good detection precision, but the processor needs to be more powerful.

The problem consists in the time needed for detection the object, processing the image, saving the data, computation of the distance based on the values saved, checking if a new object is used in front of robot.

## References


1. Happian-Smith, J.: An Introduction to Modern Vehicle Design, BH (2002)
2. Chao, E.: Automated driving system: a vision for safety, NHTSA (2017)
3. Jelena, K., Nenad, J., Vujo, D.: Sensors and sensor fusion in autonomous vehicle. In: 26th Telecommunications Dorum (TELFOR), Belgard, Serbia (2018)

4. Caveney, D., Hedrick, J.K.: Single versus tandem radar sensor target tracking in the adaptive cruise control environment. In: Proceedings of the 2002 American Control Conference, Anchorage, AK, USA (2002). https://ieeexplore.ieee.org/document/1024819. Accessed 3 Jan 2020
5. Philipp, L., Gerd, W.: 3D LIDAR processing for vehicle safety and environment recognition. In: 2009 IEEE Workshop on Computational Intelligence in Vehicles and Vehicular Systems, Nashville, TN, USA (2009). https://ieeexplore.ieee.org/document/4938725. Accessed 3 Jan 2020
6. Weinan, G., Fernando, R.G., Weitian, T., Lei, C.: Cooperative adaptive cruise control of connected and autonomous vehicles subject to input saturation, New York, NY, USA (2017). https://ieeexplore.ieee.org/document/8249022. Accessed 3 Jan 2020
7. Stein, G.P., Mano, O., Shashua, A.: Vision-based ACC with a single camera: bounds on range and range rate accuracy. In: IEEE IV2003 Intelligent Vehicles Symposium. Proceedings (Cat. No. 03TH8683), Columbus, OH, USA (2003)
8. Hermann, W., Stephan, H., Felix, L., Christina, S.: Handbook of Driver Assistance Systems - Basic Information, Components and Systems for Active Safety and Comfort. Springer, Heidelberg (2016)
9. Kim, K.-E., Lee, C.-J.: Sensor fusion for vehicle tracking with camera and radar sensor. In: 17th International Conference on Control, Automation and Systems (ICCAS 2017), Ramada Plaza, Jeju, Korea (2017)
10. Bosch, R.: Bosch Acc Adaptive Cruise Control. Bentley Pub, Auburn (2003)
11. Han, C., Yang, J., Li, M.: Robust cruise control of heterogeneous connected vehicle systems. In: 2018 Chinese Control And Decision Conference (CCDC), Shenyang, China (2018)
12. Peter, G., Jakub, F.: Maintaining vehicle speed using a mechanical cruise control. Acta Electrotechnica et Informatica **17**(2), 48–52 (2017). https://doi.org/10.15546/aeei-2017-0016
13. Idriz, A.F.: Adaptive Cruise Control with Auto-Steering for Autonomous Vehicles. Omniscriptum Gmbh & Company Kg., Saarbrücken (2015)
14. Reif, K.: Adaptive Cruise Control, Fundamentals of Automotive and EngineTechnology, pp. 202–209. Springer Fachmedien, Wiesbaden (2014)
15. Kakade, W.R.S.: Adaptive cruise control system: design and implementation, interdisciplinary programme in systems and control engineering Indian institute of technology, Bombay Mumbai (2006)
16. Ko, W., Chang, D.E.: Cooperative adaptive cruise control using turn signal for smooth and safe cut-in. In: 18th International Conference on Control, Automation and Systems (ICCAS), Daegwallyeong, South Korea (2018)
17. Evans, M.R., Peters, J.I.: Cooperative adaptive cruise control: human factors analysis, Publication No. FHWA-HRT-13-045 (2013)
18. Anayor, C., Gao, W., Odekunle, A.: Cooperative adaptive cruise control of a mixture of human-driven and autonomous vehicles. In: SoutheastCon 2018, St. Petersburg, FL, USA (2018)
19. Ghaffari, A., Panahi, A., Khodayari, A., Alimardani, F.: Neural-network based modeling for Stop&Go behavior in real traffic flow. In: 6th IEEE International Conference Intelligent Systems, Sofia, Bulgaria (2012)
20. Naranjo, J.E., Gonzalez, C., Garcia, R., de Pedro, T.: Cooperative throttle and brake fuzzy control for ACC+ Stop&Go Maneuvers. IEEE Trans. Veh. Technol. **56**, 1623–1630 (2007)
21. Kimbrough, S., Chiu, C.: A brake control algorithm for emergency stops (which may involve steering) of tow-vehicle/trailer combinations. In: 1991 American Control Conference, Boston, MA, USA (1991)
22. Alizadeh, P.: Object distance measurement using a single camera for robotic applications, The Faculty of Graduate Studies Laurentian University Sudbury, Ontario, Canada (2015)

23. Tao, Z., Changku, S., Shan, C.: Monocular vision measurement system for the position and orientation of remote object. International Symposium on Photo electronic Detection and Imaging, vol. 6623 (2007)
24. Krishnan, J.V.G., Manoharan, N., Rani, B.S.: Estimation of distance to texture surface using complex log mapping. J. Comput. Appl. **3**(3), 16 (2010)
25. Barrois, B., Wöhler, C.: 3D pose estimation based on multiple monocular cues. In: Proceeding of IEEE Conference on Computer Vision and Pattern Recognition, pp. 1–8 (2007)
26. Image formation by thin lenses. http://farside.ph.utexas.edu/teaching/302l/lectures/node141.html. Accessed 3 Jan 2020
27. Lindstrom, P.: Triangulation made easy. https://www.researchgate.net/publication/221362655_Triangulation_Made_Easy. Accessed 32020
28. 3D imaging technology - time of flight. https://www.azom.com/article.aspx?ArticleID=16003. Accessed 3 Jan 2020
29. Hahne, C., Aggoun, A., Velisavljevic, V., Fiebig, S., Pesch, M.: Baseline and triangulation geometry in a standard plenoptic camera. https://www.researchgate.net/publication/319183885_Baseline_and_Triangulation_Geometry_in_a_Standard_Plenoptic_Camera. Accessed 3 Jan 2020
30. Adi, K., Widodo, C.E.: Distance measurement with a stereo camera. Int. J. Innovative Res. Adv. Eng. (IJIRAE) **4**(11) (2017). https://ijirae.com/volumes/Vol4/iss11/05.NVAE10087.pdf. Accessed 3 Jan 2020
31. Li, L.: Time-of-flight camera – an introduction. Texas Instruments. http://www.ti.com/lit/wp/sloa190b/sloa190b.pdf. Accessed 3 Jan 2020
32. Ningthoujam, B., Ningthoujam, J.S., Namram, R.S., Nongmeikapam, K.: Image and ultrasonic sensor fusion for object size detection. In: Fifth International Conference on Image Information Processing (ICIIP -2019), Japee University, Shimla, at JUIT, India (2019)
33. https://www.jevoisinc.com/pages/hardware
34. Castanedo, F.: A review of data fusion techniques. Sci. World J. **2013** (2013). Hindawi Publishing Corporation
35. Raol, R.J.: Multi-Sensor Data Fusion with MATLAB. CRC Press, Boca Raton (2009)

# Robot Soldering Positions Correction with Cognex Vision Camera

Georgios Tsigaras[1(✉)] and Valer Dolga[2]

[1] Politehnica University Timisoara, Timişoara, Romania
georgios.tsigaras@student.upt.ro
[2] Mechatronics Department, Politehnica University Timisoara, Timişoara, Romania

**Abstract.** Inline soldering may hide many variables that can affect the soldering process especially when the pcbs are transferred on the systems' conveyor inside a custom made manufactured carrier. The manual introduction of the board inside the carrier and the placement of the components done by human operator can also be considered as a principal factor for the variations of the nominal position that the pcb must have so the soldering process can start. The duplication of the manufactured carries cannot be 100% accurate, even if the carrier is within manufacturing tolerance (0.2 mm depends on the equipment & process). This tolerance can affect the soldering process especially on very small pins >0.5 mm DIA. The offset of the PCB can lead to bad solder joints (shorts, insufficient barrel fill etc.) and make the yield of the process drop. Yields above 98% on every EMS process are considered bad, making EMS giants lose profit every second as the products require repair or in some cases need to be scraped [8]. Every retrofit or upgrade like this is realized for high demand customers who expect high yield numbers. For the above reasons the Japanese company came with relative new solution in collaboration with Mitsubishi and Cognex in order to eliminate all these position offsets using a vision camera to scan fiducials before the soldering process starts.

**Keywords:** Robot · Soldering program · Vision camera · PCB

## 1 Introduction

Over the past decades robotic arms have replaced human operators in factories all over the world [4]. EMS industry is not an exception and here we can see many different types of robots (parallel, scara, arm etc.) performing a big variety of operations (screwing, pick and place of components, soldering etc.) [3]. In line soldering was one of the most difficult processes demanding accuracy, repeatability and soldering knowledge of THT components

in order to achieve quality acceptable results [9]. As the robots started working in factories all over the world, we received feedback regarding the yield score of the robotic process. Some of the end users reported misplace of the soldering tip caused by the carrier in which the pcb is transported or the conveyor itself in case that PCBs are being transported without carrier [4]. As a first step a system with bottom guide pins


E.-C. Lovasz et al. (Eds.): MTM&Robotics 2020, MMS 88, pp. 444–456, 2021.
https://doi.org/10.1007/978-3-030-60076-1_40

was developed (Fig. 1). The retrofit had very good results but the problem for the pcbs being carried without a carrier were still a problem. As the vision cameras for different applications started to lead the market, everyone knew that sooner or later a vision system will be integrated to solve this issue [4].

## 2 Description of the Application

### 2.1 Introduction

For this application we are going to use a base Mitsubishi RH-6FH series robot specially produced for the Japanese company Japan Unix in order to perform inline soldering of pcb components. The robot arm is a RH-6FH series robot with no major modifications but the CRND controller shows us some promising futures that lead to succeed soldering of the components [9]. Despite all these futures some special applications with small pins did show us a few bad solder joints caused by the actual position of the pcb compared with the one used for teaching. Till nowadays it did seem impossible to control external factors but its time to test the new upgrade. We have received the kit with all the parts we need and our job is to install the hardware, calibrate the camera, integrate it to the process and compare the results. All the above are going to be presented in this presentation together with the results. The PCB's are transferred in and out of the system through a conveyor with belts as show in the picture below (Fig. 1).

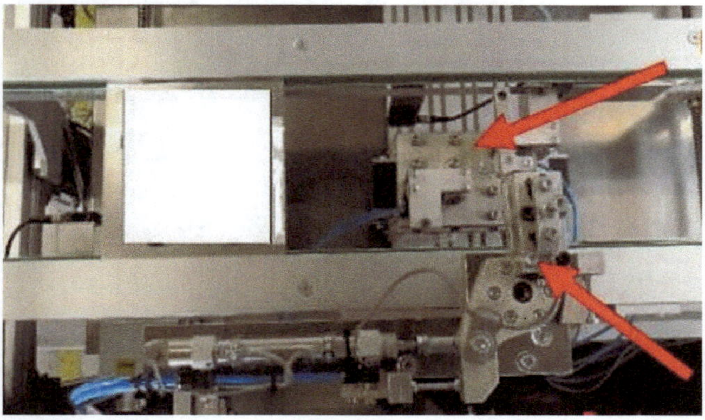

**Fig. 1.** PCB transfer conveyor. The PCB is hidden for customer- product confidentiality reasons.

Even there are top and bottom guiding mechanisms (pins, see arrows Fig. 1) triggered by air pistons the nominal position of the pcb is not always exactly the same. The soldering tip (soldering head assembly) can maintain the exactly same position by tip correction sensor (Fig. 2) [5]. Coming back to the position of the pcb inside the carrier, it is a variable that cannot be controlled by any mechanical part and the only solution is to modify the robot position calculating the angle of 2 fixed points on the PCB (fiducial marks) (Fig. 3).

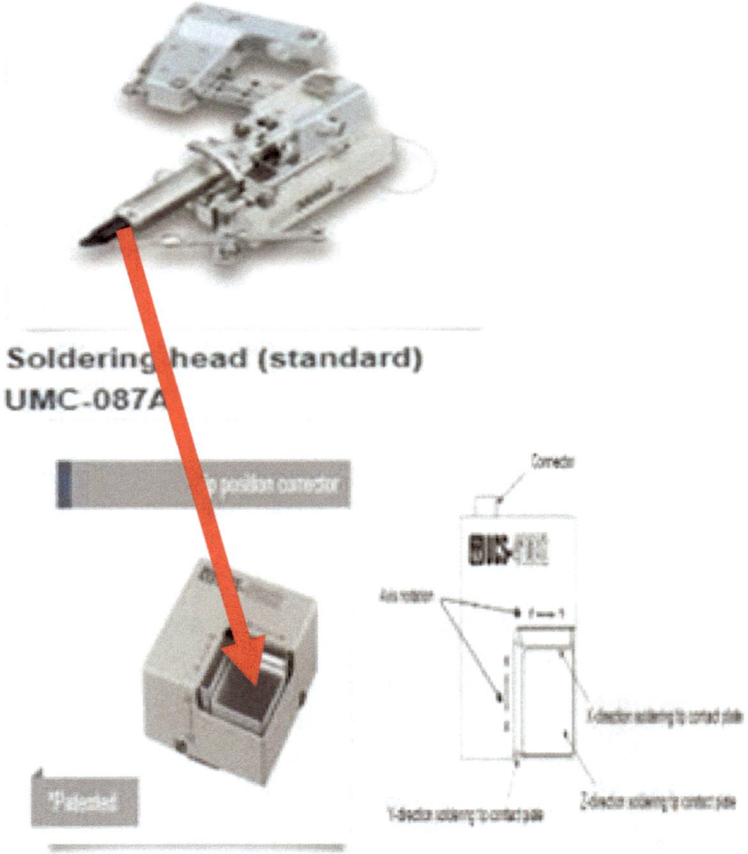


**Fig. 2.** Tip correction sensor

The pcb's fiducials will be scanned for 1st time before the soldering program gets created and the results will be saved in the memory of the robot controller. After this every pcb that is going to be soldered will be scanned and the difference of angle compared to the 1st pcb will be added, calculated and applied as X and Y value to the robot coordinates in order to obtain the correct position on the pin.

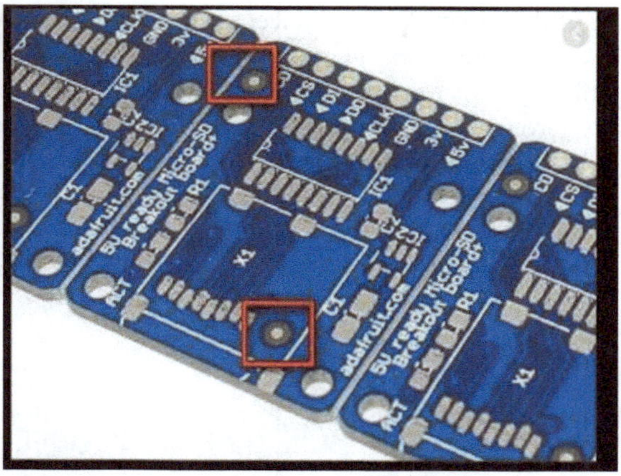


**Fig. 3.** PCB fiducial marks

## 2.2 Description of the Upgrade Kit and Installation

The installation kit contains the CognexIn-Sight 8000 Series camera, 8402 model 1600 × 1200 resolution with an alloy bracket that is going to be mounted on the soldering head of the robot, a POE switch to power up the camera and a normal switch that connects the camera LAN cable to the robot controller and the PC. For better lighting while scanning the fiducials a LED ring (Fig. 6) with red colour is also included for the camera assembly. The LED controller is going to be wired directly to the robot controller digital I/O and the power ON-OFF signal will be sent from the robot program. The camera is going to be operated by Cognex insight software so the robot cell must have a PC that sends the camera data to the robot controller. Our soldering cell is built with a Siemens S1500 PLC running on Scada (winCC runtime) so that is not going to be a problem (Fig. 4).

**Fig. 4.** Robot cell's electrical box

At the back side of the robot, inside the electrical box we mounted the new switch and we connected the 3 wires (one to the pc, one to the camera and one to the controller). The hardest part is yet to come, the lan cable of the camera has to be mounted all the way through the robot's couplings and reach the camera next to the soldering head. All the robots covers have to be removed also an additional hole has to be drilled so the cable can get out of the back-robot plate. Another obstacle to this difficult job is that the RJ45 connector is too big to go through the couplings of the robot and it has to be removed. It will be crimped again as soon as the cable management finishes (Fig. 5).

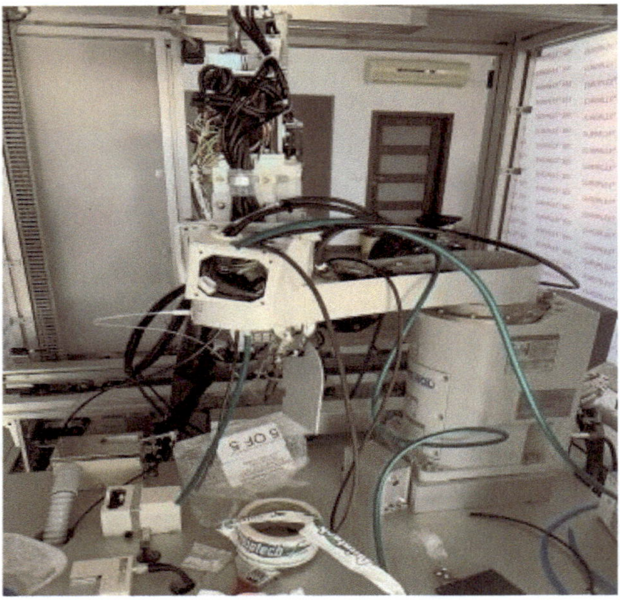

**Fig. 5.** Installation of the camera cables

As soon as we finished with the installation of the hardware, camera focus adjustment has to follow and then camera calibration as a last step before we are ready to test the upgrade. The picture below shows the final position of the camera attached next to the soldering head (Fig. 6).

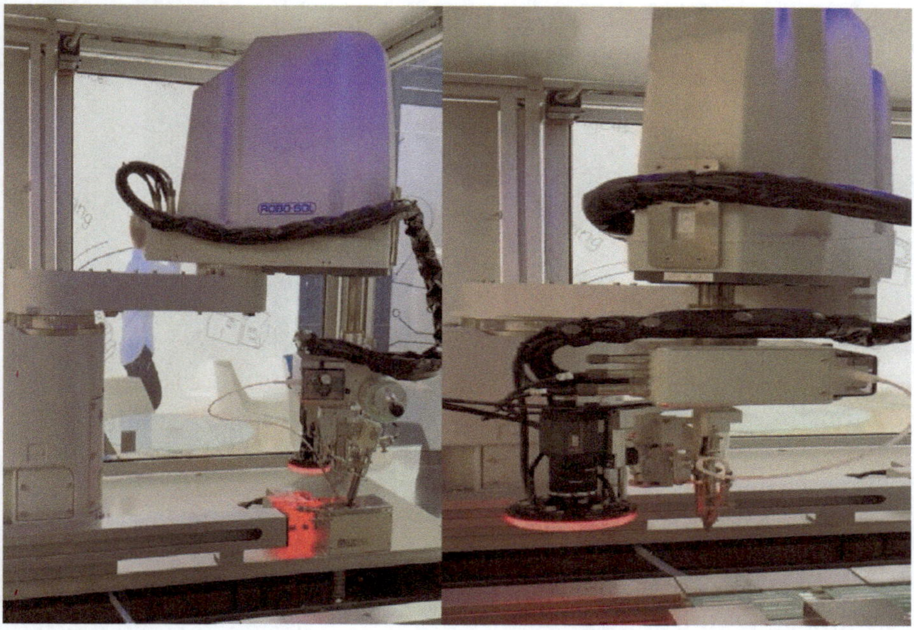

**Fig. 6.** Vision camera position

## 2.3 Camera Focus Adjustment

After all the hardware modifications are finished, I have to install the Cognex in sight software and set up the camera IP. The in-sight software can detect the camera and go to live mode although I can see no clear image because the camera focus has to be adjusted by the camera lens manually.

For this task to be completed first of all we have to find an ideal position for Z axis of the robot as the camera is attached to the head and not standstill. The position will be defined after taking in consideration all the obstacles around the robot preventing crashing the head or the camera. As the value was defined is time to run in sight software go above a fiducial mark and adjust the camera lens until we get the best results. Cognex software offers real time percentage of the image quality and makes the process easier. The next step is to calibrate the camera for the really first time (Fig. 7).

**Fig. 7.** Camera focus adjustment

# 3 Camera Calibration

This operation is very important and has to be proceeded with accuracy because all the images we are going to capture in the future are going to be transformed in data and they will be sent to the robot controller. Based on these data the controller is going to calculate the offset and add it to the robot coordinates (X, Y) in order to maintain the same position as the ones taught (Fig. 8).

**Fig. 8.** Camera focus adjustment

For this procedure we need: Calibration plate, Robot program CAMTL.prg, In-Sight Explorer JOB file "Sample_line.job" [1]. As I don't have a calibration plate and there is no time to create one, I am going to print the shapes needed on a plain A4 paper and position them below the camera where there should be if there was a plate. There are nothing more than a normal black boarder rectangle and one with a cross shape inside.

Next we will have to position the camera above the cross shape triangle and access the CAMTL program inside the robot controller which i received from Japan Unix. Inside the program i have to press function until FWD is shown for F1. Then i press FWD twice to get to TOOL P_NTool (Fig. 9).

**Fig. 9.** Robot command Tool P_NTool

The next step is to open file -> open job -> line_sample.job inside the insight explorer software. After i open the job i have to set the camera to continuous shot so i can see live image. I center the camera to the middle of the cross as shown in the picture below (Fig. 10) and I teach the position inside the cametl program “P0”.

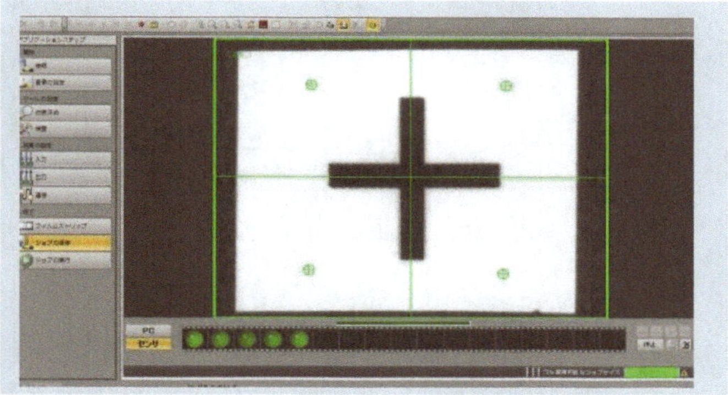

**Fig. 10.** Camera centering

After this we have to teach P90 (same position +90°) with the same procedure. Next is the second rectangle. I leave the camera in its turned position (90°) and jog it above the second calibration mark. After i center the camera to the rectangle as shown in the picture (Fig. 11) below i teach P_CALIB inside the CAMETL program

Further on its necessary to move the cross marks on the edges of the rectangle and teach them inside job points (Fig. 12).

As a last step I have to move the camera center to the edges of the rectangle with the robot JOG function and note the XY. Then I have to introduce them inside every point of the job manually (Fig. 13) [2].

Now the calib_1 shows pass and the calibration process is finished. From now on we can use the camera in production mode (Fig. 12).

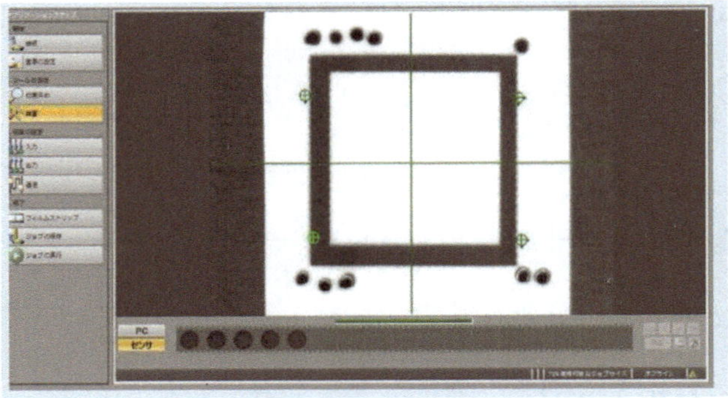

**Fig. 11.** Camera centering

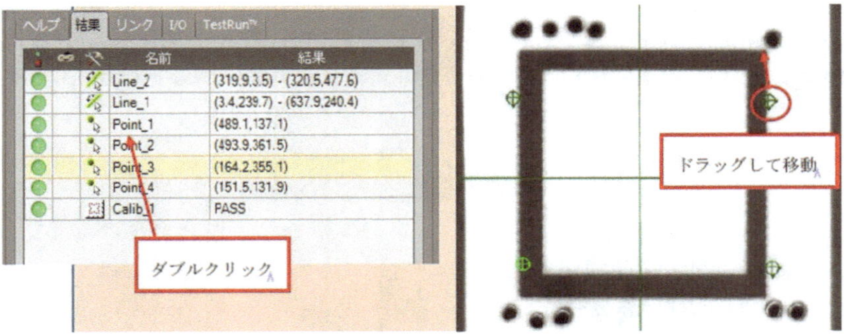


**Fig. 12.** Teaching of the edges

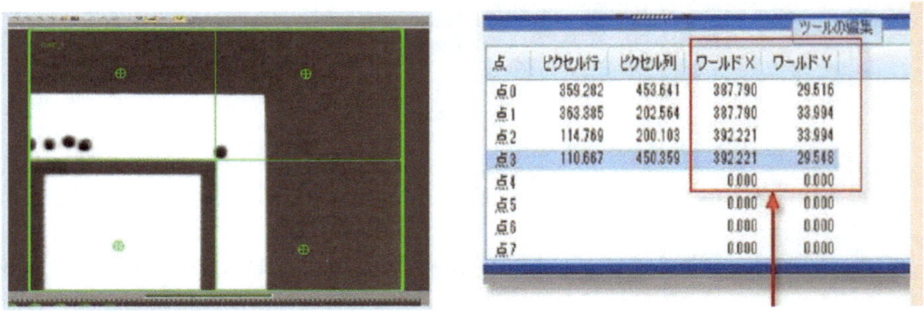


**Fig. 13.** Introducing the values

## 4 Implementation in the Soldering Program

First of all, I have to create a new job inside in sight software [7]. The software came with sample jobs pre-installed so I am going to use file Program Sample_20_E.job. I move the robot head above the fiducial and switch the camera to live image. Now I have to register the image of the fiducial inside point one and teach scan area and shape of it (Fig. 14). In the meanwhile, I have to remember the XYZ and C coordinates of the robot so I can use them inside the program when I need them.

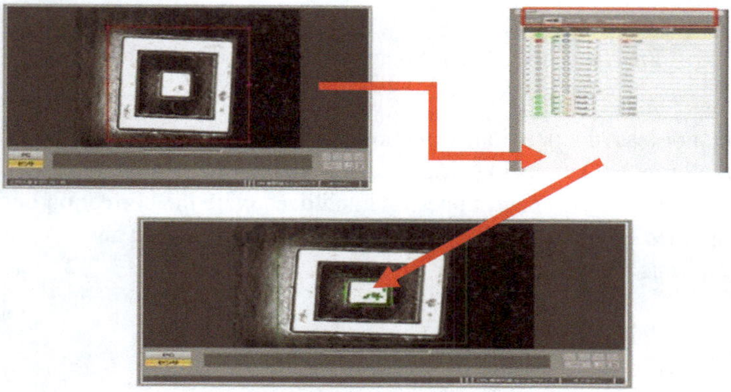

**Fig. 14.** Fiducial Scan area and shape teaching

Fiducial 1 is now registered as point one. The same procedure I am going to follow for the 2nd fiducial mark.

After both fiducials are registered inside the vision interface, I have to scan then for the 1st time with the robot. I am going to use the following program lines (Fig. 15) [6]:

```
'Initial alignment teaching
CallP "MCR ", 0, 0, 30.0, 0.0, +0.0, 0, +0.0, 0, +0.0, 0, 0,(+487.02, +58.69, +3.69, +0.00, +0.00, -0.01, +0.00, +0.00)(4, 0)
CallP "MCR ",11, 21, 0.0, 0.0, +0.0, 1, +0.5, 1, +0.0, 1, 1,(+487.02, +58.69, +3.69, +0.00, +0.00, -0.01, +0.00, +0.00)(4, 0)
If M_VSerr=1 Then GoTo *LEnd
CallP "MCR ",11, 22, 0.0, 0.0, +0.0, 1, +0.5, 1, +0.0, 2, 1,(+419.49, +18.40, +3.69, +0.00, +0.00, +89.99, +0.00, +0.00)(4, 0)
If M_VSerr=1 Then GoTo *LEnd
```

**Fig. 15.** Robot program lines

Macro 0,0 is PTP movement (evasion point) and 11,21–11,22 fiducial 1&2. The XYZ and U coordinates are inside parenthesis (X, Y, Z, additional axis, U). Lines with if are used for image acquisition error. If the vision software is not able to capture or register an image the program will finish automatically [6].

The images are now registered and we are ready to run production with fiducial correction. As a last step I am going to introduce the fiducial scan positions inside the soldering program and a special command when to start and when to end the positions correction regarding the camera readings (Fig. 16).

Lines 1 to 4 are used for the 1st fiducial and 5 to 8 for the second one. The condition of no image registered is still used.

```
CallP "MCR ", 0, 0 , 40.0 , 0.0 , +0.0, 0, +0.0, 0 , +0.0, 0 , 0 ,( +487.02, +58.69, +3.69, +0.00, +0.00, -0.01, +0.00, +0.00)(4 , 0)
'Dly 0.5
CallP "MCR ",12, 21 , 0.0 , 0.0 , +0.0, 1, +0.5, 1 , +0.0, 1 , 1 ,( +0.00, +0.00, +0.00, +0.00, +0.00, +0.00, +0.00, +0.00)(0 , 0)
If M_VSerr=1 Then GoTo *LNoWorkpiece
'Second alignment mark
CallP "MCR ", 0, 0 , 40.0 , 0.0 , +0.0, 0, +0.0, 0 , +0.0, 0 , 0 ,( +419.49, +18.40, +3.69, +0.00, +0.00, +89.99, +0.00, +0.00)(4 , 0)
'Dly 0.5
CallP "MCR ",12, 22 , 0.0 , 0.0 , +0.0, 1, +0.5, 1 , +0.0, 2 , 1 ,( +0.00, +0.00, +0.00, +0.00, +0.00, +0.00, +0.00, +0.00)(0 , 0)
If M_VSerr=1 Then GoTo *LNoWorkpiece
```

**Fig. 16.** Fiducial scan command lines in the robot program

Last but not least the position correction has to be used only for soldering positions and not for all robot positions like origin position, air clean position etc. For this reason, I use a special macro command to activate and deactivate the correction only between the program lines I want to be corrected. This macro doesn't need any XYZ coordinates inside parenthesis and looks like this (Fig. 17):

```
'Activate alignment correction
CallP "MCR ",13, 21 , 0.0 , 0.0 , +0.0, 1, +0.0, 1 , +0.0, 1 , 1 ,( +0.00, +0.00, +0.00, +0.00, +0.00, +0.00, +0.00, +0.00)(0 , 0)
'Deactivate alignment correction
CallP "MCR ",13, 21 , 0.0 , 0.0 , +0.0, 1, +0.0, 1 , +0.0, 0 , 1 ,( +0.00, +0.00, +0.00, +0.00, +0.00, +0.00, +0.00, +0.00)(0 , 0)
```

**Fig. 17.** Activation and deactivation of correction commands

## 5 Conclusions

First of all it was quite difficult task to proceed including hardware and software modifications that requires knowledge and skills. Definitely is an upgrade for existing soldering systems or new ones that seems to solve one of the biggest problems of the soldering process, the external position variables. From the advantages and disadvantages perspectives we can highlight: Its obvious a huge advantage is that the yield of the soldering process will raise, saving a lot of money for repair or scraping PCBS. The accuracy of the pcb carriers can also be reduced as there is a vision system that corrects the soldering positions, permitting end users use less resources on the manufacturing of them. As a disadvantage we see the cycle time raise, adding 2 additional points for the robot to execute and 0.5 s delay before each of them as the camera needs to be stable before scanning an image. Last but not least the cost of the kit and the installation can not be unconsidered.

For the functionality test now we are going to create a table in which we are going to add 20 measurements, 10 with camera correction and 10 without. The reference position will be taught with the tip exactly on the edges of the soldering pad and its going to be considered as point 0. We will choose a soldering tip with exactly the same width as the solder pad diameter so we can observe any deviation. We will run 10 trials with camera and 10 without. As soon as the pcb stops at the stopper and the robot starts the soldering program we will stop the robot while being at the soldering position. We will use 200 degrees temperature so we don't burn the pcb surface during the measurements. As we don't have access to a digital measuring equipment with the help of a feeler gauge we will try to measure the offset on X and Y directions compared to the position we did teach as a reference 0 position. Our feeler gauge has a scale of 50 to 1000 μm. Every value under 50 will we considered as 0. The solder pad has a diameter of 1 mm and our tip the same. Below you can observe the result of our measurements.

* *It is important also to mention the position repeatability value given by Mitsubishi for the RH-6FH45 robot is 10 μm for the X-Y gentry.*

| | With camera | | Without camera | |
|---|---|---|---|---|
| Measurement nr. | X | Y | X | Y |
| 1 | 0 | 0 | 100 | 0 |
| 2 | 50 | 0 | 150 | 50 |
| 3 | 50 | 50 | -50 | 0 |
| 4 | 50 | 0 | 150 | 100 |
| 5 | 0 | 50 | 150 | 0 |
| 6 | 0 | 0 | -100 | 50 |
| 7 | -50 | 0 | 150 | 50 |
| 8 | -50 | 0 | 150 | 100 |
| 9 | 0 | 0 | 100 | 0 |
| 10 | 50 | 50 | 100 | 0 |
| average value | 10 | 15 | 90 | 35 |
| accuracy value | 10 | 15 | 90 | 35 |
| standard deviation(σ) | 39.44053 | 24.15229 | 90.67647 | 41.1636301 |
| repeatability (3 σ) | 118.3216 | 72.45688 | 272.0294 | 123.49089 |

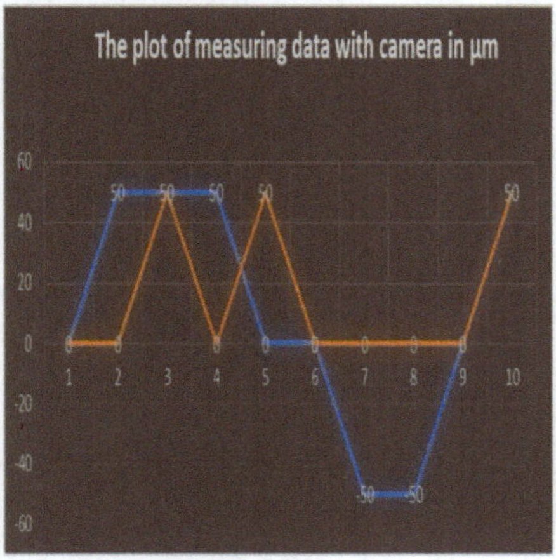


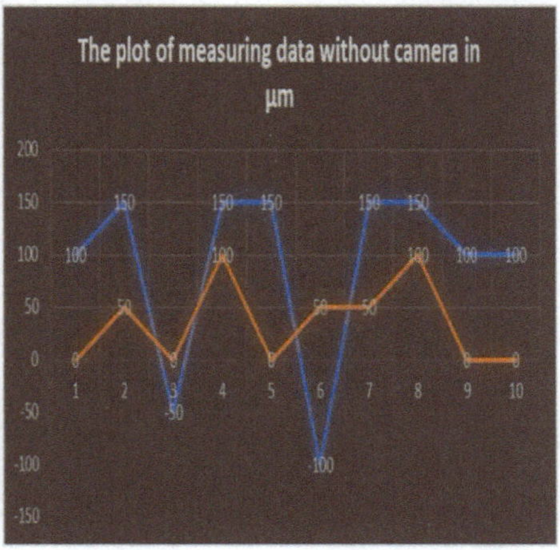


Taking a closer look to the measurement results we can observe that using the camera, deviation and repeatability values are 3 times smaller than without it. We can consider the upgrade as a successful one, achieving incredible scores such as 39.4 μm standard deviation compared to 90.67647 on X axis and almost a half value 24.15229 to 41.1636301 on Y axis. The scores without the camera cannot be considered bad in any case especially while soldering bigger pads (>2 mm diameter) with a wider tip. In our case the pcb was transferred without a carrier on the cells' conveyor and the THT component was planted by human hand each time in order to simulate a real in-production case where operators are loading the components on the PCbs that are going to be soldered by a robot. Almost the same results could be obtained using a guiding system like in Fig. 1 without using a camera but it is compatible only for production lines with carriers. Last but not least, all the equipment's' conveyors system that are placed at an EMS factory line, are adjusted up to 0.3 mm wider so the pcbs don't get stuck while being transferred from an equipment to another. The 0.3 mm can be Y or Angle offset depends how each pcb reaches the stopper position.

## References

1. Camera calibration.pdf Japan Unix. Accessed 15 Mar 2020
2. Program alignment with two groups.pdf Japan Unix. Accessed 18 Mar 2020
3. Pires, J.N.: Industrial Robots Programming: Building Applications for the Factories of the Future. Springer, Heidelberg (2007)
4. Li, X., Su, X., Liu, Y.: Cooperative robotic soldering of flexible PCBs (2017)
5. Japan Unix. Soldering robot product page. https://www.japanunix.com/en/products/automation/. Accessed 15 Mar 2020/03/15
6. Operational manual 700FH – Japan Unix
7. Cognex in sight user manual. https://support.cognex.com/en-hu/downloads/insight/training/manuals
8. Geren, N., Redford, A.H.: The significance of desoldering and resoldering methods in robotic automated rework. J. Electron. Manuf. **4**, 41–51 (1994)
9. Lehtimäki, L.: Developing a soldering robot for high-mix low-volume electronics production (2018)

# A Mechatronic System for Automatized Inspection of Powerline Transmission

Giuseppe Carbone[1,2](✉), Giuseppe Forciniti[1], Fabrizio Fuoco[1], David Rodriguez Izquierdo[1], Luigi Scarfone[1], and Domenico Viapiana[1]

[1] DIMEG, University of Calabria, Rende, CS, Italy
giuseppe.carbone@unical.it
[2] CESTER-Research Center for Industrial Simulation and Testing, Technical University of Cluj-Napoca, Cluj-Napoca, Romania

**Abstract.** This paper presents a low-cost mechatronic system for helping workers on powerline transmission management. The proposed architecture can perform remote analysis and measurements of magnetic field on powerline cables. Specific attention is addressed at the device design including kinematic and dynamic simulations. This paper also describes the 3D printing of a prototype and main software development including a graphical user interface. Preliminary tests are reported to show the soundness and engineering feasibility of the proposed design solution.

**Keywords:** Mechatronic system · Powerline transmission · Automatized control · Sensor

## 1 Introduction

Constant energy supply to the customers requires periodic inspections of powerlines to confirm the energy transmission and prevent failures. This operation is mainly achieved by direct human visual inspection or by using helicopters/drones equipped with infrared and corona cameras. However, both human and aerial inspection are expensive and risky as there can be fulgurations of humans or undesired hitting of the powerlines by helicopter/drone blades as reported for example in [1].

In the last two decades there have been several attempts to design robotic devices that can climb on powerlines to accomplish inspection tasks from close distances without needed to turn of the power supply. Some interesting examples are reported in [1–7]. Nevertheless, most of them are very complex as based on multiple rotating wheels running synchronously on multiple powerlines or they are based on flying helicopter/drone-like devices, which have a very complex control especially in presence of windy conditions, as also mentioned in [1].

This paper proposes a low-cost and user-friendly mechatronic design for the inspection of powerline transmission. The proposed design solution is at first stage simulated from kinematic and dynamic viewpoints. Afterwards a prototype is obtained


E.-C. Lovasz et al. (Eds.): MTM&Robotics 2020, MMS 88, pp. 457–467, 2021.
https://doi.org/10.1007/978-3-030-60076-1_41

by 3D printing and a dedicated software and user interface have been developed. Preliminary tests are carried out to show the soundness and engineering feasibility of the proposed design solution.

## 2 The Proposed Concept

The aim of this paper is to design and build a system able to perform inspection tasks on powerline transmission cables with key low-cost and user-friendly features.

Starting from analysis on powerline transmission and various type of cables, a preliminary study was focused at identifying a design solution with proper kinematic and dynamic behavior during motion. The proposed mechanical architecture can be driven by electric motors, specifically two servomotors and one linear actuator with a telescopic leg and a tail for balancing and sensing purposes as shown in the 3D CAD model in Fig. 1. Namely, the architecture includes a revolute joint and two prismatic joints with a total of 3 active degrees of freedom. Accordingly, the contact point between the powerline cable and the robot leg can perform a specific desired trajectory, which can be chosen at the design stage. After identifying a suitable topology further activities have been carried out such as kinematic and dynamic simulations, static analysis, the implementation of control codes and the chosen sensors to perform the analysis.

The dimensional synthesis has been carried out by starting from BS 215 Part 2 1970, where an ACSR powerline cable, [8], has been considered to identify the main technical specifications in terms of maximum acceptable weight of the robot body and the loading conditions for the powerline transmission due to presence of the robot.

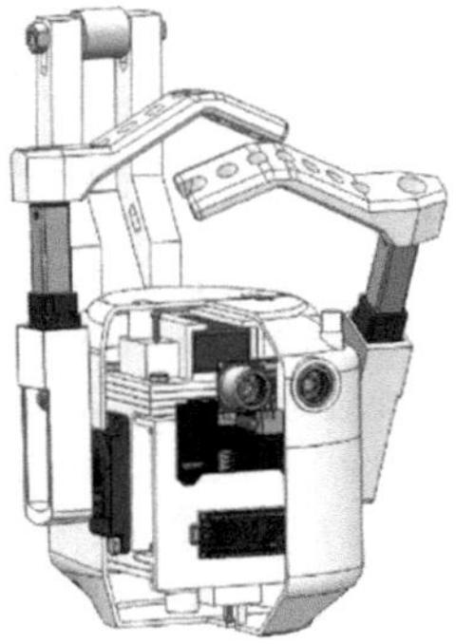

**Fig. 1.** A 3D model of the proposed design with two telescopic legs and a tail for balancing and sensing purposes.

# 3 Performance Simulations

## 3.1 Kinematics and Trajectory Planning

Direct Kinematics can be easily solved as based on Denavit-Hartenberg [9] by using the parameters in Table 1. Objective of this method is to assign a specific pose for the end effector, assumed as contact point between the system and the powerline cable, solving a matrix product as in Eq. (1). Proper joint variables will allow the end effector to reach a specific pose, by means of the Inverse Kinematic, which can be easily obtained using a geometric approach (Fig. 2) as in the Eqs. (2)–(3).

**Table 1.** DH parameters of the proposed device according to Fig. 2.

| \ | $R_z$ | $D_z$ | $D_x$ | $R_x$ |
|---|---|---|---|---|
| $T_{01}$ | $q_1 + 90°$ | $d$ | 0 | 90° |
| $T_{12}$ | 0° | $H + q_2$ | 0 | 0° |

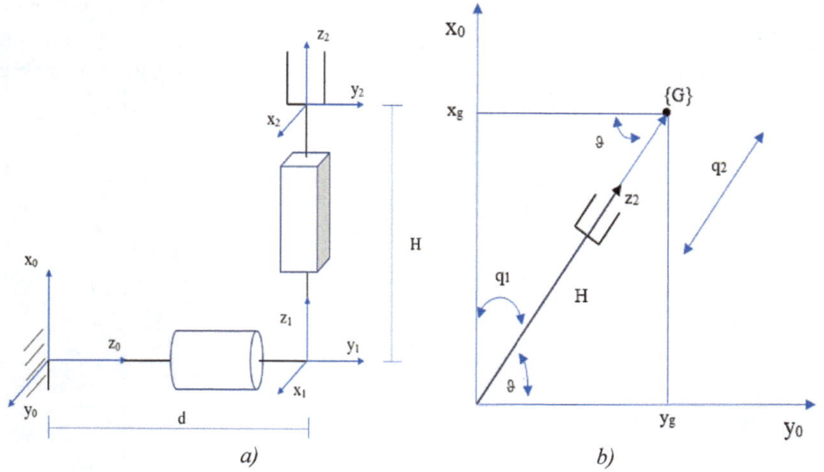


**Fig. 2.** Schemes of the proposed device for: a) Direct Kinematics; b) Inverse Kinematics.

$$T_{02} = T_{01}T_{12} \tag{1}$$

$$q_1 = \frac{\pi}{2} - \vartheta = \frac{\pi}{2} - \tan^{-1}\left(\frac{x_i}{y_i}\right) \tag{2}$$

$$q_2 = \sqrt[2]{(x_i^2 + y_i^2)} - H \tag{3}$$

To allow the movement on the powerline cable, a specific trajectory for end effector is implemented. It is crucial to avoid contact between the legs during their alternate movement, assuring at same time strong and stable contact of the end effector on the cable also during a continuous motion. The following Goals poses are set in the trajectory planning by considering the above-mentioned conditions:

- {G1} starting pose;
- {G2} middle pose (to avoid obstacles);
- {G3} final pose.

These three Goals are reached in cycling way by defining two trajectories, one for the arm during its forward motion, and one for the arm in its backward motion.

The end effector reaches a Goal using a cubic interpolation for the joint variables to avoid impulsive accelerations by satisfying four-boundary condition. For example, one can write for Goals (4):

$$\begin{cases} q_i(t_0) = q_i^{G_0} \\ q_i(t_f) = q_i^{G_f} \\ \dot{q}_i\ (t_0) = \dot{q}_{i_i}^{G_o} \\ \dot{q}_i\ (t_f) = \dot{q}_{i_i}^{G_f} \end{cases} \tag{4}$$

The Via points Goal {G2} is set with a non zero velocity condition, as final point during {G1}–{G2} trajectory, and initial point in {G2}–{G3} trajectory. The aim of this non zero condition is to give a more fluid and quick motion performance (Fig. 3).

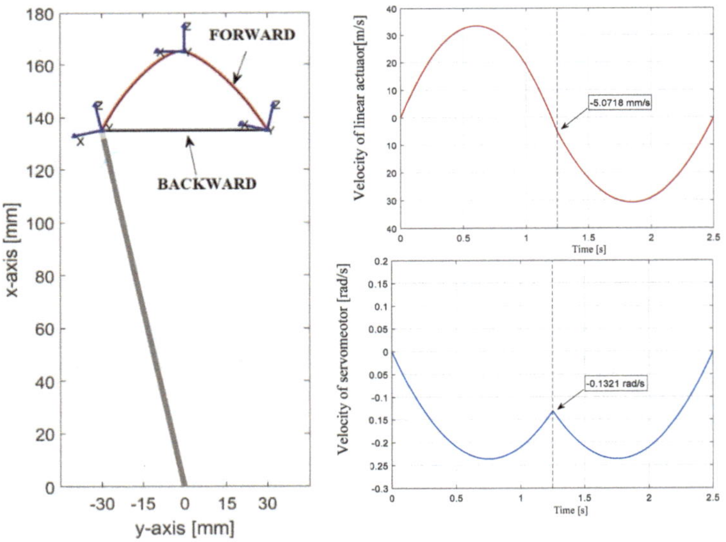


*a)* *b)*

**Fig. 3.** Simulation results: a) trajectory profile; b) velocity of prismatic and revolute joints.

### 3.2 Dynamic Analysis

In this phase the mechanical system and the trajectory are embedded into a software for obtaining data to use in real prototype, like choosing the type of motors and actuators, the best trajectory changing pace and lift, etc. At first a 2D analysis is performed: an algorithm in MATLAB, Newton-Euler approach based is solved to evaluate Force and Torque for single arm movement. The same method can find friction condition modeling the arm in reverse configuration for backward motion (Fig. 4). Other SIMSCAPE model is helpful to validate the results obtained, comparing these results with a SOLIDWORKS model (Fig. 5).

After validating the model, a 3D extension is required to introduce arm-cable contacts in the simulation. In SOLIDWORKS for modeling geometries and have a complete concept. Joint variables (displacement, velocity, acceleration) calculated from Sect. 2 are here implemented in motors and actuators. For more accurate analysis and validation, meaning contact force and force and torque of actuation, the complete robot model is exported in ADAMS View (Fig. 7). In Table 2 are shown maximum force and torque, with models of motor and actuator chosen (characteristics in Fig. 6), using a safety factor.

During dynamic analysis is necessary to adjust and improve the physical proprieties of mechanical parts. So modifying the shape of the rigid element who acts as arm can change the oscillation of the body: geometries are designed in way to obtain the maximum gain between lightness and robustness. There are some improvements in stability thanks to self-centering and self-alignment couples, generates from a correct choosing of arm shape and weights distribution.

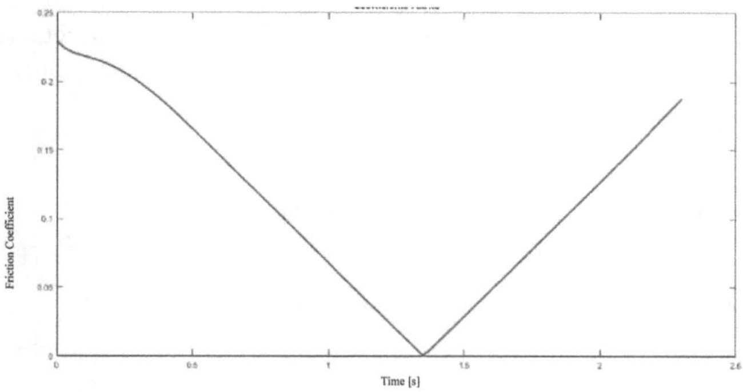


**Fig. 4.** Evolution of the dimensionless friction coefficient as obtained in SIMULINK.

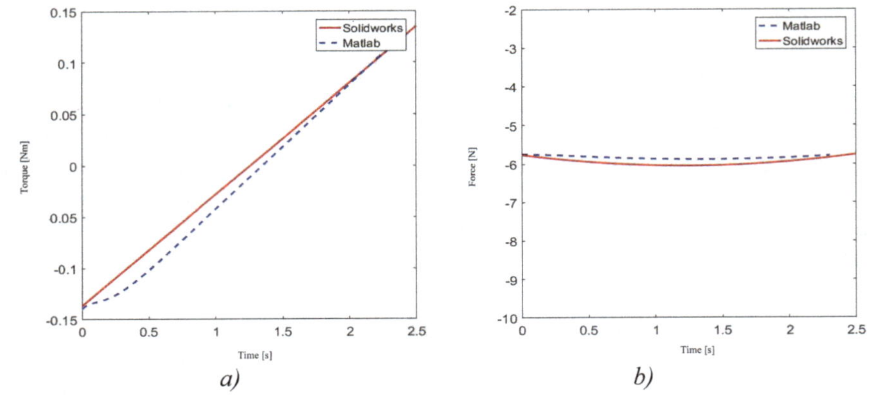


**Fig. 5.** Comparison of results versus time as obtained for: a) input torque; b) input force.

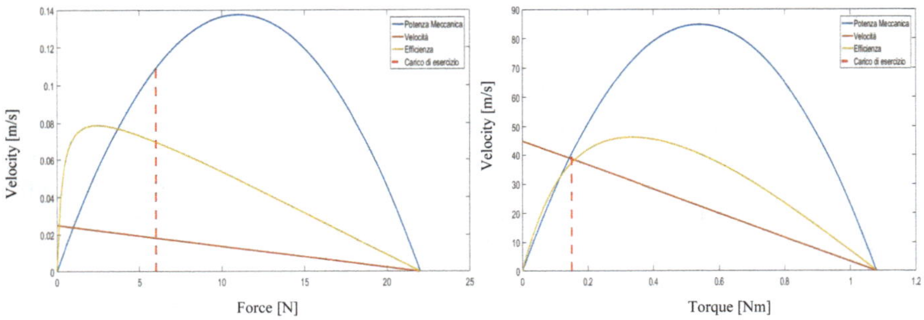


**Fig. 6.** Main characteristics of the choosen actuators: a) linear actuator; b) rotative motor (blue-mechanic power; red-speed; yellow-efficiency; dotted line selected nominal operation conditions).

**Table 2.** Maximum Force and Torque, motor and actuator chosen.

| \ | Value | SF = 3,5 | MG996R [10] | L12-30-50-6R [11] |
|---|---|---|---|---|
| Torque [N m] | 0,18 | 0,63 | 1,08 | – |
| Force [N] | 6,5 | 22,05 | – | 22 |

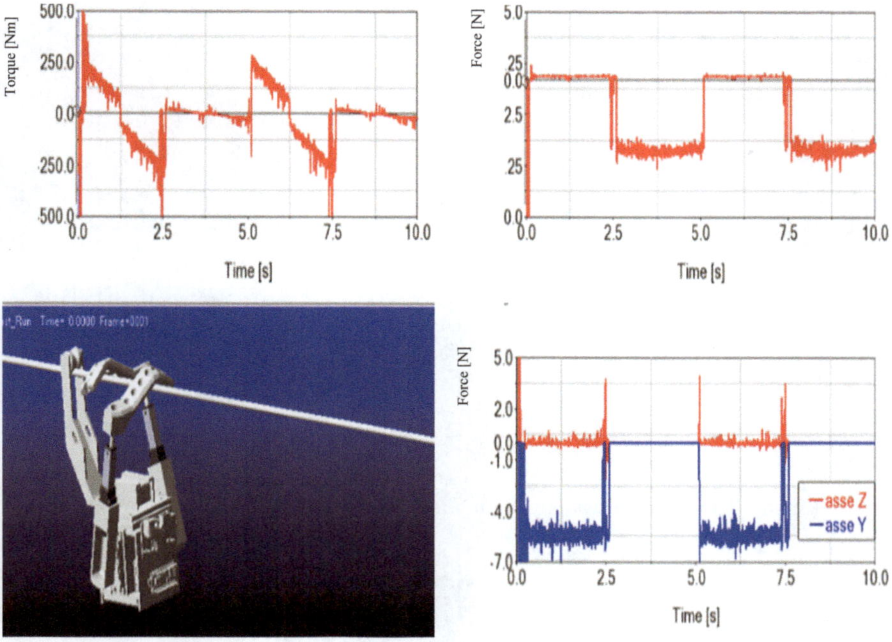


**Fig. 7.** ADAMS View results. On the top: torque and force required for each motor. On the bottom: animation and contact force between arm and the powerline cable.

## 4 Operation and Control Architecture

Operation of the proposed device is based on Arduino boards, which are an open source low-cost solution to manage both motor control and data feedbacks for a proper robot operation. Two control boards are embedded: an Arduino Nano and a NodeMCU [4]. The two control boards have different tasks. The Arduino Nano controls the path planning and motor motion of the robot receiving commands from NodeMCU, which has a Wi Fi module. Thought the Wi Fi connection a master operator can send commands and receive sensor data measurements, maintaining a safety distance from the powerline. The robot has been equipped with various sensors.

Most important sensor are an hall effect magnetic field sensor, KY-024 [5]; a GY-521 accelerometer [6], which allows to monitor the stability of the robot center of gravity while an HC-SR04 [7] ultrasonic sensor measures the presence and distance from a frontal obstacle, making the robot able to avoid impacts. Figure 8 shows the circuit configuration, including the above mentioned sensory feedbacks.

Using the remote control in any smartphone with wi fi communication one can send commands to the device maintaining a safety distance from the powerline. Figure 9 shows the developed user's interface in an Android smartphone. From this interface a user can send movement operation and can read the state of accelerations along three axes, the magnetic field from a sensor on the tail and the distance from an obstacle along forward direction.

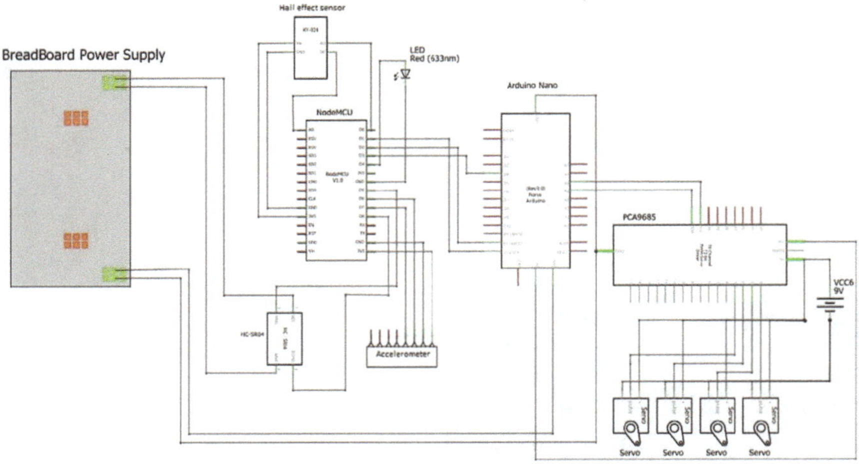


**Fig. 8.** The proposed cabling configuration set up with sensors and motors.

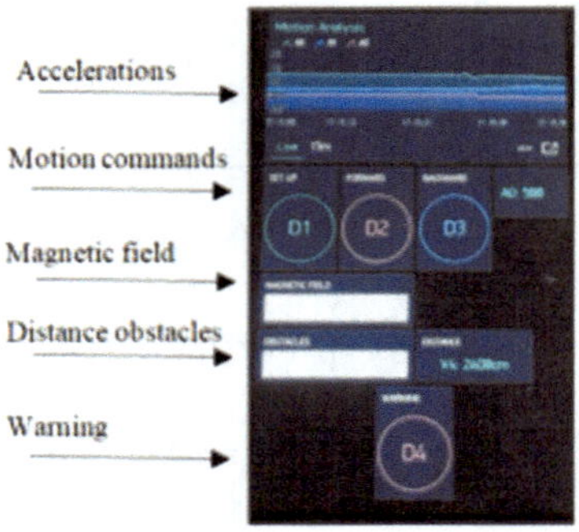


**Fig. 9.** User's interface for controlling the mechatronic system.

## 5 Preliminary Experimental Test

Mechanical components have been modelled in SOLIDWORKS (Fig. 10). Several static simulations have been carried out to verify the structural properties especially at critical points like contacts with the powerline cable, where friction can affect durability (Fig. 11). Tail and feet required very careful iterative design and refinements to achieve an effective solution. After suitable simulations, parts have been 3D printed by a Mini G2 Delta Rostock 3D printer [12]. Electronic components have been assembled and each main component has been preliminarily tested. Then, full device testing has been carried out.

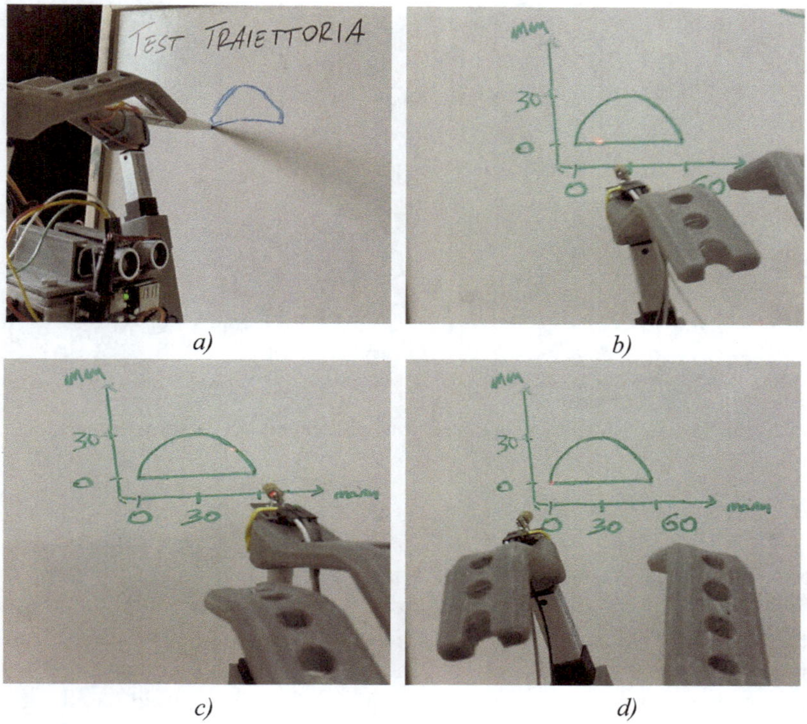


**Fig. 10.** Trajectory validation experiments: a) by drawing: b) c) d) following different trajectory poses with a laser pointer (each figure refers to a different time step).

First tests consisted in validating the end effector trajectory matching the planned trajectory. This test has been performed in two different way. Namely, by drawing the trajectory on a blackboard with a pencil and by following the same trajectory with a laser pointer module. The latter is more effective as it does not add additional loads on the device (Fig. 10). Second type of tests (Fig. 11) consisted in simulating the operation on a powerline by using the magnetic sensor to identify a defect on the powerline, which has been intentionally created in point of a powerline model. For this test, scaled model has been created to in our laboratory with a scaled frame and an aluminum cable. During these tests it has been set up a step size of 60 mm, which has been found to be the best compromise between stability and performances. Note that a step size of 100 mm has been also tested. This has been found to almost double the speed with an increase by 35%, of the oscillations, which reduces the stability and accuracy of magnetic field measurements accordingly.

**Fig. 11.** A picture of a simulated powerline monitoring test experiment.

## 6 Conclusions

This paper has addressed the design of a new mechatronic device for inspection of electric powerlines. A simple and user-friendly design solution is proposed as based on two telescopic legs. Kinematic and dynamic simulations have been carried out for a proper size synthesis of the proposed device. A simple control hardware and software has been proposed as based on Arduino and a user-friendly android based user interface. Preliminary tests have been carried out. Experiments demonstrate the feasibility of the proposed device, which can autonomously swing along a powerline while inspecting magnetic field and temperature.

**Acknowledgements.** The paper presents results from the research activities of the project ID 37_215, MySMIS code 103415 "Innovative approaches regarding the rehabilitation and assistive robotics for healthy ageing" co-financed by the European Regional Development Fund through the Competitiveness Operational Programme 2014–2020, Priority Axis 1, Action 1.1.4, through the financing contract 20/01.09.2016, between the Technical University of Cluj-Napoca and ANCSI as Intermediary Organism in the name and for the Ministry of European Funds. 20/01.09.2016, between the Technical University of Cluj-Napoca and ANCSI as Intermediary Organism in the name and for the Ministry of European Funds.

## References

1. Gonçalves, R., Carvalho, J.C.: Review and latest trends in mobile robots used on power transmission lines. Int. J. Adv. Robot. Syst. **10**, 408 (2013)
2. Nayyerloo, M., Chen, X., Wang, W., Geoffrey Chase, J.: CableClimbing robots for power transmission lines inspection, mobile robots. In: Chen, X., Chen, Y.Q., Chase, J.G. (eds.) State of the Art in Land, Sea, Air, and Collaborative Missions. IntechOpen (2009)
3. Bahrami, M.R., Abed, S.A.: Mechanics of robot inspector on electrical transmission lines conductors with unequal heights supports. Vibroeng. Procedia **25**, 60–64 (2019)
4. Aggarwal, R.K., Johns, A.T., Jayasinghe, J.A.S.B., Su, W.: An overview of the condition monitoring of overhead lines. Electric Power Syst. Res. Int. J. **53**, 15–22 (2000)

5. de Souza, A., Moscato, L.A., dos Santos, M.F., de Britto Vidal Filho, W., Ferreira, G.A.N., Ventrella A.G.: Inspection robot for high-voltage transmission lines. In: ABCM Symposium Series in Mechatronics. vol. 1, pp. 1–7 (2004)
6. Golightly, I., Jones, D.: Visual control of an unmanned aerial vehicle for power line inspection. In: Proceedings of the 12th International Conference on Advanced Robotics, (ICAR 2005), Seattle, pp. 288–295 (2005)
7. Pouliot, N., Montambault, S.: Geometric design of the LineScout, a teleoperated robot for power line inspection and maintenance. In: IEEE International Conference on Robotics and Automation, Pasadena, pp. 3970–3977 (2008)
8. Aluminum overhead conductors: CAFCA. Workington, Harare, Zimbabwe (2010)
9. Ceccarelli, M.: Fundamentals of Mechanics of Robotic Manipulation. Springer, Dordrecht (2004)
10. Sensorkit. http://sensorkit.en.joy-it.net/index.php?title=Main_Page. Accessed 20 Apr 2020
11. Actuonix. https://www.actuonix.com/category-s/1840.htm. Accessed 20 Apr 2020
12. Geeetech. http://www.geeetech.com/wiki/index.php/Delta_Rostock_mini_G2. Accessed 20 Apr 2020

# Robotic Palletization with Camera Position Determination

Jaroslav Antoš(✉) and Pavel Fišer

VÚTS, a.s., Liberec, Czech Republic
{Jaroslav.Antos,Pavel.Fiser}@vuts.cz

**Abstract.** In the serial production of automotive air filters, the last operation required is to grab and unload the cartridge set. The filter cartridge grip is always in a defined position, but the cartridge lining cannot be locked and the position can be defined in advance, so it was necessary to implement a specialized camera system. This system has to cope not only with different positions, but also with different shaped box. Furthermore, the article deals with the adaptation of the camera system and lens to the final vision processes phase and the need to meet the high requirements for the cycle time, robot movement and finally the high safety requirements of the automotive industry. Multiple exposure and a multivision process have been used for the edge detection to combine multiple images together. The whole system with camera and lights is located on the manipulation gripper, which in addition must meet the requirements for the multiple pneumatic position changing of suction cups and control vacuum valves with feedback. Furthermore, the article deals with the final implementation of the task in the now already working cooperation with the superior PLC and safety system assembly line.

**Keywords:** Robot · Gripper · Camera · PLC · Safety

## 1 Introduction

State of the art, 2D and 3D vision-guided systems are effectively bridging the gap for many applications in diverse industries, and the chances are very high that if an application challenge is present some type of robotic vision capability is now available to help real applications. Robots are good at making identical repetitive movements, such as a simple task on an assembly line. Robotic palletization with camera position determination is intended for serial, non-operator production of air filters that separate air from undesirable moisture and dirt. The aim of the paper is to solve the situation of unloading the air filters into various boxes, whose position cannot be locked and defined in advance, therefore it is necessary to introduce a specialized camera system. The first part describes about a seven-axis robotic system controlled by a PLC. A gripper is an integral part of each robot, which is why the gripper chapter follows, which in this case must meet several requirements together. The main chapter describes the vision process for selected paper and plastic boxes that require a different approach to image processing. In the automotive industry, besides high a production cycle and


E.-C. Lovasz et al. (Eds.): MTM&Robotics 2020, MMS 88, pp. 468–476, 2021.
https://doi.org/10.1007/978-3-030-60076-1_42

quality, safety matters are highly emphasized, this is why the last chapter deals with the DCS safety system for double workplaces [1, 8].

## 2 The Robotic Seven Axis PLC System

The choice of robot brand was made in advance by the end customer of the line who already uses Fanuc robots in his company. When implementing the camera system, Keyence and Fanuc Camera systems were considered. Due to the integration into the robot system and the link with the vision offset motion commands, the Fanuc camera system was chosen. The next step was the concept of the workplace. Due to continuous production and a short production cycle of 4.1 seconds per cartridge, four pallets are located at two separate workstations. The options were, the use of two independent six-axis robots, or one seven-axis robot that uses a linear seventh axis to change position between workplaces. Both variants are comparable in price, the decisive factor was the solution of safety. When palletizing it is necessary to cooperate with a person and therefore a safer option was chosen, a seven-axis robot, which brings the possibility of human entry into a safety fence, where at that moment there is no robot. However, this system requires all the energy systems to pass through the energy chain. Optimal path planning is required for a 4.1 s clock period [7].

An integral part of the pick & place operation is the accurate calibration of the concentricity of all suction cup centers and filter cartridges using a limiting mechanism to determine the exact positions of the filter cartridges on the conveyor to within one millimeter. A major problem comes in the operation place. First, the parent PLC must determine the selected workstation, check the gripper, move from any position to the home, and then call program number selected at the start the process of obtaining an unknown, different offset of the placed pallet, then select an empty row and the first vacant floor in the box offset row X and floor Z. First, the superior PLC selects the first free site with the box ready for filter cartridges, checks the gripper status, and then the robot moves from any position to the home state of the site. Then, using the robot program calling method: “Program number select” together with the parameters required row and floor, calls the selected program on the robot. According to the input parameters, the robot determines the shift in the X and Z coordinates. In addition, the resulting position must be corrected on the basis of the data obtained from the camera system and supplemented by this second offset. The entire operation of obtaining the second XY offset and R rotation begins with the home position check, followed by the robot hovering over the first corner of the box. Here the robot waits 500 ms to stabilize its movement and then the first vision process is started, after its completion the robot traverses diagonally over the second corner, as the box is 1200 mm wide and cannot be photographed completely from 1000 mm high without distorting (see Fig. 1) which was created by using a 6 mm lens not suitable for the purpose. It is therefore necessary to perform two independent series of photo shooting over each corner at a diagonal and combine these results into one image.

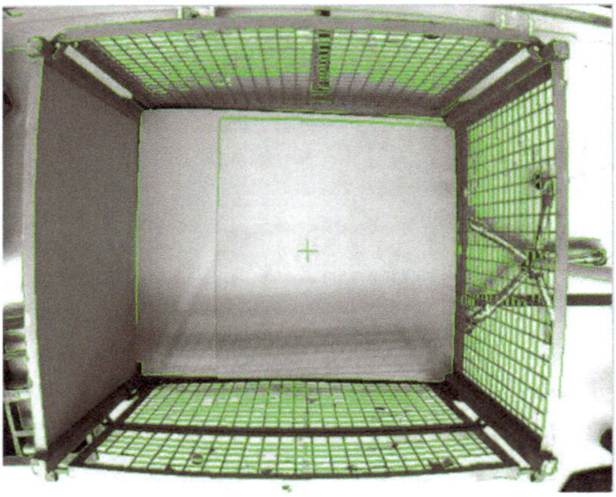

**Fig. 1.** The barrel distortion when using a 6 mm lens

Communication between the PLC and the robot is handled via the number select program with a completed hand shake at the end to confirm the execution of the selected program. The logic of these calls allows the programming of the main program in the PLC, which also communicates with the Sick safety system and other parts of the line such as the conveyor and control of the actuators or buttons in the cage of the station with the robot.

## 3 Multi-function Gripper Designed for Gripping and Simultaneous Vision Process

The first phase was the selection of the suction cup itself, different types from several manufacturers were tested, but none of them met the defined requirements for repeatability, dynamics during movement and unloading accuracy, therefore unique suction cups were developed by the Czech company Bohemia Vakuum. These aluminum suction cups with seals meet all criteria, the only prerequisite remains the aforementioned concentricity of the filter cartridge and the suction cup, which must not exceed 1 mm in the X and Y directions, and at the same time their perpendicularity must be met. For this purpose, calibration mandrels were produced and calibrating cartridges were marked on the lathe. Another problem was to shorten the suction time and suction time of the filter cartridges. Shortening the suction time is achieved not only by switching off the vacuum, but also by instantly switching on the pressure again gradually from one suction to the fifth. In addition to valves, the vacuum/overpressure piping is equipped with air pressure sensors for monitoring and control. Each suction tube is equipped with a damping spring in the direction of the Z axis and supplemented with a limit sensor of the spring compression position, which brings the possibility of detecting a unique box bottom or a gripper collision with the alarm [11].

Because the packing boxes have different filter cartridge spacing, the gripper has the ability to change them to three different positions using eleven pneumatic cylinders.

A special requirement is when placing four filter cartridges in a narrow box. For this choice the gripper is complemented by tilting the fifth suction cup by 45°. The design of the gripper was inspired by the human left hand with five fingers, and the folding fifth sucker represents similar possibilities to the human thumb. During the movement process, the span is automatically reduced to a minimum for safer transport.

The gripper must also support the camera and be positioned so that the vacuum cone is not obstructed by the vacuum hoses and electrical wiring, but must also be in a safe place against collision with the metal box. Two inclined lights with a wavelength of 650 μm in the red visible spectrum were chosen for the lighting inside the box. Lenses went through a large selection, the original 12 mm was not satisfactory, because the robot could not be placed in such a position that the entire photograph could be seen in the resulting photo, so gradually tested lenses with a smaller focal length of 6 mm and 8 mm were tested. Six-millimeter occupied a sufficient field, but at the cost of a large distortion of the image barrels, therefore, a medium size of 8 mm was chosen. After focusing the lens mechanically on the working area and selecting the appropriate mechanical aperture, a special lens protection frame was prepared on the 3D printer, as those with a lower focal length extend beyond the length of the camera housing [10].

The next step is to calibrate the camera, where the two options are planar or orthogonal. Because the boxes may be at different Z levels, an orthogonal calibration was chosen. Since the line will be moved to another building after completion, it will be necessary to recalibrate the camera as well. Using orthogonal calibration, we detect the positions of the box at different Z levels (see Fig. 2) [3, 9].

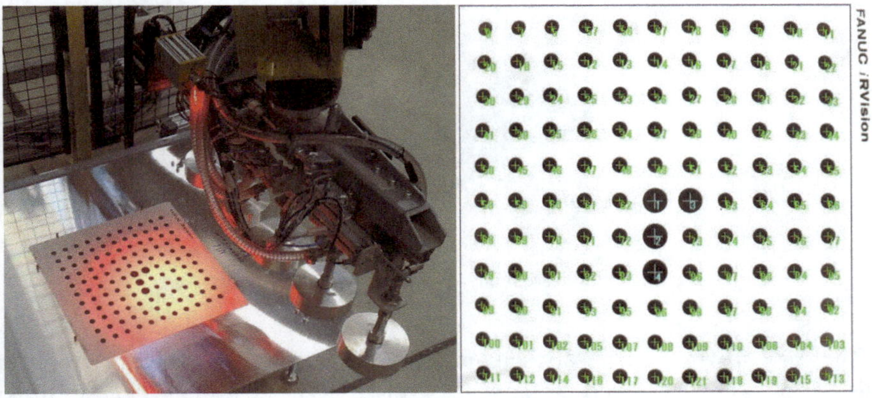


**Fig. 2.** The gripper, camera calibration with auxiliary table and camera view

## 4 Vision Processes

When the robot is in the shooting position, in focus and with the aperture and light level set, the vision process is self-evaluated and evaluated separately. It consists of multiple operations sorted into sequences.

## 4.1 Paper Box with the Method of Joining Multiple Images Together

The aim of vision processes is to determine the offset of position X, Y and R around the Z axis of the newly placed box against the defined position from the program to the unloading space, which is complemented by mechanical stops that ensure position delimitation in centimeters, but only the camera is in charge of its precise focus in millimeters. The first process is created by a 2D multi-view vision process, he basic program setting method is merging two camera views. There can be as many camera views as the number of measurement points. For each measurement point, a camera view is named camera view 1 or camera view 2. Command tool comes under each camera view (see Fig. 3) [7].

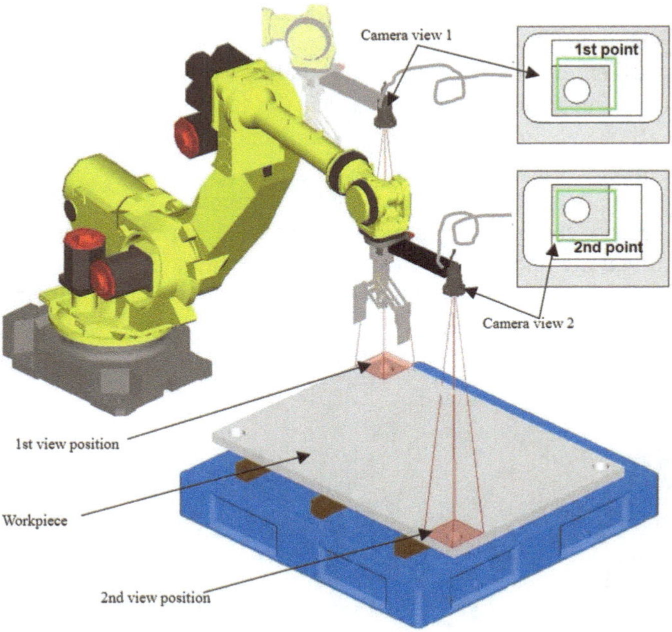


**Fig. 3.** Overview of 2D multi view vision process [2]

The camera was set with a resolution of 1024 × 1280 pix, triple exposure time 0.62–2.5 ms. Threefold exposure is at the cost of a grayer resulting image, that is, smaller edge derivatives and longer shooting times, but with greater stability against extreme changes in illumination, such as a skylight. A major problem occurred in the detection of the box edges (see Fig. 4), indefinite search for the derivative occurred.

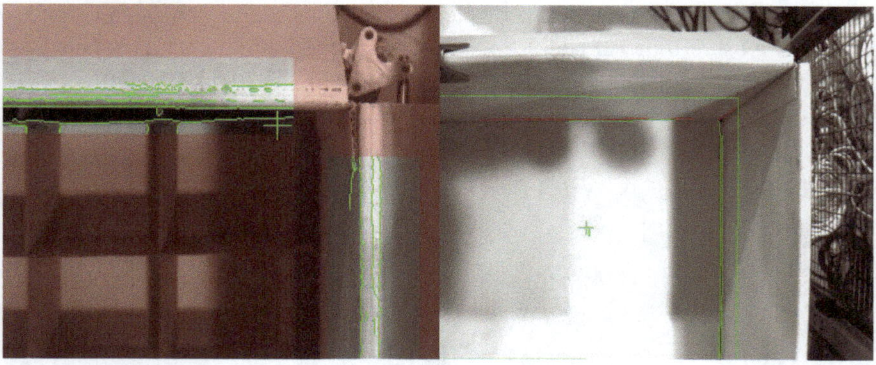

**Fig. 4.** The indefinite boundaries of derivation in a box

The edge search option at the bottom of the box was also tested, but also unsuccessfully. Therefore, detection using supplemented black and white contrast marks was used. First there was a problem due to the reflectance of shiny materials, where the black-shiny surface under the camera appears to be pure white, so the method of matt 3D printing was chosen, which proved to be better. However, the shape of the symbol was detected by the camera system, due to similarities in the mesh occupied by the protective mesh (see Fig. 5) [5].

**Fig. 5.** The mark and unwanted detection of the sought mark in the surrounding fence

Therefore, we proceeded to develop a new unique shape, which could not be confused with the surrounding objects captured by the camera. The best variant was shown to be a circular shape mark (see Fig. 6), which already shows repeatability with scorings greater than 95% throughout the day when light conditions change. This is important in cases where the line cannot be obscured to create constant light conditions.

For our case, the camera was complemented by two 20 W lights, which can reduce the maximum exposure time. In addition, the robot's motion program was supplemented with triple repetition of shooting attempts two more attempts to take a picture, and then it issues an alarm not taken [6].

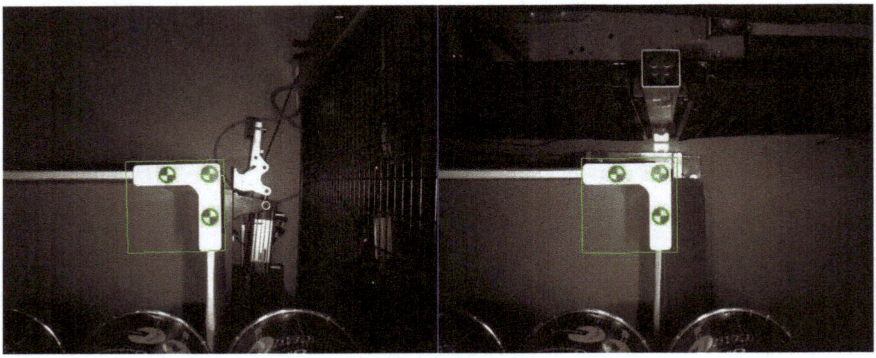

**Fig. 6.** The Round mark from a 3D printer made of matt plastic

## 4.2 Plastic Box with the Method of Determining the Offset Z Using the Scale

The second task is to unload the filter cartridges into a plastic box, the difference is not only in material and less space, when it is necessary to reduce the gripper by pneumatic cylinders, but especially in that it is necessary to determine Z levels together with the XY shift by the R rotation. Therefore we will only deal with the determination of the Z-axis levels, the determination of the other offsets is described above. The placement of the plastic inserts between the individual trays of the established filter cartridges can vary up to several centimeters. In this case, the box cannot be locked to the exact position on the workplace. The main problem is that positioning auxiliary symbols cannot be used, so it is only necessary to rely on the detection of natural shapes in the field of view of the camera, which are already there at the factory. Specifically, it is a plastic lining. In the derivative tests, it exhibited multiple circular portions (see Fig. 7), since the surface is scanned entirely from one point, distortion occurs. Illumination of space proved to be an essential parameter in capturing and subsequently evaluating the image. After many tests with different levels of lighting, circles in the corners of the box were discarded due to their poor illumination. Another problem was the impossibility of detecting the derivative due to the small narrow space. The main goal, however, remains to determine the individual Z levels, for this purpose, two levels were taught as a reference through the Vision process, and a reference a distance in the Z axis direction was assigned to each of them at intermediate positions orthogonally.

**Fig. 7.** The plastic lining with circular derivatives

## 5 Safety Dual Check System

Dual check safety limits the movement space or speed of the robot according to input parameters. It primarily serves to protect the operator (interaction between the machine and human) and secondarily between individual machine parts (for example, a robot versus a fence). Three spaces have been modeled for the palletizing workstation, two are restricted zones and the third is an auxiliary crossing zone called "tunnel", which is used for a special HOME in case of stopping the robot during its crossing. In addition, it also serves to detect the position of the robot, which must be sent internally to the safety PLC, from this information can be precisely set the opening of the electronic locks at the door of the protective cage can be precisely set and thus allow access to the space when there is no robot and. The most dangerous situation occurs when unloading to a repository where it is necessary to meet the robot and the human in the fence in the automatic operation, for this purpose, a working zone for positioning and a restricted zone for safe movement of the robot was created (see Fig. 8) [4].

**Fig. 8.** The dual check safety workplace

## 6 Conclusion

Robotic palletization with camera position determination was achieved according to specifications. We managed to solve the situation of unloading the air filters into various types of boxes, where it is not possible to determine the exact position in advance, but thanks to a camera system it is possible to continuously unload a large number of cartridges, with orders of over 3 000 cartridges a day. Special detection systems for positioning were designed and implemented for paper boxes and the whole system was re-programmed to these shapes. For plastic boxes with interlay, the matrix was taught in the limit positions and the detection layer is then scaled to which, in addition to the XY and the R rotation, the Z level is assigned. A separate chapter of the system was a gripper that complies to a combination of more features and together with a robot on a linear guide, a seven-axis system is created, which is complete with dual check safety. Further development at this station will be in an additional extension with new types of storage boxes of different materials and different dimensions.

## References

1. OPERATOR'S MANUAL: R-30+B Plus/R-30+B Mate Plus CONTROLLER, Fanuc robot series. https://www.fanuc.eu/cz/cs?gclid=EAIaIQobChMIiNaLo9rO6AIVlOR3Ch0CeQVvEAAYASAAEgLFnPD_BwE. Accessed 04 Apr 2020
2. OPERATOR'S MANUAL: Vision 2D Camera Application, Fanuc robot series. https://www.fanuc.eu/cz/cs?gclid=EAIaIQobChMIiNaLo9rO6AIVlOR3Ch0CeQVvEAAYASAAEgLFnPD_BwE. Accessed 04 Apr 2020
3. CAMERA PACKAGE FOR R-30iB Plus Controller, Fanuc robot series. https://www.fanuc.eu/cz/cs?gclid=EAIaIQobChMIiNaLo9rO6AIVlOR3Ch0CeQVvEAAYASAAEgLFnPD_BwE. Accessed 04 Apr 2020
4. SAFETY HANDBOOK: B-80687EN/15, Fanuc robot serie. https://www.fanuc.eu/cz/cs?gclid=EAIaIQobChMIiNaLo9rO6AIVlOR3Ch0CeQVvEAAYASAAEgLFnPD_BwE. Accessed 04 Apr 2020
5. Sciencedaily: New filter enhances robot vision on 6D pose estimation (2019).https://www.sciencedaily.com/releases/2019/07/190710131932.htm. Accessed 22 July 2020
6. Qualitymag: Design, CMS, Hosting & Web Development: ePublishing. Qualitymag: The Details of Vision Guided Robotics (2020). https://www.qualitymag.com/articles/96012-the-details-of-vision-guided-robotics. Accessed 22 July 2020
7. Koubaa, A.: Robot Operating System (ROS): The Complete Reference, vol. 4. Springer, Cham (2020). ISBN 978-3-030-14984-0. ISSN 1860-949X
8. Mazal, J., Fagiolini, A., Vasik, P. (eds.): Modelling and Simulation for Autonomous Systems: The Complete Reference, vol. 4. 5th International Conference, MESAS 2018, Prague, Czech Republic, Springer, Cham (2020). ISBN 978-3-030-20189-0. ISSN 1860-949X
9. Sergiyenko, O., Flores-Fuentes, W., Mercorelli, P. (eds.): Machine Vision and Navigation. Springer, Cham (2020). ISBN 978-3-030-22586-5
10. Villegas, V., Osiris, O., Alfaro, N., de Jesús, M., Marrufo, S., Israel, A. (eds.): Advanced Topics on Computer Vision, Control and Robotics in Mechatronics. Springer, Cham (2018). ISBN 978-3-319-77769-6
11. Sethi, I.K.: Computational Vision and Robotics. Proceedings of ICCVR 2014. Springer, New Delhi (2015). ISBN 978-81-322-2195-1. ISSN 2194-5357

# Implementation of Manipulator with Rotary and Translational Axis Using Electronic Cams

Pavel Dostrašil(✉) and Aleš Richter

VÚTS, a.s, Liberec, Czech Republic
{pavel.dostrasil,ales.richter}@vuts.cz

**Abstract.** This article deals with the complete implementation of the control system for a newly developed manipulator. The system is based on electronic cams defining the movement of the rotary and translation axes of the manipulator. The endpoint trajectory is only partially defined. The key point is the design of suitable displacement diagrams for individual drives. A specific software tool has been developed for this purpose, allowing detailed adjustments of the dynamic parameters of displacement diagrams while also checking the endpoint trajectory.

The control system logic had to adapt to the design constraints of the task while allowing the most efficient operation for maximum machine cycle. A side effect of increasing the machine cycle is increasing the positional error of the resulting trajectory. At the end of the article, it is demonstrated how this effect can be partially eliminated due to the applied control concept.

**Keywords:** Electronic cam · Displacement diagram · PLC

## 1 Introduction

Within the implementation of the assembly line in VÚTS company, a special manipulator for a feeding seaming machine was developed. This is the part of the assembly line where the products come in two belts and are placed in one position in seaming machine. This is a key point in terms of overall cycle time. At the same time, it is a place where the frame of the first part of the line is not connected to the second part of the line. It must be possible to easily change the start and end point of the movement. The manipulator excels in especially very small built-up areas and short time cycle. It consists of two arms. The movement of each arm is achieved by rotation and translation. In detail, its mechanical construction is devoted in the patent [1]. This paper deals mainly with data preparation and its implementation in the manipulator control system.

The project implementation can be divided into two stages. In the first stage, we basically take the ideal endpoint trajectory data and achieve this by using electronic cams. From the beginning it was clear that this was not an ideal procedure, but due to the enormous time pressure to the start of the line, this procedure was first implemented.

In the second stage the problem was analyzed in detail. From the original trajectory, only the key parts relevant to the associated mechanisms have been preserved. The rest was recreated for the ideal dynamic properties of the individual drives.


E.-C. Lovasz et al. (Eds.): MTM&Robotics 2020, MMS 88, pp. 477–487, 2021.
https://doi.org/10.1007/978-3-030-60076-1_43

## 2 Manipulator Design

The manipulator consists of two curved arms whose shape is designed to make maximum handling space (see Figs. 1 and 2). Each arm has its own linear guide with belt portal axis which allows it to translate. Each moving platform is fitted with its own servomotor with gearbox providing rotary motion. Four 400 W Omron servo drives provide the movement. The translation axis achieves a speed of 968 mm/s at rated drive speed. The rotary axis at the rated speed achieves an output speed of 900°/s.

Later, it turned out that the choice of used reducers was applicable to the desired tact, but for further acceleration, this would be limiting. In the case of a linear axis, this would mean exceeding the rated speed of the drive. In the case of the rotary axis, the situation was worse because the moment of inertia of the loaded manipulator on the motor axis exceeded the manufacturer's recommended values. To maintain system stability, the controller had to be set to a relatively low stiffness. The result was a higher following error, i.e. deviation of the actual and desired position.

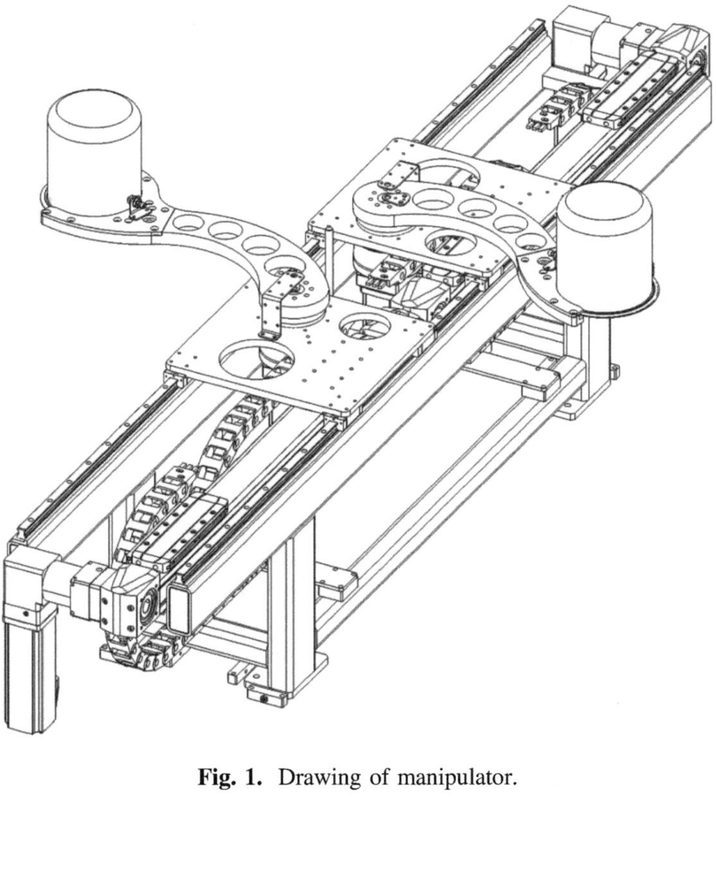

**Fig. 1.** Drawing of manipulator.

**Fig. 2.** Implementation of the manipulator in the assembly line.

For further calculations, it is important to define the conversions between the Cartesian coordinate system (x, y) and the manipulator axes ($\varphi$, $y_2$). The Cartesian coordinate system originates in the center of the rotary mechanism on the right (see Fig. 3). The linear axis of the manipulator is shifted by a distance $x_{off}$ = 795 mm. Arm length L = 400 mm is defined from the axis of rotation to the center of the transferred cylindrical product. Equations (1), (2), (3) and (4) describe the relationships for mutual conversions.

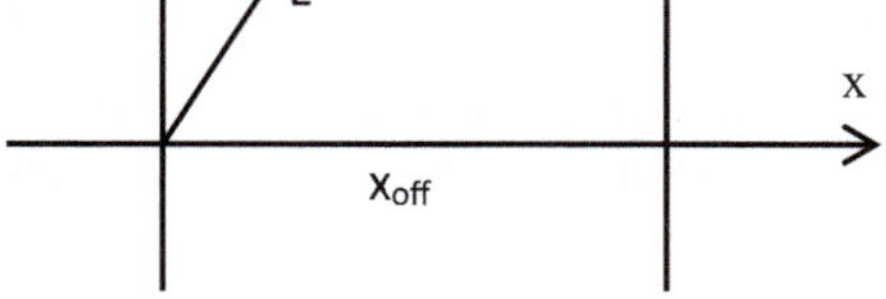


**Fig. 3.** Coordinate systems.

$$x = x_{off} + L \sin \varphi \tag{1}$$

$$y = y_2 - L \cos \varphi \tag{2}$$

$$\varphi = arcsin\, \frac{x - x_{off}}{L} \tag{3}$$

$$y_2 = y + L \cos \varphi \tag{4}$$

## 3 Movement Along the Original Trajectory

The project was taken over in progress with minimal time for implementation. So we continued with the original plan and made movements according to the original trajectory, as shown in the picture below (Fig. 4).

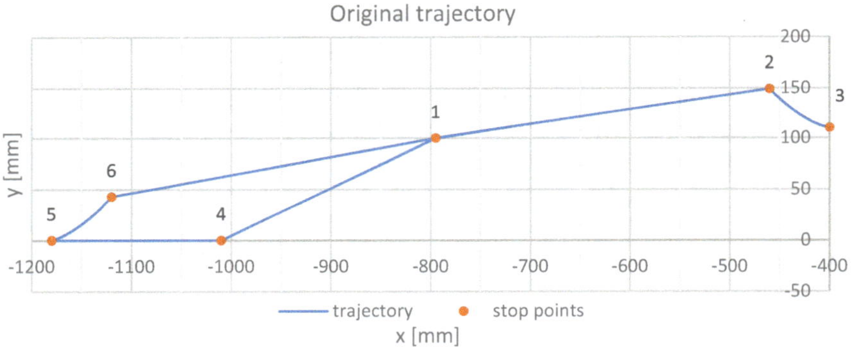


**Fig. 4.** Trajectory of the original concept.

The movement always starts and ends at point 1. Upon request of the higher-level PLC, the manipulator loads or unloads the product.

Loading: 1 => 2 => 3 (waiting for confirmation) => 2 => 1.
Unloading: 1 => 4 => 5 (waiting for confirmation) => 6 => 1.

The graph shows that the motion curve is not smooth and therefore cannot be implemented without stopping at individual points. Moreover, the trajectory itself does not say anything about binding to the virtual master axis, respectively to the time. In this first phase, we chose the simplest approach, dividing the trajectory after constant distance steps and assigning them a virtual master with a constant step. Subsequent smooth control of the virtual master can also achieve smoothness in the movement of individual drives. Due to the transformation of motion to translation and rotation, it is not possible to achieve effective control of individual axes in terms of the dynamics of the individual drives.

The cooperating seaming station cannot produce during the unloading of the product, so it is desirable to minimize this operation. In order to reach the production cycle, blocking by the manipulator should take a maximum of 1 s. Waiting in position 1 and stopping at each point proved to be a very inefficient approach. The manipulator was put into test operation at the required time and the resulting tact was sufficient for to the running-in of the line. Significantly higher cycle times are required for use in the manufacturing process, which would not be possible under the above conditions.

## 4 Movement Along a Modified Trajectory

The first step was a detailed analysis of the hand's path. First we identified the parts of the trajectory which had to be followed precisely for the correct loading and unloading of the product. We identified new key points of the trajectory more precisely. That is, the closest position where the manipulator can safely wait without limiting the working processes of the cooperating mechanisms. And points beyond which cooperating mechanisms can start working safely.

The next step was to prepare displacement diagrams for the individual drives. We have several tools available for the preparation of the electronic cam data [2] and [3], which are always used to prepare data for independent drives. Based on my previous experience, we were able to develop a software tool in C# for designing the displacement diagrams for the manipulator in a short space of time. We divided the displacement diagram into a shorter section, which we replaced with a selected function. We chose a 5$^{th}$ degree polynomial, because it can easily be defined by the boundary conditions 0$^{th}$, 1$^{st}$ and 2$^{nd}$ derivatives. Neighbouring polynomials always share their boundary conditions, which ensures the continuity of the resulting displacement diagram to the second derivative. The function is determined by a conventional notation [4].

$$y = c_5x^5 + c_4x^4 + c_3x^3 + c_2x^2 + c_1x^1 + c_0 \tag{5}$$

For implementation in the control system, notation is used according to Horner scheme.

$$y = ((((c_5x + c_4)x + c_3)x + c_2)x + c_1)x + c_0 \tag{6}$$

This implementation guarantees the most effective method of calculation. It used only 5 additions and 5 multiplications. The calculation of individual coefficients is more complex as described in [4].

The tool also implements a graphical component to display the manipulator endpoint trajectory (see Fig. 5). Displacement diagrams are composed of fifth degree polynomials that allow an intuitive design using the boundary conditions of the zero, first and second derivatives [4].

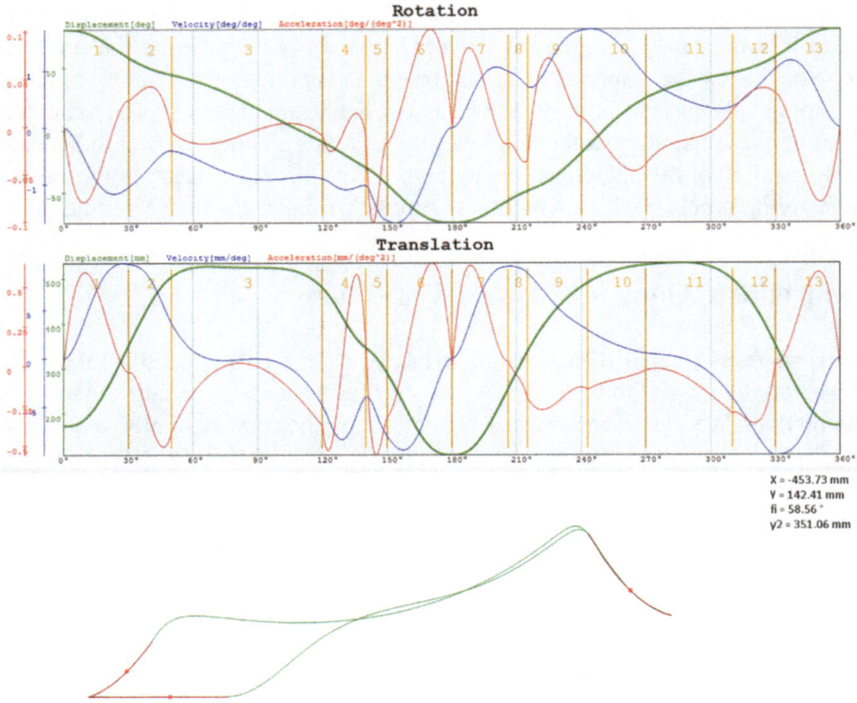


**Fig. 5.** Software tool for displacement diagram design.

The design of the displacement diagrams was then as follows: First, we defined the end positions for both drives, i.e. zero derivatives for the virtual master points 0°, 180° and 360°. We defined the same for points where it is still necessary to follow the prescribed trajectory (red points in Fig. 5). Subsequently, we began to adjust other boundary conditions (first and second derivatives) to achieve the best possible dynamic parameters of the drive and still kept the desired trajectory. Initially, this process seemed non-intuitive due to the transformation of motion into a rotational and translational axis, but after a short time it ceased to be a problem. After defining these critical parts, it was sufficient to prepare the rest of the trajectory where the exact trajectory was not strictly defined and the design process was easier. A collision check with the cooperating mechanisms was performed. Further adjustments were made to reduce the collision space with the other arm of the manipulator. The resulting trajectory with the new key points is shown in Fig. 6 and Fig. 7. The meaning of each key point is described in Table 1.

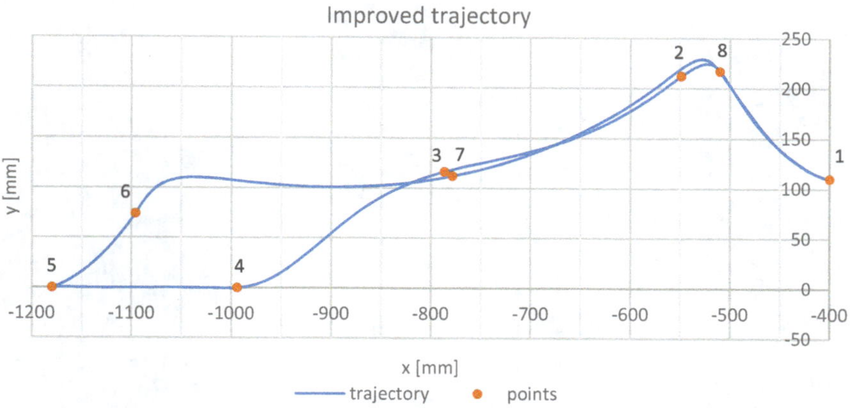


**Fig. 6.** Improved trajectory with key points.

**Table 1.** Meaning of key points.

| Id | VM [°] | Description |
|---|---|---|
| 1 | 0 | Loading position |
| 2 | 64 | Boundary where the loaded arm stops blocking the workspace of the carousel |
| 3 | 108 | Waiting point where the loaded arm enters the collision area with the other arm |
| 4 | 140 | Waiting point where the loaded arm enters the workspace of the seaming mechanism |
| 5 | 180 | Unloading position |
| 6 | 215 | Boundary where the empty arm leaves the workspace of the seaming mechanism |
| 7 | 252 | Boundary where the empty arm leaves the collision space and releases it for the second loaded arm |
| 8 | 310 | Waiting point in front of the carousel workspace |

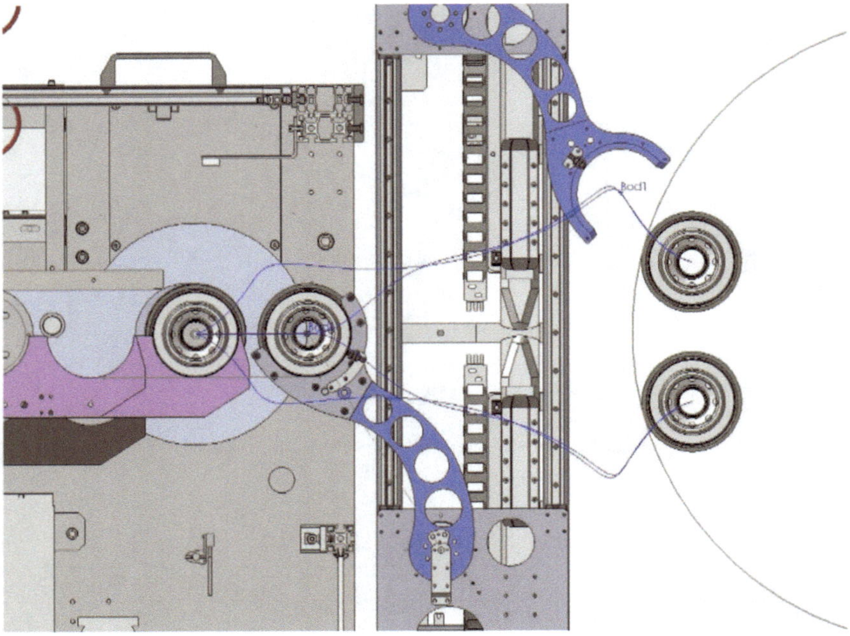

**Fig. 7.** Drawing of mechanism and trajectories of both arms.

## 5 Implementation of the Control System

The control system is based on an Omron platform, specifically the compact PLC NX1P2-1140DT. The PLC supports the synchronous control of four real axes with a primary task cycle of 2 ms. Directly on the controller body 24 digital inputs and 16 digital outputs are implemented, which are used for communication with the superior PLC. The superior PLC sends only basic commands, such as: load product and unload product. PLC programming was performed in Sysmac Studio [5], which supports IEC 61131-3 standard. The program was written in ST (structured text) as a two-layer state machine for both arms. In the right part of the work area, the arms can move independently, but in the left part they must cooperate with each other to avoid a collision. Slowing down or completely stopping the delivery of products on one of the belts is not a problem. Overall, this is a product with a very good price performance ratio.

As mentioned above, displacement diagrams are composed of fifth degree polynomials. It has already been shown that it is appropriate to maintain this method for PLC implementation [2]. In terms of memory consumption and computational complexity is a very effective way. The displacement diagrams were composed of 13 polynomials. Each polynomial is described by 6 coefficients and a polynomial length on the virtual master. Overall, 182 coefficients are required for both drives.

## 6 Reduction of Positional Errors by Adjusting the Displacement Diagram

Due to the low rigidity of the arm rotation servo driver, the resulting movement exhibited a positional error. The error was repeatable at a given speed. Thanks to the developed tool for design of the displacement diagram, it was possible to adjust the shape very easily to counteract the measured error [6]. A small modification of the shape, led to an adjustment of speed and acceleration (this is done automatically by composing displacement diagram from polynomials). From the perspective of the dynamics of the whole system, it is a negligible change, which leads to a significant improvement in the accuracy of product placement. The graphs below (Figs. 8 and 9) shows the effect of this correction on the measured data.

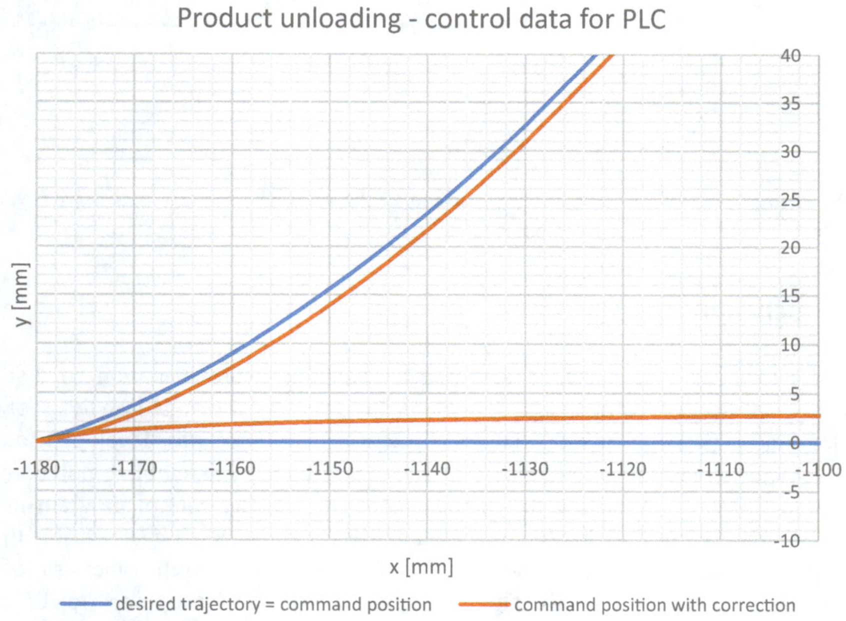


**Fig. 8.** Comparison of the original trajectory and the trajectory with the correction.

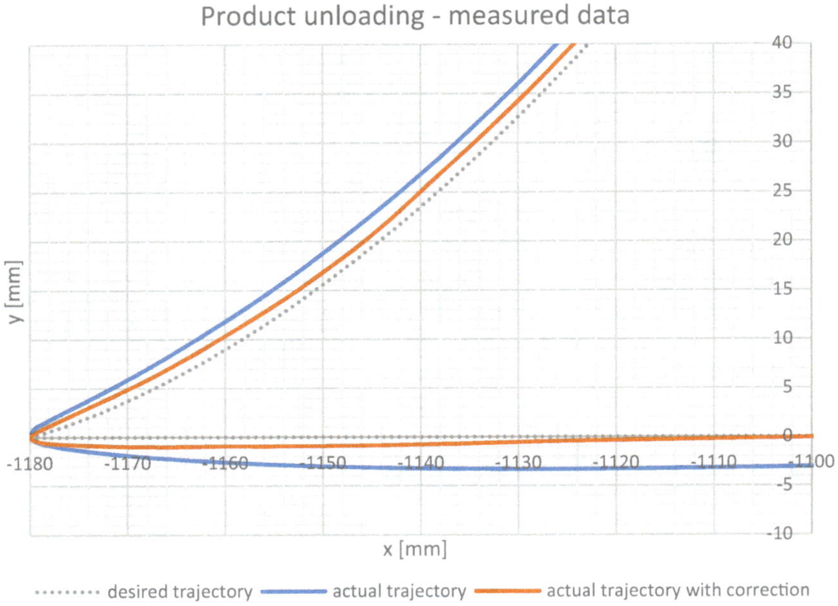


**Fig. 9.** Comparison of measured data according to the original trajectory and trajectory with correction.

# 7 Conclusion

Within a few days we managed to put the manipulator into operation so that the assembly line testing could begin. This was followed by a detailed analysis of the problem and a redesign of the solution. A new software tool was developed for the design of the displacement diagrams. The design was done by assembling polynomials of the 5th degree by adjusting the boundary conditions. The control program in the PLC has been modified so that the manipulator always waits as close as possible to the target position and at the same time does not allow collision with the other arm of the manipulator. The trajectory was designed smoothly between the endpoints (loading and unloading). Under ideal conditions, the manipulator can travel the entire path without stopping.

The total movement time is 4 s (only the movement without technological processes). However, this number is not as important as the time of entry and exit of the seaming machine. The setpoint was 1000 ms. In practice, we reached 924 ms (total unproductive time of seaming process caused by manipulator).

The position error was corrected by adjusting the displacement diagrams. The error was reduced, as shown in Fig. 8. Significantly better results could be achieved using an iterative algorithm [7]. However, we did not proceed with further modifications because the manipulator movement was within the required tolerances.

## References


1. Richter, A., Palaštuk, P.: Manipulátor pro podávání výrobků ze dvou větví výrobní linky do následné společné větve, Patent application
2. Dostrašil, P.: Implementation of specific displacement diagrams for the control of kinetic sculptures with Yaskawa electronic cams. In: Beran, J., Bílek, M., Žabka, P. (eds.) Advances in Mechanism Design II. MMS, vol. 44, pp. 377–382. Springer, Cham (2017). https://doi.org/10.1007/978-3-319-44087-3_51
3. Jirásko, P., et al.: Mechatronika pohonů pracovních členů mechanismů. VÚTS, Liberec (2015)
4. Koloc, Z., Václavík, M.: Cam Mechanisms. Elsevier, Amsterdam (1993)
5. Sysmac Studio. https://industrial.omron.cz/cs/products/sysmac-studio. Accessed 31 Mar 2020
6. Dostrašil, P.: Finding the parameters of the compensation pulse with the use of iterative Nelder-Mead method for suppressing residual vibrations of electronic cam. In: Advances in Robotics, Mechatronics and Circuits: Proceedings of the 18th International Conference on Circuits (part of CSCC 2014), Greece, pp. 155–159 (2014)
7. Čermák, L., Hlavička, R.: Numerické metody. Akademické nakladatelství CERM, Brno (2008)

# Study on the Fuel Consumptions Using Traffic Simulation with Example of City Duisburg

Xiaowei Hu, Xiaoyi Ma(✉), and Dieter Schramm

Chair of Mechatronics, Universität Duisburg-Essen, Duisburg, Germany
{xiaowei.hu, xiaoyi.ma}@uni-due.de

**Abstract.** With the development of microscopic traffic simulation and computer performance, traffic simulation is no longer limited to solving problems in the field of transportation. Microscopic traffic simulation can provide massive details, which can provide more possibilities for many studies. For studying traffic emissions, fuel consumptions and how the vehicle dynamic characteristics will affect traffic flow, a series of driver model and vehicle model for traffic simulation were built in this work. A traffic scenario of part of City Duisburg was also built in Simulation of Urban MObility (SUMO), which took advantage of the road network from Open Street Map (OSM) and traffic counter data provided by the municipality of Duisburg. As the vehicle models in SUMO are quite abstract and simplified, more detailed driver model and vehicle models were developed and applied in the simulation. The driver model is a fuzzy logic model based on behavior data of human drivers collected from driving simulator. In the vehicle models the characteristics of powertrains, braking systems, and the changes of driving resistances are all considered. These models make the simulated vehicles more similar to human drivers and realistic vehicles. The average fuel consumptions in this scenario was studied and the results with different models were also compared.

**Keywords:** Microscopic traffic simulation · SUMO · Driver modeling · Vehicle modeling

## 1 Introduction

Energy conservation and emission reduction is a continuous discussed topic in this age of increasingly scarce energy. Many kinds of technologies have been development to improve the efficiency of vehicles and reduce the energy consumption. There have been many researches about the energy consumption of vehicles [1–3], however, the energy consumption in a traffic environment and the impact of vehicle dynamic characters on traffic flow have been paid less attention. Large-scale field tests are economically and spatially restricted, traffic simulation in computer make the large-scale experiment possible. With the help of powerful computer, massive data can be calculated in a short time, which makes it possible for traffic simulation to develop from macroscopic to microscopic. Microscopic traffic simulation can reproduce the specific state of each traffic participant at each simulation step. More details in simulation mean more possibilities in application.


E.-C. Lovasz et al. (Eds.): MTM&Robotics 2020, MMS 88, pp. 488–495, 2021.
https://doi.org/10.1007/978-3-030-60076-1_44

In this work, a 24-h traffic scenario of part of City Duisburg was firstly established in Simulation of Urban MObility (SUMO). The road network of City Duisburg was obtained from Open Street Map (OSM). Then, a driver model was built based on the driving behaviors of realistic humanity drivers. The tested drivers completed a same driving test in a driving simulator. Their operating behaviors of gas/brake pedal in specific driving situations were collected. Based on these driving data, a fuzzy logic model of driver has been built. In addition, a model of internal combustion engine vehicle (ICEV) has also been made. The combination of driver model and vehicle model was applied as a new kind of car-following model into the traffic simulation scenario of City Duisburg. With the help of more detailed powertrain model, the fuel consumptions of vehicle in different kinds of driving situation could be more accurate calculated. The final simulation results with proposed models were also compared with the results using original models in SUMO.

## 2 Traffic Scenario of City Duisburg

In this work, a traffic scenario of part of Duisburg was built. The simulated area is the center part of the city, which is also called Duisburg inner ring, as shown in Fig. 1. As mentioned in introduction, the scenario was built and simulated in SUMO, which is an open source, microscopic and continuous road traffic simulation software [4]. It is able to provide space-continuous and time-discrete microscopic traffic flow simulation [5]. As mandatory inputs, a road network and a set of routes are needed for a simulation in SUMO. The way of obtaining and generating the necessary inputs from available data would be introduced in following.

### 2.1 Road Network

Road networks are now available in many ways. Here the network file of corresponding area was downloaded from OSM. Using an additional tool of SUMO, the OSM map file can be converted into XML (Extensible Markup Language) and be directly used in SUMO. After correcting the false information in the map, we got a road network as shown in Fig. 1.

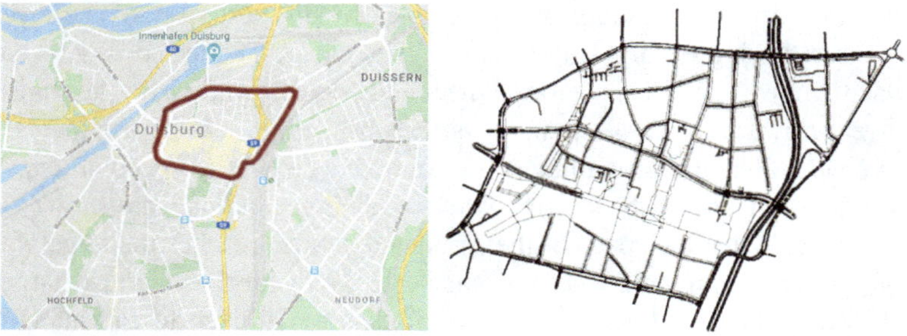


**Fig. 1.** Duisburg inner ring in Google Map (left) and imported network from OSM (right).

### 2.2 Traffic Demand

Besides the network file, another necessary input of SUMO is a set of routes, i.e., traffic demand. Traffic demand means set of all vehicles in the traffic system, with their associated routes [6]. SUMO supports many kinds of data for generating the traffic demand, including OD-matrices, population statistics, and detector data. In this work the traffic demand was generated from detector data. The traffic detectors include 76 induction loops and camera counters, which locate around 12 intersections in the simulated area. A 24-h traffic counter data of a normal workday were used for generating the traffic demand. The detector data and locations were provided by municipality of Duisburg. With the help of corresponding tools of SUMO, 29918 vehicles in total were generated in the simulation scenario and were distributed to different routes and depart time according to the detector data. The following simulations with different models were all based on the scenario built in this chapter.

## 3 Modeling of Driver and Vehicle

Only the car-following process is considered in this work, the lane-changing process is ignored. Therefore, the steering operation of drivers and the lateral movement of vehicles are not considered in modeling.

### 3.1 Driver Modeling

Generally, in microscopic traffic simulation software, the car-following model and lane-changing model determine how a vehicle drives in the simulation scenario. These models are mathematically abstracted from the traffic observation data. The basic element of the traffic observation and microscopic modeling is the "driver-vehicle-unit". Therefore, the car-following model and lane-changing model represent the combined characters of driver and vehicle. When more detailed vehicle models are applied in simulation, it is impossible to adjust the vehicle part of car-following model. Before the modeling of vehicle, a driver model, which can control the vehicle model as a humanity driver, is necessary in this work.

The mentioned driver model mainly reflects the operation of human driver on the acceleration/brake pedal. For gathering the operation data of humanity drivers, a driving test in a driving simulator was performed by 34 participants. The driving test scenario is free driving on highway without other traffic. The tested drivers can drive the simulated vehicle with real steering wheel and pedals, which is in the same way of driving a real car. The driving simulator is shown in Fig. 2. The participants need to drive the vehicle to the corresponding speed following a series of speed orders. There is an LCD on the right side of the steering wheel. The target speeds were shown on it in a certain sequence. When the speed of the simulated vehicle reaches the target value and the change of speed is maintained in a range of 5 km/h for 5 s, the next target speed will be shown on the LCD, meanwhile the simulator will beep to call attention. The profile of acceleration/brake pedal position of the simulated vehicle was record during the whole test. The data recording frequency is no lower than 63 Hz. In this way, the operation behavior of acceleration and brake pedal for humanity drivers by corresponding current speed and desired speed can be collected. These data were used for a fuzzy logic modeling.

**Fig. 2.** Driving simulator and its cockpit.

The human driver is a very complex and nonlinear control system. It is difficult to simulate such a system using classical control methods. Therefore, the fuzzy logic was selected to simulate human driver behavior. In realistic car-following situation many factors will be considered, such as the speed of leading car, current gap and minimum gap, current speed, etc. It will make the fuzzy model extremely complex if make all these factors as the model inputs. The human driver was simplified to a model with dual inputs and single output. Current speed is one of the inputs, the difference of desired speed and current speed is the other one. The output of the model is pedal position, positive means acceleration while negative means braking. The membership functions of fuzzy driver model are shown in Fig. 3.

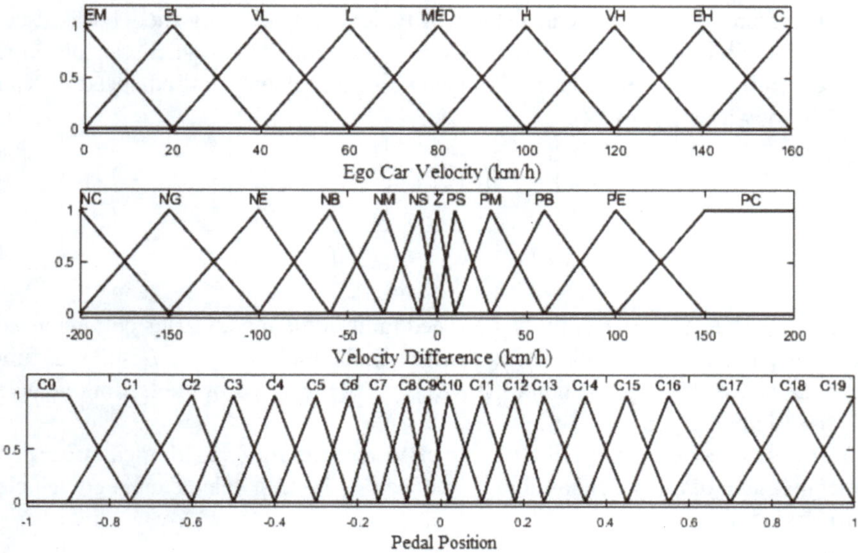


**Fig. 3.** Membership functions of fuzzy driver model.

By analyzing the collected data from driving simulator, the relationship of current speed, difference of desired speed and current speed, pedal position was tabulated according to the membership functions. This table contains the fuzzy rules, which are listed in Table 1.

**Table 1.** Gear ratios of gearbox and differential.

| V_d | V | | | | | | | | |
|---|---|---|---|---|---|---|---|---|---|
| | EM | EL | VL | L | MED | H | VH | EH | C |
| NC | C0 | C0 | C0 | C0 | C0 | C0 | C0 | C0 | C0 |
| NG | C1 | C1 | C1 | C1 | C1 | C1 | C1 | C1 | C1 |
| NE | C2 | C2 | C2 | C2 | C2 | C2 | C1 | C2 | C1 |
| NB | C3 | C3 | C3 | C1 | C2 | C2 | C2 | C2 | C2 |
| NM | C5 | C5 | C3 | C2 | C2 | C2 | C3 | C3 | C3 |
| NS | C6 | C4 | C3 | C3 | C3 | C4 | C4 | C4 | C5 |
| Z | C14 | C14 | C14 | C15 | C16 | C17 | C17 | C17 | C18 |
| PS | C14 | C14 | C14 | C14 | C14 | C14 | C15 | C15 | C15 |
| PM | C17 | C14 | C15 | C16 | C16 | C16 | C16 | C16 | C16 |
| PB | C17 | C17 | C17 | C16 | C17 | C17 | C17 | C17 | C17 |
| PE | C18 | C18 | C18 | C18 | C17 | C17 | C17 | C17 | C18 |
| PC | C18 | C18 | C18 | C18 | C18 | C18 | C19 | C19 | C19 |

In car-following process, the maximum speed, gap to leading car and other factors must be considered, while the desired speed, as one of the necessary input parameters of fuzzy model, cannot be collected directly from the driver of vehicle during driving tests. Therefore, a mathematical method of typical car-following model was used. It makes it possible to calculate desired speed with parameters, which can be directly retrieved during traffic simulation. The method of calculating desired speed in Krauss model is shown as follow [7].

$$V_{desired} = \min\left[v_{safe}, v_{max}, v_n(t) + aT\right] \tag{1}$$

$$v_{safe} = -\tau b + \sqrt{(\tau b)^2 + v_{leading}(t)^2 + 2b\, g(t)}, \tag{2}$$

where $v_{safe}$ is the safe speed, $v_{max}$ is the speed limit of the road, $v_n(t)$ is current speed at time $t$, a is the maximum acceleration, $T$ is the simulation step length, $b$ is the maximum deceleration, $\tau$ is driver's reaction time, $v_{leading}(t)$ is the speed of the leading vehicle, $g(t)$ is the following gap.

The driver model, consists of this function and the fuzzy model, can operate the acceleration and brake pedal based on parameters of leading vehicle and ego vehicle in car-following situations.

## 3.2 ICEV Powertrain and Braking System Modeling

In SUMO a vehicle is represented with an object with length and width moving in two-dimensional space. Since this study focuses on the emissions and energy consumptions of vehicle, the vehicle model consists of powertrain model and longitudinal mechanical model of body.

In an ICEV, the powertrain consists of internal combustion engine (ICE), gearbox, differential and wheels. These structures determine the traction of a vehicle.

The internal combustion engines applied in vehicles are also nonlinear systems. In simulation they could be replaced with lookup tables. The steady state data of engine working at several specific operating points are tabulated. During simulations, the variables' value of engine is determined based on the lookup tables by interpolation. In this study, the ICE model was divided to mechanical part and fuel consumption part. The mechanical part determines the output torque of engine at all engine speeds and all acceleration pedal position. The fuel consumption model gives out the fuel consumption value at all engine operating points (Figs. 4 and 5).

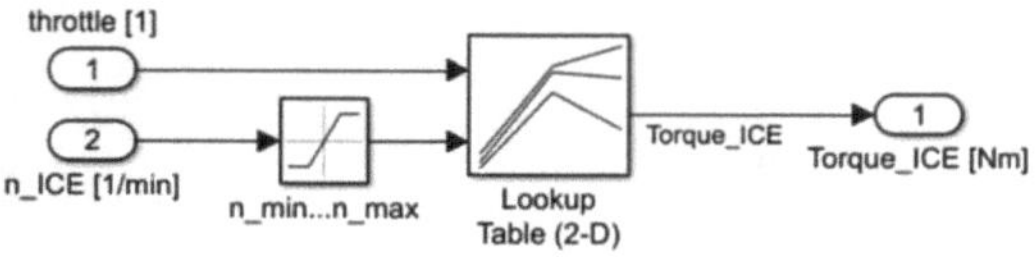


**Fig. 4.** Schematic of engine mechanical model.

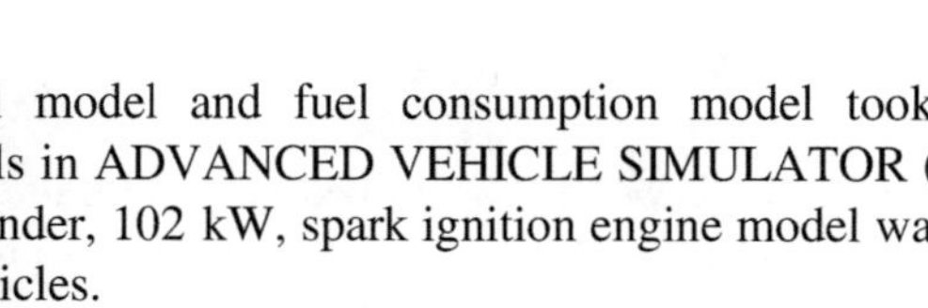


**Fig. 5.** Schematic of fuel consumption model.

The mechanical model and fuel consumption model took advantage of the parameters of models in ADVANCED VEHICLE SIMULATOR (ADVISOR). In this work a 3.0 L, 6-cylinder, 102 kW, spark ignition engine model was used as the engine of all simulated vehicles.

Gearbox and differential can help ICE to enlarge the range of output torque and trie to make engine work at more efficient area. As in a 24-h traffic simulation the simulation step is usually set to 1 s, the gear shifting process is ignored here. The gear will be selected to the optimal gear according to vehicle speed.

To sum up, the tractive force of vehicle in each simulation step can be calculated with following function.

$$F_t = \frac{T_m i \eta_G \eta_D}{r_{dyn}}, \tag{3}$$

where $F_t$ is the tractive force of vehicle, $T_m$ is the output torque of engine, $i$ is gearbox transmission ratio, $r_{dyn}$ is the dynamic radius of wheel.

The model of braking system is also a one-dimensional lookup table,which reflects the relationship between brake pedal position and total braking torque on wheels.

### 3.3 Vehicle Modeling

The entire vehicle is simplified to a particle with one-dimensional motion along the lane as the steering and lane-changing processes are not considered. The particle is affected by tractive/braking force, rolling resistance, wind drag and acceleration resistance. Because the road network in SUMO is two-dimensional and without altitude information, the grade force on vehicle is always zero.

The acceleration of vehicle $a$ in each simulation step can be obtained according to Newton's second law. As the vehicles in SUMO can only drive forward, there will be only two conditions should be considered.

When acceleration pedal is pressed:

$$F_t - F_f - F_w - F_j = ma. \tag{4}$$

When brake pedal is pressed:

$$-F_b - F_f - F_w - F_j = ma, \tag{5}$$

where $F_f$ is rolling resistance, $F_w$ is wind drag, $F_j$ is acceleration resistance, $F_t$ is tractive force and $F_b$ is braking force.

The vehicle speed in simulation step $i$ can be obtained by following function, the direction is along the lane.

$$v_i = v_{i-1} + aT. \tag{6}$$

## 4 Simulation with SUMO and MATLAB

As an open source software, SUMO allows embedding any traffic models in C++ for simulation. However, the driver model and vehicle models in previous chapter have used fuzzy logic and lots of interpolation operations. In MATLAB there are corresponding functions that can be called directly. Therefore, both the driver model and vehicle model are programmed in MATLAB.

TraCI, abbreviation for "Traffic Control Interface", is an interface of SUMO. With TraCI, SUMO can work as a library. MATLAB or other software can through TraCI retrieve all required variables value, calculate the vehicles state in next simulation step using external models and give the results back to control the simulated objects. Then SUMO updates the positions of vehicles and displays on the SUMO-GUI.

The Scenario was simulated two times, with the default model in SUMO and with the proposed model, respectively. Because the proposed engine model includes fuel consumption information, it is possible to achieve the fuel consumption of each vehicle in each simulation step. Thus, it is also possible to calculate the total and average fuel consumption after simulation. Additionally, fuel consumption is also one of the emission outputs for simulation in SUMO. The emission model in SUMO is modified based on Handbook Emission Factors for Road Transport (HBEFA).

As a result, in the simulation with SUMO default model, the average fuel consumption of all simulated vehicles is 25.56 L/100 km. When using the proposed driver and vehicle model, however, with the SUMO emission model, the average fuel consumption of all simulated vehicles is 24.30 L/100 km. When totally using the proposed models in Sect. 3, the average fuel consumption dropped to 18.49 L/100 km.

## 5 Conclusion

The average fuel consumptions from three methods are all obviously higher than common experience about passenger cars. However, the engine model used here is a 3.0 L, 1991 model. When driving in pure urban traffic, the result seems more reasonable. If more kinds of engine model can be inserted to the simulation, the results can be more reasonable.

The proposed modeling and simulate methods aim to study not only the traffic energy consumptions, but also the effect of vehicle dynamic and driving behavior on traffic flow. Therefore, the modeling of electric vehicles and self-driving and applying into simulation will be the future work.

## References

1. Chen, C., Xu, X.: Modeling the conducted EMI emission of an electric vehicle (EV) traction drive. In: IEEE EMC Symposium 1998. International Symposium on Electromagnetic Compatibility. Symposium Record (Cat. No. 98CH36253), vol. 2, pp. 796–801. IEEE (1998)
2. Arslan, O., Karasan, O.E.: Cost and emission impacts of virtual power plant formation in plug-in hybrid electric vehicle penetrated networks. Energy **60**, 116–124 (2013)
3. Ao, G.Q., Qiang, J.X., Zhong, H., et al.: Fuel economy and NOx emission potential investigation and trade-off of a hybrid electric vehicle based on dynamic programming. Proc. Inst. Mech. Eng. Part D J. Automobile Eng. **222**(10), 1851–1864 (2008)
4. Lopez, P.A., et al.: Microscopic traffic simulation using SUMO. In: 21st International Conference on Intelligent Transportation Systems (ITSC), Maui, HI, pp. 2575–2582 (2018)
5. SUMO Documentation. https://sumo.dlr.de/docs/index.html. Accessed 30 May 2020
6. Baquela, E.G., Olivera, A.C.: Optimization of traffic network design using nature-inspired algorithm: an optimization via simulation approach. In: Handbook of Research on Modern Optimization Algorithms and Applications in Engineering and Economics, IGI Global, pp. 266–287 (2016)
7. Krauß, S: Microscopic modeling of traffic flow: investigation of collision free vehicle dynamics. Ph.D. dissertation (1998)

# The Analysis of the Dynamic Behavior of a Tire Pressure Sensor

Paul Florin Druţa, Calin Gozman-Pop, Dorin Simoiu, Ion Crâştiu, and Liviu Bereteu(✉)

Mechanics and Materials Strength Department, Politehnica University of Timişoara, Bd. Mihai Viteazul, nr. 1, 300222 Timişoara, Romania
paul.druta@yahoo.com, calin.gozman@gmail.com, ion_crastiu@yahoo.com, {dorin.simoiu,liviu.bereteu}@upt.ro

**Abstract.** Acceleration, braking and steering of a vehicle depends largely on the contact between the tire and the road, and it depends on the internal pressure. Higher than recommended pressure creates an overheating of the tires and is often the cause of accidents. A lower pressure leads to lower tire forces increasing the braking distance, decreases the vehicle's maneuverability, leads to tire wear and consequently to a shorter service life. To all this must be added an additional fuel consumption. All this, and not only, has led to the development of tire pressure monitoring systems (TPMS). The active element through which the pressure is monitored is the sensor. It can be mounted in the tire or on the tire valve. Even for masses of a few tens of grams, the developed inertial forces reach hundreds of Newton. Therefore, it is important to know the behavior of these sensors in dynamic regimes, this being the purpose of this paper.

**Keywords:** Pressure sensor · Mathematical model · Dynamic behavior analysis

## 1 Introduction

The Tire Pressure Monitoring System (TPMS) was introduced for new vehicles from 1 October 2005 in the US and from 1 November 2012 in the EU in order to increase road safety by limiting the number of tire explosions, but also to protect environment by reducing $CO_2$ emissions. According to some studies performed [1], with the decrease of the tire pressure, the rolling resistance increases, respectively the fuel consumption. Thus, for a tire with a nominal pressure of 2 bar a decrease of 0.6 bar of the pressure determines an increase of the fuel consumption by approximately 4%. In 2018, approximately 90 million vehicles were produced, and according to studies 19% of vehicles have inadequate pressure, which means that 17 million vehicles have inadequate tire pressure at some point. Due to these aspects, the additional fuel consumption is estimated at 4 billion liters per year, which has a value of approximately 5.2 billion Euros and generates an additional 9.3 million tons of $CO_2$ [2]. For these reasons, the market for TPMS sensors is growing rapidly. In order to be competitive, they must be produced with low prices and small dimensions in order to reduce the dynamic forces


E.-C. Lovasz et al. (Eds.): MTM&Robotics 2020, MMS 88, pp. 496–503, 2021.
https://doi.org/10.1007/978-3-030-60076-1_45

and increase their lifespan [3]. The variation of the tire pressure during the movement of the vehicle can create dangerous problems by changing the position of its center of gravity and implicitly of the contact forces with the road [4]. Techniques for increasing the efficiency of signal analysis in TPMS are presented in [5].

This paper, given the role that pressure sensors play in the structure of new types of smart tires, aims at their dynamic modeling and numerical simulation for different speeds and different types of wheels.

## 2 Mathematical Modeling of the Dynamic Behavior of a Pressure Sensor

The analyzed sensor is with a metal valve and fixed to it with a variable fixing angle. The fixing with a variable angle is made so that it can be used for different types of rims. Otherwise, in the event of a tire explosion, the sensor is quickly destroyed. In the dynamic analysis of this sensor it will be considered that the rim is a rigid body and that each point of it, including the orifice through which the valve passes, i.e. the place where the sensor is fixed, will have a cycloidal trajectory (Fig. 1).

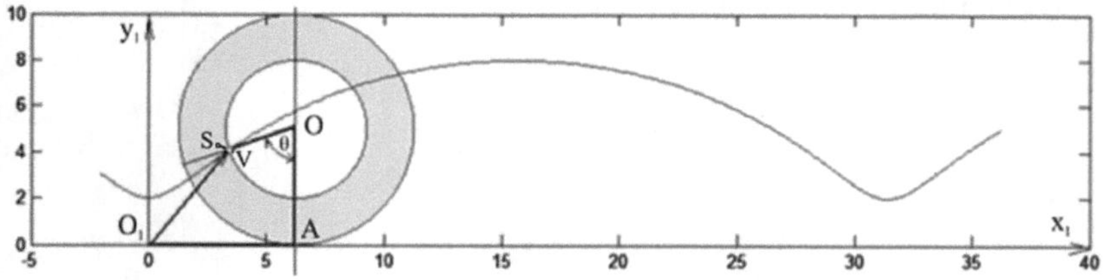


**Fig. 1.** Cycloidal motion of a pressure sensor

In Fig. 1, the fixed reference system was denoted by $O_1x_1y_1$, point O is the center of the wheel, A is the point of contact between the wheel and the road, and OA = R is the radius of the wheel. The position of the valve is denoted by V, and OV = r, the inner radius of the rim. With $\theta = \theta(t)$ the angle described by the OV radius of the rim was denoted. In relation to the fixed system $O_1x_1y_1$ the coordinates of the valve are:

$$X_V = R\theta - r\sin\theta, \quad (1)$$

$$Y_V = R - r\cos\theta. \quad (2)$$

Considering that the movement of the vehicle is done with constant speed, it is possible to write

$$O_1A = R\theta = vt \text{ and } \omega = \dot{\theta} = v/R, \quad (3)$$

It is denoted by C $(X_{C0}, Y_{C0})$ the center of mass of the sensor considered as a rigid body, and its coordinates are given with respect to point V, for fixing it on the rim in the plane of wheel movement. Under these conditions it will have a cycloidal motion having the parametric equations:

$$X_C = R\theta - X_{C0} - r\sin\theta, \tag{4}$$

$$Y_C = R - Y_{C0} - r\cos\theta. \tag{5}$$

Except for the battery, printed circuit board (PCB) and sensor, the other components are plastic. This fact will cause the sensor, as a whole, under the effect of the inertial forces generated by the cycloidal movement of the point V in which it is fixed, to be subjected to deformations. Under these conditions Newton's fundamental law is applied:

$$m\overrightarrow{a_S} = -\overrightarrow{F_e} - \overrightarrow{F_a}, \tag{6}$$

where with m was denoted the mass of the sensor, with $\overrightarrow{a_S}$ the acceleration of the center of mass in dynamic regime, i.e. under the action of the forces of inertia, elastic and internal friction (damping). With $\overrightarrow{F_e}$ the result of the elastic forces was noted, and with $\overrightarrow{F_a}$ the result of the damping forces. The dynamic model of a pressure sensor whose V fixing point on the rim moves on a cycloid is shown in Fig. 2.

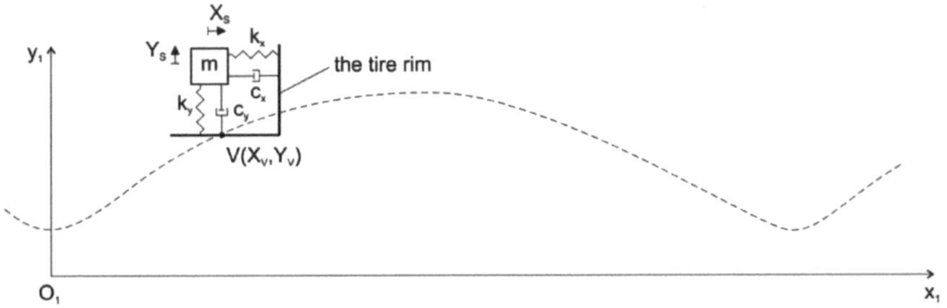


**Fig. 2.** Dynamic model of a pressure sensor

The dynamic displacements of the center of mass of the sensor are denoted by, $X_S$, and $Y_S$. By projecting Newton's law on the axes of the fixed reference system, it was obtained:

$$m\ddot{X}_S = -c_x(\dot{X}_S - \dot{X}_C) - k_x(X_S - X_C); \tag{7}$$

$$m\ddot{Y}_S = -c_y(\dot{Y}_S - \dot{Y}_C) - k_y(Y_S - Y_C), \tag{8}$$

where $c_x$, $c_y$, $k_x$, and $k_y$ are the damping constants, respectively the elastic constants of the sensor in the two directions. As, in a first analysis, the dynamic behavior is interesting from a qualitative point of view, it will be considered that the behavior of the sensor after the two directions is the same, that is, the constants will be equal.

If, it is taken into account that $X_{Sr} = X_S - X_C$ and $Y_{Sr} = Y_S - Y_C$ are displacements of the sensor (deformations) with respect to the static position of its center of mass, on the two axes, the above equations become:

$$m\ddot{X}_{Sr} + c\dot{X}_{Sr} + kX_{Sr} = -m\ddot{X}_C; \tag{9}$$

$$m\ddot{Y}_{Sr} + c\dot{Y}_{Sr} + kY_{Sr} = -m\ddot{Y}_C. \tag{10}$$

After deriving Eqs. (4) and (5) and replacing them in Eqs. (9) and (10) is obtained:

$$m\ddot{X}_{Sr} + c\dot{X}_{Sr} + kX_{Sr} = mr\left(\frac{v}{R}\right)^2 \sin\frac{vt}{R}; \tag{11}$$

$$m\ddot{Y}_{Sr} + c\dot{Y}_{Sr} + kY_{Sr} = mr\left(\frac{v}{R}\right)^2 \cos\frac{vt}{R}. \tag{12}$$

If it is considered the real axis of the complex plane $O_1y_1$ axis and the imaginary axis $O_1x_1$ axis, the center of mass of the sensor in dynamic mode corresponds to affix

$$z_{Sr}(t) = Y_{Sr}(t) + iX_{Sr}(t). \tag{13}$$

Multiplying Eq. (11) by complex number $i = \sqrt{-1}$ and adding by the Eq. (1) is obtained:

$$m\ddot{z}_{Sr} + c\dot{z}_{Sr} + kz_{Sr} = mr\left(\frac{v}{R}\right)^2 e^{i\omega t}. \tag{14}$$

It was found that the movements of the sensor in the two directions are produced by harmonic forces therefore the affix of the center of mass of the sensor in dynamic regime will have the law of motion given by:

$$z_{Sr}(t) = Z_{Sr}e^{i\omega t}. \tag{15}$$

Substituting this solution in Eq. (14) it is obtain equation presented below:

$$\left(-m\omega^2 + i\omega c + k\right)Z_{Sr} = m\omega^2 r. \tag{16}$$

From here the modulus of the relative displacement in dynamic regime of the center of mass of the sensor with respect to its rest position is determined:

$$Z_{Sr} = \frac{r(\omega/\omega_n)^2}{\left[\left(1 - (\omega/\omega_n)^2\right)^2 + (2\zeta\omega/\omega_n)^2\right]^{1/2}} \tag{17}$$

where r is the rim radius, $\omega_n = (k/m)^{1/2}$ is the natural circular frequency of the sensor, $\omega = v/R$ is the circular frequency, and $\zeta = c/2m\omega_n$ is the damping ratio.

## 3 Numerical Simulations

Given the large number of pressure sensors produced annually in the world, over 200 million, with the reduction of their cost price, there is generally a reduction in their mass. At high speeds this is very important, given that the sensors are subject to forces of inertia that can destroy. To illustrate this aspect, the variation of inertia forces will be simulated for four types of sensors having masses of 30 g, 35 g, 40 g and 45 g depending on the running speeds for a 225/70/R15 tire. For this purpose in Fig. 3 are represented the diagrams of the inertia forces resulted by the graphical representation according to the formula:

$$F_i = m\left(\frac{v}{R}\right)^2 r = 2m\left(\frac{v}{D}\right)^2 d, \tag{18}$$

where m is the mass of the sensor, v is the speed of the car, r is the radius of the rim and R is the radius of the tire, respectively d is the diameter of the rim and D is the diameter of the tire.

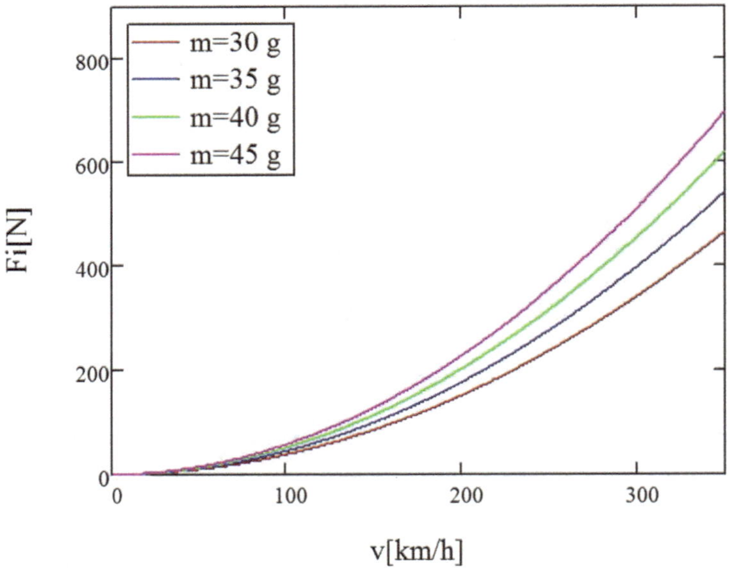


**Fig. 3.** Variation of inertia forces for different masses of the pressure sensor

In the automotive field there are a multitude of types and sizes of wheels. It is found that for the most common types of rims, their diameters can vary between 14″ and 18″ without the tire diameter (outer diameter) varying within a few millimeters [6].

For these reasons, considering the fact that the sensors are fixed on the rim, it is important to analyze the variation of inertia forces for the same type of sensor mounted on different types of wheels. Figure 4 shows the variations of inertial forces acting on a pressure sensor if it is placed on wheels having a diameter from 14″ to 18″.

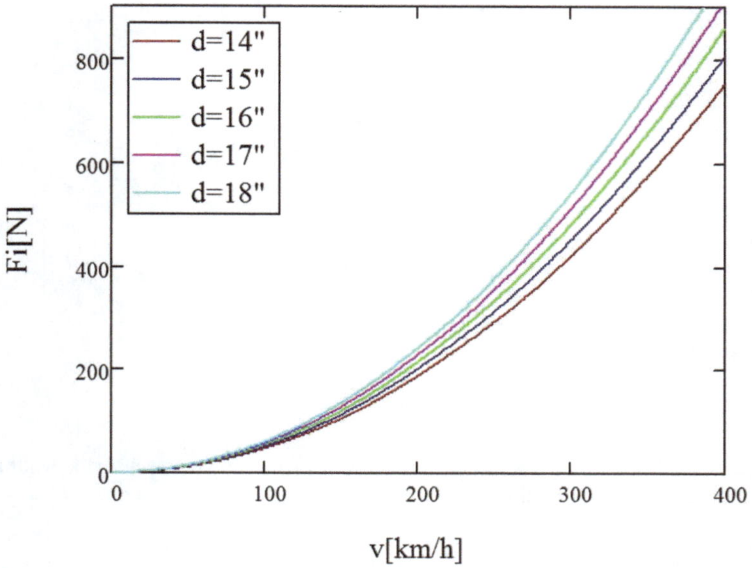


**Fig. 4.** Variation of inertia forces for a sensor placed on different rim sizes

From the relative dynamic displacement of the position of the center of mass of a pressure sensor, it is possible to follow the frequency domains in which it can operate without suffering deformations that would endanger its integrity. It was found [7] that sensors that in principle have in the composition the same materials and close dimensions have the first natural frequency after 2000 Hz. To obtain small deformations the sensor should work in the range for which $\omega/\omega_n \ll .15$ or vehicle speed will be $V < .15R\omega_n = .3\pi f_n R$. This relationship gives much higher speeds than usual. In conclusion, the deformations of the pressure sensor in the range of current speeds are small (Fig. 5).

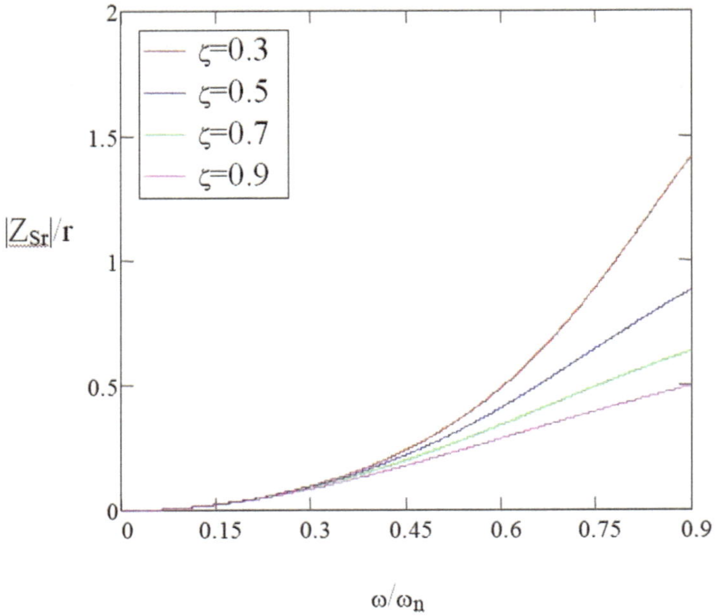


**Fig. 5.** Relative displacement of the center of mass of the pressure sensor depending on the damping ratio and frequency ratio

## 4 Conclusions

The paper aims to determine the source that produces the vibrations of a pressure sensor. For this purpose, the equations of motion were deduced in an original way, in the plane of motion of the wheel. It has been shown that the movement of the sensor on a cycloid is the source of vibrations. Some conclusions can be drawn from the analysis of the dynamic behavior of a pressure sensor:

1. Even for masses of the order of tens of grams on a pressure sensor can act, at very high running speeds, forces of several hundred newtons.
2. Given that these inertial forces produce vibratory movements, the pressure sensors fatigue the valves to which they are attached. Therefore, wear-resistant materials must be chosen for the valves.
3. If the same type of sensor is used for different rim sizes, it must be taken into account that it is dynamically stressed on larger diameter rims much stronger than on smaller diameter rims.
4. Regarding the relative movements of the sensor in dynamic regimes, taking into account the fact that its natural frequency is somewhere near 2000 Hz, and the running speeds are below 400 km/ h, results for the ratio $\omega/\omega_n$ very low values, in conclusion small deformations.
5. The dynamic behavior of the sensor depends on the mechanical properties of the component materials, but also on its geometry. A detailed analysis of these aspects will be followed in a future paper in which, by Modal Analysis using the Finite

Element Method, both the dynamic forms of the vibration modes and the natural frequencies of several types of pressure sensors will be obtained. The influence of the material properties on the sensitivity or reliability of the sensor can be given only after these analyzes.

## References

1. Yang, A.: TPMS Presentation, Sr.Auto FAE, Freescale Confidential Proprietary, 1 April 2014
2. Mustafić, I., Klisura, F., Berković, M.: Regulation and tire pressure monitoring system. In: 14th International Research/Expert Conference, Trends in the Development of Machinery and Associated Technology, TMT 2010, Mediterranean Cruise, 11–18 September 2010
3. Wei, C., Zhou, W., Wang, Q., Xia, X., Li, X.: TPMS (tire-pressure monitoring system) sensors: Monolithic integration of surface-micromachined piezoresistive pressure sensor and self-testable accelerometer. Microelectron. Eng. **91**, 167–173 (2012)
4. Shraim, H., Rabhi, A., Ouladsine, M., M'Sirdi, N.K., Fridman, L.: Estimation and analysis of the tire pressure effects on the comportment of the vehicle center of gravity. In: Proceedings of the 2006 International Workshop on Variable Structure Systems, Alghero, Italy, 5–7 June, pp 268-273 (2006)
5. Marton, Z., Fodor, D., Enisz, K., Nagy, K.: Frequency analysis based tire pressure monitoring. In: Proceedings of the IEEE International Electric Vehicle Conference (IEVC 2014), IEEE, Florence, Italy, pp. 1–5 (2014)
6. Tire size calculator. https://tiresize.com/calculator/. Accessed 15 June 2020
7. Gozman-Pop, C., Simoiu, D., Crâştiu, I., Nyaguly, E., Oanţă, E.N., Bereteu, L.: Dynamic analysis of satellites used for side-airbag system. In: 7th International Conference on Advanced Materials and Structures - AMS 2018, IOP Conference Series: Materials Science and Engineer, vol. 416, pp 1–7 (2018)

# Biomechanics

# Investigations on the Human Body and Seat Suspension Response Using Quarter, Half and Full Car Models

Raj Desai[1(✉)], Anirban Guha[1], and P. Seshu[2]

[1] Department of Mechanical Engineering, Indian Institute of Technology, Mumbai, India
rajdesai@iitb.ac.in
[2] Department of Mechanical Engineering, Indian Institute of Technology, Dharwad, India

**Abstract.** A number of human body and vehicle models have been developed to study vehicle and human dynamics. Which combination of these models is appropriate is an open question since it requires a compromise between modeling complexity and simulation accuracy. This paper attempts to answer this question by combining previously developed 12 degrees of freedom (DoF) human body and nonlinear cushion contact force models with quarter car (2-DoF), half-car (4-DoF) and full car (7DoF) models through a seat suspension system with an inclined multi-compression damper. The seat suspension parameters are obtained using a genetic algorithm (GA) based optimization exercise which minimizes seat effective amplitude transmissibility (SEAT) factor for random and bump road profiles. These systems are then analyzed using MATLAB-SIMULINK for pitch and roll. Various response parameters such as seat acceleration, cushion contact force, head fore-aft and vertical acceleration have been studied with different vehicle models and road profiles. The results indicate clear inferior outcomes of the quarter car model and thus advises avoiding them for integrated driver-vehicle ride comfort analysis.

**Keywords:** Biodynamics · Human body model · Genetic algorithm · Seat suspension · Vehicle model

## 1 Introduction

It is widely established that there is a high risk of health impairment due to prolonged exposure to whole-body vibration [1]. This situation is commonly experienced by vehicle occupants. So, an understanding of the human body and vehicle dynamics can help in lowering the heath risk of vehicle occupants. Vehicle ride comfort is an essential and important criterion in vehicle dynamics [2]. The comfort experienced by humans has considerable variations as different people respond in different ways to the same situation. Studies have shown that the effects of vibrations on humans could be quite detrimental and, in certain cases, may give rise to physical and psychological health impairments [3]. Research has shown that intervertebral disc degeneration [4] and chronic low-back pain as a result of constant exposure to vibrations occur more


E.-C. Lovasz et al. (Eds.): MTM&Robotics 2020, MMS 88, pp. 507–516, 2021.
https://doi.org/10.1007/978-3-030-60076-1_46

frequently among drivers than in the general population [5]. Human body response to vibration depends on posture, magnitude, frequency, the direction of excitation and individual anthropometry.

The vehicle seat suspension system plays a significant role in the separation of the vibrations from road disturbances [6]. The suspension system should be able to dissipate vibrational energy with minimal suspension travel. Also, the system should be able to perform effectively over a wide range of frequency and large amplitude vibrations. Often the seat suspension system is evaluated using seat effective amplitude transmissibility (SEAT) factor [7]. This is the ratio of the vibration experienced on top of the seat to that of the vibrating floor. Different kinds of vehicle models have been used to study vehicle dynamics and human body response. For the integrated human-vehicle analysis it is important to have a model that is both reasonable in its complexity and accurate enough to give predictions close to the experimental results. The quarter car model is the simplest for analysis but does not allow pitch and roll motion of vehicle chassis to be investigated. A half-car model is able to capture pitching effect but not roll. With a full car model, these limitations are eliminated. However, the corresponding increase in modeling complexity makes parameter identification difficult. Thus, it is important to investigate differences in the response of the human body with these three different vehicle models and take a decision on the appropriate model for this analysis. This is what has been attempted in this work.

A previously developed 12 DoF multibody human body model and a non-linear cushion contact force model [8] are integrated with full, half and quarter car models through a multi compression seat suspension system. The seat suspension parameters are obtained for the three different integrated human-vehicle models using a genetic algorithm (GA) by minimizing the SEAT factor. Both random and bump road profiles have been studied. The effect of different vehicle models on the outcome (SEAT value, cushion contact force, head vertical and fore-aft motion) of an integrated system is used to judge and decide on the appropriate model to use for this analysis.

## 2 Mathematical Modelling

Detailed descriptions of the human body and nonlinear cushion contact force models used in this study and their integration with the full vehicle model are presented in previously published work [8, 9]. A 12 degree of freedom (DoF) multibody two-dimensional human body model ($m_1 - m_4$) and a nonlinear seat human contact force model ($F_{cu}$) are integrated with full [10]/half/quarter car vehicle models as represented in Fig. 1 through a seat suspension system. The force developed by the seat suspension system is represented by $F_{SS}$. The human body performance parameters are represented by cushion contact force $F_{cu}$, head fore-aft ($\ddot{x}_4$) and head vertical ($\ddot{z}_4$) accelerations. The road input to the vehicle model occurs at $z_{r1} - z_{r4}$. The Newtonian method is used to formulate the equations of motion. The human body model and nonlinear cushion models are validated using experimental data. The response function for the human body model was direct vertical seat to head transmissibility and for the cushion model it was contact force corresponding to three different levels of excitation [8]. The road disturbance input is at $z_{r1} - z_{r4}$ and each human body mass ($m_1 - m_4$) has three degrees of freedom (vertical fore-aft and rotational). The human body masses are interconnected using rotational and translational springs and dampers. For the quarter car model, only one fourth of chassis mass and vehicle suspension system are engaged (Fig. 1a). For half car vehicle model, only one side of the vehicle suspension and tire systems is considered. For chassis, only half of the mass and inertia are taken into account (Fig. 1b).

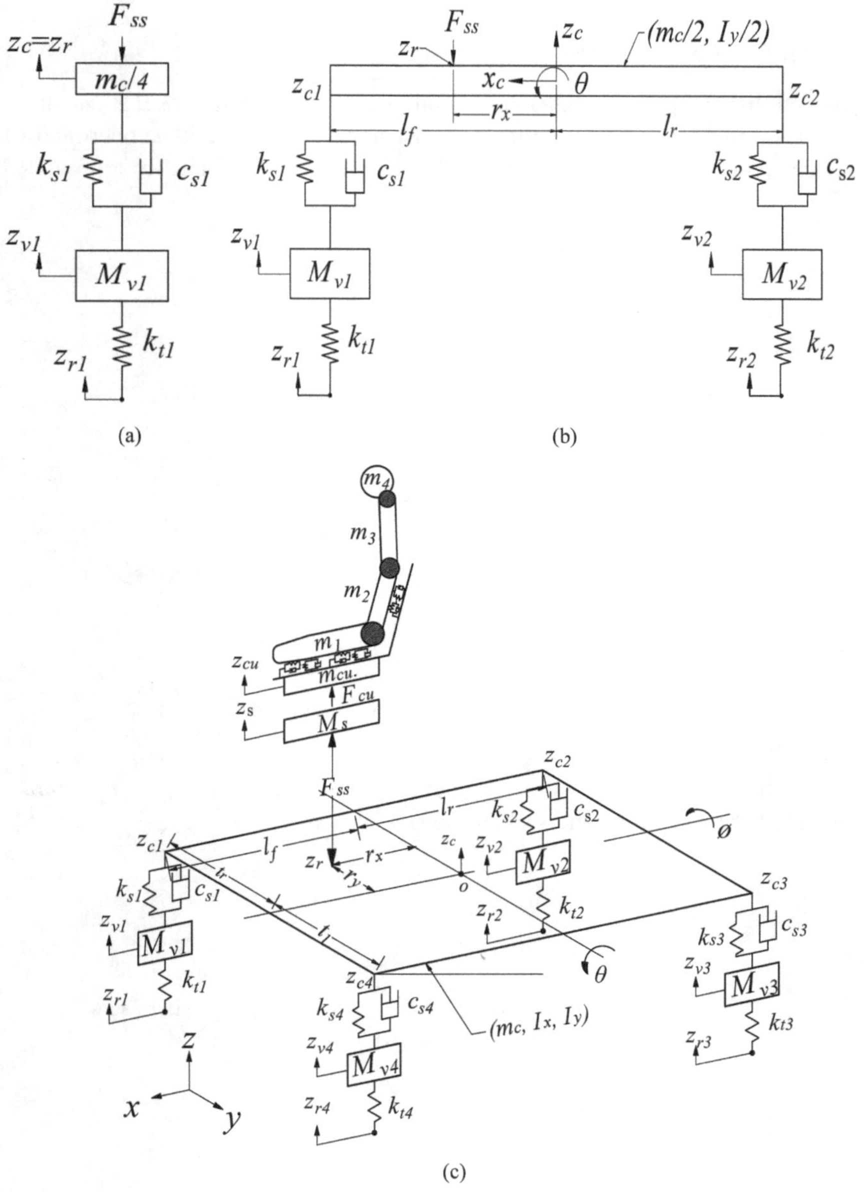


**Fig. 1.** Model representation: (a) quarter car (b) half car (c) 21 DoF full car vehicle human model integration.

## 3 Seat Suspension System

The suspension system consists of various combinations and configurations of spring, damper and linkage systems. Spring and damping coefficients are key factors involved in the attenuation of vibrations. It is important to have optimal values of spring and damper coefficients to provide best possible comfort over a wide range (frequency and amplitude) of road disturbances.

### 3.1 Seat Suspension Configuration

A coupled linear spring parallel to a multi-stage damper inclined at an angle is investigated in this work. The optimum suspension parameters and damper inclination angle are obtained. The configuration of seat suspension system is represented in Fig. 2.

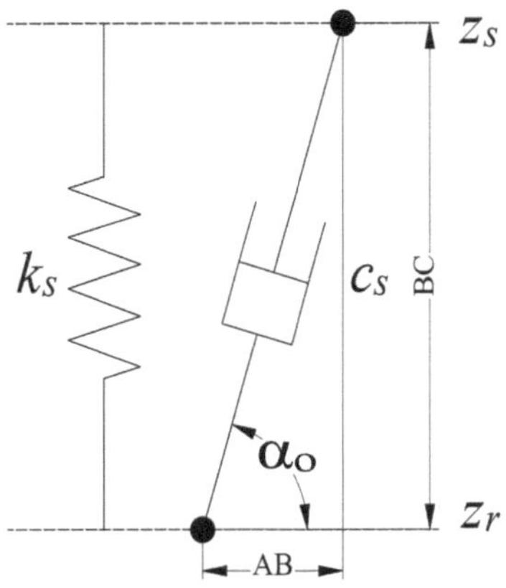


**Fig. 2.** Seat suspension configuration.

The vertical distance between damper mounting points (*BC*) is 220 mm. The vertical force developed by the seat suspension system is given by:

$$F_{susp} = k_s(z_r - z_d) + F_{damp,i} \tag{1}$$

$$F_{damp,i} = \begin{cases} c_{s1} v \sin^2 \alpha_s & 0 \le v \sin \alpha_s \le v_1 \\ [c_{s1} v_1 + c_{s2}(v \sin \alpha_s - v_1)] \sin \alpha_s & v \sin \alpha_s > v_1 \\ c_{s3} v \sin^2 \alpha_s & v_2 \le v \sin \alpha_s < 0 \\ [c_{s3} v_2 + c_{s4}(v \sin \alpha_s - v_2)] \sin \alpha_s & v \sin \alpha_s < v_2 \end{cases} \tag{2}$$

Here:

$$\alpha_s = \tan^{-1}\left[\tan\alpha_0 - \frac{z_r - z_s}{BB}\right] \quad (3)$$

### 3.2 Performance Evaluation

The integrated human-cushion-vehicle system is simulated using MATLAB-SIMULINK. The suspension design parameters are optimized to maximize comfort. The objective function formulated in the optimization process for performance estimation of seat suspension was based on seat effective amplitude transmissibility (SEAT) factor with the excitation reflecting different road disturbances. Suspension system parameters were optimized for minimizing the SEAT value. The mathematical expression is:

$$SEAT = \frac{\left(\ddot{z}_{s,w}\right)_{RMS}}{\left(\ddot{z}_{r,w}\right)_{RMS}} \quad (4)$$

$\ddot{z}_{s,w}$ represents frequency-weighted acceleration. Frequency-weighting $W_k$ [11] was applied to account for human sensitivity to vibration. The optimization objective function is summarised as follows:

$$\text{To minimize:} f = SEAT \ (5)$$

$$\text{Subject to: } (z_r - z_s)_{max} - (z_r - z_s)_{min} \leq z_{rs,allow.}$$

The maximum allowable seat travel $z_{rs,allow.}$ was assumed to be 45 mm [12].

## 4 Road Input Modelling

Automobiles moving on a random, irregular-profiled road give rise to vibration generated due to tire-road interactions. These are harmful for the occupants' health and comfort. The disturbances observed by the occupant when a vehicle passes over a bump or pothole is different from that observed while driving over an inconsistent uncertain road. The primary task of a better seat suspension system is to ensure good ride comfort and minimize the vehicle body roll, pitch and yaw motion for a variety of road conditions and vehicle manoeuvres.

### 4.1 Random Road Disturbance

Road roughness is described by displacement profile along the path over which vehicle tyre moves [13]. The differential equation for stationary road roughness is given by Eq. (6) and is simulated in MALAB-SIMULINK. Figure 3 illustrates the random road profile corresponding to grade C [14, 15] and vehicle moving at a constant speed of 72 km/h using white noise.

$$\dot{q}(t) + 2\pi n_0 v q(t) = \sqrt{2\pi G_r v}\omega(t) \tag{6}$$

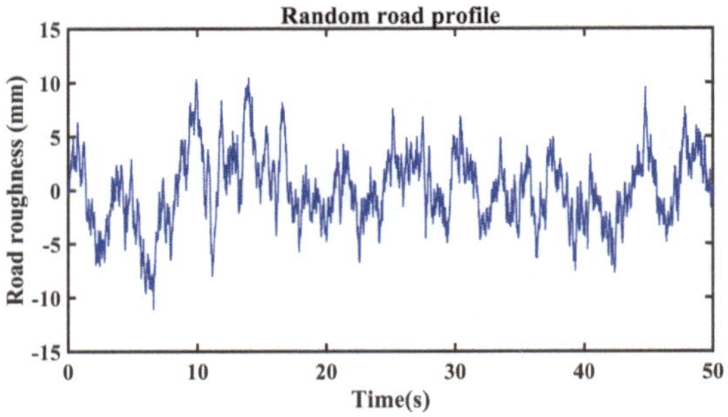


**Fig. 3.** Random constant velocity road profile.

### 4.2 Bump Road Profile

The seat suspension performance is examined when a vehicle moving at a constant speed of $v = 15\,\mathrm{km\,h^{-1}}$ is subjected to a single bump road disturbance as revealed in Fig. 4. This road profile is employed to the fore and rear wheels, with the same peak amplitude but a time delay of $(l_f + l_r)/v$ [16]. Here $v$, $l_f$ and $l_r$ represent vehicle speed and distance of front end and rear end from centre of gravity of chassis.

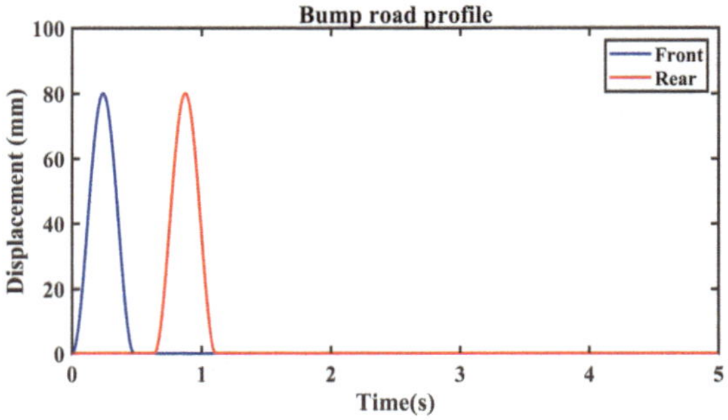


**Fig. 4.** Bump road profile.

## 5 Results and Discussion

Seat suspension parameters are obtained for full, half and quarter car models in accordance with Sect. 3.2 using genetic algorithm optimization technique. The 2D human body model is integrated with full, half and quarter car models to investigate the suspension and human body dynamics in a random road profile. The seat suspension optimization parameters for C-class random road profile obtained by minimization of SEAT factor with constraint on suspension travel are reported in Table 1.

**Table 1.** Seat suspension system results using a random road profile (C-class).

| | Optimized value | | |
|---|---|---|---|
| | Full car | Half car | Quarter car |
| $k_s(N/m)$ | 10578.19 | 10507.08 | 10507.00 |
| $c_{s1}(Ns/m)$ | 2648.09 | 3112.39 | 741.02 |
| $c_{s2}(Ns/m)$ | 2397.45 | 120.34 | 997.83 |
| $c_{s3}(Ns/m)$ | 2963.15 | 3397.09 | 929.48 |
| $c_{s4}(Ns/m)$ | 1469.68 | 394.31 | 960.00 |
| $\alpha_0(°)$ | 21.03 | 26.90 | 26.87 |

Comparative analysis of human body related parameters for half and quarter car models are listed in Fig. 5. The quarter car model shows lower SEAT value and higher suspension travel compared to full and half car models. The acceleration $(\ddot{z}_{s,w})$ at seat is higher for the quarter car model compared to the full-half car model as road disturbance gets directly transferred to the base of the seat. Consequently, higher head acceleration is observed for quarter car vehicle models compared to full and half car models whereas contact force is almost similar as seen in Fig. 5.

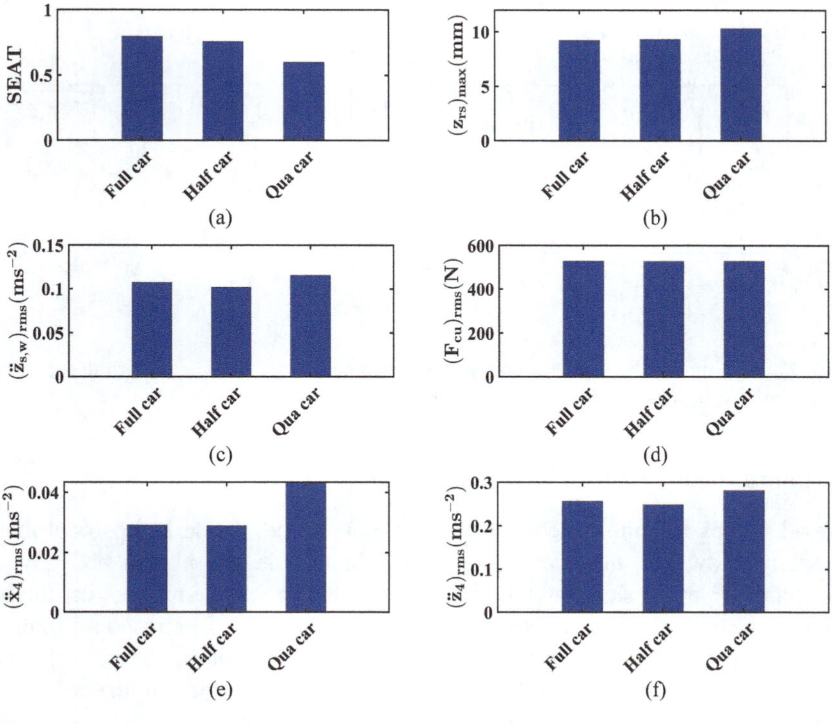


**Fig. 5.** Seat suspension and human body performance assessment parameters for random road profile (Road class C-units): (a) SEAT value, (b) suspension travel, (c) seat wt. RMS accel., (d) cushion contact RMS force (e) head fore-aft and (f) head vertical RMS acceleration.

This variation is observed due to the integration of the whole human body model with a quarter/half system of the vehicle. The total mass of the human body in full/half/quarter vehicle system remains the same whereas the mass of the vehicle system differs, thus altering the total integrated system dynamics. The quarter vehicle model is seen to be underestimating the SEAT value significantly. Thus, it is recommended that the full vehicle model be used for human-vehicle system analysis.

## 5.1 Random Road Profile

The fluctuations in the seat and human body and seat dynamic parameters are presented in Fig. 6. Half and full car models are able to give consistent response for head vertical and seat acceleration while the quarter car model is not able to match the results of the others. Although RMS of head vertical acceleration is higher with the quarter car model, the peaks in acceleration response are lower. Similar results are seen in seat response of acceleration.

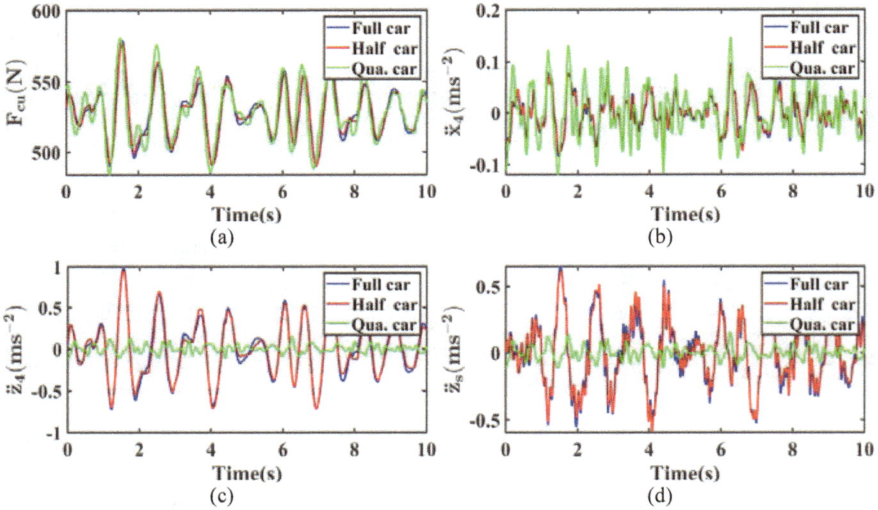


**Fig. 6.** Fluctuations in the seat and human body parameters for bump road profile: (a) cushion contact force (b) head fore-aft (c) head vertical accel. (d) seat acceleration.

## 5.2 Bump Road Profile

The road bumps are often used to control vehicle speed. These bumps contribute to road safety. However, they cause high peaks in vehicle and human body dynamic parameters. The bump shape and the speed with which a vehicle moves over the bump influences the discomfort perceived by the human body. The time response in the seat and human body parameters on a single bump are presented in Fig. 7. As with random road profile, half and full car model results are close to each other but the head vertical and seat acceleration of quarter car model is quite different.

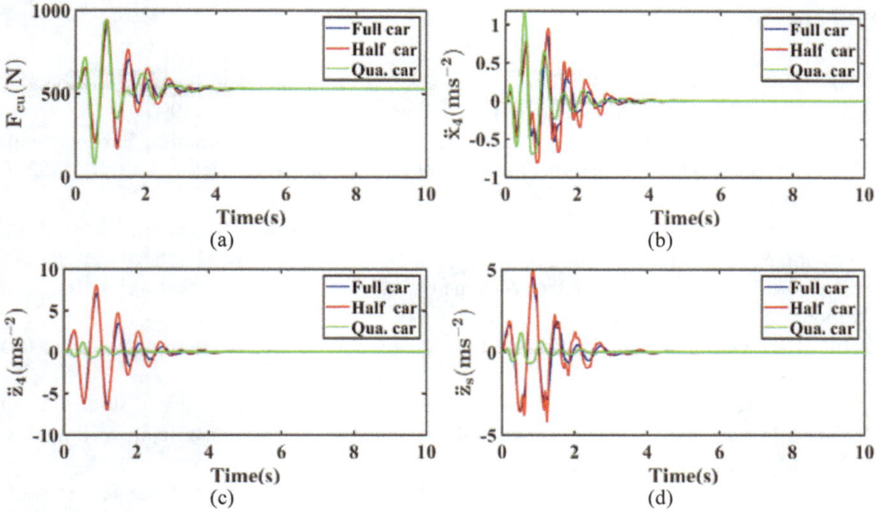


**Fig. 7.** Variation in seat and human body parameters for random road profile: (a) cushion contact force (b) head fore-aft (c) head vertical accel. (d) seat acceleration.

## 6 Conclusions

This work attempts to answer the question: which amongst models - full, half and quarter car - is suitable for analysing rider comfort for random and bump road profiles. A comparison of these models when the human body and seat cushion models have been integrated with them indicates that the quarter car model has lower SEAT value and higher suspension travel compared to full and half car models. The acceleration value at the seat is also higher as road disturbances get directly transferred to the base of the seat. Higher head acceleration value is also observed for the quarter car model even though the contact force value is almost similar. This is probably because the total mass of the human body in full/half/quarter vehicle system remains the same whereas the mass of the vehicle system differs, thus altering the total system dynamics.

Thus, we recommend that the use of a quarter car model be avoided for the analysis of human comfort in a human-seat-vehicle integrated model. The half car model seems to be good enough for most situations and may be used in place of the full car model to reduce computational complexity unless the road/drive conditions generate significant rolling motion in the vehicle.

Some driving conditions, e.g. a hilly terrain with embankments and hairpin bends, may generate significant rolling and yawing motion in the vehicle. What level of change in road and driving conditions create a significant difference between the results predicted by the half and full car models is an interesting question. This may be addressed by future research in this domain.

## References


1. Magnusson, M., Pope, M., Wilder, D., Hansson, T.: Vibrations as the cause of low back pain disorders. Professional drivers are at risk. Lakartidningen **92**, 1711 (1995)
2. Gad, S., Metered, H., Bassuiny, A., Abdel Ghany, A.M.: Ride comfort enhancement of heavy vehicles using magnetorheological seat suspension. Int. J. Heavy Veh. Syst. **22**, 93–113 (2015)
3. Diyana, M.Y.A., Karmegam, K., Shamsul, B.M.T., Irniza, R., Vivien, H., Sivasankar, S., Syahira, M.J.P.A., Kulanthayan, K.C.M.: Risk factors analysis: Work-related musculoskeletal disorders among male traffic policemen using high-powered motorcycles. Int. J. Ind. Ergon. **74**, 102863 (2019)
4. Hill, T.E., Desmoulin, G.T., Hunter, C.J.: Is vibration truly an injurious stimulus in the human spine? J. Biomech. **42**, 2631–2635 (2009)
5. Bovenzi, M., Hulshof, C.T.J.: An updated review of epidemiologic studies on the relationship between exposure to whole-body vibration and low back pain. J. Sound Vib. **215**, 595–611 (1998)
6. Patil, M.K., Palanichamy, M.S., Ghista, D.N.: Man-tractor system dynamics: towards a better suspension system for human ride comfort. J. Biomech. **11**, 397–406 (1978)
7. Mansfield, N.J.: Human Response to Vibration. CRC Press, Boca Raton (2004)
8. Desai, R., Guha, A., Seshu, P.: Modelling and simulation of an integrated vehicle-driver vibrational system using multibody human and nonlinear cushion contact force model-Under review. Simul. Model. Pract. Theory. (2020)
9. Desai, R., Guha, A., Seshu, P.: Multibody biomechanical modelling of human body response to direct and cross axis vibration. Procedia Comput. Sci. **133**, 494–501 (2018)
10. Zuo, L., Nayfeh, S.A.: Structured H2 optimization of vehicle suspensions based on multi-wheel models. Veh. Syst. Dyn. **40**, 351–371 (2003)
11. Vibration, M.: Shock—Evaluation of Human Exposure to Whole-Body Vibration—Part 1: General Requirements, ISO Stand. 2631 (1997)
12. Wang, Y.H., Shih, M.C.: Design of a genetic-algorithm-based self-tuning sliding fuzzy controller for an active suspension system. Proc. Inst. Mech. Eng. Part I J. Syst. Control Eng. **225**, 367–383 (2011)
13. Zhang, L.-J., Lee, C.-M., Wang, Y.S.: A study on nonstationary random vibration of a vehicle in time and frequency domains. Int. J. Autom. Technol. **3**, 101–109 (2002)
14. Liang, C.-C., Chiang, C.-F., Nguyen, T.-G.: Biodynamic responses of seated pregnant subjects exposed to vertical vibrations in driving conditions. Veh. Syst. Dyn. **45**, 1017–1049 (2007)
15. Nigam, N.C., Narayanan, S.: Applications of Random Vibrations. Springer, Cham (1994)
16. Du, H., Li, W., Zhang, N.: Integrated seat and suspension control for a quarter car with driver model. IEEE Trans. Veh. Technol. **61**, 3893–3908 (2012)

# Efficient FEM Based Optimization of a Parallel Robotic System for Upper Limb Rehabilitation

Doina Pisla[1], Nicoleta Pop[1(✉)], Bogdan Gherman[1], Ionut Ulinici[1], Iosif Luchian[1], and Giuseppe Carbone[1,2]

[1] CESTER, Technical University of Cluj-Napoca, Cluj-Napoca, Romania
{Doina.Pisla,Nicoleta.Pop,Bogdan.Gherman, Iosif.Luchian}@mep.utcluj.ro, Ionut.Ulinici@omt.utcluj.ro

[2] DIMEG, University of Calabria, Rende, Italy
Giuseppe.Carbone@unical.it

**Abstract.** Cardiovascular stroke (CVS) is one of the leading causes of motor disabilities worldwide, and unfortunately the number of cases keeps increasing, and will continue to increase until personnel shortages will make the motor rehabilitation procedure to be more challenging. The main solution for this is the automation of the rehabilitation process through the use of robotic technologies capable of providing high dosage and intensity training with minimal interference from the kinesiotherapy specialist. In this paper, the authors present a parallel robotic solution for the rehabilitation of the wrist joint. FEM based simulations are carried out on the most stressed/strained components to identify the reaction forces acting on them during the execution of a rehabilitation exercises. Furthermore the mechanical structure of the targeted components is optimized and placed under FEM analysis again to demonstrate the improvements that have been brought, while tests in medical environment are presented to validate the rehabilitation robotic system.

**Keywords:** Upper limb rehabilitation · Parallel robot · Robotic rehabilitation · FEM · Siemens NX

## 1 Introduction

The robotic rehabilitation field has seen increased attention within the last few decades, despite the first International Conference on Rehabilitation Robotics being held in 1989. This continuous increase in interest towards the kind of robotic devices used mostly in the rehabilitation of patients that suffer from neurological disorders, has been mostly a response to the unavoidable personnel crisis that will occur within the physical rehabilitation field. The said crisis, refers to the fact that by 2030 [1], due to the ever increasing elderly population (a direct result of increasing quality of life), which is the main target group for stroke occurrence, the rate of appearance of qualified medical rehabilitation personnel on the market won't be able to keep up with the demand.

As a result of the incoming challenges in rehabilitation, a more logical and efficient solution would be the automation of the physical exercises that patients need to execute


E.-C. Lovasz et al. (Eds.): MTM&Robotics 2020, MMS 88, pp. 517–532, 2021.
https://doi.org/10.1007/978-3-030-60076-1_47

in their post-stroke stage. Given the simplistic nature of the limb motions that need to be executed during therapy, especially during acute post-stroke rehabilitation, it should come as no surprise that introducing robotic rehabilitation devices within the process, would provide several advantages over classic human-assisted rehabilitation, both for the patient and especially for the therapist.

First and foremost, the implementation of rehabilitation robots (RR) would greatly reduce the physical stress that is placed upon the kinetotherapist during the rehabilitation procedure, as the physician is responsible with manually manipulating the afflicted limbs of the patient (which in the acute stage of rehabilitation, can act as dead weights). Not only does this aspect place unwanted fatigue upon the specialist, but due to the human body's biological limitations, continuous physical effort does decrease in time the efficiency of manipulation and ability to concentrate of the therapist. Secondly, the increasing number of patients will also mean that it will be impossible for the therapist to cater to every patient's need for rehabilitation (one human can only treat one patient at a time).

A review of neurorehabilitation robotics, showcases several such devices for upper and lower limb rehabilitation, [2]:

- ACT-3D [3], is a rehabilitation device that targets the shoulder and elbow joints, and executes task specific reach and payload manipulation training, through VR, haptic and auditory robot to patient interaction, the device requires as input Kinematic data and force parameters.
- ARM GUIDE [4], is also a shoulder/elbow rehabilitation device that employs task specific training, through passive, active assisted and active resistance intervention methods. Patient interaction is VR-assisted and haptic, while as measurement parameters, kinematic data, force, smoothness and accuracy are within its capabilities
- ARMEOSPRING [5], is a shoulder, elbow and forearm rehabilitation device, with task specific training and passive/gravity compensation intervention methods. The device uses VR and haptic interaction methods with the patient, and registers kinematic data and force grasping patterns.
- SUPINATOR-EXTENDER [6], is a forearm and wrist rehabilitation device, with task oriented training programming, active assisted interaction methods, VR and haptic patient interaction, that registers kinematic data and force parameters.
- ASPIRE [7], is a parallel robotic system that can provide rehabilitation training tasks for shoulder. The robotic structure of ASPIRE is based on a spherical mechanism.
- RAISE [8] is a parallel robotic system dedicated for the lower limb rehabilitation of bed-ridden patients. The device is meant for the acute rehabilitation of all the lower limb joints, and can serve a large range of people, with different anthropometric dimensions.

As a result of the aforementioned, our team has decided to introduce a novel robotic upper limb rehabilitation device, and to demonstrate through the finite element method (FEM) the viability of the system in providing safe and efficient wrist rehabilitation with minimum part wear.

The ParReEx robotic system is a parallel modular robot used for upper limb rehabilitation. The module dedicated for elbow rehabilitation focuses on flexion/extension and pronation/supination motions and the module for wrist rehabilitation executes the flexion/extension, respectively the adduction/abduction [9–11]. As it is mentioned above, the robotic device adopts a parallel structure. This has been chosen due to the increased precision that parallel devices present when relating to serial structure, and reduced loads to which the components of the device are subjected to (as a result of the parallel architecture), which increase the safety factor to which the patient is subjected when using one such device. The device itself has been developed to be able to reproduce the basic motions of the human wrist, in the case of patients that have little to no control of their affected limb, and present increased frailness of the body, as such ParReEx – wrist is capable of reproducing it's targeted motions at low velocities and with increased smoothness thus eliminating the prospect of inducing unwanted forces within the patient's hand or wrist during operation.

The aim of this paper is to analyze the design of the robotic part which supports the weight of the hand, considered to be with the largest mechanical load.

In this paper, the authors analyzed the static behavior of one module of a parallel robotic system used for upper limb rehabilitation in order to identify the optimal design solutions for the more heavily strained components of the system.

The finite element method (FEM) is a numerical technique used to perform finite element analysis (FEA) in both static and dynamic conditions for any structure or any given physical phenomenon [12] including human-mounted robotic orthoses [13].

## 2 The ParReEx-Wrist Robot for the Medical Recovery of the Wrist Joint

Figure 1 presents the ParReEx – wrist rehabilitation robotic system [14]. The patient's hand (and wrist) is positioned above the robot circular rail (Fig. 2), which leads to a more compact design and reduced dimensions, as one of the target design criteria. The inverse kinematics is straightforward, as shown in Eq. (1) and (2).

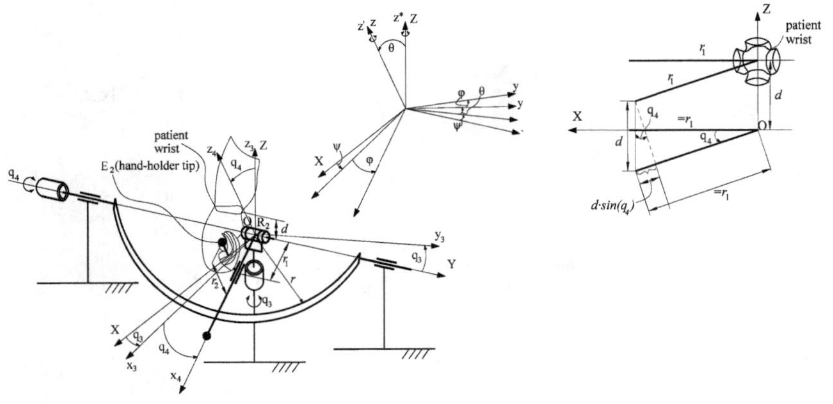


**Fig. 1.** ParReEx-wrist rehabilitation robot: a. kinematic scheme; b. position during rehabilitation

$$q_4 = \pm \text{asin}\left(\frac{r_1 + \sqrt{4 \cdot d \cdot Ze_2 + r_1^2}}{2d}\right) \tag{1}$$

$$q_3 = \frac{Ye_2}{\cos(q_4) \cdot (r_1 - d \cdot \sin(q_4))} \tag{2}$$

This architecture yields a singularity configuration (Fig. 2) [15], but this position is out of the operational workspace of the robot, limited by the natural motion limits presented in [13].

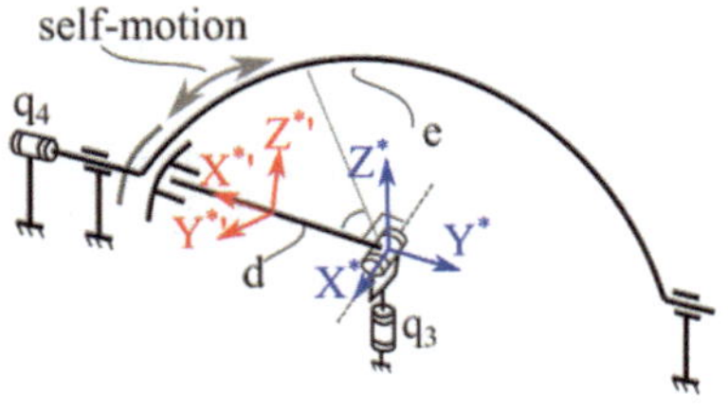


**Fig. 2.** ParReEx wrist – singular pose

Figure 3 presents the initial CAD model of the ParReEx – wrist robot [9] for the medical rehabilitation of the adduction-abduction and flexion-extension motions. The two servomotors are placed inside the box acting as a base for the circular rail and the hand support. The motion transmission is achieved using bevel gears for the flexion-extension motions and a timing belt for adduction-abduction.

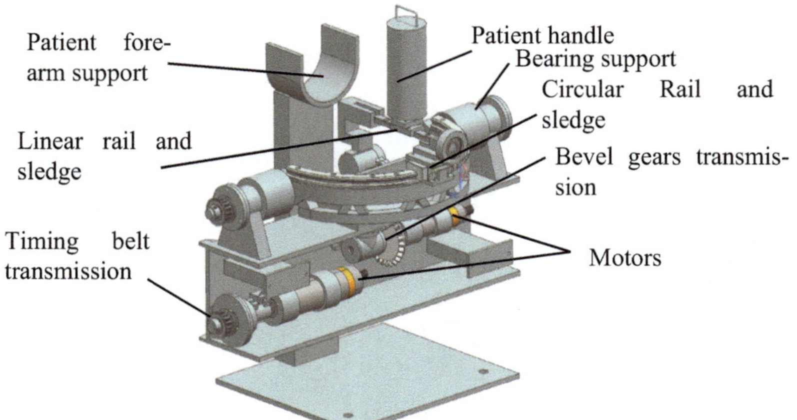


**Fig. 3.** ParReEx – wrist robotic system for wrist motions rehabilitation: flexion/extension and adduction/abduction – initial CAD design

During operation, the patient's hand is safely secured to a joystick/handle placed on a rail directly connecting the circular guide (used for flexion/extension) and the rotation joint responsible for guiding the arm (during adduction-abduction). As a result of this the weight of the hand is sustained from multiple anchors, which greatly reduces the stress placed upon the system's components. It's also necessary to mention that the weight of the arm is also picked up by a separate anchor that has no tie in to the module used for the reproduction of the wrist motions and manipulation of the hand. Consequently all of this means that the only forces that are actively being transmitted through the system's components are those induced by the motors and those induced by the human hand which on average weights under 1 kg for people with bodyweights between 70 and 120 kg as shown in [7].

Most of the elements have been designed to be made of plastic (ABS). In addition, steel components may be used, mainly the circular guide and the ball bearings (THK, type HCR12A), as well as the linear guide and the slider from TBI Motion and a series of bearings of different sizes. The weights of the assembly elements (in most cases screws with a 3 mm nominal diameter head) have not been taken into account. In the FEM simulations performed, only a series of parts subjected to high stresses were considered and whose geometry can lead to unusually large, undesirable deformations. Based on the FEM analysis, a structural optimization of the parts will be carried out later for maximum rigidity under the conditions of a minimum mass.

## 3 FEM Analysis

The FEM analysis presented in this paper focuses on the optimization of several critical components of the robotic structure. Figure 4 presents the design flowchart main steps approached in this paper.

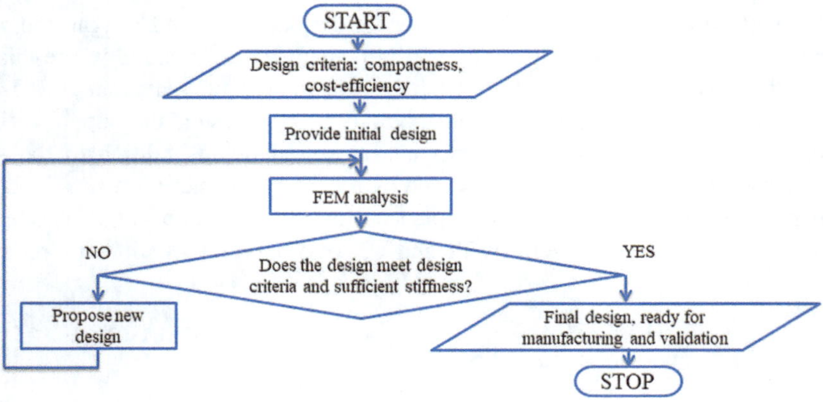


**Fig. 4.** ParReEx – wrist design flowchart

The authors have chosen FEM analysis since the accelerations used during rehabilitation exercises are small (roughly $1°/s^2$ for both flexion/extension and adduction/abduction motions), so the robot operates in the quasi-static domain, which means that a static analysis can be successfully applied, considering the most disadvantageous configurations of the robot. The authors have used FEM analysis to obtain rigid and cost-effective plastic components (as the strategy is to build them in-house). The Fig. 5 presents the initial version of this component, made entirely of plastic material (ABS-M30) using the Stratasys Fortus 380mc production system. Its role is to rigidly support the guide rail (HCR12A-60-100R) on which the slide will slide (HCR12A-60-100R-UU - Fig. 6).

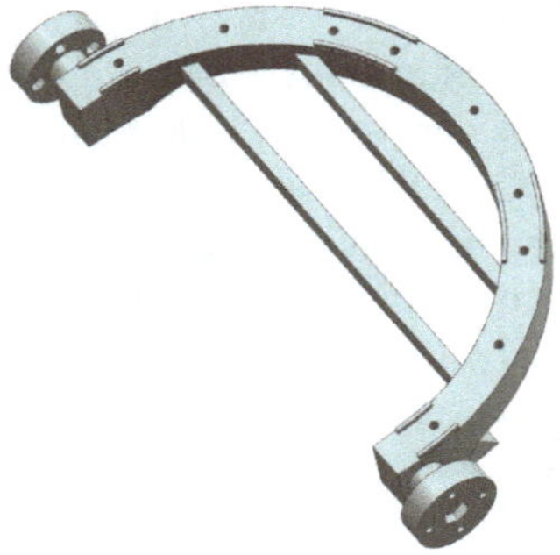

**Fig. 5.** The circular guide support of the ParReEx-wrist module for wrist flexion/extension

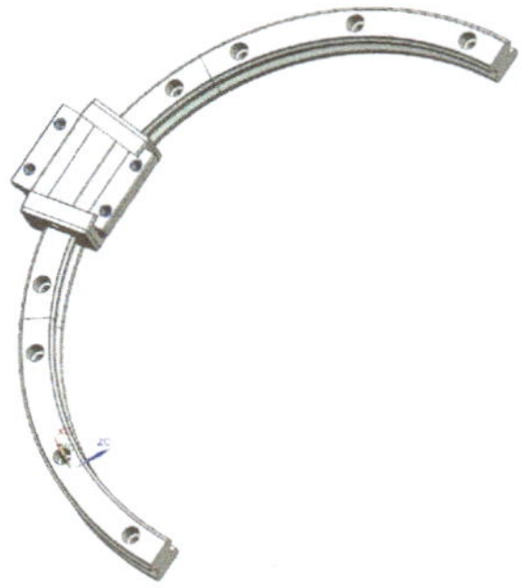

**Fig. 6.** The guide rail and sled

Figure 7 presents the FEM analysis of this component, using Siemens NX. An estimated (using Siemens NX RecurDyn) concentrated force F1 = 11 N was applied (the force was determined by estimating the weight of the other elements of the robot that the part must support, with the addition of the patient's hand weight, the acceleration the robotic system being negligible) in point A, in the position where this part sustains the most stress. The result shows a deformation (arrow) maximum of 0.52 mm in point A, which is considered as unacceptable. As a result, the support will be modified by adding a structure for mechanical stiffening. In addition to the strain induced from the vertical plane (represented by force F1), the piece is also subjected to a (much smaller) force in the horizontal plane (in the BB' direction - Fig. 7). To cancel the effect of this force, the elements T1 and T2 were provided as stiffening rods.

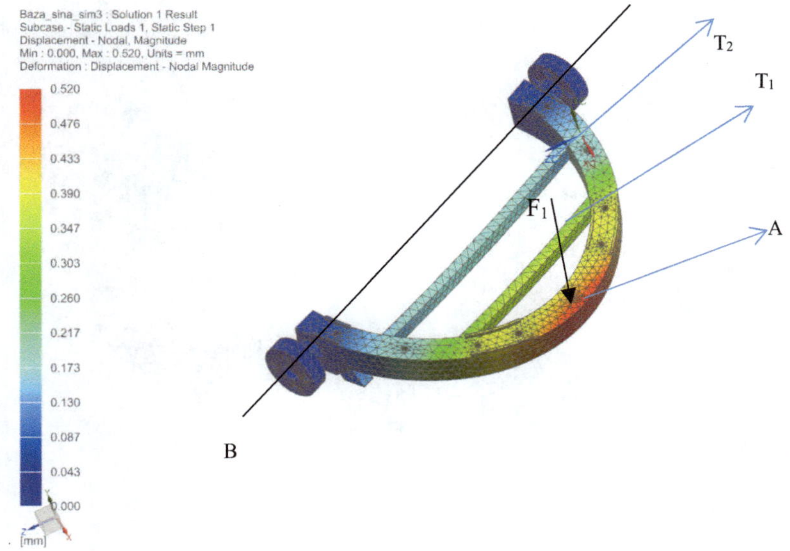


**Fig. 7.** The circular guide support of the ParReEx-wrist

In order to reduce the size of the piece deformation, an attempt was made to create a beam-like structure with lattices (Fig. 8). The FEM analysis shows a reduction of the deformation to 0.42 mm, which is not a substantial enough reduction.

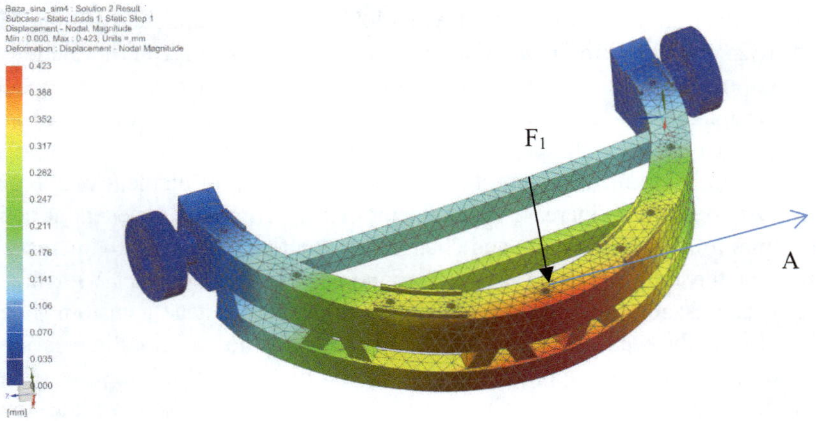


**Fig. 8.** The circular guide support with a lattice like structure (FEM analysis)

In the next stage of optimization it was decided to increase the thickness of this support, from 15 mm to 25 mm, which obviously should lead to an improvement of the bending strength of this element. In Fig. 9 the result of this simulation is presented.

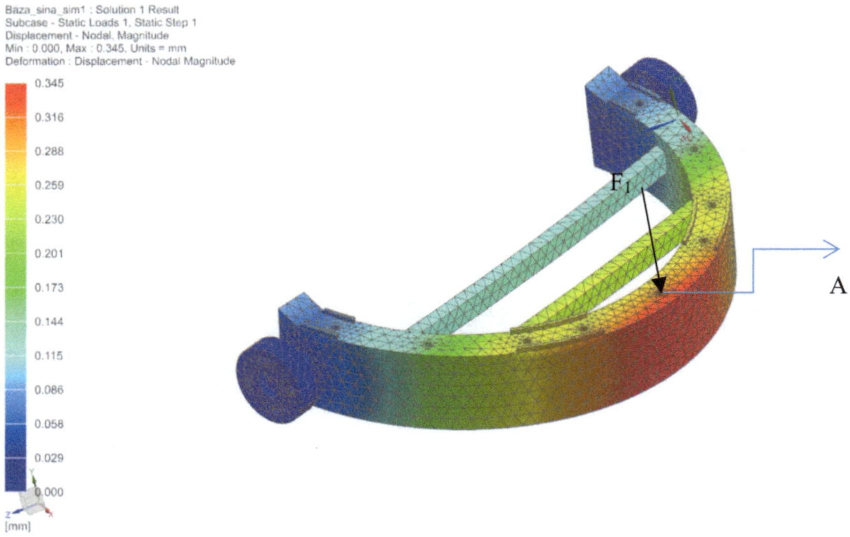


**Fig. 9.** The circular guide support of the ParReEx-wrist robot at a 25 mm thickness (FEM analysis)

As it can be seen, the maximum deformation is 0.345 mm, so an improvement of 150% compared to the initial situation, therefore the component has been considerably optimized at this point.

Figure 10 presents the result of the simulation at a 35 mm thickness of the rail support (circular guide), thus an increase of 140%. The maximum deformation is in this case of 0.325 mm, resulting in a reduction of about 6%, insignificant in relation to the percentage increase of the thickness of the element. Consequently, it was decided to return to a thickness of 25 mm, which ensures an acceptable deformation of the element, negligible for the robot's load.

In order to study further the deformability of the support element of the circular guide, it was decided to increase the diameter of the cylindrical fasteners at their end from 15 mm to 25 mm (Fig. 11) and change the structure of the arc, by increasing the thickness at the top, where the circular support comes into contact with the slider (Fig. 12). The usage of a guide rail was also removed and instead a custom slider was developed, mainly because of cost efficiency reasons. The custom slider is in direct contact with the circular support and achieves the motion through the use of several metal roller wheels that enter in direct contact with the circular support that now also fulfills the role of a guide rail. It was also discovered that in this case the reinforcement bars were no longer necessary. The result shows a considerable improvement: a maximum deformability of 0.1104 mm, so a 360% improvement, which shows that the support area of this element is the key to reducing the deformability of the robot element.

For cost-effectiveness reasons, the guide rail HCR12A-60-100R on which slides the (HCR12A-60-100R-UU - Fig. 6 component has been removed entirely and a custom made slider with 7 ball bearings has been designed. It has to fit the in-house designed circular rail and thus assure a smooth execution of the flexion-extension motions rehabilitation.

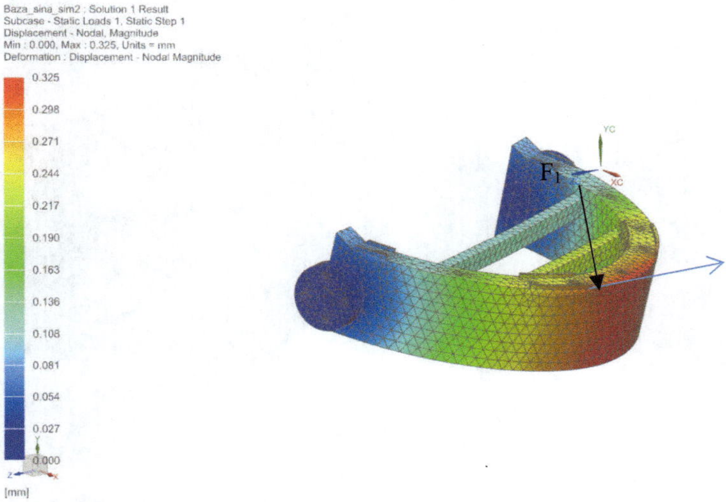


**Fig. 10.** The circular guide support of the ParReEx-wrist robot at a 35 mm thickness (FEM analysis)

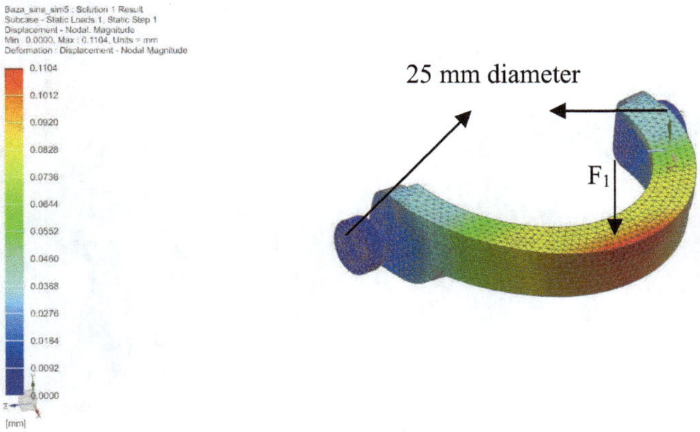


**Fig. 11.** The circular guide support at a thickness of 35 mm and a diameter of the connection elements of 25 mm

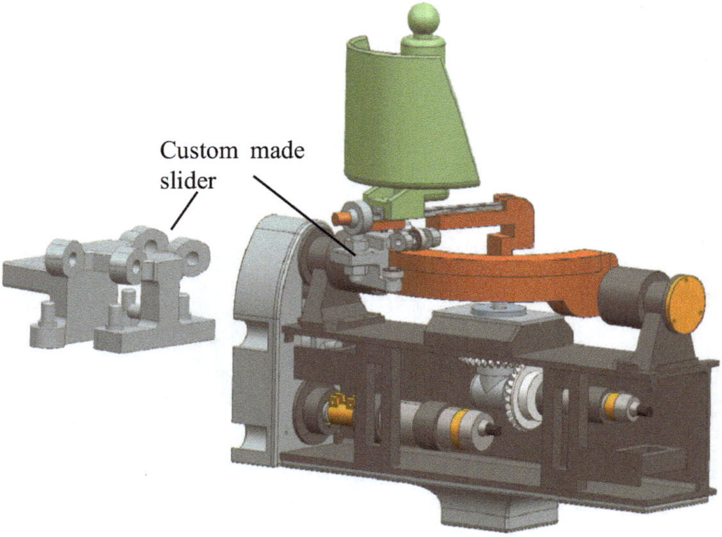


**Fig. 12.** Final CAD design

Another component of interest is the bearing housings that support the support element of the circular guide. The simulation of the two housings is shown in Fig. 13 and 14. This shows a negligible bearing type deformation at a force distributed by F2 = 15 N, on the side of the circular guide and F3 = 10 N on the side of the motor (driven).

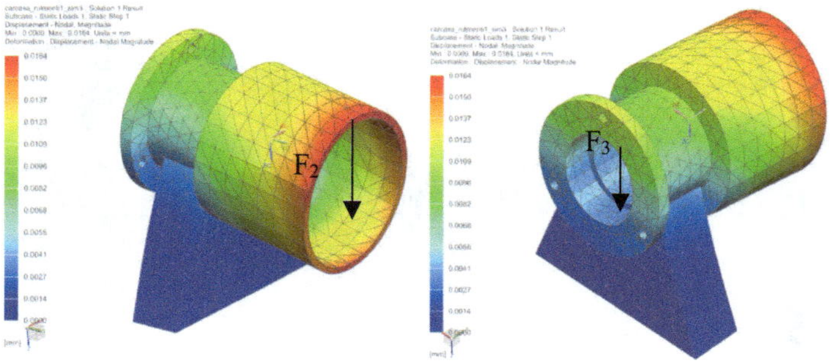


**Fig. 13.** Bearing cover

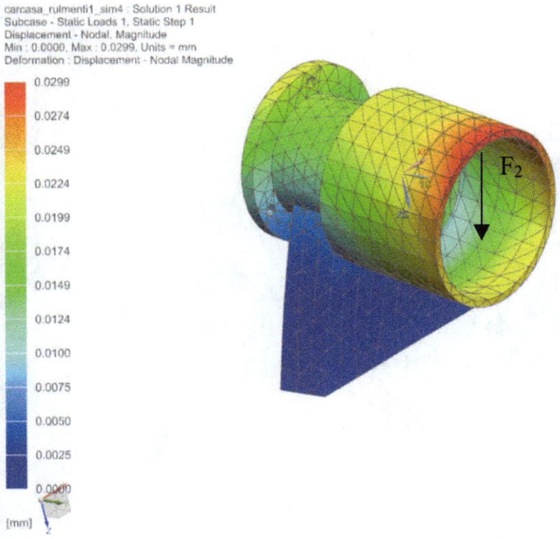


**Fig. 14.** Free bearing support

Another part undergoing major stress is the one that supports the weight of the hand and all the other elements (plastic related parts - ABS and bearings), resulting in a total force of F4 = 20 N, evenly distributed. The piece will be fastened using the screws in the robotic module housing. Figure 15a presents the simulation result, with a deformation of approximately 0.02 mm, therefore more than acceptable.

Another piece under increased tension is the one that supports the linear guide under the patient's hand. The maximum expected load is 15 N evenly distributed on the surface shown in Fig. 15b. The simulation result shows a maximum deflection of 0.217 mm, acceptable under the conditions where the other end of the linear guide is supported by the support element of the circular guide.

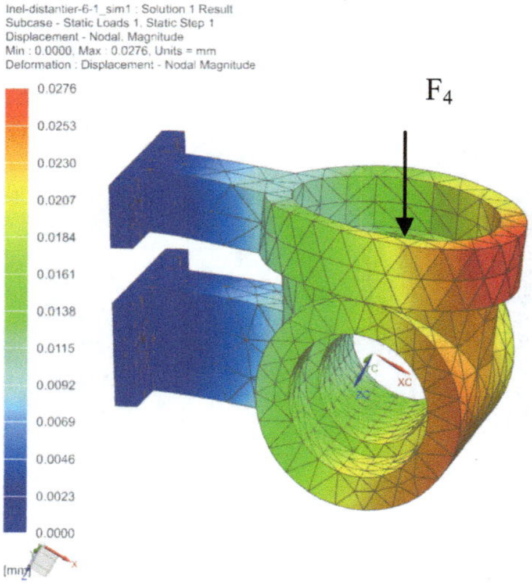


Fig. 15a. FEM analysis of the part

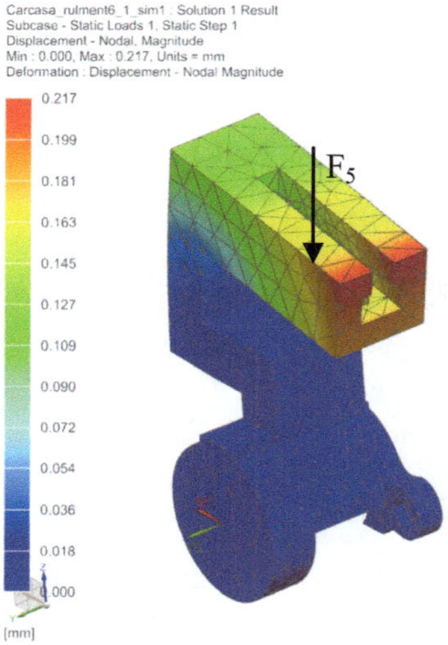


**Fig. 15b.** FEM analysis of the wrist support component

## 4 Experimental Tests

The presented design steps have been used in the development of the robotic system experimental model (Fig. 16). This has been tested using healthy subjects in laboratory and initial set of measurements showed very little deformation with practically no influence in the achievement of the training exercises. Another set of experimental tests [7] have been performed on 18 patients at the Municipal Cluj-Napoca Hospital, Romania, in order to validate the functionality of ParReEx parallel robotic system. All the subjects used for the testing were patients aged between 55 and 76 with various neurological affections and motor-deficit of the right upper limb.

Out of the 18 patients, 8 patients suffered a stroke, 8 suffered from Parkinson's disease, 1 underwent brain surgery and 1 had gouty arthritis. The table also includes the affection phase (meaning time past from initial diagnosis). Each subject was hospitalized for 7 days which means a total time of ParReEx-wrist robotic rehabilitation module (Fig. 16) functioning of 1260 min.

The experimental data is synthetized in the Table 1:

**Table 1.** Experiment subject parameters

| Subject no. | Affection phase [months] | Time on ParReEx-wrist rehabilitation device daily [min] |
|---|---|---|
| Males | | |
| 1 | 2 | 10 |
| 2 | 9 | 10 |
| 3 | 24 | 10 |
| 4 | 6 | 10 |

*(continued)*

**Table 1.** (*continued*)

| Subject no. | Affection phase [months] | Time on ParReEx-wrist rehabilitation device daily [min] |
|---|---|---|
| 5 | 30 | 10 |
| 6 | 6 | 10 |
| 7 | 6 | 10 |
| 8 | 12 | 10 |
| Females | | |
| 9 | 24 | 10 |
| 10 | 12 | 10 |
| 11 | 6 | 10 |
| 12 | 36 | 10 |
| 13 | 12 | 10 |
| 14 | 12 | 10 |
| 15 | 1 | 10 |
| 16 | 18 | 10 |
| 17 | 20 years | 10 |
| 18 | 16 | 10 |

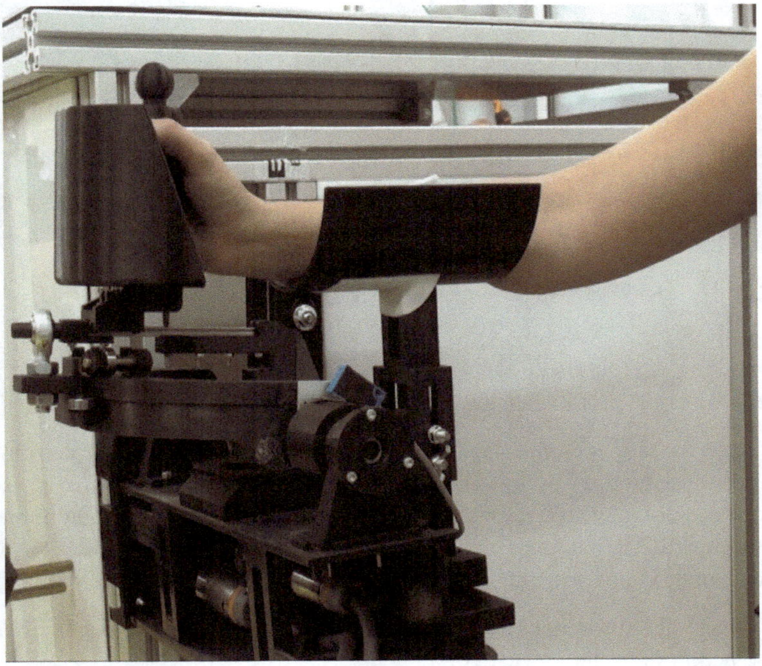

**Fig. 16.** ParReEx wrist rehabilitation system – experimental model

The wrist rehabilitative motions, wrist flexion/extension, respectively wrist abduction/adduction (Fig. 17) were performed with a number of 6 series for each motion with 10 repetitions per motion. The minimum amplitudes values for each wrist motion have been calculated along with the average values and the maximum ones as it is shown in Table 2.

**Table 2.** The amplitudes values for wrist motions

| Rehabilitative motions | Amplitudes values [°] | | |
|---|---|---|---|
| | Minimum | Average | Maximum |
| Wrist flexion | 0 | 3 | 20 |
| Wrist extension | 70 | 99 | 110 |
| Wrist abduction | 20 | 45 | 50 |
| Wrist adduction | 1 | 7 | 10 |

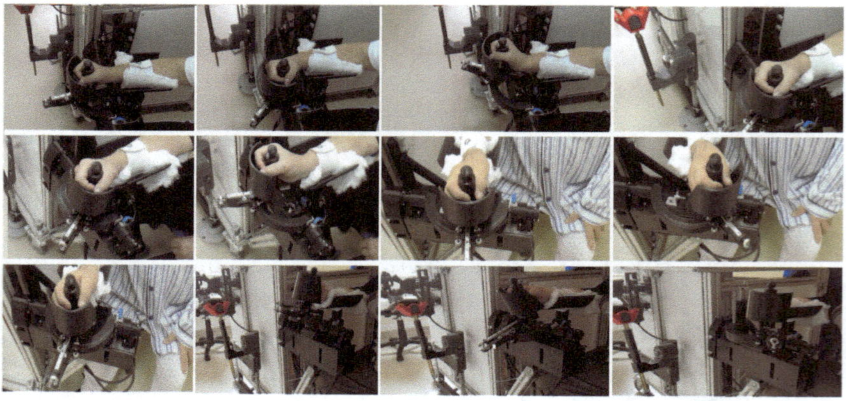

**Fig. 17.** The wrist rehabilitative motions performed with ParReEx wrist system

Following the experimental tests carried out under the conditions presented above, we were able to determine that the ParReEx robotic system:

- Was able to carry out the rehabilitation procedure of the wrist according to the requirements imposed by the kinetotherapist
- Was able to complete the testing phase in a hospital environment without presenting any hazards that could have immerged as a result of mechanical, electrical, thermal, noise, vibration or structural causes.

The device did not present any traces of mechanical wear under the effect of forces that were acting upon the device components.

## 5 Conclusions

This paper focuses on ParReEx-wrist, a parallel robotic system used for the rehabilitation of the wrist. The design of ParReEx requires careful attention regarding the possible deformations, which might occur and affect its operation. Accordingly, the FEM analysis have been carried out at specific configurations and loading conditions, which refer to the worst case scenarios of wrist flexion/extension and abduction/adduction rehabilitation. Results have been used for improving the ParReEx design in order to achieve optimal stress. Experimental tests demonstrate that the optimized design is suitable for the intended wrist rehabilitation tasks.

**Acknowledgment.** The paper presents results from the research activities of the project ID 37_215, MySMIS code 103415 "Innovative approaches regarding the rehabilitation and assistive robotics for healthy ageing" co-financed by the European Regional Development Fund through the Competitiveness Operational Programme 2014–2020, Priority Axis 1, Action 1.1.4, through the financing contract 20/01.09.2016, between the Technical University of Cluj-Napoca and ANCSI as Intermediary Organism in the name and for the Ministry of European Funds.

## References

1. Robotics Multi-Annual Roadmap 2014–2020
2. Oña, E.D., Cano-de la Cuerda, R., Sánchez-Herrera, P., Balaguer, C., Jardón, A.: A review of robotics in neurorehabilitation: towards an automated process for upper limb. J. Healthc. Eng. 9758939 (2018). https://doi.org/10.1155/2018/9758939. PMID: 29707189; PMCID: PMC5901488
3. Ellis, M.D., Sukal-Moulton, T.M., Dewald, J.: Impairment-based 3-D robotic intervention improves upper extremity work area in chronic stroke: targeting abnormal joint torque coupling with progressive shoulder abduction loading. IEEE Trans. Robot. **25**(3), 549–555 (2009). https://doi.org/10.1109/tro.2009.2017111
4. Kahn, L.E., Zygman, M.L., Rymer, W.Z., Reinkensmeyer, D.J.: Robotassisted reaching exercise promotes arm movement recovery in chronic hemiparetic stroke: a randomized controlled pilot study. J. Neuroeng. Rehabil. **3**(1), 12 (2006). https://doi.org/10.1186/1743-0003-3-12
5. Sanchez, R.J., Liu, J., Rao, S., et al.: Automating arm movement training following severe stroke: functional exercises with quantitative feedback in a gravity-reduced environment. IEEE Trans. Neural Syst. Rehabil. Eng. **14**(3), 378–389 (2006). https://doi.org/10.1109/tnsre.2006.881553
6. Allington, J., Spencer, S.J., Klein, J., Buell, M., Reinkensmeyer, D.J., Bobrow, J.: Supinator extender (sue): a pneumatically actuated robot for forearm/wrist rehabilitation after stroke. In: 2011 Annual International Conference of the IEEE Engineering in Medicine and Biology Society; Boston, MA, USA. pp. 1579–1582 (2011)
7. Tucan, P., Vaida, C., Plitea, N., Pisla, A., Carbone, G., Pisla, D.: Risk-based assessment engineering of a parallel robot used in post-stroke upper limb rehabilitation. Sustainability **11** (10), 2893 (2019)
8. Vaida, C., Birlescu, I., Pisla, A., Ulinici, I., Tarnita, D., Carbone, G., Pisla, D.: Systematic design of a parallel robotic system for lower limb rehabilitation. IEEE Access **8**, 34522–34537 (2020)

9. Gherman, B., Carbone, G., Plitea, N., Ceccarelli, M., Banica, A., Pisla, D.: Kinematic design of a parallel robot for elbow and wrist rehabilitation. In: Doroftei, I., Oprisan, C., Pisla, D., Lovasz, E. (eds.) New Advances in Mechanism and Machine Science. Mechanisms and Machine Science, vol. 57. Springer, Cham (2018)
10. Tucan, P., Gherman, B., Major, K., et al.: Fuzzy logic-based risk assessment of a parallel robot for elbow and wrist rehabilitation. Int. J. Environ. Res. Public Health **17**(2), 654 (2020). https://doi.org/10.3390/ijerph17020654
11. Gherman, B., Pisla, D., Plitea, N., Vaida, C., Carbone, G., Pisla, A., Banica, A.: A family of robotic systems for the rehabilitation of the upper limb, Patent no. 132234 (2017)
12. Sahu, S., Choudhury, B.B.: Dynamic behaviour analysis of an industrial robot using FEM. In: Nayak, J., Abraham, A., Krishna, B.M., Chandra Sekhar, G.T., Das, A.K. (eds.) Soft Computing in Data Analytics. AISC, vol. 758, pp. 193–202. Springer, Singapore (2019). https://doi.org/10.1007/978-981-13-0514-6_20
13. Tarnita, D., Pisla, D., Geonea, I., et al.: Static and dynamic analysis of osteoarthritic and orthotic human knee. J. Bionic Eng. **16**, 514 (2019)
14. Birlescu, I., Pisla, D., Gherman, B., Pisla, A., Vaida, C., Carbone, G., Plitea, N.: On the singularities of a parallel robotic system used for elbow and wrist rehabilitation. In: Lenarcic, J., Parenti-Castelli, V. (eds.) ARK 2018. SPAR, vol. 8, pp. 203–211. Springer, Cham (2019). https://doi.org/10.1007/978-3-319-93188-3_24
15. Vaida, C., Carbone, G., Major, K., Major, Z., Plitea, N., Pisla, D.: On human robot interaction modalities in the upper limb rehabilitation after stroke. Acta Tech. Napocensis. Ser. Math. Mech. Eng. **60**(1), 91–102 (2017)

# Automatic Arterial Puncture Sensorial Device for Fast Arterial Blood Gas Sampling from Radial Artery During Covid-19 Pandemic

Bogdan Mocan[1(✉)], Mircea Fulea[1], Mircea Murar[1], Mihai Steopan[1], and Mihaela Mocan[2]

[1] Technical University of Cluj-Napoca, Cluj-Napoca, Romania
bogdan.mocan@muri.utcluj.ro
[2] Iuliu Haţieganu University of Medicine and Pharmacy, Cluj-Napoca, Romania

**Abstract.** The main objective of the current paper was to design an automatic arterial puncture sensorial device to facilitate a fast extraction of arterial blood gas sampling from radial artery during Covid-19 pandemic. COVID 19 (Coronavirus) pandemic has generated surge demand for essential healthcare equipment, medicines along with the requirement for medical examination equipment. Cannulation of the radial artery can sometimes be difficult and require multiple attempts, which prolongs the time required for the procedure and this is not indicated in a COVID-19 pandemic. Within this paper we provide a detailed description of a novel device for fast radial artery puncture, developed for use during the COVID-19 Pandemic, when the contact of medical staff with the infected patient must be minimal. The proposed concept of the device allows the immobilization of the upper limb of the patient in the appropriate position, facilitates the precise identification of the position of the radial artery even in a situation of low peripheral pulse, allows visualization of the subcutaneous area of the vein plane in the analyzed area to avoid puncturing them and maintains needle orientation at a precise angle to the central horizontal plane of the patient's forearm. Thus, the proposed of AUSmart-Astrup concept creates pre-requisites for reducing the time needed for arterial blood sampling, increased the efficiency of the medical procedure, provides ergonomics for the medical staff, and increased safety in the context of COVID-19 pandemic.

**Keywords:** Medical device design · Automatic arterial puncture sensorial device · Arterial blood gas sampling during Covid-19 pandemic · Fast extraction of arterial blood gas sampling

## 1 Introduction and Related Works

COVID 19 (Coronavirus) pandemic has affected almost all countries from the world and has made a significant impact on the available healthcare facilities and treatment systems. There is a requirement for the introduction of various advance technologies to tackle various problems related to this viral pandemic [1]. In Covid-19 pandemic context infection is a major risk for the tandem patient - medical staff. Infections occur mainly through prolonged contact with the infected patient, so the World Health


E.-C. Lovasz et al. (Eds.): MTM&Robotics 2020, MMS 88, pp. 533–542, 2021.
https://doi.org/10.1007/978-3-030-60076-1_48

Organization (WHO) recommendations are to reduce patient - medical staff contact. In this context, all medical procedures must be shortened. Thus, the idea of this device appeared. So, the efficiency of the current procedure of arterial blood sampling is not optimal at the moment, having a relatively high rate of failure depending on the medical skills of the examiner and hemodynamic status of the patient [2].

So, arterial puncture (AP) is an invasive procedure frequently used in adult patients hospitalized for respiratory diseases or metabolic disorders, especially in Emergency Rooms (ER) and intensive care units (ICU). It is the only reliable determination of ventilation success as evidenced by $CO_2$ content. It constitutes a more precise measure of successful gas exchange and oxygenation. ABG sampling is the only way of accurately determining the alveolo-arterial oxygen gradient [3]. Arterial blood gases (ABG) sampling provides valuable information on the acid base balance at a specific point in the course of a patient's illness [4]. AP is usually performed on the radial artery, which is easily accessible because the superficial anatomic presentation, the absence of large veins aside, and with sufficient blood supply through collateral circulation of the hand. This technique must be acquired by al interns, no matter the specialization domain [5].

The main indications for ABG sampling are stipulated in national and international guidelines of many diseases. In brief, they comprise the following: the identification of respiratory, metabolic, and mixed acid-base disorders, measurement of the partial pressures of respiratory gases involved in oxygenation and ventilation; monitoring of acid-base status, as in patient with diabetic ketoacidosis on insulin infusion; assessment of the response to therapeutic interventions such as mechanical ventilation in a patient with respiratory failure; quantification of the levels of dyshemoglobins (e.g., carboxyhemoglobin and methemoglobin) [6].

There are absolute and relative contraindication to AP. The absolute contraindications include [4]:

- An abnormal blood supply through collateral circulation validated by modified Allen test [7]; in this case the recommendation is to use a different AP site;
- The presence of local infection or scars with distorted anatomy at the puncture site (e.g., previous surgical interventions, congenital or acquired vascular malformations, or burns)
- The presence of arteriovenous fistulas or vascular grafts, in which case arterial vascular puncture should not be attempted.

The relative contraindications are referring to systemic disorders such as those associated with high risk of hemorrhage (anticoagulant treatment, thrombocytopenia etc.) or thrombosis [3].

The current procedure consists of puncturing the radial artery with a needle attached to a heparinized syringe: After the patient has been informed and have given his consent (when is possible), the Allen test is performed. Then, the radial artery is identified by palpating the pulse and determining the point of maximum impulse [3] in patients with wrist hyperextension. When the radial artery cannot be identified properly, Doppler ultrasound device might be used to find the optimum puncture site [8]. Holding the arterial blood gas syringe with the dominant hand, aim the needle away from the patient's hand toward the upper arm. The skin is punctured at a 30-to-45-degree angle at a point just below the index and middle fingers of the performer's hand. The needle should be slowly advanced until the syringe easily and passively fills with bright red, pulsating blood. Ideally, at least 1 to 2 cc of blood should be obtained. If no blood is obtained, the procedure is repeated. After the blood sample is collected, the syringe is withdrawn and pressure is applied for at least 5 min to assure correct hemostasis and avoid local hematoma [3, 9]. The sample should be rapidly examined, in less than 15 min, and should not be air contaminated to avoid false results.

There is a diversity of solutions for the ABG sampling from radial artery. Literature research revealed some interesting patents and scientific articles that are described below:

- the development of new syringe and needles - US 4187860 A "Arterial blood collection device"; CN 204500730 U – "Artery blood sampling device"; US 4215702 A – "Arterial blood extraction device"; US 4657028 A – "Blood sampling device"; US 20120215124 A1 – "Non-invasive arterial blood gas determination"; US 5483973 A – "Needle stopper and needle removal device"; US 5947932 A – "Closed system blood sampling device"; EP 2768393 B1 – "Syringe with removable plunger for arterial blood gas sample collection".
- devices meant to ensure the blood sample extraction and radial artery compression:- US 20040039413 A1 "Radial artery compression system"; CN 202078309 U "Wrist support pad for radial artery puncture"; CN 204121123 U "Improved radial artery puncture fixing device"; CN 204133790 U "Radial artery puncture arm support for catheter bed".

In conclusion, most products on the market facilitate the sampling of venous blood, not arterial blood, and none of the above-mentioned patents or devices on the market that facilitate the sampling of radial artery blood do not fully address the medical procedure -integrated approach. By integrated approach we mean: precise identification of the position of the radial artery, visualization of the superficial venous circulation in order to avoid their puncture, guiding and maintaining the orientation of the needle throughout the puncture and, last but not least, ensuring arterial stasis, avoiding bleeding the puncture site, after removing the needle from the artery.

The technical problem solved by the present invention is to make a device and a method of using the device to maintain and immobilize in position the patient's upper limb, to accurately identify the radial artery and the distance between it and the sensor. even in the case of a low peripheral pulse, to facilitate the visualization of the subcutaneous area of the vein plane in the analyzed area and to maintain the orientation of the needle with which the puncture is made at a precise angle (33°) to the central horizontal plane of the patient's forearm.

The structure of this paper is as follows: Sect. 1 makes the introduction and presents different related researches addressing blood sampling procedure, then Sect. 2 presents the developed design methodology, Sect. 3 describe the generated concept and the virtual prototype of the automatic arterial puncture sensorial device prototype for fast arterial blood gas sampling from radial artery Sect. 4 highlights the conclusions and discussions.

## 2 Design Methodology

Due to the absence of technology specific regulation or national standards for designing medical devices in Romania, a methodology for designing such equipment was proposed hereinafter. For developing the current proposed methodology international standards, technical regulations and different design methods have been consulted [10, 11]. Particular attention was given to safety issues regarding invasive implications of the arterial blood sampling medical procedure [12].

Figure 1 outlines the proposed design methodology for automatic arterial puncture sensorial device. The entire development consists of five phases:

- **Phase 1**: Defining/Understanding the vision of the automatic arterial puncture sensorial device (AUSmart-Astrup).
- **Phase 2**: Identify the safety, clinical and technical requirements applicable to medical devices.
- **Phase 3**: Conceptual design of the AUSmart-Astrup device.
- **Phase 4**: AUSmart-Astrup device risk assessment.
- **Phase 5**: Concept adjustments/final solution/Testing and validation.

**Phase 1**:: ***Defining/Understanding the vision of the automatic arterial puncture sensorial device (AUSmart-Astrup)***

**AUSmart-Astrup** must allow fast and efficient radial arterial puncture. The radial artery is the preferred place of arterial puncture because it is easily accessible, being located superficially, it is not adjacent to large veins and has collateral supply through the palmar arch.

**Phase 2**:: ***Identify the Safety, Clinical and Quality Requirements Applicable to AUSmart-Astrup device***

The requirements identification means to determine the specific technical requirements with which a medical device must comply. In general, these technical requirements fall under one of the following three areas: 1) product safety; 2) clinical safety; and 3) quality systems compliance.

Product safety is one of the key issues in designing a medical device. Medical devices and equipment are used in medical settings for patients and concern human life. The safety issues of medical devices are more stringent, dedicated, and critical than the other types of devices whatever they are. Depending on the specificity of the medical devices, three categories of product safety requirements are considered: a) Hardware safety (from the mechanical and electro- mechanical points of view), b) Software safety, c) Operational safety.

To evaluate the product safety of a given medical device someone should consider firstly the following international safety standards or regulations: IEC 60601-1; IEC 60601-1-2; IEC 60601-1-6; ISO 10993; ISO 14971; ISO 13482.

**Phase 3**:: ***Conceptual design of the AUSmart-Astrup device***

The main objective of designing a medical device is to keep it simple and functional. On the other hand, designing medical equipment is not enough to guarantee that it is safe for patients and for medical staff. Under these circumstances, standards cannot hope to cover all risks. So, designers must make up for the area's standards do not cover by conducting a comprehensive risk analysis.

Designers classically use both redundancy and diversity as safety features. Redundancy is simply duplicating the same feature while diversity is the use of two different methods to bring the same function [13].

Finally, the ergonomic component should not be neglected; medical equipment must be finally easy to use – ergonomic.

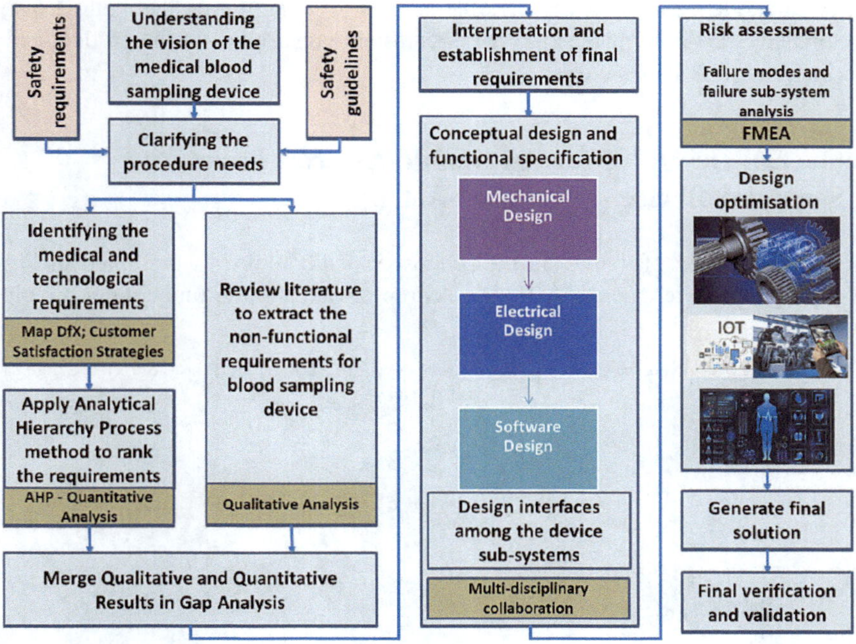


**Fig. 1.** Design framework for automatic arterial puncture sensorial device [adapted from [12]].

**Phase 4**:: ***AUSmart-Astrup device risk assessment***

Once the general safety parameters have been identified, it is necessary to perform a Risk Assessment (also called Hazard Analysis). This is one step of an overall Risk Management process, as required by ISO 14971 (Application of risk management to medical devices). The requirements of ISO 14971:2007 are applicable to all stages of the life cycle of a medical device.

Designers must judge the severity of potential harm and the probability that the harm occurs. Once designers have identified the unacceptable risks, their next step is to define safety measures to mitigate them.

**Phase 5**:: ***Concept tests and adjustments/Final solution/Final validation***

Prototype testing must be done to gather information concerning how the developed concept could be improved as well as highlighting if any critical aspect has not been given enough attention during the concept development phase.

Thus, only a methodical approach based on a methodology, in the stage of product conception, allows to design a mature solution for a mechatronic product. The methodology presented serve for a better understanding of the disciplines and constrains involved in the design process of complex mechatronic products and provides an overview of the entire conception process. Vital conceptual decisions can be made and withdrawn based on interdisciplinary and integrated approaches to the proposed methodology. Certainly, the current realities must also be taken into consideration. Thus, the common truth nowadays is that "what was true and novel yesterday might be not true and obsolete today". Hence any mechatronic product needs to be crafted and launched in a very flexible and fast way. However, the methodology of development should be matched to the circumstances, specifics of the product and the field in which it is launched.

## 3 Overall Design of the Automatic Arterial Puncture Sensorial Device - AUSmart-Astrup

The generated concept, based on the proposed methodology, is presented in Figs. 2 and 3 - a general view of the AUSmart-Astrup device and the functioning principles.

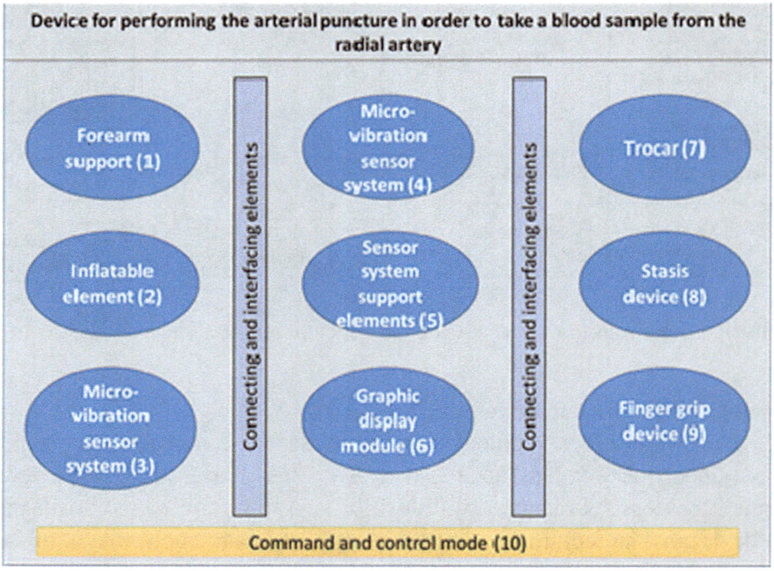


**Fig. 2.** Structural and control scheme of the AUSmart-Astrup device components.

Figure 2 shows the structural and control scheme of the components of the AUSmart-Astrup device for performing the arterial puncture to take a blood sample from the radial artery. These components are forearm support 1, inflatable element 2, micro-vibration sensor system 3, vein visual sensory system 4, sensor system support elements 5, graphic display device, trocar 7, stasis device 8, device 9. The connection of the various component modules of the device is facilitated by quick-connect interfaces; thus, facilitating their coupling and detachment.

Thus, Fig. 3 highlights also the modules of the AUSmart-Astrup device. The structure of the device is a modular one, and each module can be detached for sterilization [14]. The component modules explicitly indicated in Fig. 3 are: forearm support 1; the inflatable element 2, the micro-vibration sensory system 3 and the sensory system for subcutaneous visualization of the plane of the superficial veins 4; the sensor and graphic display system support elements 5, the graphic display modulus 6, the trocar 7 (which has an orientation of 33° with respect to the horizontal median plane of the forearm), the stasis device 8; and finger grip modulus 9. It can also be seen from Fig. 3 that the AUSmart-Astrup device ensures an ergonomic position of the arm throughout the procedure and, at the same time, ensures the immobilization of the arm, so as to prevent its involuntary withdrawal by the patient (especially in children or adolescents). The forearm will be fixed in the AUSmart-Astrup device by the angular movement of the graphic display system support elements 5 (see Fig. 3).

The operation mode of the AUSmart-Astrup device is as follows: the device provides an optimum position and orientation of the patient's forearm-palm tandem; the palm should be flexed towards the forearm, so as to facilitate radial artery puncture arm in the device. The positioning of the fingers and their proper orientation is done with the help of the finger gripping elements. Thus, to identify the position of the radial artery, the internal face of the distal region of the forearm is swept transversely, and when the two micro-vibrations sensors identify independently, but both at the same time, a maximum of the arterial pulse this information is posted on the desired screen and the "acoustic signaling element" is activated emitting an audible signal of medium

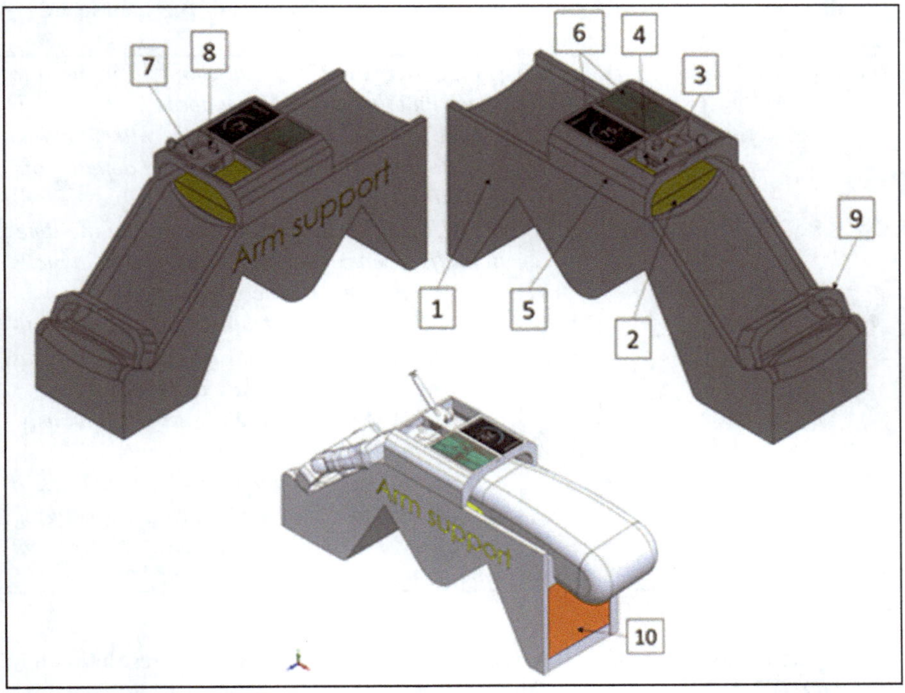


**Fig. 3.** The overview of the AUSmart-Astrup device.

intensity “beep” for 0.1 s. After identifying the radial artery (maximum of the arterial pulse - value 5, on the scale from 1 ÷ 5) and the distance to it, the position of the sensory module is blocked to perform the arterial puncture.

Therefore, the sensor module for identifying the radial artery integrates two micro-vibration and ultrasonic sensors, for the precise identification of the area with maximum arterial pulse and the distance between the sensor and the maximum point of the vibration of the artery and two infrared proximity sensors, which facilitate the precise identification of the route of the veins in the analyzed area. The two micro-vibration sensors (no 3 highlighted in Fig. 3), integrated in the device, are arranged on the same median axis with the syringe and the puncture needle. The distance between the two micro-vibration sensors is 15 mm. The puncture needle guidance is in the middle of the distance between the two sensors; in this configuration we make sure that the needle can be guided precisely towards the radial artery, identified by the two micro-vibration sensors.

As for the purpose of identifying the route of the superficial veins in the analysed area, it is to avoid puncturing the veins; one of the risks of the arterial puncture procedure is to collect venous blood instead of arterial blood. By identifying and visualizing the route of the veins in real time, the risk of puncturing a vein instead of the radial artery is much reduced [5]. Also, the AUSmart-Astrup device will be capable to guide very precisely the puncture needle in the radial artery and to realise the stasis after the extraction of the needle from the artery.

For calibrating the micro-vibrating sensor, we will use the following pseudocode:

| **Pseudocode for calibrating the micro-vibrating sensor module** | **Explanations and details of the pseudocode** |
|---|---|
| *(1) AUS1X_BootState* | *(1)* ***AUS1X_BootState*** *function will be used to check that the sensors have booted.* |
| *(2) AUS2X_SensorInit* | *(2)* ***AUS2X_SensorInit*** *function will be called once to initialize the sensor with a default configuration.* |
| (3) AUS3X_SetTimingBudget | *(3)* ***AUS3X_SetTimingBudget*** *will be the time, in milliseconds, required by the sensor to make one measurement of micro-vibration* |
| (4) AUS5X_CheckforDataReady | *(4)* ***AUS5X_CheckforDataReady*** *receive notification that ranging data are ready - returns "1" when new ranging data are ready.* |
| (5) AUS8X_Stop | *(5)* ***AUS8X_Stop*** *ranging occurs continuously. To stop the current ranging operation, the user can use the stop command, AUS8X_Stop (). If the stop command is issue during the ranging operation, the device completes the current ranging operation before stopping.* |

The graphic display module (see Fig. 3) consisting of the two high-resolution mini-screens (OLEDs) - the information that will be displayed refers to the identification of the area with maximum arterial pulse, to the display of the heart rate, to the display of the $O_2$ blood saturation level and to the display of the vein route in the area of interest. One of the two high-resolution mini screens (OLEDs) is dedicated to displaying the vein path. Therefore, the display of information on the two screens can be decided by the doctor who performs the puncture maneuver depending on the arm at which the puncture is made and the position of the sensitive module.

## 4 Conclusions and Discussions

The proposed concept of the AUSmart-Astrup device allows the immobilization of the upper limb of the patient in the appropriate position, facilitates the precise identification of the position of the radial artery even in a situation of low peripheral pulse, allows visualization of the subcutaneous area of the vein plane in the analyzed area to avoid puncturing them and maintains needle orientation at a precise angle to the central horizontal plane of the patient's forearm. Thus, the proposed concept creates prerequisites for reducing the time needed for arterial blood sampling, increased the efficiency of the medical procedure, provides ergonomics for the medical staff, and increased safety, both for medical staff and patients, in the context of COVID-19 pandemic.

## References

1. Mocan, M., et al.: Cardiac rehabilitation protocols outcome in frail patients undergoing transcatheter aortic valve implantation. BALNEO Res. J. **9**(4), 401–405 (2018). https://doi.org/10.12680/balneo.2018.220
2. Mocan, M., Rahaian, R., Mocan, B., Blaga, S.N.: Multi-marker evaluation of the cardiometabolic risk. Atherosclerosis **263**, E201–E202 (2017). https://doi.org/10.1016/j.atherosclerosis.2017.06.650
3. Dev, S.P., Hillmer, M.D., Ferri, M.: Arterial puncture for blood gas analysis. NEJM **364**, e7 (2011). https://doi.org/10.1056/nejmvcm0803851
4. Danckers, M., Fried, E.D.: Arterial Blood Gas Sampling. pp. 1–3, 8 September 2014
5. Brzezinski, M., Luisetti, T., London, M.: Radial artery cannulation: a comprehensive review of recent anatomic and physiologic investigations. Anesth. Analg. **109**(6), 1763–1781 (2009)
6. Davis, M.D., Walsh, B.K., Sittig, S.E., Restrepo, R.D.: AARC clinical practice guideline: blood gas analysis and hemoximetry. Respir. Care **58**(10), 1694–1703 (2013). https://doi.org/10.4187/respcare.02786
7. Asif, M., Sarkar, P.: Three-digit Allen's test. Ann. Thorac. Surg. **84**(2), 686–687 (2007)
8. Laursen, C.B., Pedersen, R.L., Lassen, A.T.: Ultrasonographically guided puncture of the radial artery for blood gas analysis: a prospective, randomized controlled trial. Ann. Emerg. Med. **65**(5), 618–619 (2015). https://doi.org/10.1016/j.annemergmed.2015.01.016
9. McSwain, S.D., Yeager, B.E.: Is there an easy, effective, efficient, and inexpensive technique to reduce pain of arterial punctures? Respir. Care **60**(1), 141–143 (2015). https://doi.org/10.4187/respcare.03865
10. Mocan, M., Mocan, B.: Cardiac rehabilitation for older patients with cardiovascular pathology using robotic systems - a survey. BALNEO Res. J. **10**(1), 33–36 (2019). https://doi.org/10.12680/balneo.2019.236

11. Bintintan, V., et al.: New inductive proximity sensor platform for precise localization of small colorectal tumors. Mater. Sci. Eng. C-MATERIALS Biol. Appl. **106** (2020). https://doi.org/10.1016/j.msec.2019.110146
12. Mocan, B., Fulea, M., Mocan, M., Bintintan, V.: A robotic helping hand to the detection of small colorectal tumours in laparoscopic surgery. Acta Tech. Napocensis. Ser. Math. Mech. Eng. **62**(1), 91–98 (2019)
13. Moldovan, C.E., Perju, D., Lovasz, E.-C., Modler, K.-H., Maniu, I.: On the kinematic analysis of a sixth class mechanism. In: Corves, B., Lovasz, E.-C., Hüsing, M., Maniu, I., Gruescu, C. (eds.) New Advances in Mechanisms, Mechanical Transmissions and Robotics. MMS, vol. 46, pp. 47–57. Springer, Cham (2017). https://doi.org/10.1007/978-3-319-45450-4_5
14. Javaid, M., Haleem, A., Vaishya, R., Bahl, S., Suman, R., Vaish, A.: Industry 4.0 technologies and their applications in fighting COVID-19 pandemic. Diabetes Metab. Syndr. Clin. Res. Rev. **14**(4), 419–422 (2020). https://doi.org/10.1016/j.dsx.2020.04.032

# On the Measurement of Dynamic Stability of Normal and Osteoarthritic Human Knee During Ascending and Descending the Stairs

Daniela Tarnita(✉), Marius Georgescu, and Alin Petcu

Faculty of Mechanics, University of Craiova, Craiova, Romania
tarnita.daniela@gmail.com

**Abstract.** The aim of this study is to measure the variation of the flexion-extension angles of the knee and to study the stability of the knees movements during ascending and descending the stairs for healthy subjects and for patients suffering of osteoarthritic knee (OAK). Tools of nonlinear dynamics are used in order to quantify the stability of human healthy knee and OAK joints during ascending and descending the stairs. The values of Lyapunov exponents (LE), which represent the measure of instability of knee movements, are computed based on the experimental time series collected for both knee joints movements: flexion-extension (fl-ext) in sagittal plane and varus-valgus rotation (var-valg) in frontal plane. LEs obtained for the OAK are associated with more divergence, more variability and less stability, being larger than those are obtained for the knees of healthy subjects.

**Keywords:** Ascending stairs · Descending stairs · Osteoarthritic knee · Stability · Lyapunov exponents · Phase planes · State space reconstruction

## 1 Introduction

The experimental evaluation and measurement of the biomechanical parameters of human gait has an increased importance, and it is more and more recognized and used in biomechanics [1–7], in robotics [8–14], in rehabilitation and in clinical research field [11, 15–18]. In the medical field, the monitoring of human gait parameters provide important information about quantitative measurement of different gait characteristics [4–7, 19, 20], about the evolution of different diseases and about physical therapy involving rehabilitation [11–18]. Ascending and descending the stairs are daily activities of people, both healthy and with musculoskeletal disorders, one of the most known disease being osteoarthritis of the joints [5–7, 15, 17, 19, 20]. Compared with normal walking, ascending and descending the stairs are activities which imply more effort, especially in people with musculoskeletal diseases [19–28]. Several studies were performed to investigate normal human stair ascent and descent [29–31]. Some papers are focused on the analysis of joint kinematics [21–24], joint moments [32] that occur during staircase walking. Some studies investigate ascending the stair by patients with knee implants [33] and hip implants [34], or amputees with artificial limbs [35, 36].


E.-C. Lovasz et al. (Eds.): MTM&Robotics 2020, MMS 88, pp. 543–555, 2021.
https://doi.org/10.1007/978-3-030-60076-1_49

In the last years, new mathematical tools have been used to better characterize the stability and the nonlinear features of gait variability [37–50]. Local dynamic stability is quantified by Lyapunov exponents which represent a measure of the ability of the walking subject to maintain continuous motion and to attenuate the effects of the local perturbations [37–50]. Many papers studied chaotic dynamics in human gait over-ground and on treadmill [37–44]. To our knowledge, however, the use of nonlinear dynamic analysis in order to investigate the influence of the osteoarthritic process in the knee on the dynamic stability of human ascending and descending the stairs has not been reported. The stability of is usually studied in other papers by using the centre of mass or the centre of pressure, but almost no nonlinear tools were used. In this paper, tools of nonlinear dynamics are used in order to quantify the stability of human healthy knee and OAK knee joints during ascending and descending stairs. In order to design and develop different bio-inspired robotic structures that must maintain their stability, researchers used the results obtained during the experimental evaluation of human gait variability and stability, depending on different parameters [17, 39–45].

The objective of this paper study is to measure the variation of flexion-extension angle and to study, by using tools of nonlinear dynamics, as Lyapunov exponents, the influence of the OAK on the stability of movements, during ascending and descending the stairs for patients suffering of osteoarthritis of knee (OAK) and compare them with the results obtained for healthy subjects.

## 2 Experimental Study

The experimental method which allows obtaining the kinematic parameters diagrams for the human knee joint uses a Biometrics Ltd data acquisition system based on electrogoniometers, which are wearable sensors successfully used in biomechanics and medical applications [2, 4, 17, 18, 20, 24, 40, 42, 43, 51].

For data acquisition six electrogoniometers, one for each of six joints of both lower limbs, connected with two 8-channel DataLOG devices at a frequency of 500 Hz (MWX 8 Biometrics Ltd) were used. The data were transferred from DataLOG to computer in real time via Bluetooth communication. In Fig. 1 the data acquisition system mounted on the subject is shown.

Two samples of 7 healthy subjects and, respectively, of 5 patients suffering by knee osteoarthritis, second stage of gravity, were considered for experimental tests. The average anthropometric data for both samples are presented in Table 1. The healthy subjects were pain-free and had no evidence or known history of motor and skeletal disorders or record of surgery to the lower limbs. The Ethical Committee of the University of Craiova approved the study.

The experimental tests consist of ascending stairs and descending the stairs. A staircase with a number of 12 stairs, with the sizes: height = 0.18 m, width = 0.3 m, length = 1.5 m is considered for experimental tests.

**Fig. 1.** The subject with acquisition system mounted on lower limbs joints

**Table 1.** Anthropometric data of the healthy subjects and patients

| | | Age | Weight [Kg] | Height [cm] | Lower limb length [cm] | Length hip-knee [cm] | Length knee-ankle [cm] |
|---|---|---|---|---|---|---|---|
| Subjects | Average | 26,86 | 71,29 | 174,29 | 79,93 | 40,93 | 39 |
| | StdDev | 1,46 | 12,50 | 8,27 | 9,52 | 4,83 | 4,72 |
| Patients | Average | 62,46 | 73,46 | 169,33 | 81,00 | 41,33 | 39,67 |
| | StdDev | 2,37 | 2,65 | 2,08 | 2,00 | 1,15 | 1,15 |

The angular amplitudes of human knee flexion-extension during ascending and descending stairs were obtained for each test as data files. In Fig. 2 the diagram of experimental flexion-extension joint angles, in sagittal plane, for the three joints of the left leg of Subject 1 in respect with time [s], during ascending and, respectively, descending stairs, collected by Biometrics system, are presented.

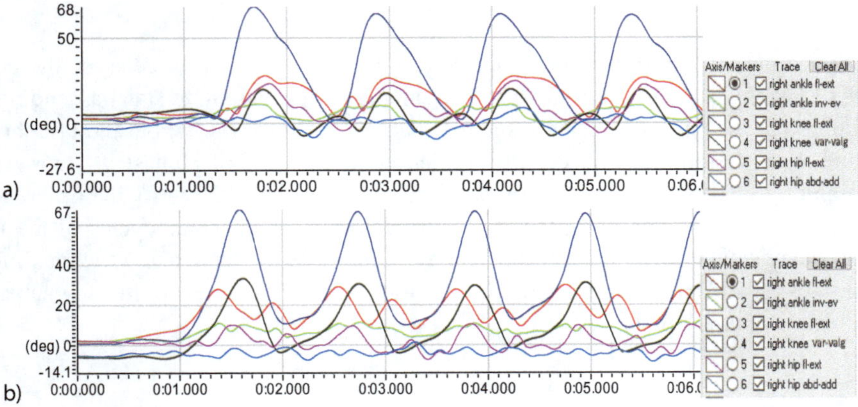


**Fig. 2.** Experimental flexion-extension angles, in sagittal plane, for left leg of subject 1: a) ascending stairs; b) descending stairs

# 3 Results

Human gait variability from one cycle to another for the same subject and from one subject to another imposes the normalization of gait cycles. Normalization is done using SimiMotion software where the data files are transferred [52]. In Fig. 3 the medium cycle of flexion-extension movement of right knee for entire samples of healthy subjects and of patients during ascending and descending the stairs are presented.

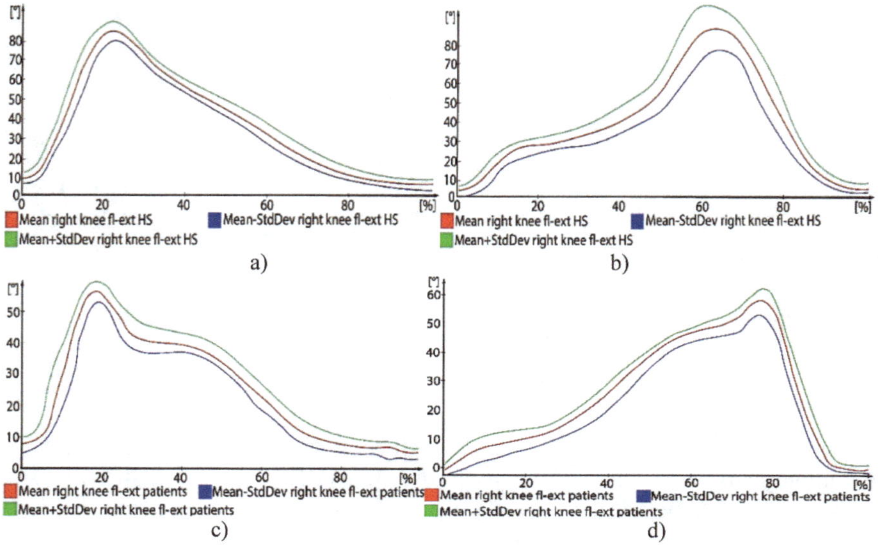


**Fig. 3.** Medium cycle of flexion-extension movement of right knee for the sample of healthy subjects: a) ascending stairs; b) descending stairs and of patients: c) ascending stairs; d) descending stairs

By comparing the diagrams of the average flexion-extension cycles of the right knee of healthy sample and of OA patients sample corresponding to stairs ascending, it is found that the curves of healthy subjects and patients are similar, but the last shows disturbances, tremors, which denote oscillations during walking on stairs caused by pain and by the instability of the joints affected by osteoarthritis. Also, the differences in amplitude can be observed, the maximum values of the average cycles of the OA knees being smaller than those of the healthy subjects. In Tables 2 and 3 the maximum flexion-extension angles and maximum reaction forces of the knee for healthy subjects and for patients are presented.

**Table 2.** The maximum flexion-extension angles of the knee for healthy subjects

| Subjects | Descending | Ascending |
|---|---|---|
| | Max. angle [deg] | Max. angle [deg] |
| 1 | 97 | 95 |
| 2 | 101 | 98 |
| 3 | 97 | 94 |
| 4 | 96 | 94 |
| 5 | 99 | 96 |
| 6 | 96 | 94 |
| 7 | 100 | 98 |
| Medium cycle | 98 | 95.57 |

**Table 3.** The maximum flexion-extension angles of the knee for patients

| Patients | Descending | Ascending |
|---|---|---|
| | Max. angle [deg] | Max. angle [deg] |
| 1 | 66 | 64 |
| 2 | 61 | 60 |
| 3 | 66 | 63 |
| 4 | 63 | 61 |
| 5 | 65 | 62 |
| Medium cycle | 64.20 | 62.00 |

The variation curves of flexion-extension knee angle have a similar pattern for patients and for healthy subjects. The flexion-extension angle values are higher with about 2–3° for descending than for ascending the stairs for both samples: healthy subjects and patients. Experimental tests have shown that the maximum values of flexion-extension angles are lower by about 20–26° for the osteoarthritic knees comparing with the healthy knees for both t tests: ascending and descending the stairs. The maximum values of medium cycle of flexion-extension angle is lower with about 24° for the patients' sample than for healthy subjects, in both cases: ascending and descending the stairs. An explanation of the angular differences could be that the OA patients decreased the amount of knee flexion because of increased pain and joint stiffness. Previous studies on gait showed that a reduction in knee flexion represents a compensatory effect to minimize the compressive forces across the knee. It has been reported that the affected knees of OA patients do exhibit less dynamic range of motion compared with healthy subjects [5, 19, 22, 23, 27, 28].

## 4 Dynamic Analysis

In this paper, tools of nonlinear dynamics are used in order to quantify the stability of human healthy knee and OAK knee joints during ascending and descending stairs. The values of Lyapunov exponents (LE), which represent the measure of instability of knee movements, are computed based on the experimental time series collected for both knee joints movements: flexion-extension (fl-ext) in sagittal plane and varus-valgus rotation (var-valg) in frontal plane. The algorithm used for estimating the knee non-linear system stability is shown in Fig. 4.

The state space reconstruction S is done by using the method that consists in to generate the delay coordinates vectors [46]:

$$x_n = \left[s(t_0 + nT_s);\ s(t_0 + nT_s + T), \ldots.\ s(t_0 + nT_s + (d_E - 1)T)\right] \qquad (1)$$

Where the integer $d_E$ is the embedding dimension, the notation s(.) is a measured scalar function, $T_s$ is the sampling time, n = 1, 2,…, $d_E$, while $T = kT_s$ is an appropriately time delay. In [47] is demonstrated that the dynamics in the reconstructed state space and the original dynamics are equivalent, the two spaces have the same invariants, such as LE. Phase plane portraits are used, in order to characterize the kinematics of the system, by correlating the knee angular rotations with the knee angular velocities [39–43]. In Fig. 5 the state-space reconstruction and the phase-plane portraits corresponding to knee flexion-extension movement of Subject 4 and Patient 1 are shown, while in Fig. 6 the state-space reconstruction and the phase-plane portraits corresponding to knee varus-valgus movement of Patient 1 are shown. Similar diagrams are obtained for the other subjects and patients. It can be seen that for healthy knee the cycle's curves show less divergence in their trajectories, are more compact, the amplitudes tend to be constant, while the curves traced for OAKs show more divergence and increased spread in their trajectories.

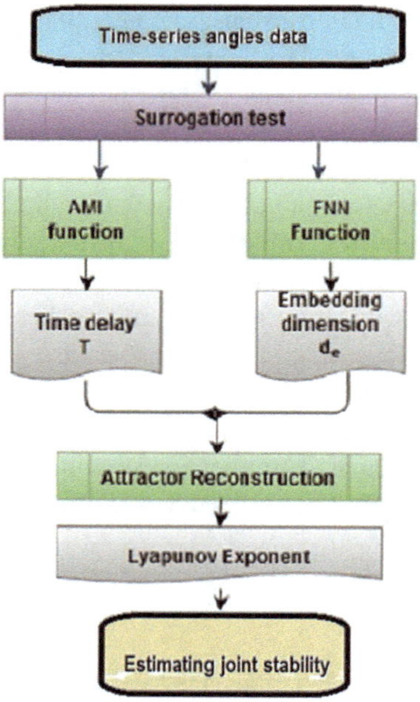


**Fig. 4.** Algorithm for study the joint stability

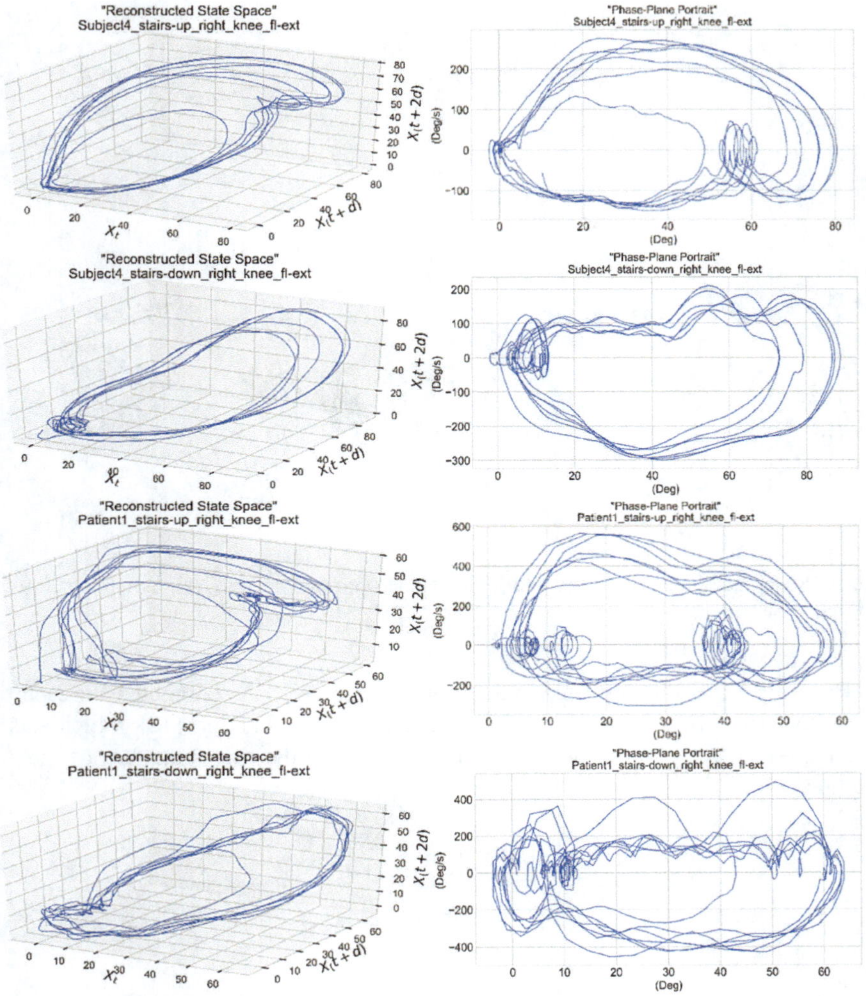


**Fig. 5.** State space reconstruction and Phase plane portrait for flexion-extension (f-e) movement of the right knee of Subject 4' and of Patient 1

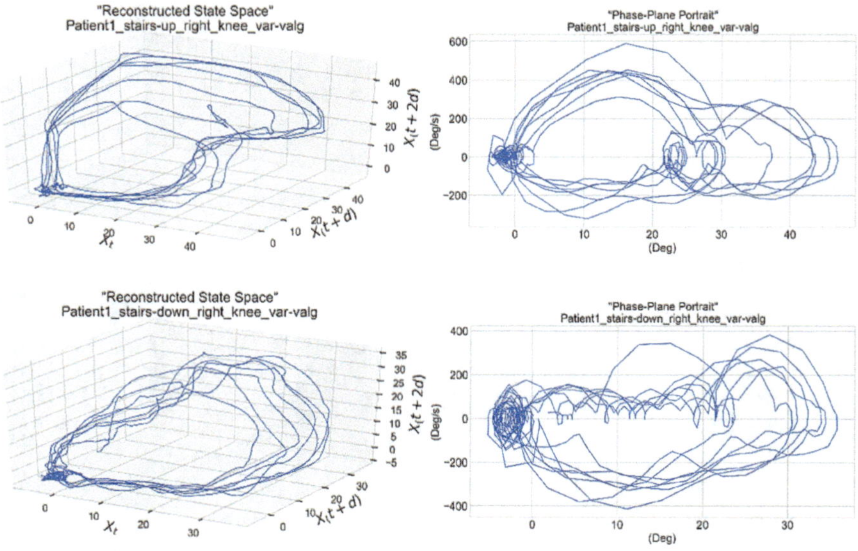


**Fig. 6.** State space reconstruction and Phase plane portrait for varus-valgus movement of the right knee of Patient 1

By using TISEAN software, an appropriate time lag, T, is computed for each time series, by using the average mutual information (AMI) function [48], which sets the time lag equal to the value of delay corresponding to the first minimum of the AMI function (Fig. 7).

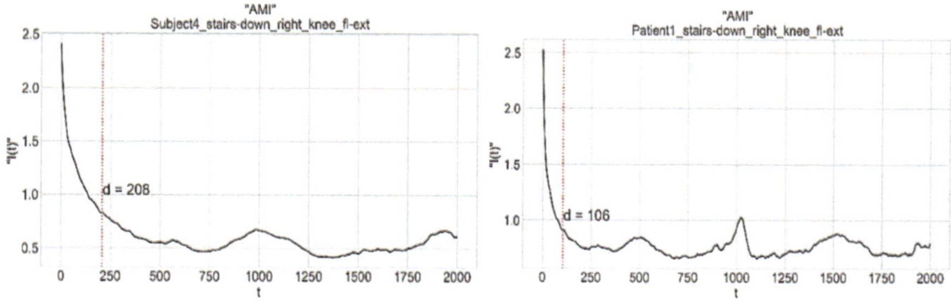


**Fig. 7.** AMI function diagram for stairs-down Subject 4 and Patient 1

A suitable embedding dimension, which is the minimum value that trajectories of the reconstructed state vector may not cross over each other in state space, was chosen by using the false nearest neighbour method (FNN) [49] (Fig. 8).

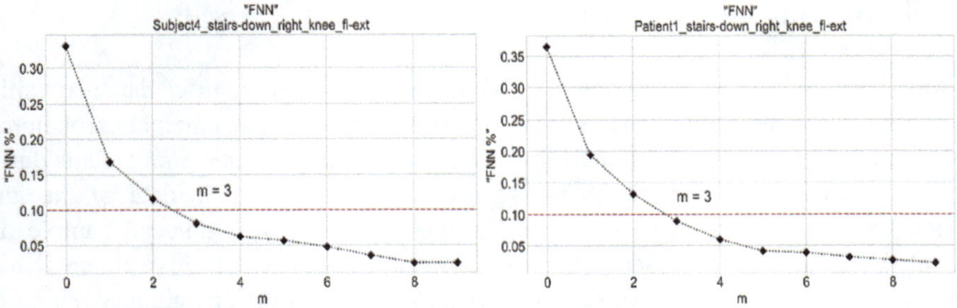


**Fig. 8.** FNN diagram for Subject 4 and Patient 1- descending stairs

The LEs are computed using 'lyap_r' routine, available in TISEAN package, based on the Rosenstein algorithm [50]. If LEs increase, then the local dynamic stability decreases [40–49]. The LEs calculated for all time series were positive, that means the human knee joint is a deterministic chaotic system, in accordance with [47]. In Table 4 the average values of the computed LEs for both knee movements of healthy subjects' sample and patients' sample are presented, while in Fig. 9 these values are plotted.

**Table 4.** Average values of Lyapunov exponents

| | P left | P right | HS left | HS right |
|---|---|---|---|---|
| stairs-down f-e | 2.416547317 | 2.572345909 | 1.774434920 | 1.847812011 |
| stairs-up f-e | 2.101810936 | 2.283159951 | 1.707204798 | 1.653789153 |
| stairs-down v-v | 1.824528492 | 2.094635268 | 1.100643761 | 0.916221168 |
| stairs-up v-v | 1.249618601 | 1.396367137 | 0.735011699 | 0.618985014 |

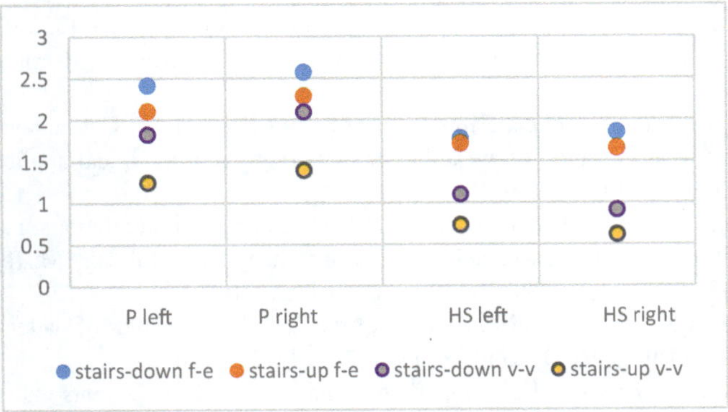


**Fig. 9.** Average Lyapunov exponents values for ascending and descending stairs, for: Healthy subjects (HS) and Patients (P), both movements:flexion-extension (f-e) and varus-valgus (v-v)

## 5 Discussions and Conclusions

A study based on the tools of nonlinear dynamics and on a Biometrics data acquisition system based on wearable sensors to visualize the steady state kinematics of human knee joint movement and to estimate the stability of the human knee movements during ascending and descending the stairs is presented. The kinematic data of the knee flexion–extension angles and varus-valgus angles for healthy subjects' sample and patients' sample were analysed. The purpose of this study was to investigate chaotic characteristics of movements of the knee joint. We applied the chaotic analysis to the rhythmic movements of human knee and showed that there existed the chaotic feature of angle positions in the rhythmic knee movements using the chaotic measure such as LE. The LE obtained for each test of human knee were positive values. The mean LE for human knee joint are smaller for ascending the stairs compared to the descending the stairs. The observation is valuable for healthy subjects as well as for the patients. We also observed the increasing of the LE for patients than the healthy subjects for both movements. For the OAK (the right knee), the values of LE are bigger than those obtained the other knee of the patients. The values of LEs obtained for the OAK are associated with more divergence and less stability. Larger values of LEs obtained for OAKs are explained by the influence of the pain, while smaller values obtained for healthy subjects reflect a local stability, less divergence and less variability. A limitation of this study is the relatively low number of subjects in the two samples. Although there are published articles that were based on a similar number of subjects (7–10 people) [23, 38], for future work we aim to deepen our research, based on an increased number of participants in the samples (17–20 persons).

## References

1. Tao, W., Liu, T., et al.: Gait analysis using wearable sensors. Sensors **12**, 2255–2283 (2012)
2. Tarnita, D.: Wearable sensors used for human gait analysis. Rom. J. Morphol. Embryol. **57** (2), 373–382 (2016)
3. Muro-de-la-Herran, A., Garcia-Zapirain, B., Mendez-Zorrilla, A.: Gait analysis methods: an overview of wearable and non-wearable systems highlighting clinical applications. Sensors **14**, 3362–3394 (2014)
4. Tarnita, D., Geonea, I., Petcu, A., Tarnita, D.N.: Numerical simulations and experimental human gait analysis using wearable sensors. In: Husty, M., Hofbaur, M. (eds.) MESROB 2016. MMS, vol. 48, pp. 289–304. Springer, Cham (2018)
5. Al-Zahrani, K.S., Bakheit, A.M.: A study of the gait characteristics of patients with chronic osteoarthritis of the knee. Disabil. Rehabil. **24**, 275–278 (2002)
6. McGibbon, C.A., Krebs, D.E.: Compensatory gait mechanics in patients with unilateral knee arthritis. J. Rheumatol. **29**, 2410–2419 (2002)
7. Tarnita, D., Catana, M., Tarnita, D.N.: Experimental measurement of flexion-extension movement in normal and osteoarthritic human knee. Rom. J. Morphol. Embryol. **54**(2), 309–313 (2013)
8. Barbu, I., Alexandru, C.: Virtual prototyping tools applied in mechanical engineering. Mecatronica **3**, 9–12 (2004)

9. Tarnita, D., Catana, M., Tarnita, D.N.: Contributions on the modeling and simulation of the human knee joint with applications to the robotic structures. In: Rodić, A., Pisla, D., Bleuler, H. (eds.) New Trends on Medical and Service Robotics: Challenges and Solutions Mechanisms and Machine Science, vol. 20, pp. 283–297. Springer, Cham (2010)
10. Enescu, M.L., Alexandru, C.: Modeling and simulation of A 6 DOF robot. Adv. Mater. Res. **463**, 1116–1119 (2012)
11. Tarnita, D., Tarnita, D.N., Bizdoaca, N., Popa, D.: Contributions on the dynamic simulation of the virtual model of the human knee joint. Mater. Sci. Eng. Tech. **40**(1–2), 73–81 (2009). Special Edition Biomaterials Wiley-Vch.
12. Gherman, B., Birlescu, I., Nicolae, P., Carbone, G., Tarnita, D., Pisla, D.: On the singularity-free workspace of a parallel robot for lower-limb rehabilitation. Proc. Rom. Acad. Ser. A **20** (4), 383–391 (2019)
13. Vaida, C., Birlescu, I., Pisla, A., Ulinici, I., Tarnita, D., et al.: Systematic design of a parallel robotic system for lower limb rehabilitation. IEEE Access **8**, 34522–34537 (2020)
14. Plitea, N., et al.: Kinematic analysis of an exoskeleton-based robot for elbow and wrist rehabilitation. In: Carvalho, J., Martins, D., Simoni, R., Simas, H. (eds) Multibody Mechatronic Systems. MuSMe 2017. Mechanisms and Machine Science, vol. 54. Springer, Cham (2018)
15. Tarnita, D., Catana, M., Tarnita, D.N.: Design and simulation of an orthotic device for patients with osteoarthritis. In: Bleuler, H., Bouri, M., Mondada, F., Pisla, D., Rodic, A., Helmer, P. (eds.) New Trends in Medical and Service Robots, pp. 61–77. Springer, Cham (2016)
16. Mateas, M., Moldovan, C., et al.: Novel rehabilitation system for the lower limb. IOP Conf. Ser. Mater. Sci. Eng. **444**(5), 052021 (2018)
17. Tarnita, D., Pisla, D., Geonea, I., Tarnita, D.N., Vaida, C.: Static and dynamic analysis of osteoarthritic and orthotic human knee. J. Bionic Eng. **16**, 514–525 (2019)
18. Geonea, I., Tarnita, D.: Design and evaluation of a new exoskeleton for gait rehabilitation. Mech. Sci. **8**(2), 307–322 (2017)
19. Hicks-Little, C., et al.: Lower extremity joint kinematics during stair climbing in knee osteoarthritis. Med. Sci. Sports Exer. **43**(3), 516–524 (2011)
20. Petcu, A., Georgescu, M., Calafeteanu, D., Tarniţă, D.: Kinematics and kinetics of healthy and osteoarthric knee during walking stairs. Bull. Transilvania Univ. Brasov Ser. I Eng. Sci. Spec. Issue **9**, 203–208 (2016)
21. Tarnita, D., Popa, D., Tarnita, D.N., Grecu, D.: CAD method for three-dimensional model of the tibia bone and study of stresses using the finite element method. Rom. J. Morphol. Embryol. **47**(2), 181–186 (2006)
22. Nadeau, S., McFadyen, B.J., Malouin, F.: Frontal and sagittal plane analyses of the stair climbing task in healthy adults aged over 40 years: what are the challenges compared to level walking. Clin. Biomech. **18**, 950–959 (2003)
23. Riener, R., Rabuffetti, M., Frigo, C.: Stair ascent and descent at different inclinations. Gait Posture **15**(1), 32–44 (2002)
24. Tarnita, D., Geonea, I., Petcu, A., Tarnita, D.N.: Experimental characterization of human walking on stairs applied to humanoid dynamics. In: International Conference on RAAD, pp. 293–301 (2016)
25. Della, C.U., Bonato, P.A.: Novel design for an instrumented stairway. J. Biomech. **40**(3), 702–704 (2007)
26. Stacoff, A., Diezi, C., Luder, G., Stussi, E., Kramers-de Quervain, I.A.: Ground reacţion forces on stairs: effects of stair inclination and age. Gait Posture **21**(1), 24–38 (2005)

27. Protopapadaki, A., Drechsler, W.I., Cramp, M.C., Coutts, F.J., Scott, O.M.: Hip, knee, ankle kinematics and kinetics during stair ascent and descent in healthy young individuals. Clin. Biomech. **22**, 203–210 (2007)
28. McFadyen, B.J., Winter, D.A.: An integrated biomechanical analysis of normal stair ascent and descent. J. Biomech. **21**, 733–744 (1988)
29. Andriacchi, T.P., et al.: A study of lower-limb mechanics during stair climbing. J. Bone Joint Surg. **62A**, 749–757 (1980)
30. Costigan, P.A., Deluzio, K.J., Wyss, U.P.: Knee and hip kinetics during normal stair climbing. Gait Posture **16**, 31–38 (2002)
31. Zachazewski, J.E., Riley, P.O., Krebs, D.E.: Biomechanical analysis of body mass transfer during stair ascent and descent of healthy subjects. J. Rehabil. Res. Dev. **30**, 412–422 (1993)
32. Kowalk, D.L., Duncan, J.A., Vaughan, C.L.: Aduction–adduction moments at the knee during stair ascent and descent. J. Biomech. **29**, 383–388 (1996)
33. Andriacchi, T.P., et al.: The influence of total knee-replacement design on walking and stair climbing. J Bone Joint Surg. **64A**, 1328–1335 (1982)
34. Bergmann, G., Graichen, F., Rohlmann, A.: Is staircase walking a risk for the fixation of hip implants. J. Biomech. **5**, 535–553 (1995)
35. Ferreira, C.R., Barauna, M.A., da Silva, K.C.: Analysis of the performance of above-knee amputees in climbing stairs. In: Proceedings of the Fourteenth International Symposium on Biomechanics in Sports. Lisbon, Portugal, pp. 533–536 (1996)
36. Powers, C.M., Boyd, L.A., Torburn, L., Perry, J.: Stair ambulation in persons with transtibial amputation: an analysis of the Seattle LightFoot. J. Rehabil. Res. Develop. **34**, 9–18 (1995)
37. Dingwell, J.B., Cusumano, J.P., et al.: Local dynamic stability versus kinematic variability of continuous overground and treadmill walking. J. Biomech. Eng. **123**, 27–32 (2001)
38. Miyoshi, T., Murata, A.: Chaotic characteristic in human hand movement. In: Proceedings of the 2000 IEEE International Workshop on Robot and Human Interactive Communication, Osaka, Japan, 27–29 September, pp. 194-199 (2000)
39. Tarnita, D., Georgescu, M., Geonea, I., Petcu, A., Tarnita, D.-N.: Nonlinear analysis of human ankle dynamics. In: Carbone, G., Ceccarelli, M., Pisla, D. (eds.) New Trends in Medical and Service Robotics. MMS, vol. 65, pp. 235–243. Springer, Cham (2019)
40. Tarnita, D., Georgescu, M.: Applications of nonlinear dynamics to human knee movement on plane & inclined treadmill. In: Wenger, P., Chevallereau, C., Pisla, D., Bleuler, H., Rodić, A. (eds.) New Trends in Medical & Service Robots, vol. 39, pp. 59–73. Springer, Cham (2016)
41. Hurmuzlu, Y., Basdogan, C., Stoianovici, D.: On the measurement of dynamic stability of human locomotion. ASME J. Biomech. Eng. **116**(1), 30–36 (1994)
42. Tarnita, D., Catana, M., Tarnita, D.N.: Nonlinear analysis of normal human gait for different activities with application to bipedal locomotion. Rev. Roum. J. Tech. Sci. Appl. Mech **58** (1–2), 177–192 (2013)
43. Tarnita, D., Marghitu, D.: Nonlinear dynamics of normal and osteoarthritic human knee. In: Proceedings of the Romanian Academy, pp. 353–360 (2017)
44. Dingwell, J.B., Cusumano, J.P.: Nonlinear time series analysis of normal and pathological human walking. Chaos **10**(4), 848–863 (2000)
45. Yang, C., Wu, Q.: On stabilization of bipedal robots during disturbed standing using the concept of Lyapunov exponents. Robotica **24**(5), 621–624 (2006)
46. Packard, N.H., Crutchfield, J.P., Farmer, J.D., Shaw, R.S.: Geometry from a time. Ser. Phys. Rev. Lett. **45**, 712–716 (1980)
47. Nayfeh, A.H.: Introduction to Perturbation Techniques. Wiley, New York (1981)
48. Fraser, A.M., Swinney, H.L.: Independent coordinates for strange attractors from mutual information. PhysRev **A33**, 1134–1140 (1986)

49. Kennel, M.B., Brown, R., Abarbanel, H.D.I.: Determining minimum embedding dimension using a geometrical construction. PhysRev **A45**, 3403–3411 (1992)
50. Rosenstein, M.T., Collins, I.J., Deluca, C.J.: A practical method for calculating largest Lyapunov exponents from small data sets. Physics D **65**, 117–134 (1993)
51. www.biometricsltd.com/. Accessed 08 Apr 2020
52. www.simi.com. Accessed 20 Mar 2020

# Modeling and Simulation of Biomechanical Behavior of Ankle Ligaments with Applications in Vibration Therapy

Adela Neamţu Popescu, Ion Crâştiu, Dorin Simoiu, Andreea Raluca Ursu, and Liviu Bereteu(✉)

Mechanics and Materials Strength Department, Politehnica University of Timişoara, Bd. Mihai Viteazul, nr. 1, 300222 Timişoara, Romania
adelaneamtu@yahoo.com, ion_crastiu@yahoo.com, {dorin.simoiu, liviu.bereteu}@upt.ro

**Abstract.** Ankle sprain is one of the most common injuries. Tens of thousands of ankle sprains are recorded daily, especially in athletes, but not only. This injury consists of stretching of ligaments or partial or complete rupture of ligaments. In this study it is proposed, based on a rheological model of the tissue of a ligament, to explain from a biomechanical point of view the effect of vibration on the body, which it have in the post-traumatic rehabilitation phase of an ankle sprain. The Bingham type viscoelastoplastic model with transition to the sliding phase at elongation and elastic return in the relaxation phase was analyzed. A vibrating platform was used for the case study in which it was possible to act on the vibration frequency, amplitude and exposure time.

**Keywords:** Modeling · Simulation · Mechanical behavior · Ankle ligaments · Vibration therapy

## 1 Introduction

Ligaments and tendons are tissues with similar structures, made up of a bunch of solid fibers. The essential differences between them are that while the tendons connect the muscle to the bone, the ligaments connect two bones together, on the one hand, and on the other hand the tendons provide strength and transmit muscle power to the bone ensuring movement, ligaments ensure stability of movement.

Unlike tendons that can rupture, totally or partially, or become inflamed, the ligaments are strong and resistant to overload. This is why most ligament problems are acute in the form of mild, moderate or severe sprains. The ankle is the joint most frequently affected by sprains. Anterior talofibular ligament (ATFL), posterior talofibular ligament (PTFL) and calcaneofibular ligament (CTL) are involved in the ankle sprain. The most common incident is a sprain by inversion when the sole of the foot is brought inside, resulting in stretching or rupture of the ATFL and CFL [1, 2]. After the acute pain has disappeared, the normal rehabilitation begins, which involves first of all physiotherapy.


E.-C. Lovasz et al. (Eds.): MTM&Robotics 2020, MMS 88, pp. 556–565, 2021.
https://doi.org/10.1007/978-3-030-60076-1_50

Ultrasound is frequently used in combination with other forms of treatment in the management of lateral ligament sprains of the ankle. Ultrasound is said to have an analgesic, decontracting and decongestant effect on affected tissue. Despite the widespread use of this method, the critical analysis. Despite the widespread use of this method, the critical analysis shows that there is little scientific evidence to support its role in the management of dislocated ankles [3].

Also, electrotherapy offers the possibility of obtaining the effects of muscle excitation, hyperemic and decontracting effect. Studies show that electrotherapy can improve short-term recovery outcomes in athletes with acute ankle sprain, and that it can speed up the resumption of sports activities [4, 5].

In recent years, whole body vibration (WBV) has become a method of neuromuscular training with widespread popularity in health and fitness centers. The method can be considered as the substitution for the classic training or as an additional one. The main purpose of the method is to increase muscle strength and power. In [6] a critical analysis of the effect of WBV on muscle performance, strength, flexibility and muscle power is made.

Recently, the training of athletes through the whole body vibration method (WBV) was introduced as a preventive and rehabilitation after ankle sprains [7].

Almost half of patients with a lateral ankle sprain develop chronic ankle instability, characterized by recurrent injuries and sensations of failure. It is suggested that WBV has the ability to ameliorate injuries caused by mechanoreceptor damage following an accident, and to improve proprioception in people with ankle instability [8].

The effect of vibration therapy can also be quantified through gait analysis programs, after the achievement of the post-trauma rehabilitation stage [9–12].

The aim of this paper is to model the biomechanical behavior of the tissues that make up the ligaments through rheological elements whose behavior over time can be analyzed based on the constitutive equation. A case study comes to confirm some theoretical results.

## 2 Elasto-Viscoplastic Models with Elastic Return

There are many rheological models that characterize the sliding or plastic behavior of biological tissues such as ligaments or tendons. To explain the effects of vibrations on these tissues, the most important rheological models are those that show purely elastic behavior during recovery after reaching maximum stress. Such a rheological model must have a spring inserted with a group of two elements connected in parallel, one of which must be the so-called slipper, as is show in Fig. 1. This friction element corresponds to a plastic behavior and slides when the stress reaches a threshold $\sigma_P$.

The group of elements connected in parallel is locked on the increasing domain of stress, ($\dot{\sigma} > 0$) for all values for which, $\sigma \leq \sigma_P$. For values of the stress, $\sigma > \sigma_P$, the slip will appear and the model from Fig. 1 will behave elasto-viscoplastic.

When the stress decreases, so for, $\dot{\sigma} < 0$, the group of elements connected in parallel remains blocked, and the behavior of the model will be purely elastic.

Regardless of the behavior of the X element connected in parallel with the slipper, the rheological model in Fig. 1 will have an elastic return characteristic. This essential property is important for the simplifications it brings in the study of the effect of vibrations on some tissues, which can lead to mechanical stresses in the field of slipping.

One can imagine a significant number of rheological models that have the structure corresponding to the model described above.

First it can be assumed that instead of the element X there is an elastic element characterized by the modulus of elasticity $E_2$. In this case the model will have an elastoplastic behavior.

Similarly, if the element denoted by X is replaced by a viscous element, characterized by the $\eta$ coefficient, the so-called Bingham model is obtained. This distinguishes the Prandtl model by including the viscous element that will act only after overcoming the slip stress.

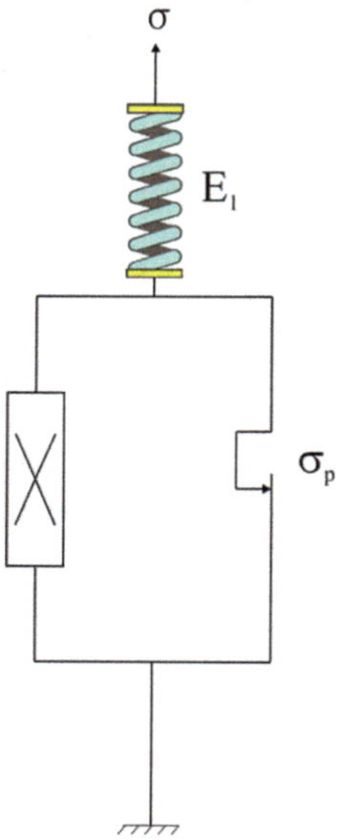


**Fig. 1.** Rheological model with elastic return

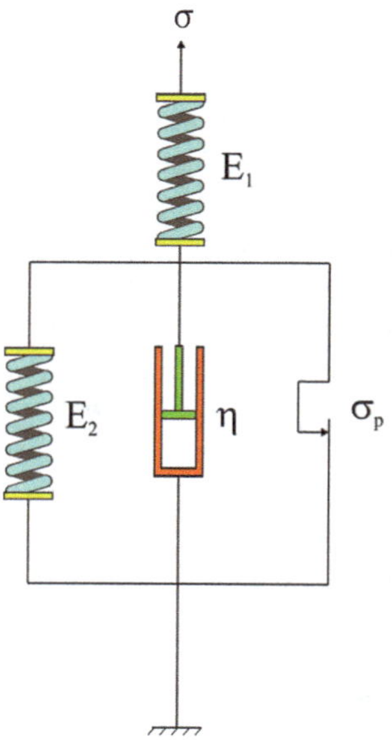


**Fig. 2.** The Goldschmit model

Finally, the element X can be considered as being formed by two elements, one elastic having the modulus of elasticity $E_2$ and the other dashpot having the viscosity coefficient $\eta$, connected in parallel as shown in Fig. 2. The so-called Goldsmith model is obtained, which has an elastoviscoplastic behavior if the load exceeds the sliding stress, $\sigma_p$. As can be easily seen, all the tissues that correspond to the model in Fig. 2 are essentially viscoplastics with viscous properties above the slip stress limit. As a result, the general model given in Fig. 2, when discharged behaves elastically, and the following equation can be written

$$\sigma = \sigma_1 - E_1(\varepsilon_1 - \varepsilon) \tag{1}$$

where $\sigma$ is the value of the stress at a certain point on the diagram ($\sigma$, $\varepsilon$), (Fig. 3), $\varepsilon$ is the value of the corresponding strain, and $\sigma_1$ is the maximum value of the stress at the point $M_1$.

## 3 The Principle of Relaxation Through Vibration

The explanation of the decrease of the stress state of a tissue by vibration can be made starting from the analysis of the behavior of this material in the sliding field. By applying vibrations it reaches the sliding domain, also called plastic domain, and for the average state of vibration which reaches the end of several cycles, this state will correspond to a stress below the initial value. The following reasoning is based on the model given in Fig. 2, which has a purely elastic characteristic on return. Therefore, it will be assumed that in the initial state of equilibrium, the tissue is loaded at stress $\sigma_0$ which corresponds to $\varepsilon_0$ strain.

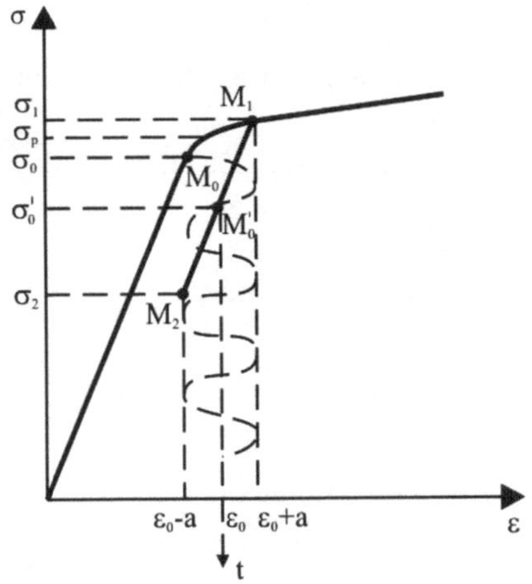


**Fig. 3.** Diagram of elastoplastic model under vibrations

Over this deformation is applied a vibration deformation with amplitude a, transmitted from the vibrating platform, so that the total deformation will have the following variation law

$$\varepsilon = \varepsilon_0 + a.\sin\omega t, \tag{2}$$

where $\omega = 2\pi f$ is the circular frequency of the vibrations produced by the platform, and f is their frequency. By *a,* was noted the dimensionless amplitude which in fact is the ratio between the amplitude of the vibrations of the biomechanical model and its initial elongation.

According to Eq. (2) the extreme deformations of the tissue will be

$$\varepsilon_1 = a + \varepsilon_0 \text{ and } \varepsilon_2 = a - \varepsilon_0, \tag{3}$$

and the corresponding values of stress will be $\sigma_1$, respectively $\sigma_2$, as shown in the diagram from Fig. 3. Starting from the initial state of deformation $\varepsilon_0$ and stress, $\sigma_0$, the maximum deformation $\varepsilon_1$ and stress $\sigma_1$ are reached. The elastoviscoplastic model was considered to behave as in Fig. 3, i.e. the maximum stress is higher than that at which the slip is reached ($\sigma_1 > \sigma_p$). The discharge will be made until the minimum deformation $\varepsilon_2$ and stress $\sigma_2$ on the right segment $M_1M_2$, so an elastic return, and in this return the modulus of elasticity remains $E_1$.

A number of n loading-unloading cycles follows, $n = f * t$, where t is the time of exposure to vibrations, in seconds, until the vibration stops, which will take place at the initial deformation $\varepsilon_0$. Due to the elastic return of the biomechanical model, stress will be found on the $M_1M_2$ segment. Therefore, on the segment $M_1M_2$ at the initial deformation, $\varepsilon_0$ it will correspond to a stress value equal to the average of the extreme values, given by the relation:

$$\sigma_0' = \frac{\sigma_1 + \sigma_2}{2} < \sigma_0. \tag{4}$$

Thus, a reduction of the initial stress given by the expression is achieved.

$$\Delta\sigma = \sigma_0 - \frac{\sigma_1 + \sigma_2}{2}. \tag{5}$$

Taking into account that the return is made elastic, it is possible to write the slope of the $M_1M_2$ segment

$$\sigma_1 - \sigma_2 = E_1(\varepsilon_1 - \varepsilon_2), \tag{6}$$

and considering relationships (3)

$$\sigma_2 = \sigma_1 - 2E_1 a, \tag{7}$$

if the relation given by Eq. (5) is introduced in the Eq. (7), the expression of the absolute decrease of the stress state is obtained.

$$\Delta\sigma = \sigma_0 - \sigma_1 + E_1 a. \tag{8}$$

Obviously, once the parameters of the biomechanical model are known, the relative decrease of stress can be determined.

$$\frac{\Delta\sigma}{\sigma_0} = 1 - \frac{\sigma_1}{\sigma_0} + \frac{E_1}{\sigma_0} a. \tag{9}$$

This ratio can be seen as a measure of the effect of vibration relaxation.

## 4 Vibration Analysis of Tissues with Elastoviscoplastic Behavior

The paper [13] analyzed a tissue whose behavior was considered according to the Bingham model. It behaves perfectly elastic for stress values lower than the slip limit ($\sigma \leq \sigma_p$) and for values higher than this limit ($\sigma > \sigma_p$) it behaves viscoplastic. Next, the Goldschmit model will be analyzed, which will have an elastic behavior for lower stress values than the slip limit value ($\sigma \leq \sigma_p$), but above this value ($\sigma > \sigma_p$) the behavior will be viscoelastoplastic. Applying the same calculation scheme for the Goldschmit model as for the Bingham model, in case of loading in the sliding domain, the following constitutive equation is obtained:

$$\frac{1}{E_1}\frac{d\sigma}{dt} + \frac{1}{\eta}\left[\left(1 + \frac{E_2}{E_1}\right)\sigma - \sigma_p\right] = \frac{d\varepsilon}{dt} + \frac{E_2}{\eta}\varepsilon. \tag{10}$$

If the law of vibration given by Eq. (2) is introduced in this constitutive equation, it is obtained

$$\dot{\sigma} + h\sigma = A\cos(\omega t - \varphi) + B, \tag{11}$$

where the following notations were made:

$$h = \frac{E_1 + E_2}{\eta}; A = \frac{E_1 a}{\eta}\sqrt{E_2^2 + \eta^2\omega^2}; tg\varphi = \frac{E_2}{\eta\omega}; B = \frac{E_1\sigma_p + E_2\sigma_0}{\eta}. \tag{12}$$

The solution of the differential Eq. (10) is of the form:

$$\sigma = Ce^{-ht} + \frac{A\omega}{\omega^2 + h^2}\left[\sin(\omega t - \varphi) + \frac{h}{\omega}\cos(\omega t - \varphi)\right] + \frac{B}{h}, \tag{13}$$

where C is an integration constant that can be determined from the continuity conditions imposed on the stress when crossing the sliding limit.

From Eq. (2), at the moment $t = t_p$, the stress passes through the slip limit value, moment for which it is possible to write

$$\sin\omega t_p = \frac{\sigma_p - \sigma_0}{E_1 a}. \tag{14}$$

At this moment the value of the stress is at the limit of the slip value:

$$\sigma(t_p) = \sigma_p. \tag{15}$$

Imposing the continuity condition to the solution given by Eq. (13), the integration constant C is determined, having the following expression:

$$C = e^{ht_p}\left\{\sigma_p - \frac{A\omega}{h^2+\omega^2}\left[\sin\left(\omega t_p - \varphi\right) + \frac{h}{\omega}\cos\left(\omega t_p - \varphi\right)\right] - \frac{B}{h}\right\}. \tag{16}$$

Analyzing the stress given by Eq. (13) it is found on the one hand that the $Ce^{-ht}$ component decreases exponentially over time and after a number of vibration cycles can be neglected, and on the other hand it will be taken into account that the maximum stress is obtained at times in which, $\sin\omega t = 1$. Based on these considerations from Eq. (12), the maximum value of stress is obtained

$$\sigma_1 = E_1 a\frac{(E_1+E_2)E_2 + \omega^2 h^2}{(E_1+E_2)^2 + \omega^2 h^2} + \frac{E_1\sigma_p + E_2\sigma_0}{h\eta}. \tag{17}$$

From Eqs. (8), and (17) can be determined the absolute variation of the stress from the biomechanical model in the situation where the deformation given by the vibrations is maximum

$$\Delta\sigma = \frac{E_1^2(E_1+E_2)a}{(E_1+E_2)^2 + \omega^2\eta^2} - \frac{E_1\left(\sigma_p - \sigma_0\right)}{E_1+E_2}. \tag{18}$$

Consequently, the measure of the relative decrease of the stress, as a result of the applied vibrations, will characterize the degree of relaxation by vibrations and is given by the relation:

$$\frac{\Delta\sigma}{\sigma_0} = \frac{E_1}{E_1+E_2}\left[1 - \frac{\sigma_p}{\sigma_0} + \frac{1}{1+\frac{\omega^2\eta^2}{(E_1+E_2)^2}}\frac{E_1 a}{\sigma_0}\right]. \tag{19}$$

It can be seen that in the slip field the two elastic elements $E_1$ and $E_2$ are connected in series, and the value of this equivalent modulus of elasticity will characterize the biomechanical model in the slip field, so it will represent the modulus of elasticity $E_p$. It can write:

$$\frac{1}{E_2} = \frac{1}{E_p} - \frac{1}{E_1}. \tag{20}$$

If it is taken into account that $E_1$ and $E_p$ can be determined experimentally, it is preferable to replace the value of the modulus of elasticity $E_2$ from Eq. (20) to Eq. (19) and obtain:

$$\frac{\Delta\sigma}{\sigma_0} = \frac{E_1 - E_p}{E_1}\left[1 - \frac{\sigma_p}{\sigma_0} + \frac{1}{1+\frac{\omega^2\eta^2}{E_1^4}\left(E_1 - E_p\right)^2}\frac{a}{\varepsilon_0}\right]. \tag{21}$$

Based on this formula, some conclusions can be given regarding the effect of vibrations in decreasing the relative state of stress in a ligament.

## 5 Case Study

This case study describes the effect of vibration training on recovering an athlete's ankle sprain. The patient was 29 years old and, following a jump, suffered a second degree lateral ankle sprain. The study aims to highlight the fact that the vibration therapy treatment procedure has improved the degree of movement and extensibility of the soft tissues of the ankle and stimulates proprioception.

After the radiological consultation that confirmed the presence of a sprain in which the ATFL and CTL ligaments were involved and that there was no presence of a fracture, a three-phase treatment program was performed.

In the first phase, during the three days after the installation of the ankle sprain, the protocol was one, namely rest, compression by orthotics, ice three times a day for 10 min on the ankle and elevation, when the foot must be higher than the heart. For swelling, an anti-inflammatory cream is used, which is applied with gentle lymphatic drainage maneuvers to reduce edema.

The second phase begins 3–4 days after the installation of the ankle sprain, and the sub acute treatment protocol includes some light exercises to tone the joint, but without loading the ankle. An elastic band is used through which, from the sitting position, the leg is placed on the vibrating platform (Fig. 4). The exercise program continued with the flexion movement, the extension or dorsiflexion movement, the lateral movement and the rotational movement. All these movements were performed with the leg placed on the vibrating platform and with the elastic band maintained on the forefoot. The regime chosen for the vibrating platform was one corresponding to the small amplitude and low frequencies.

The third phase of the ankle sprain treatment protocol corresponds to the rehabilitation program. At this stage it was possible to load the ankle with the entire weight of the body and it can be applied after the sixth day after the installation of the ankle sprain (Fig. 5). The rehabilitation phase is a time in the treatment program when the proprioceptive part and the recovery of strength at the ankle are very important. A series of exercises followed on the vibrating platform, in orthostatic position, then from orthostatic the weight is moved from the heel to the toes, respectively from the same position the foot is positioned at the lateral edge of the platform. All exercises lasted 60 s and were repeated three times. The exercises are designed to tone the calf muscles and stimulate proprioception. The vibrating platform regimes were chosen in the low frequency and high amplitude range compared to the second phase.

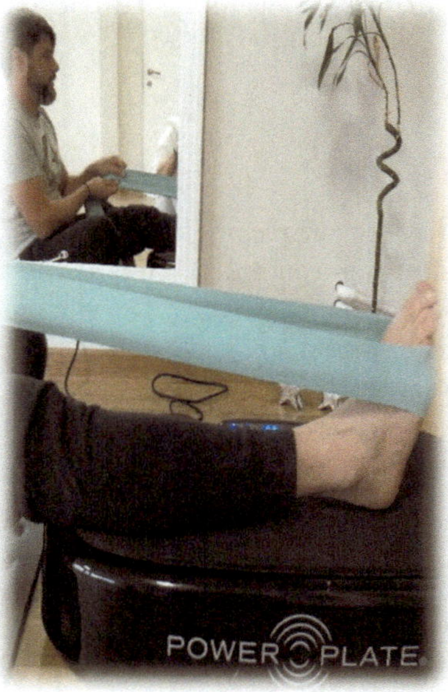


**Fig. 4.** A patient during second phase

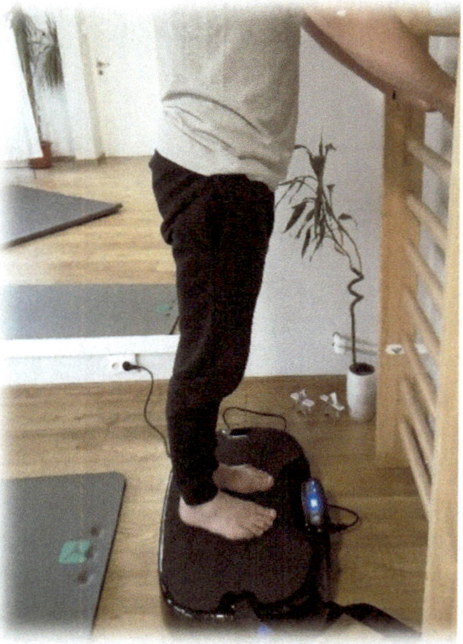

**Fig. 5.** The third phase of protocol

## 6 Conclusions

Based on the theoretical results and the experimental results obtained in the case study, the following conclusions and observations can be formulated:

1) The Eq. (21) which gives the measure of the relative decrease of stress for the rheological model of a tissue with viscous properties generalizes the case of the rheological model with elastoplastic behavior, which becomes a particular case [13].
2) The magnitude of the relative decrease of stress increases with the value of the relative amplitude of vibrations.
3) The presence of viscosity diminishes the effect of vibrations on the relative reduction of stress.
4) Increasing the frequency of vibrations also has a reducing effect on decrease the level of stress.
5) From the case study it was found that in situation of an ankle sprain it is important to control, in the first phase, pain and swelling. Through vibration therapy, the mobility of the articulation was restored, as a result of the stress relaxation, the muscular strength was restored, as a result of the vibratory actions, and to a certain measure the proprioception was stimulated.

## References

1. Dubin, J.C., Comeau, D., McClelland, R.I., Dubin, R.A., Ferrel, E.: Lateral and syndesmotic ankle sprain injuries: a narrative literature review. J. Chiropractic Med. **10**(3), 204–219 (2011)
2. Hertel, J.: Functional anatomy, pathomechanics, and pathophysiology of lateral ankle instability. J. Athl. Train. **37**(4), 364–375 (2002)
3. Williamson, J.B., George, T.K., Simpson, D.C., Hannah, B., Bradbury, E.: Ultrasound in the treatment of ankle sprains. Injury **17**(3), 176–178 (1986)
4. Razzano, C., Izzo, R., Savastano, R., Colantuoni, C., Carbone, S.: Noninvasive interactive neurostimulation therapy for the treatment of low-grade lateral ankle sprain in the professional contact sport athlete improves the short-term recovery and return to sport: a randomized controlled trial. J. Foot Ankle Surg. **58**(3), 441–446 (2019)
5. Feger, M.A., Goetschius, J., Love, H., Saliba, S.A., Hertel, J.: Electrical stimulation as a treatment intervention to improve function, edema or pain following acute lateral ankle sprains: a systematic review. Phys. Ther. Sport. **16**(4), 361–369 (2015)
6. Mohd Mukhtar, A., Abid Ali, K., Mohd, F.: Effect of whole-body vibration on neuromuscular performance: a literature review. Work **59**(4), 571–583 (2018)
7. Sierra-Guzmán, R., Jiménez-Diaz, F., Ramírez, C., Esteban, P., Abián-Vicén, J.: Whole-body-vibration training and balance in recreational athletes with chronic ankle instability. J. Athl. Train. **53**(4), 355–363 (2018)
8. Otzel, D.M., Hass, C.J., Wikstrom, E.A., Bishop, M.D., Borsa, P.A., Tillman, M.D.: Motoneuron function does not change following whole-body vibration in individuals with chronic ankle instability. J. Sport. Rehabil. **28**(6), 614–622 (2019)
9. Suciu, O., Onofrei, R.R., Totorean, A.D., Suciu, S.C., Amaricai, E.C.: Gait analysis and functional outcomes after twelve-week rehabilitation in patients with surgically treated ankle fractures. Gait Posture **49**, 184–189 (2016)
10. Onofrei, R.R., Amaricai, E., Petroman, R., Surducan, D., Suciu, O.: Preseason dynamic balance performance in healthy elite male soccer players. Am. J. Mens. Health **13**(1), 1–7 (2019)
11. Suciu, O., Bereteu, L., Draganescu, G., Ioanovici, T.: Determination of the elastic modulus of hydroxiapatite doped with magnesium through nondestructive testing. In: Advanced Materials Research, vol. 814, pp. 115–122, Timisoara, Romania (2013)
12. Suciu, O., Bereteu, L., Draganescu, G.: Determination of mechanical properties of hydroxyapatite doped with magnesium. In: Bunoiu, M, Biris, C.G., Avram, N. TIM 2012, Physics Conference, Timisoara, vol. 1564, pp. 132–137 (2012)
13. Druta, P.F., Neamtu Popescu, A., Crâştiu, I., Gozman Pop, C., Simoiu, D., Bereteu, L. : Vibration effect on some tissues biomechanical models. In: IEEE 18th International Symposium on Intelligent Systems and Informatics (SISY 2020), Subotica, Serbia, pp. 1–6 (2020)

# Author Index





E.-C. Lovasz et al. (eds.): MTM&Robotics 2020, MMS 88, pp. 567–569, 2021.
https://doi.org/10.1007/978-3-030-60076-1